COCCIDIOIDOMYCOSIS
Sixth International Symposium

ANNALS OF THE NEW YORK ACADEMY OF SCIENCES
Volume 1111

COCCIDIOIDOMYCOSIS
Sixth International Symposium

Edited by Karl V. Clemons, Rafael Laniado-Laborin, and David A. Stevens

Published by Blackwell Publishing on behalf of the New York Academy of Sciences
Boston, Massachusetts
2007

Library of Congress Cataloging-in-Publication Data

International Symposium on Coccidioidomycosis (6th : 2006 : Stanford University)
  Coccidioidomycosis : Sixth International Symposium / edited by Karl V. Clemons ... [et al.].
  p. ; cm. – (Annals of the New York Academy of Sciences, ISSN 0077-8923 ; v. 1102)
  Includes bibliographical references.
  ISBN-13: 978-1-57331-688-0 (alk. paper)
  ISBN-10: 1-57331-688-1 (alk. paper)
  1. Coccidioidomycosis–Congresses. I. Clemons, Karl V. II. Title. III. Series.
  [DNLM: 1. Coccidioidomycosis–Congresses. W1 AN626YL v. 1102 2007 / WC 460 I61c 2007]

  RC136.3.I58 2006
  362.196'936–dc22

                                                        2007005112

The *Annals of the New York Academy of Sciences* (ISSN: 0077-8923 [print]; ISSN: 1749-6632 [online]) is published 28 times a year on behalf of the New York Academy of Sciences by Blackwell Publishing with offices at 350 Main St., Malden, MA 02148 USA; 9600 Garsington Road, Oxford, OX4 2ZG UK; and 600 North Bridge Rd, #05-01 Parkview Square, 18878 Singapore.

Information for subscribers: For new orders, renewals, sample copy requests, claims, changes of address and all other subscription correspondence please contact the Journals Department at your nearest Blackwell office (address details listed above). UK office phone: +44 (0)1865 778315, fax +44 (0)1865 471775; US office phone: 1-800-835-6770 (toll free US) or 1-781-388-8599; fax: 1-781-388-8232; Asia office phone: +65 6511 8000, fax; +44 (0)1865 471775, Email: customerservices@blackwellpublishing.com

Subscription rates:
Institutional Premium   The Americas: $4043   Rest of World: £2246
The Premium institutional price also includes online access to full-text articles from 1997 to present, where available. For other pricing options or more information about online access to Blackwell Publishing journals, including access information and terms and conditions, please visit www.blackwellpublishing. com/nyas
*Customers in Canada should add 6% GST or provide evidence of entitlement to exemption.
**Customer in the UK or EU: add the appropriate rate for VAT EC for non-registered customers in countries where this is applicable. If you are registered for VAT please supply your registration number.

Mailing: The *Annals of the New York Academy of Sciences* is mailed Standard Rate. Mailing to rest of world by DHL Smart & Global Mail. Canadian mail is sent by Canadian publications mail agreement number 40573520. **Postmaster:** Send all address changes to *Annals of the New York Academy of Sciences*, Blackwell Publishing Inc., Journals Subscription Department, 350 Main St., Malden, MA 02148-5020.

Membership information: Members may order copies of *Annals* volumes directly from the Academy by visiting www.nyas.org/annals, emailing membership@nyas.org, faxing 212-298-3650, or calling 800-843-6927 (US only), or 212-298-8640 (International). For more information on becoming a member of the New York Academy of Sciences, please visit www.nyas.org/membership. Claims and inquiries on member orders should be directed to the Academy at email: membership@nyas.org or Tel: 212-298-8640 (International) or 800-843-6927 (US only).

Printed in the USA. Printed on acid-free paper.

Disclaimer: The Publisher, the New York Academy of Sciences and the Editors cannot be held responsible for errors or any consequences arising from the use of information contained in this publication; the views and opinions expressed do not necessarily reflect those of the Publisher, the New York Academy of Sciences, or the Editors.

Annals are available to subscribers online at the New York Academy of Sciences and also at Blackwell Synergy. Visit www.blackwell-synergy.com or www.annalsnyas.org to search the articles and register for table of contents e-mail alerts. Access to full text and PDF downloads of *Annals* articles are available to nonmembers and subscribers on a pay-per-view basis at www.blackwell-synergy.com and www.annalsnyas.org.

The paper used in this publication meets the minimum requirements of the National Standard for Information Sciences Permanence of Paper for Printed Library Materials, ANSI Z39.48_1984.

ISSN: 0077-8923 (print); 1749-6632 (online)
ISBN-10: 1-57331-688-1 (alk. paper); ISBN-13: 978-1-57331-688-0 (alk. paper)

A catalogue record for this title is available from the British Library.

ANNALS OF THE NEW YORK ACADEMY OF SCIENCES

Volume 1111
September 2007

COCCIDIOIDOMYCOSIS
Sixth International Symposium

Editors

KARL V. CLEMONS, RAFAEL LANIADO-LABORIN, AND
DAVID A. STEVENS

This volume is the result of a meeting entitled **Sixth International Symposium on Coccidioidomycosis,** held on August 23–26, 2006 at Stanford University. It was jointly sponsored by the Coccidioidomycosis Study Group, Valley Fever Center for Excellence; and the University of California, San Diego, School of Medicine.

CONTENTS

Part III. Vaccine Research

Part IV. Laboratory Diagnosis

Part V. Clinical Disease

Part VI. Veterinary Infection

Part VII. Therapy

Preface

The Sixth International Symposium on Coccidioidomycosis was convened on August 23–26, 2006 at Stanford University. This symposium is held as a decennial meeting and this year constituted the 50th gathering of the Coccidioidomycosis Study Group (CSG), an informal group comprising interested physicians, researchers, and students, all with an interest in various aspects of coccidioidomycosis. The CSG includes members who have attended all 50 meetings, as well as those whose first meeting was in 2006. Attendance included 185 persons from 17 different countries, and with educational backgrounds that included medical doctors, PhDs, veterinarians, registered nurses, osteopathic practitioners, and other interested parties.

The intent of the international symposium is to bring together experts from around the world to provide the latest information available on coccidioidomycosis to those health-care workers and scientists with an interest in this disease, to foster collegiality among the attendees, and to cross-fertilize the medical and basic sciences to stimulate innovative collaborations. Developed under the guidance of John Galgiani, with the other twelve members of the organizing committee, the program included 70 talks in clinical and basic science sessions, as well as over 30 poster presentations.

Three half-day workshops preceded the official beginning of the meeting and were followed by a welcome reception for all attendees and guests. Thursday morning signaled the official start of the meeting, with John Galgiani giving Medicine Grand Rounds for the Department of Medicine, Stanford University Medical School and its community. Rounds were open to the public and filled Fairchild Auditorium. The scientific sessions followed, each focused on topics related to issues raised in the grand rounds. The evening session was a satellite symposium sponsored by Astellas Pharma US. Friday morning began the second day of sessions, as well as the poster presentations, presided over by "sage squads" of senior faculty members. The scientific sessions were followed in the evening by the conference dinner held at the Stanford Golf Course. Saturday sessions were limited to a half day, and the meeting was adjourned in the early afternoon. An overview of the scientific program, as well as poster abstracts from the meeting, can be found at http://www.vfce.arizona.edu. The program and this volume together underscore the important advances made toward understanding this pathogen, reducing infection with it, and treating those unfortunate to have developed the disease.

In addition to the science, other points were noteworthy. Mitch Magee presented a short eulogy, with a moment of silence to honor the passing of our colleague Rebecca Cox, who was a valued contributor to our knowledge of the

Ann. N.Y. Acad. Sci. 1111: xi–xii (2007). © 2007 New York Academy of Sciences.
doi: 10.1196/annals.1406.052

immunology of coccidioidomycosis. Two awards were presented by Richard Hector, on behalf of an awards subcommittee of the organizing committee, for lifetime achievement and contributions in basic and applied research in the field of coccidioidomycosis. The awardees were Antonino Catanzaro (Emmet Rixford Memorial Award) and David A. Stevens (Charles E. Smith Memorial Award). These awards are only given every decade and reflect the long-term commitment of these individuals to both clinical and basic research aspects of this disease, as well as to the mentoring of young clinicians and researchers.

The editors of this volume and as members of the organizing committee thank the following sponsors of the symposium, whose financial support and assistance allowed the meeting to be held: Astellas Pharma US, Inc.; Schering-Plough; Centocor, Inc; Enzon Pharmaceuticals, Inc.; and Merck & Co., Inc. We particularly thank the New York Academy of Sciences for their interest in publishing these proceedings. Their expert staff and responsiveness have made the preparation of this volume possible. We thank those individuals who gave freely of their time to critically review the submitted manuscripts. Last, we thank all of the authors who responded in short time frames with the authoritative contributions published in this volume.

We look forward to the Seventh International Symposium on Coccidioidomycosis in 2016.

—KARL V. CLEMONS
Stanford University
California Institute for Medical Research, and
Santa Clara Valley Medical Center

—RAFAEL LANIADO-LABORIN
Universidad Autónoma de Baja California

—DAVID A. STEVENS
Stanford University
California Institute for Medical Research, and
Santa Clara Valley Medical Center

Organizing Committee

John N. Galgiani, Chair
University of Arizona

Neil M. Ampel
University of Arizona

Janis Blair
Mayo Clinic, Scottsdale

Antonino Catanzaro
University of California, San Diego

Karl V. Clemons
Stanford University
California Institute for Medical Research
Santa Clara Valley Medical Center

Richard Hector
University of California, San Francisco

Royce H. Johnson
Kern County Medical Center

Peter C. Kelly
Arizona Department of Health Services

Rafael Laniado-Laborin
Universidad Autónoma de Baja California

Hillel B. Levine
Coccidioidomycosis Study Group

Laurence F. Mirels
Santa Clara Valley Medical Center

Demosthenes Pappagianis
University of California, Davis

David A. Stevens
Stanford University
California Institute for Medical Research
Santa Clara Valley Medical Center

Reviewers

The Editors wish to thank the following individuals for their contributions as reviewers.

Coccidioidomycosis: Changing Perceptions and Creating Opportunities for Its Control

JOHN N. GALGIANI

Valley Fever Center for Excellence, College of Medicine, University of Arizona, and the Southern Arizona VA Health Care System, Tucson, Arizona, USA.

ABSTRACT: The perceptions of coccidioidomycosis as a medical problem has undergone sequential and dramatic metamorphoses since its first description more than a century ago. First thought to be rare and lethal, coccidioidomycosis was subsequently found to be common and often mild. During World War II, its overall impact upon large populations came sharply into focus and the consequences for public health became clearer. Early treatments had significant limitations and toxicities, and therefore treatment of coccidioidomycosis was reserved for only the sickest patients. Since then, safer oral therapies have become commonplace. Despite their availability, there has been no investigation of their use in the less severe and much more common early infections. Even newer drugs such as nikkomycin Z, which might actually cure infections, until very recently have had trouble finding a sponsor to move it through clinical trials. Perceptions once formed by the understanding of coccidioidomycosis as a medical problem now appear to hinder the future study of newer therapeutic opportunities. It is suggested in this review that it is time to revisit and possibly change these perceptions if we are to improve our care of patients.

KEYWORDS: public health; medical history; antifungal therapies

INTRODUCTION

I delivered this presentation to Stanford's Department of Internal Medicine for its weekly Grand Rounds and as an introduction to the Sixth International Symposium on Coccidioidomycosis, published in this volume. Twelve years previously, Dr. Stanley Deresinski and Dr. Richard Hector used the same occasion ahead of the Fifth International Symposium on Coccidioidomycosis

Address for correspondence: John N. Galgiani, M.D., Medical Service (1–111INF), Southern Arizona VA Health Care System, 3601 South Sixth Avenue, Tucson, AZ 85723. Voice: 520-792-1450, ext: 6793; fax: 520-529-4738.
  spherule@u.arizona.edu

Ann. N.Y. Acad. Sci. 1111: 1–18 (2007). © 2007 New York Academy of Sciences.
doi: 10.1196/annals.1406.041

1

to review the personalities and accomplishments of several early Stanford faculty members that figured prominently in the history of this interesting disease.[1] For my lecture, I also took an historical approach, but for a different purpose.

During the past century, there have been successive waves of new understandings about coccidioidomycosis, first in its epidemiology and later in its treatment with antifungal therapies. With each successive set of revelations, the medical community has shifted its perception about the importance of coccidioidomycosis as a medical problem and how best to manage it. Such changes over time are perhaps not surprising because they generally occur with nearly all diseases. However, the point to be emphasized here is that once these perceptions are formed they in themselves can also have profound effects on the future direction of research and progress in medical care. In particular, I would like to suggest that the very progress we have made in treating the most serious complications of coccidioidomycosis may now be preventing us from addressing the most common forms of the disease, the initial pulmonary syndrome. As a result of current perceptions, we may miss opportunities to develop cures for infected patients rather than to simply continue to manage their illnesses as a chronic disease.

COCCIDIOIDAL GRANULOMA: FIRST PERCEPTIONS

The first patient identified in 1893 with what was to be known as coccidioidomysosis was an Argentinian soldier with wide-spread cutaneous lesions.[2,3] The following year, a similar patient was found in San Francisco by Dr. Emmet Rixford, a faculty member of Cooper Medical College, which a decade later was to become Stanford's School of Medicine.[4-6] Beginning with these initial reports and continuing for the next 30 years, occasional other patients, such as the one shown in FIGURE 1, were identified in California; most of these patients had extensive and ultimately fatal illnesses. Histologically, these lesions demonstrated granulomatous inflammation and within the lesions were large multicellular organisms. These were first thought to be protozoan, but within a few years they were discovered to be fungal structures by the Stanford pathologist, Dr. W. Ophuls.[7] Until 1935, patients identified with coccidioidal granuloma remained fairly rare, there being fewer than 500 cases reported to the California Department of Public Health.

COCCIDIOIDAL GRANULOMA AND VALLEY FEVER ARE CAUSED BY THE SAME FUNGUS

Based on these earliest observations, the first perception of the nature of coccidioidomycosis was that it was a rare and often very severe infection. However,

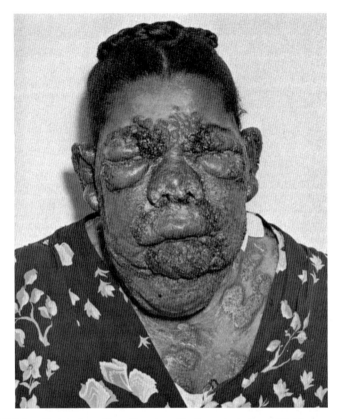

FIGURE 1. Patient from Kern County with disseminated coccidioidomycosis. (From Conant, *et al. Manual of Clinical Mycology.* [1971]. Reprinted by permission from W.B. Saunders.)

this perception was going to change radically because of new epidemiologic observations.

An important key to the true spectrum of coccidioidal disease was the dermatologic condition known as erythema nodosum. Erythema nodosum is caused by subcutaneous inflammation that produces tender, violacious areas, usually of the shins and lower legs. These lesions are not specific for any one disease, but they are so distinctive and discomforting that when they occur they are hard to miss.

In the mid 1930s, Dr. Myrnie Gifford, another Stanford Medical School–trained physician, became a public health officer in the Kern County, California, Department of Health. While there, she became interested in the epidemiology of a common, local disease called San Joaquin Fever.[8] This was a self-limited pulmonary illness of unknown etiology and one of its manifestations was the presence of Erythema nodosum.

In other studies, Dr. Gifford also noted that Erythema nodosum was an occasional finding in persons with coccidioidal granuloma. For example, in one series she found 3 of the 15 such patients manifest the skin rash.

A few years earlier, the finding of Erythema nodosum after coccidioidal infection also had been noted at Stanford, when Harold Chope, a medical student being trained in the study of coccidioidal granuloma, accidentally suffered an intensive laboratory exposure when he opened a petri dish containing an old culture of *Coccidioides*.[9] Nine days later, he developed a respiratory illness and as part of his syndrome also developed *E. nodosum*. To everyone's great surprise, Chope did not die from the coccidioidal infection, but rather his illness was completely resolved.

In 1938, Dr. Ernest Dickson, whose laboratory was the site of Chope's accident, visited Dr. Gifford in Kern County. His visit was due in part to the fact that coccidioidal granuloma had become a reportable disease in California, and many reports of new cases of this disease were coming out of the Central Valley. During that visit, it became clear that Erythema nodosum was associated with both coccidioidal granuloma and San Joaquin Valley Fever, and this soon led to the discovery that both diseases were caused by the same fungus.

The epidemiologic studies of Valley Fever were then picked up by one of Dr. Dickson's junior faculty, Dr. Charles Smith. Also, at this time, a skin testing reagent became available which stimulated delayed-type hypersensitivity early after the onset of Valley Fever and persisted long after the illness had resolved. Dr. Smith used skin testing to show that a very high proportion of Central Valley residents exhibited a positive reaction.[10,11] This finding indicated that many persons in the Central Valley, even without apparent illness, had undergone a coccidioidal infection.

These clinical observations were being developed in the context of other information about coccidioidomycosis. For example, after coccidioidal granuloma became a reportable disease, the numbers of cases reported in California substantially increased, but surprisingly the proportion of patients dying from the infection decreased.[8] Also, in 1931, it was reported that 18% of livestock actually were found to have histopathologic evidence of coccidioidal granuloma, but displayed no illness.[12] Thus, Gifford and Dickson's discovery that *Coccidioides immitis* could be isolated from patients with Valley Fever and Smith's skin testing results fit nicely with other information that was accumulating from other sources.

Recalling this time, Charles Smith subsequently wrote "Coccidioidomycosis is an exceedingly common infection, rather than a rare one, and that its disseminated form, coccidioidal granuloma, is the exception rather than the rule."[13] That is, the earlier perception of coccidioidomycosis as a rare and often fatal disease had shifted to the new perception that it was a common and benign infection.

IMPACT OF COCCIDIOIDOMYCOSIS ON THE WORLD WAR II MILITARY EFFORT

Perceptions were going to change again with the advent of World War II. As part of the war effort, the southwest desert regions were used for extensive military training. During these maneuvers, Valley Fever had a major impact on troop health. For example, over a 1-month period hundreds of infections were identified during training exercises in Kern County.[13] Most of these illnesses were self-limited, uncomplicated pneumonia, but none the less they incapacitated the service personnel. Similar outbreaks occurred in military camps across the southwest.

Later, when German prisoners were housed in Florence, Arizona, skin test conversion rates were as high as 50% over a 6-month period.[13] The problem became so serious that there were protests that these POW camps violated the Geneva Convention.

Once again the perception about Valley Fever shifted. No longer was it a disease of little impact, but rather one that caused considerable disability within large populations and one which required a response. Because there was not yet an effective antifungal treatment available, the focus centered on avoidance and dust control.

ROLE OF AMPHOTERICIN B AS THE FIRST TREATMENT FOR COCCIDIOIDOMYCOSIS

Although not yet available during World War II, effective antifungal therapy was on the way. By 1960, the focus on managing coccidioidomycosis had shifted again because of the breakthrough therapeutic agent, amphotericin B.[14-18] Because much of coccidioidomycosis was affecting those living in the agricultural and rural areas of California, it was the physicians who lived in the farming communities of the Central Valley that were responsible for managing these infections. Although many clinicians were involved, two that taught us the most about how to use amphotericin B were William Winn (FIG. 2) and Hans Einstein (FIG. 3). Not only were they among the first to use amphotericin B intravenously, but they also initiated the practice of injecting amphotericin B intrathecally into the cerebrospinal fluid as the first effective treatment of coccidioidal meningitis.[19-22] Because of their studies, amphotericin B became widely regarded as the most effective therapy for treating coccidioidomycosis.

However, its therapeutic benefit needed to be balanced with its toxicity. The main problems with amphotericin B are pretty well known. It requires intravenous administration. Treatment is often associated with significant untoward reactions, most notably fever, nausea, and muscle aches. Of course,

FIGURE 2.   William Winn. (Photo courtesy of William Winn, Jr.)

its most limiting complication is nephrotoxicity which is both dose-related
and cumulative, in some patients producing complete renal failure.[23] So, despite its effectiveness, amphotericin B was necessarily reserved for the most
seriously ill, that is, for persons who might benefit from such a toxic treatment. As a result, the perception that evolved by the 1970s was "If you
didn't need amphotericin B treatment, you had the milder or mild form of the
disease."

FIGURE 3. Hans Einstein.

MICONAZOLE AS AN ALTERNATIVE TREATMENT TO AMPHOTERICIN B

After amphotericin B, two decades would pass until a second drug was identified to be useful in treating coccidioidomycosis and this only because of a fortunate collaboration between Hillie Levine (FIG. 4), a mycologist at UC Berkeley, and David Stevens (FIG. 5), a new Stanford faculty member based at Santa Clara Valley Medical Center. Dr. Levine was at that time involved with animal testing of coccidioidal vaccines and in developing new skin testing reagents for Valley Fever.[24–29] Dr. Stevens, on the other hand, had previously studied the immunology of herpes virus infections,[30–35] but was looking for a

FIGURE 4.   Hillel B. Levine.

new direction for his research. David became interested in the cellular immune responses to *Coccidioides* and began working with Hillie in the field trials of a new skin-testing antigen, spherulin.[27,36–39]

As part of his other interests, Dr. Levine also tested antifungal drugs for pharmaceutical firms because his lab was well versed in the murine model of coccidioidal infection. One of the drugs tested by Dr. Levine was miconazole, sent to him by Janssen Pharmaceutical Company.[40] In these studies, untreated mice given pulmonary infections with *Coccidioides* all succumbed to infection, but with increasing doses of miconazole, survival improved to 100%. On the basis of these preclinical results, David Stevens started to obtain miconazole to treat patients with coccidioidomycosis, first as compassionate individual use, but eventually under an FDA IND, which eventually led to a New Drug Application for use in coccidioidomycosis and other mycoses.[41–43]

The overall response rates to miconazole treatment for the main areas of complications from initial infection were 67% for chronic pneumonia, 50% for skin, 41% for the skeleton, and 35% for the meninges.[42] Whether miconazole

FIGURE 5. David A. Stevens.

was as effective as amphotericin B is unknown because there was never a comparative study. What was clear, however, was that miconazole did not cause nephrotoxicity. On the other hand, miconazole, like amphotericin B, required parenteral administration. Also, miconazole produced its own set of untoward effects, which included hyperlipidemia, anemia, and thrombocytosis. Thus, when miconazole became available for clinical use, it was perceived as an alternative therapy for basically the same patient group, the small subset of all infected patients that had progressively severe infections.

ADVENT OF ORALLY ABSORBED AZOLE
ANTIFUNGAL DRUGS

After miconazole was approved by the FDA for clinical use, it did not enjoy extensive use. Even so, miconazole was a milestone in antifungal drug development because it demonstrated that the azole class of drugs was effective. In future screening, other azole congeners were discovered that were both orally effective and possessed fewer untoward reactions.

The three oral azole antifungals introduced in the 1980s and 1990s were ketoconazole, itraconazole, and fluconazole. Each has been the subject of multiple clinical trials as therapy for chronic pulmonary and extrapulmonary coccidioidomycosis including coccidioidal meningitis.[44–62] Many of these studies were carried out by a single, multicenter collaboration known as the Mycoses Study Group, and most were phase II, open-label studies of individual drugs; none were randomized studies in comparison to amphotericin B. However, one study by this group compared fluconazole and itraconazole in a prospective, randomized, double-blinded protocol.[62] In that study there is a trend suggesting that itraconazole is more effective against skeletal lesions. The overall results indicated a response rate of approximately 50–70%, depending upon the site of infection for both drugs. Similar results have also been found with these therapies for meningitis.[58,60] Again, because of the lack of comparative studies, it is not known how these response rates would compare to those achieved by amphotericin B. However, even without formal comparisons, it is very clear that oral azole antifungals, especially fluconazole and itraconazole, as oral medications are much easier to administer and much less toxic than what was available before.

On the basis of these numerous studies, the recently revised Infectious Diseases Society of America Practice Guidelines state that "Azole therapy (fluconazole, itraconazole) has largely supplanted amphotericin B as treatment of progressive coccidioidal infections," with polyenes largely being reserved for the most fulminant or progressive infections.[63] The thinking here is that safety and the ability to administer long, even life-long courses of therapy makes azole therapy the preferable first-line treatment.

PERCEPTIONS AND FUTURE DEVELOPMENT OF NEW
TREATMENTS FOR COCCIDIOIDOMYCOSIS

Now with convenient and remarkably safe antifungal drugs in hand, one might expect that there would have developed a wealth of information regarding the use of oral azole therapies for the less severe but greatly more common syndrome associated with the initial coccidioidal infection. However, in the same IDSA Guidelines there is also the following statement: "How best to manage primary respiratory coccidioidal infections is an unsettled issue because of the lack of prospective controlled trails." This categorical statement begs the following question: How could such an achievable and potentially useful direction for clinical trials been left so completely unexplored?

For me, the answer hinges on a reversal in the relationship between therapeutic technology and our perception of the disease we our treating. Up until this time, treatment was so difficult and toxic that only the most seriously ill would benefit. This situation existed for so long, some 30 or 40 years, that it engendered the fixed perception that treatment should be reserved for only the most seriously ill. As a result, despite the availability of safe and convenient drugs, they have not been studied for their usefulness in helping patients with early coccidioidal infections. For the remainder of this discussion, I will examine three underpinnings of this perception in more detail to see if perhaps they might be worth changing.

Perception 1: Uncomplicated Coccidioidal Pneumonia Is Not Really a Problem for Most People

While it is certainly true that the primary pulmonary syndrome is not nearly as serious as the less frequent complications, it is often anything but mild. Coccidioidal pneumonia is similar to lower respiratory illnesses caused by many other pathogens. It is characterized by cough, chest pain, fever, and frequently drenching night sweats. Weight loss of 5–10% of body weight is common. Bone and joint pain is typical as are skin rashes such as we have discussed already. For many persons, fatigue is a strikingly prominent symptom, which often prevents patients from going to work or carrying out other routine activities. This illness often lasts for weeks to months. For example, in a study at a university campus health center, one-quarter of otherwise healthy college students required medical care for at least four months.[64]

Although Valley Fever is greatly underreported, the trend of these statistics is clearly on the rise. Shown in Figure are the figures for reported cases of coccidioidomycosis in Arizona over the past 16 years and as you can see, there is a striking increase over this time period. Actually, in Arizona, the cases in 2006 exceeded those of the previous year by 46%.

The progressive increases in incidence have not gone unnoticed by the public and the media. An article from the Phoenix newspaper, the *Arizona Republic*, reads "We don't know why this vexing fungus is rising." From another story in a Tucson newspaper: "There's no sure way to avoid exposure, and there's no quick recovery from the disease." Tourists can develop Valley Fever as well. Here is a consumer update from the Chicago Tribune, which reads: "What can be worse than coming home from a lovely vacation, becoming seriously ill and not knowing what you have." And again, this time from the *National Geographic Adventure* magazine: "The CDC labels it an epidemic—So why doesn't your doctor recognize Valley Fever?" I think these popular journalis

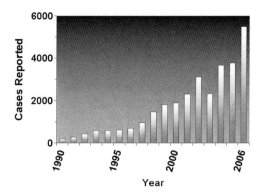

FIGURE 6. Cases of coccidioidomycosis reported to the Arizona Department of Health Services by year.

stories illustrate the fact that the public is becoming increasingly aware that this disease exists and its perception is changing to think that something should be done about it.

Perception 2: It Is Not Medically Useful to Consider Valley Fever as a Cause of Community-Acquired Pneumonia

What do physicians think about coccidioidal pneumonia? As one of the articles just mentioned suggests, the reason that Valley Fever is so underreported is that many if not most clinicians are simply not trying to diagnose it. And for many, the reason they are not diagnosing Valley Fever is because of their perception that it is not medically useful to consider *Coccidioides* as a cause of community-acquired pneumonia.

The logic of this perception goes something like this: because most persons get better without therapy, why not ignore patients with this self-limited illness and, instead, only identify those who develop complications because they are the ones you need to treat. Reinforcing this logic is the fact that the diagnosis of Valley Fever requires laboratory testing, most frequently serologic. In one study from the University of Arizona Student Health Service, 300 persons needed to be tested in order to identify 50 new infections.[64] At approximately $80 per patient tested, this would suggest a cost of nearly $500 to diagnose each case.

It is now clear that clinicians within the endemic region who take this approach are likely to be missing an etiologic diagnosis in a sizable proportion of their patients with community-acquired pneumonia. In a recently reported observational study, 29% of 55 patients in Tucson, Arizona with a clinical diagnosis of CAP were found to be serologically positive for coccidioidal antibodies.[65] Even with this relatively small sample, the 95% confidence interval ranged from 16% to 44%. An important correlate to this finding is that tourists who develop pneumonia in the weeks after leaving an endemic area have this same high likelihood of having Valley Fever. In addition to knowing more completely what is causing patients' illnesses, there are at least four additional very good reasons why making a diagnosis of Valley Fever improves patient care, whether or not specific antifungal therapy is begun.

First, by making an etiologic diagnosis, the clinician can give the disease a name. This moves the illness out of the realm of the unknown and reduces a patient's anxiety and fear about what is causing the symptoms. This is especially true for older patients, who may be afraid that their illness is a symptom of cancer. Also, once a diagnosis is made, there is an opportunity for patient education about what can be expected from the disease in the future.

Second, with a diagnosis in hand, there is much less need to pursue other possibilities, thereby avoiding costly and sometimes invasive additional testing such as CT scans, bronchoscopy, or biopsy.

Third, if a diagnosis of a fungal infection is made, then it becomes clear that antibacterial therapies are not going to be of much help and can be discontinued. Because Valley Fever is a common cause of CAP, this may well have a significant impact on needless antibacterial use.

Finally, because many of the complications of coccidioidomycosis occur within weeks or months of the initial infection, recognizing the early symptoms for what it is increases the chances that complications might be detected earlier and thereby decreases the consequences by earlier initiation of treatment.

So all of these reasons could be used in support of early diagnosis and I would suggest would be of benefit to our patients. Given a more systematic approach to diagnosing early coccidioidal infections, enrollment of such patients into clinical trials to identify therapeutic value of antifungal drugs would be a logical next step.

Perception 3: New Therapies for Coccidioidomycosis Will Not Be Developed Because Valley Fever Is an Unprofitable Orphan Disease

An orphan disease is defined by the FDA as having a prevalence of less than 200,000. By that definition, Valley Fever is, in fact, an orphan disease. Spread out evenly across the entire U.S. population that would be a prevalence of approximately 7 per 10,000 residents. However, Valley Fever is extremely concentrated in the southwest, and the actual prevalence in that region is approximately 50-fold greater. Either way, the small market size has not been attractive for pharmaceutical to develop new therapies. As an example, lets review attempts to develop nikkomycin Z.

Nikkomycin Z inhibits the synthsis of chitin, one of the principle polymers of the fungal cell wall.[66,67] Basing conclusions on *in vitro* testing, many fungi were seen to be resistant to nikkomycin Z. However, *Coccidioides* was strikingly sensitive to the drug, leading to its study in animals. Richard Hector and his colleagues were the first to show that nikkomycin Z was an effective therapeutic agent for coccidioidomycosis.[68] Mice infected intranasally with spores of *Coccidioides* and treated orally with varying doses of nikkomycin Z from days 2–12 showed a dose-response from 100% lethality in untreated animals to 100% survival in animals receiving 20 mg/kg or greater per day. Even more significantly, in other similarly infected animals given the same treatment regimen which were sacrificed on day 14, the quantitative cultures of the lungs from control animals showed more than 2 million cfu. In contrast, 7 of the 8 treated animals demonstrated sterile lungs and the eighth only had a single colony of fungus identified.

These results raise for the first time the possibility of a therapy that might result in a cure. If nikkomycin Z is going to be curative in patients, it likely would have its best chance of doing so in the early stages of illness, before fibrosis and other anatomic barriers might interfere with drug bioavailability.

This line of reasoning leads to a new paradigm: By treating the early infection, the disease is cured and complications will not occur. In this context, the idea of treating the early infections is now not simply to make patients feel better more quickly, although that might also occur. Rather, treating the early infection might be the best approach to eliminating the problem altogether instead of simply managing the complications when they arise, as we do today, as chronic disabilities.

To summarize the chronology of the nikkomycin Z story, the drug was discovered in the 1970s, and the studies reported by Hector and colleagues were carried out in the 1980s. In the 1990s a small Bay Area company, Shaman Pharmaceuticals, completed a preclinical package, filed an IND with the FDA, and conducted a single-dose safety trial in humans, which was completed successfully with no toxicity observed. However, shortly after that, Shaman was dissolved, and at that point further development of nikkomycin Z stalled because no other pharmaceutical sponsor has yet to come forward to resume clinical trials and commercialization.

Although the fate of nikkomycin Z to date would seem to reinforce the perception that new therapies for Valley Fever will not emerge, it is possible that other factors may also come into play. Since Shaman Pharmaceuticals dissolved, the sponsorship of the FDA application for nikkomycin Z was assumed by the University of Arizona in May of 2005. Since then, federal, state, and philanthropic support has been sought to allow for the resumption of clinical trials with the currently existing nikkomycin Z drug. The success of these efforts would allow for a multidose drug safety trial to begin as early as 2007.

For other drugs with broader spectra of antifungal activity, the larger mycologic diseases such as candidiasis or aspergillosis have provided the commercial incentive to conduct the needed clinical trials. When the activity spectra also includes coccidioidomycosis, patients with this disease may benefit when the drug is made commercially available, even if FDA indication is not obtained.

This was the case for fluconazole and itraconazole in the past and may also be the case for voriconazole and posaconazole, two new triazoles antifungals that are now on the market.[69]

SUMMARY

Over the past century, information and limited therapeutics have influenced perceptions about coccidioidomycosis. Now the situation has been reversed in that the existing perceptions about coccidioidomycosis and its importance appear to be limiting the development of new therapeutics. Thus, in the future, whether there will be new therapies for Valley Fever is likely to hinge upon whether or not they are perceived to be needed.

REFERENCES

1. DERESINSKI, S.C. & R.F. HECTOR. 1996. The history of coccidioidomycosis. *In* Coccidioidomycosis. Proceedings of the 5th International Conference on Coccidioidomycosis.H.E. Einstein & A. Catanzaro, Eds. 48–76. Washington, DC. National Found on for Infectious Diseases.

2. POSADA, A. 1892. Un nuevo caso de micosis fungoidea con psorospermias. Anales del Circulo Médico Argentino **15:** 585–597.

3. WERNICKE, R. 1892. Ueber einen protozoenbefund bei mycosis fungoides (?). Zentralblatt fur Bakteriologie **12:** 859–861.

4. RIXFORD, E. 1894. A case of protozoic dermatitis. Occidental Med. Times **8:** 704–707.

5. RIXFORD, E. 1894. Case for diagnosis presented before the San Francisco Medico-Chirurgical Society. Occidental Med. Times **8:** 326.

6. RIXFORD, E. & T.C. GILCHRIST. 1896. Two cases of protozoan (coccidioidal) infection of the skin and other organs. Johns Hopkins Hosp. Rep. **1:** 209–268.

7. OPHULS, W. & H.C. MOFFITT. 1900. A new pathogenic mould. (formerly described as a protozoon: *Coccidioides immitis* pyogenes): Preliminary report. Philadelphia Med. J. **5:** 1471–1472.

8. GIFFORD, M.A., W.C. BUSS, R.J. DOUDS, *et al.* 1936. Data on *coccidioides* fungus infection, Kern County. Kern County Dept. Pub. Health Ann. Rep.: 39–54.

9. DICKSON, E.C. 1938. Primary coccidioidomycosis. The initial acute infection which results in coccidiodal granuloma. Am. Rev. Tuberc. **38:** 722–729.

10. FABER, H.K., C.E. SMITH & E.C. DICKSON. 1939. Acute coccidioidomycosis with erythema nodosum in children. J. Pediatr. **15:** 163–171.

11. SMITH, C.E. 1940. Epidemiology of acute coccidioidomycosis wth erythema nodosum. Am. J. Pub. Health **30:** 600–611.

12. BECK, M.D., J. TRAUM & E.S. HARRINGTON. 1931. Coccidioidal granuloma: occurrence in animals: reference to skin tests. J. Am. Vet. M. A. **78:** 490–499.

13. SMITH, C.E. 1958. Coccidioidomycosis. *In* Communicable Diseases Transmitted Chiefly Through Respiratory and Alimentary Tracts. J.B. Coates & E.C. Hoff, Eds. 285–316.Washington, DC. Office of the Surgeon General, Medical Department, US Army.

14. FIESE, M.J. 1957. Treatment of disseminated coccidioidomycosis with ampho-
    tericin B. Report of a case. Calif. Med. **86:** 119–120.
15. HOROWITZ, L. & E.M. BERGLUND. 1957. Amphotericin B: a new antifungal an-
    tibiotic in the treatment of coccidioidomycosis. *In*: Transactions of the 16th
    Conference on Chemotherapy in Tuberculosis, Washington D.C. U.S. Veterans
    Administration p345–347.
16. LITTMAN, M.L. 1957. Preliminary observations on the intravenous use of ampho-
    tericin B, an antifungal antibiotic, in the therapy of acute and chronic coccidioidal
    osteomyelitis. Proc. Symposium on Coccidioidomycosis Phoenix, AZ **575:** 86–
    94.
17. HUNTER, R.C. & E.S. MONGAN. 1958. Disseminated coccidioidomycosis treated
    with amphotericin B. U.S. Armed Forces Med. J. **9:** 1474–1486.
18. HARDENBROOK, M.H. & S.L. BARRIERE. 1982. Coccidioidomycosis: evaluation of
    parameters used to predict outcome with amphotericin B therapy. Mycopatholo-
    gia **78:** 65–71.
19. WINN, W.A. 1959. The use of amphotericin B in the treatment of coccidioidal
    disease. Am. J. Med. **27:** 617–635.
20. EINSTEIN, H.E., C.W. HOLEMAN, JR, L.L. SANDIDGE & D.H. HOLDEN. 1961. Coc-
    cidioidal meningitis. The use of amphotericin B in treatment. Calif. Med. **94:**
    339–343.
21. WINN, W.A. 1963. Coccidioidomycosis and amphotericin B. Med. Clin. North Am.
    September: 1131–1148.
22. WINN, W.A. 1964. The treatment of coccidioidal meningitis. The use of ampho-
    tericin B in a group of 25 patients. Calif. Med. **101:** 78–89.
23. BINDSCHADLER, D.D. & J.E. BENNETT. 1969. A pharmacologic guide to the clinical
    use of amphotericin B. J. Infect. Dis. **120:** 427–436.
24. LEVINE, H.B., J.M. COBB & C.E. SMITH. 1960. Immunity to coccidioidomycosis
    induced in mice by purified spherule, arthrospore, and mycelial vaccines. Trans.
    N. Y. Acad. Sci. **22:** 436–447.
25. LEVINE, H.B., J.M. COBB & C.E. SMITH. 1961. Immunogenicity of spherule-
    endospore vaccines of *Coccidioides immitis* for mice. J. Immunol. **87:** 218–227.
26. LEVINE, H.B., R.L. MILLER & C.E. SMITH. 1962. Influence of vaccination on respi-
    ratory coccidioidal disease in cynomolgus monkeys. J. Immunol. **89:** 242–251.
27. LEVINE, H.B., D. PAPPAGIANIS & J.M. COBB. 1970. Development of vaccines for
    coccidioidomycosis. Mycopath. Mycol. Appl. **41:** 177–185.
28. LEVINE, H.B., G.M. SCALARONE, J.M. COBB & D. PAPPAGIANIS. 1972. Spherule
    phases coccidioidin (spherulin) in delayed dermal hypersensitivity responses
    to *Coccidioides immitis* infection. *In* Yeasts and Yeast-Like Microorganisms in
    Medical Science. Second International Specialized Symposium on Yeasts. 229–
    236. University of Tokyo Press, Tokyo.
29. LEVINE, H.B., A. GONZALEZ-OCHOA & D.R. TEN EYCK. 1973. Dermal sensitivity
    to *coccidioides immitis*: a comparison of responses elicitied in man by spherulin
    and coccidioidin. Am. Rev. Respir. Dis. **107:** 379–386.
30. STEVENS, D.A., T. PINCUS, M.A. BURROUGHS & B. HAMPAR. 1968. Serologic rela-
    tionship of a simian herpes virus (SA8) and Herpes simplex virus: heterogeneity
    in the degree of reciprocal cross-reactivity shown by rabbit 7S and 19S antibod-
    ies. J. Immunol. **101:** 979–983.
31. STEVENS, D.A., T.W. PRY, E.A. BLACKHAM & R.A. MANAKER. 1970. Immuno-
    diffusion studies of EB virus (herpes-type virus)-infected and-uninfected hemic
    cell lines. Int. J. Cancer **5:** 229–237.

32. STEVENS, D.A., T.W. PRY & E.A. BLACKHAM. 1970. Prevalence of precipitating antibody to antigens derived from Burkitt lymphoma cultures infected with herpestype virus (EB virus). Blood **35:** 263–275.

33. STEVENS, D.A., T.W. PRY, E.A. BLACKHAM & R.A. MANAKER. 1970. Comparison of antigens from human and chimpanzee herpes-type virus-infected hemic cell lines. Proc. Soc. Exp. Biol. Med. **133:** 678–683.

34. STEVENS, D.A., S.D. KOTTARIDIS & R.E. LUGINBUHL. 1971. Investigation of antigenic relationship of Marek's disease, herpes virus and EB virus (herpes-type virus). J. Comp. Pathol. **81:** 137–140.

35. STEVENS, D.A. & T.C. MERIGAN. 1972. Interferon, antibody, and other host factors in herpes zoster. J. Clin. Invest. **51:** 1170–1178.

36. DERESINSKI, S.C., H.B. LEVINE & D.A. STEVENS. 1974. Soluble antigens of mycelia and spherules in the *in vitro* detection of immunity to *Coccidioides immitis.* Infect. Immun. **10:** 700–704.

37. STEVENS, D.A., H.B. LEVINE & D.R. TENEYCK. 1974. Dermal sensitivity to different doses of spherulin and coccidioidin. Chest **65:** 530–533.

38. DERESINSKI, S.C., H.B. LEVINE, L.J. BLAINE & D.A. STEVENS. 1975. Use of spherulin in clinical coccidioidomycosis. Am. Rev. Respir. Dis. **111:** 916.

39. LEVINE, H.B., M.A. RESTREPO, D.R. TEN EYCK & D.A. STEVENS. 1975. Spherulin and coccidioidin: cross-reactions in dermal sensitivity to histoplasmin and paracoccidioidin. Am. J. Epidemiol. **101:** 512–516.

40. LEVINE, H.B., D.A. STEVENS, J.M. COBB & A.E. GEBHARDT. 1975. Miconazole in coccidioidomycosis. I. Assays of activity in mice and *in vitro.* J. Infect. Dis. **132:** 407–414.

41. STEVENS, D.A., H.B. LEVINE & S.C. DERESINSKI. 1975. Miconazole therapy of human coccidioidomycosis. Am. Rev. Respir. Dis. **111:** 950.

42. STEVENS, D.A. 1977. Editorial. Miconazole in the treatment of systemic fungal infections. Am. Rev. Respir. Dis. **116:** 801–806.

43. STEVENS, D.A. 1983. Miconazole in the treatment of coccidioidomycosis [review] [26 refs]. Drugs **26:** 347–354.

44. GRAYBILL, J.R., D. LUNDBERG, W. DONOVAN, *et al.* 1980. Treatment of coccidioidomycosis with ketoconazole: clinical and laboratory studies of 18 patients. Rev. Infect. Dis. **2:** 661–673.

45. CATANZARO, A., H. EINSTEIN, B. LEVINE, *et al.* 1982. Ketoconazole for treatment of disseminated coccidioidomycosis. Ann. Intern. Med. **96:** 436–440.

46. DEFELICE, R., J.N. GALGIANI, S.C. CAMPBELL, *et al.* 1982. Ketoconazole treatment of nonprimary coccidioidomycosis. Evaluation of 60 patients during three years of study. Am. J. Med. **72:** 681–687.

47. CATANZARO, A., P.J. FRIEDMAN, R. SCHILLACI, *et al.* 1983. Treatment of coccidioidomycosis with ketoconazole: an evaluation utilizing a new scoring system. Am. J. Med. **74**(1B): 64–69.

48. CRAVEN, P.C., J.R. GRAYBILL, J.H. JORGENSEN, *et al.* 1983. High-dose ketoconazole for treatment of fungal infections of the central nervous system. Ann. Intern. Med. **98:** 160–167.

49. STEVENS, D.A., R.L. STILLER, P.L. WILLIAMS & A.M. SUGAR. 1983. Experience with ketoconazole in three major manifestations of progressive coccidioidomycosis. Am. J. Med. **74**(1B): 58–63.

50. PHILLIPS, P., R. FETCHICK, I. WEISMAN, *et al.* 1987. Tolerance to and efficacy of itraconazole in treatment of systemic mycoses: preliminary results. Rev. Infect. Dis. **9**(Suppl. 1): S87–S93.

51. GALGIANI, J.N., D.A. STEVENS, J.R. GRAYBILL, *et al.* 1988. Ketoconazole therapy of progressive coccidioidomycosis. Comparison of 400- and 800-mg doses and observations at higher doses. Am. J. Med. **84**:603–610.

52. GRAYBILL, J.R., D.A. STEVENS, J.N. GALGIANI, *et al.* 1988. Ketoconazole treatment of coccidioidal meningitis. Ann. N. Y. Acad. Sci. **544**: 488–496.

53. TUCKER, R.M., P.L. WILLIAMS, E.G. ARATHOON & D.A. STEVENS. 1988. Treatment of mycoses with itraconazole. Ann. N. Y. Acad. Sci. **544**: 451–470.

54. CATANZARO, A., J. FIERER & P.J. FRIEDMAN. 1990. Fluconazole in the treatment of persistent coccidioidomycosis. Chest **97**: 666–669.

55. GRAYBILL, J.R., D.A. STEVENS, J.N. GALGIANI, *et al.* NAIAD Mycoses Study, Group. 1990. Itraconazole treatment of coccidioidomycosis. Am. J. Med. **89**: 282–290.

56. TUCKER, R.M., J.N. GALGIANI, D.W. DENNING, *et al.* 1990. Treatment of coccidioidal meningitis with fluconazole. Rev. Infect. Dis. **12**(Suppl. 3): S380–S389.

57. TUCKER, R.M., D.W. DENNING, E.G. ARATHOON, *et al.* 1990. Itraconazole therapy for nonmeningeal coccidioidomycosis: clinical and laboratory observations. J. Am. Acad. Dermatol. **23**(Suppl.): 593–601.

58. TUCKER, R.M., D.W. DENNING, B. DUPONT & D.A. STEVENS. 1990. Itraconazole therapy for chronic coccidioidal meningitis. Ann. Intern. Med. **112**: 108–112.

59. DIAZ, M., R. PUENTE, L.A. DE HOYOS & S. CRUZ. 1991. Itraconazole in the treatment of coccidioidomycosis. Chest **100**: 682–684.

60. GALGIANI, J.N., A. CATANZARO, G.A. CLOUD, *et al.* 1993. Fluconazole therapy for coccidioidal meningitis. Ann. Intern. Med. **119**: 28–35.

61. CATANZARO, A., J.N. GALGIANI, B.E. LEVINE, *et al.* 1995. Fluconazole in the treatment of chronic pulmonary and nonmeningeal disseminated coccidioidomycosis. Am. J. Med. **98**: 249–256.

62. GALGIANI, J.N., A. CATANZARO, G.A. CLOUD, *et al.* 2000. Comparison of oral fluconazole and itraconazole for progressive, nonmeningeal coccidioidomycosis. A randomized, double-blind trial. Mycoses Study Group. Ann. Intern. Med. **133**:676–686.

63. GALGIANI, J.N., N.M. AMPEL, J.E. BLAIR, *et al.* 2005. Coccidioidomycosis. Clin. Infect. Dis. **41**(9):1217–1223.

64. KERRICK, S.S., L.L. LUNDERGAN & J.N. GALGIANI. 1985. Coccidioidomycosis at a university health service. Am. Rev. Respir. Dis. **131**: 100–102.

65. VALDIVIA, L., D. NIX, M. WRIGHT, *et al.* 2006. Coccidioidomycosis as a common cause of community-acquired pneumonia. Emerg. Infect. Dis. **12**:958–962.

66. CABIB, E. 1991. Differential inhibition of chitin synthetases 1 and 2 from *Saccharomyces cerevisiae* by polyoxin D and nikkomycins. Antimicrob Agents Chemother. **35**: 170–173.

67. R.F. HECTOR, B.L. ZIMMER & D. PAPPAGIANIS. 1991. Inhibitors of cell wall synthesis. Nikkomycins. *In* Recent Progress in Antifungal Chemotherapy. H. Yamaguchi, G.S. Kobayashi & H. Takahashi, Eds. 341–353. Marcel Dekker, Inc. New York.

68. HECTOR, R.F., B.L. ZIMMER & D. PAPPAGIANIS. 1990. Evaluation of nikkomycins X and Z in murine models of coccidioidomycosis, histoplasmosis, and blastomycosis. Antimicrob. Agents Chemother. **34**: 587–593.

69. DERESINSKI, S.C. 2001. Coccidioidomycosis: efficacy of new agents and future prospects. Curr. Opin. Infect. Dis. **14**: 693–696.

Expanding Understanding of Epidemiology of Coccidioidomycosis in the Western Hemisphere

RAFAEL LANIADO-LABORIN[a,b]

[a]Facultad de Medicina Tijuana, Universidad Autónoma de Baja, California, Mexico

[b]Hospital General de Tijuana, ISESALUD de Baja California, Tijuana, Mexico

ABSTRACT: Coccidioidomycosis is a disease of both national and worldwide importance that is most often diagnosed in nonendemic regions. The endemic region for *Coccidioides* spp. lies exclusively in the Western Hemisphere. *Coccidioides* spp. has long been identified in semiarid areas of the United States and Mexico, and endemic foci have been described in areas of Central and South America. Infection is usually the result of activities that cause the fungus to become airborne and inhaled by a susceptible host. Underlying medical diseases that affect T cell function are known to increase the risk of disseminated disease and include human immunodeficiency virus, cancer, and disease processes requiring transplantation and its subsequent immunosuppressive agents. In recent years the incidence of the coccidioidomycosis has increased in California and Arizona, which may be partially due to the massive migration of Americans to the Sunbelt states. To date the highest number of cases reported in Arizona was in 2004, when a total of 3,665 cases of coccidioidomycosis was reported, representing a 281% increase since 1997. Statistics on the prevalence and incidence of coccidioidomycosis in Latin America either are fragmentary or simply are not available.

KEYWORDS: coccidioidomycosis; epidemiology; Western hemisphere

INTRODUCTION

Coccidioidomycosis is the oldest of the major mycoses.[1] The disease was described in 1892 and was first thought to be parasitic in nature.[2] It is caused by two nearly identical species, *Coccidioides (C.) immitis* and *C. posadasii*, generally referred as the "Californian" and non-Californian" species, respectively.[3] These two organisms are genetically different, but at this time they

Address for correspondence: Rafael Laniado-Laborin, M.D., M.P.H, F.C.C.P., Facultad de Medicina Tijuana, Universidad Autónoma de Baja, P.O. Box 436338, San Ysidro, CA 92143-6338. Voice/fax: 011-52-664-686-5626.
rafaellaniado@gmail.com

Ann. N.Y. Acad. Sci. 1111: 19–34 (2007). © 2007 New York Academy of Sciences.
doi: 10.1196/annals.1406.004

cannot be distinguished phenotypically nor is the disease or immune response to the organisms distinguishable.[4] This article discusses up to date issues in the epidemiology of coccidioidomycosis, as presented at the Sixth International Symposium on Coccidioidomycosis.

ECOLOGY

The endemic region for *Coccidioides* spp. lies exclusively in the Western Hemisphere, nearly all of it between the 40° latitudes north and south. This life zone corresponds with the hot deserts of the southwestern United States and northwestern Mexico (the Mojave, Sonoran, and Chihuahuan deserts). This region is situated below 4,500 feet where creosote (*Larrea tridentata*), jojoba, paloverde, mesquite, bursage, and cacti abound. The climate is arid with a yearly rainfall ranging from 10 to 50 cm, with extremely hot summers, winters with few freezes and alkaline, sandy soil.[5,6]

In the United States this semiarid zone encompasses the southern parts of Texas, Arizona, New Mexico, and much of central and southern California. Endemic regions have long been identified also in semiarid areas of Mexico and endemic foci have been described in areas of Central and South America (FIG. 1).[3]

Cases of coccidioidomycosis may also arise outside endemic areas. Such cases also occur because of a recent visit to an endemic area or infection through exposure to fomites from such an area.[7] In this setting the diagnosis is often delayed because the infection is not considered initially.[6]

RISK FACTORS FOR INFECTION AND DISEASE

Infection is usually the result of activities that cause the arthroconidia to become airborne and inhaled by a susceptible host. Coccidioidomycosis is not spread from person to person except in extraordinary circumstances. The main risk factors for acquiring the infection or developing active disease are discussed in the following subsections.

Exposure to Dust

Environmental conditions appear to have an important impact on coccidioidomycosis incidence. Some studies have identified associations linking climate and other factors to seasonal patterns of coccidioidomycosis and to interannual variability and trends in the disease. Significant variables included drought indices, precipitation, temperature, wind speed, and dust during the preceding one or more years.[8,9] Infection usually occurs during the dry season.

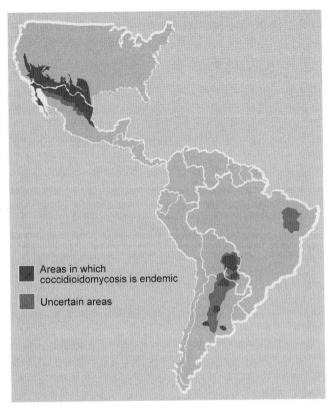

FIGURE 1. Geographic distribution of coccidioidomycosis. (From Hector and Laniado-Laborin.[80] Reproduced by permission.)

Because *Coccidioides* infects humans by the respiratory route, exposure to dust is one critical factor determining the risk of infection. The main risk factors for acquiring infection from *Coccidioides* spp. are activities that bring one into contact with dust from undisturbed soil in the endemic areas.[4] *Coccidioides* spp. are distributed unevenly in the soil and a majority of positive sites seem to be concentrated around animal burrows and ancient Indian burial sites. It is usually found 10 to 30 cm below the surface of the soil.[10,11]

Existing *Coccidioides* mycelia present in dry soil need increased soil moisture to grow, followed by a dry period during which fungal hyphae desiccate, mature, and form arthroconidia. Wind or other disturbance is required to fragment the hyphae and disperse the spores for inhalation by a host. On average, peaks in exposure to the fungal spores occur during the drier and dustier months of the year. Fewer exposures occur during the wetter and less dusty months.[12,13]

From 1991 through 1992 there was a dramatic increase in the number of cases of coccidioidomycosis reported from Kern County in the San Joaquin Valley, California, with 995 cases reported in 1991 and 3,027 cases in 1992.[14] After a 5-year drought in this region heavy rains fell in March 1991 and in February and March 1992. This increased precipitation may have brought on the germination of arthroconidia from mycelia accumulated over 5 years.

Dust storms in the endemic area are often followed by outbreaks of coccidioidomycosis. One particularly severe dust storm in 1977 carried dust from the San Joaquin Valley up to the San Francisco Bay area and resulted in hundreds of cases of nonendemic coccidioidomycosis in areas north of the San Joaquin Valley.[15]

Above this ambient risk occupational and recreational dust exposure as well as natural phenomena has occasionally caused outbreaks. Outbreaks of coccidioidomycosis have been described under several different circumstances: military maneuvers, construction work,[16] earthquakes,[17] model airplane competitions, and hunting (armadillo) expeditions.[18]

Coccidioidomycosis has long been and continues to be a threat to military personnel who reside or train in areas where *Coccidioides* spp. is endemic as the Army, Navy, Marines, and Air Force have traditionally deployed large numbers of personnel to endemic areas.[19] During World War II, when several training airfields were built in the San Joaquin Valley, California, coccidioidomycosis was the most common cause of hospitalization at many airbases in the southwest.[20] More recently, there was an outbreak of coccidioidomycosis among Navy SEALs during training exercises in Coalinga, California. Ten (45%) of 22 men had serologic evidence of acute coccidioidomycosis, the highest attack rate ever reported for a military unit. All patients were symptomatic, and 50% had abnormal chest radiographs.[19] Coccidioidomycosis must be considered an occupational disease that occurs with increased frequency among personnel exposed to the soil in endemic areas during military training.[21]

A coccidioidomycosis outbreak occurred in Ventura County, and was directly linked to dust clouds that emanated from landslides in the Santa Susanna Mountains caused by the Northridge earthquake in January 1994. In all, 170 cases were reported in a 7-week period following the earthquake. This outbreak is unusual in that Ventura County is not typically considered a hyperendemic area of coccidioidomycosis.[17]

Gender

Males are more often infected, which is likely related to occupational dust exposures; however, males also appear to be at a higher risk for dissemination, suggesting a hormonal or genetic component.[21] Drutz *et al.* studied the direct effect of human sex hormones and related compounds on the growth and maturation of *C. immitis in vitro.* 17β-estradiol, progesterone, and testosterone were highly stimulatory for the parasitic phase of *Coccidioides* spp. growth, whereas cholesterol, ergosterol, and 17α-estradiol (a physiologically inactive stereoisomer of 17β-estradiol), lacked such effects. Rates of spherule maturation and endospore release were accelerated, in a dose-dependent fashion, with the most striking effects seen at levels encountered in advanced pregnancy. A stimulatory effect of 17β-estradiol on the saprobic phase of fungal growth was also detected. This suggests that direct stimulation of *Coccidioides* spp. by human sex hormones may help to account for sex- and pregnancy-related predisposition to dissemination of coccidioidomycosis.[22]

Race

There is no known racial predilection for the acquisition of disease; however, disseminated disease occurs 10–175 times more often among Filipinos and African Americans. Whether Native Americans, Hispanics, or Asians have a higher risk is debated.[23] The 1977 dust storm in California provided a natural means of confirming this increased risk. The incidence of disseminated coccidioidomycosis in the non-Caucasian population was disproportionate to its overall representation.[1] During this wind-borne outbreak of coccidioidomycosis in the nonendemic disease region of Sacramento County, California, the rate per 100,000 of disseminated coccidioidomycosis among African American men compared with Caucasian men was 23.8 versus 2.5 (ratio 9.1:1). This difference could not be explained by differential exposure.[15] More recently, in the endemic area of Kern County, California, African American men had an adjusted odds ratio for disseminated coccidioidomycosis 28 times higher than that of any other ethnic group. The apparent variation in susceptibility among ethnic groups suggests that genetic factors influence the development of disseminated coccidioidomycosis.[1]

Although little is known about the role of T cells in eliminating *Coccidioides* spp., activated T cells elicit a delayed-type hypersensitivity (DTH) inflammatory response, indicating a Th1-type response. While DTH reactivity is regulated by class II HLA interactions with T cells, the host immune response to intracellular pathogens is primarily regulated by class I HLA molecules. Deresinski *et al.* found a significant association of blood group B and disseminated coccidioidomycosis. HLA-A9 and blood group B are both more common in persons of black and Filipino ancestry.[24]

Louie *et al.*[25] examined host genetic influences on coccidioidomycosis severity among class II HLA loci and the ABO blood group. Participants included African American, Caucasian, and Hispanic persons with mild or severe disseminated coccidioidomycosis. Among Hispanics, predisposition to symptomatic disease and severe disseminated disease is associated with blood types A and B, respectively. The HLA class II DRB1∗1301 allele marks a predisposition to severe disseminated disease in each of the three groups. Reduced risk for severe disease is associated with DRB1 ∗ 0301-DQB1 ∗ 0201 among Caucasians and Hispanics and with DRB1 ∗ 1501-DQB1 ∗ 0602 among African Americans. These data support the hypothesis that host genes, in particular HLA class II and the ABO blood group, influence susceptibility to severe coccidioidomycosis.

Immunosuppression

Underlying medical diseases that affect T cell function are known to increase the risk of disseminated disease including human immunodeficiency virus (HIV), cancer (particularly Hodgkin's disease), and disease processes requiring transplantation and subsequent immunosuppressive agents.

Dissemination among patients with cancer appears to be related to the immunosuppressive effect of the chemotherapy rather than radiation therapy or the nature of the disease itself.[23]

Coccidioidomycosis is a recognized opportunistic infection among persons infected with HIV. The first reports of coccidioidomycosis associated with acquired immunodeficiency syndrome (AIDS) occurred just a few years after the initial reports of AIDS.[26]

A prospective study in the late 1980s revealed that almost 25% of a cohort of HIV-infected individuals living in coccidioidal-endemic region developed symptomatic coccidioidomycosis within 3.5 years of follow-up.[27] Two predictive variables for the development of coccidioidomycosis were a peripheral blood CD4 lymphocyte count of <250 cells/μL and a diagnosis of AIDS.[27]

Although nearly 50% of the cases of coccidioidomycosis occurring in persons with AIDS were found to be from the coccidioidal endemic area (>90% from Arizona or California), the rest were from all other regions in the United States.[28] Therefore, the diagnosis of coccidioidomycosis should be considered in any immunosuppressed HIV-infected patient presenting with a compatible clinical syndrome.[26]

Early in the HIV epidemic, most cases presented as overwhelming diffuse pulmonary disease with a high mortality rate.[29] The incidence of severe symptomatic coccidioidomycosis has declined dramatically since the advent of potent antiretroviral therapy. Although these cases are still seen, they are typically in patients with previously undiagnosed HIV infection and extremely low peripheral blood CD4 cell counts.[26]

Pregnancy

Pregnant women have long been considered to be at increased risk of developing severe or disseminated coccidioidomycosis, presumably because of a general depression in cell-mediated immunity or because of changes in the levels of hormones that stimulate the growth of the fungus.[22]

A recent review of the literature identified 81 cases of coccidioidomycosis in pregnancy. Disseminated disease was strongly associated with the trimester of pregnancy: 50% of the cases diagnosed in the first trimester, 62% of the cases diagnosed in the second trimester, and 96% of the cases diagnosed in the third trimester had dissemination. In addition, African American women had a 13-fold increased risk of dissemination compared to that of white women.[30]

However, another viewpoint suggests that higher dissemination and mortality rates in pregnancy are contrary to the experience of practitioners and academic physicians in endemic areas and further, that maternal death is rare. It has been hypothesized that reports of increased maternal morbidity and mortality rates might be artifacts of reporting bias, which have led to an inaccurate portrayal of the natural history of coccidioidomycosis in pregnancy.[31]

Age

Coccidioidomycosis occurs in all age groups. In general, the incidence rate increases with age; the extremes of age carry a higher risk for complicated disease, including chronic pulmonary infection and dissemination.[23]

Solid-Organ Transplantation

Coccidioidomycosis is the most common endemic mycosis to cause disease in solid-organ transplant patients in North America.[32] Underlying renal and liver disease, T lymphocyte suppression from antirejection medication, and activation of immunomodulating viruses, such as cytomegalovirus, all increase the risk for coccidioidomycosis among these patients. About one-half of all cases are the result of reactivation of previously acquired coccidioidal infection and occur during the first year after transplantation. Although disseminated disease is common, most of these patients manifest with pulmonary symptoms.[32] Coccidioidomycosis has been reported in patients who receive organs from donors infected with the fungus.[33]

Hemodyalisis for Chronic Renal Failure

Dialysis patients are at increased risk for fungal infections compared to the general population, which substantially decreases patient survival. In a study by Abbott et al. dialysis patients had an age-adjusted incidence ratio for fungal infections of 9.8 compared to the general population, with candidiasis accounting for 79% of all fungal infections, followed by cryptococcosis (6.0%) and coccidioidomycosis (4.1%).[34]

RECENT TRENDS OF COCCIDIOIDOMYCOSIS
IN THE UNITED STATES

An estimated 150,000 new infections occur annually in areas of the southwestern United States. However, since coccidioidomycosis is not a nationally reportable disease (reportable only in Arizona and California), the exact incidence is unknown.

In recent years the incidence of the disease has increased in California and Arizona, which may be partially due to the massive migration of Americans to the Sunbelt states and, in particular, to Arizona, one of the fastest-growing states in the United States. The regions in Arizona in which *C. immitis* is most intensely endemic were previously sparsely populated and now contain major population centers, filled primarily with persons who have moved from areas where *C. immitis* was not endemic.[35] For example, Maricopa County (Phoenix) in 1950 had a population of 0.1 million;[35] in 2005 the estimated population had reached 3.6 million;[36] for Pima county (Tucson) population in 1950 was 0.1 million;[35] in 2005 it was estimated at 924,000.[36] Similar population expansion has also occurred in central California and west Texas.[35] As these populations have expanded in endemic areas, a growing segment of persons unusually susceptible to the most serious consequences of infection has also emerged.

In 1997 laboratory reporting of coccidioidomycosis became mandatory in Arizona. This was followed by a marked increase in the number of reported cases. To date the highest number of cases reported in Arizona was in 2004, when a total of 3,665 cases of coccidioidomycosis was reported (62.7 cases per 100,000 population), which represents a 281% increase since 1997 (958 cases).[37] From January to July 2006 the Arizona Department of Health Services reported 3,510 cases of coccidioidomycosis (compared to 1,425 cases during the same period in 2005; FIG. 2).[37]

Cases have recently been discovered outside areas previously identified as endemic, suggesting the endemic region may be wider than originally described.[23] In 2001 an outbreak of acute respiratory disease occurred among persons working at a Native American archeological site at Dinosaur National Monument in northeastern Utah. Ten workers met the clinical case definition; 9 had serologic confirmation of coccidioidomycosis, and 8 were hospitalized. All 10 were present during sifting of dirt through screens. This outbreak documents a new endemic focus of coccidioidomycosis, which extends northward its known geographic distribution in Utah by approximately 200 miles.[38]

COCCIDIOIDOMYCOSIS OUTSIDE THE ENDEMIC AREAS

Coccidioidomycosis is a disease of both national and worldwide importance that is often diagnosed in nonendemic regions, typically related to travel.[39] It

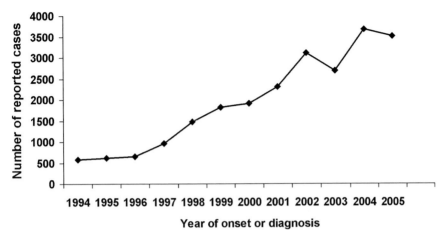

FIGURE 2. Reported cases of coccidioidomycosis, Arizona, 1994–2005. (Source: Arizona Department of Health Services, Infectious Disease Epidemiology Section.)

is usually diagnosed when individuals who live in a nonendemic region return home from visiting an endemic area.

For example, in July 1996 the Washington State Department of Health in Seattle was notified of a cluster of a flu-like, rash-associated illness in a 126-member church group. The group had recently returned from Tecate, Mexico, where members had assisted with construction projects at an orphanage. Eventually there were 21 serologically confirmed cases of coccidioidomycosis (attack rate, 17%) among this group.[16]

Chaturvedi et al.[39] reported that during a 5-year period (1992–1997), 161 persons in New York State had hospital discharge diagnoses of coccidioidomycosis, and from 1989 to 1997, 49 cultures from patients were confirmed as C. immitis; 26 of these patients had traveled to disease-endemic areas. Sixteen patient isolates were available for multilocus genotyping; all these patients had a history of travel to the Southwest, with 12 of 16 traveling to Arizona. Furthermore, while information on travel history was limited, all 16 patients from whom information was obtained had traveled to disease-endemic areas before becoming ill.

Coccidioidomycosis can create a clinical dilemma even in countries far away from the endemic areas. A 60-year-old Israeli resident traveled to Arizona, developed influenza-like infection, and returned to Israel with an airspace-occupying lesion in the lung. Since the patient was a heavy smoker, lung cancer was suspected and he was operated on. A granuloma with spherules was reported on stain preparations and C. immitis was isolated by culture.[40]

Diagnosis is often delayed because the infection is not considered initially.[6] Travelers visiting regions where Coccidioides spp. is endemic should be made aware of the risk of acquiring coccidioidomycosis, and health care providers should be familiar with the presenting signs and symptoms of this disease.

COCCIDIOIDOMYCOSIS IN LATIN AMERICA

Statistics on the prevalence and incidence of coccidioidomycosis in Latin America are either fragmentary or simply not available.

Mexico

Skin test surveys carried out in Mexico indicate that *Coccidioides* spp. infections are as prevalent there as in the endemic areas of the United States.[41] The studies by González-Ochoa (*Encuesta Nacional 1961–1965*) on skin testing with coccidioidin defined the epidemiologic distribution of coccidioidomycosis infection in three endemic zones in the country: the Northern zone, the Pacific Coast zone, and the Central zone, with variable rates of infection in the states of Baja California, Chihuahua, Colima, Coahuila, Durango, Guanajuato, Guerrero, Jalisco, Michoacán, Nayarit, Nuevo León, San Luis Potosí, Sinaloa, Tamaulipas, and Zacatecas.[42] More recently, coccidioidin skin test regional surveys for prevalence of infection have shown rates of 10% (Tijuana, Baja California, 1991[43]), 40% (Torreón, Coahuila, 1999[44]), and 93% (12 communities in the state of Coahuila, 2005[45]).

As mentioned, coccidioidomycosis is caused by two nearly identical species. To determine the prevalent species in northern Mexico, Bialek *et al.*,[46] through conventional nested PCR and real-time PCR assay, tested 120 clinical strains isolated within 10 years in Monterrey, Nuevo Leon, Mexico. All the strains corresponded to the Silveira strain (now known to be *C. posadasii*), as expected from the previous geographical studies by Fisher *et al.*[47]

In Mexico most clinical case reports originate in the northern region of the country. Since coccidioidomycosis is not a reportable disease, its true incidence is unknown.[48]

Tuberculosis and coccidioidomycosis share epidemiological, clinical, radiographic, and even histopathological features. Since tuberculosis is also endemic in Mexico, coccidioidomycosis and tuberculosis can coexist, making the correct diagnosis of both entities extremely difficult in such cases.[49]

Central America

In Central America coccidioidin surveys conducted more than 40 years ago, showed that 21% of children tested at the Motagua River Valley in Guatemala, gave positive reactions, and in the Comayagua Valley of Honduras, skin test surveys revealed an overall prevalence of 25% positivity among the subjects tested.[41] The first human case of coccidioidomycosis in Nicaragua was reported in 1979[50]; there are no published reports of prevalence of infection in that country.

South America

Argentina

Historically, Argentina is of the greatest interest because the first known case was reported by Posadas from that country in 1892.[2] Coccidioidomycosis is one of the three endemic systemic mycoses in Argentina (histoplasmosis and paracoccidioidomycosis being the other two). The endemic area includes the semidesert regions from Puna to Patagonia.[51,52]

Few coccidioidin skin test surveys have been carried out in Argentina, and thus the magnitude of infections in the endemic areas is unknown. In Santiago del Estero, a skin test survey by Negroni et al. revealed a prevalence of 19% positive reactions among 2,213 children aged 6 to 16 years.[41] In a more recent skin test survey in the Catamarca province, another skin test survey conducted by Negroni et al. in 827 children 6 to 15 years of age revealed a prevalence of infection of 16%. In 1979 in the province of San Luis, which included 1,609 school children and adults, Bonardello et al. reported a prevalence of 14.8%.[53]

Brazil

The first autochthonous cases of coccidioidomycosis in Brazil were reported in 1978 and 1979, the first case being from the State of Bahia[54] and the second one from Piauí. About 15 years later, the first micro-outbreak of this mycosis in Brazil was also reported in the State of Piauí.[55] Since then, the number of published cases has increased considerably. The association between this infection and the digging of armadillo (Dasypus novemcinctus) burrows has been described.[56] The fungus has already been isolated from tissues of this animal, from dogs, and from soil samples collected in armadillo burrows.[56] Currently, this systemic mycosis is considered endemic in the Northeast Brazilian States of Bahia, Ceará, Piauí, and Maranhão.[57]

Other Countries in South America

Little is known about areas of coccidioidomycosis in Paraguay and Bolivia. The probable endemic areas are in the Gran Chaco region, which both countries share with Argentina.[41] In Paraguay, Gomez[58] reported that coccidioidomycosis was endemic in the departments of Boqueron and Olimpo. He found that 44% of a group of Guazurangue Indians living in the department of Boqueron had positive reactions to coccidioidin.

There have been case reports of coccidioidomycosis from Colombia.[41,59] There are no reports of coccidioidin skin test surveys, and therefore the extension of the endemic area and the prevalence of infection remain unknown.

The states of Falcon, Lara, and Zulia in Venezuela have long been considered an endemic area for coccidioidomycosis on the basis of case reports and skin test surveys.[41,60]

COCCIDIOIDOMYCOSIS IN NONHUMAN HOSTS

The organism has been described in a wide spectrum of mammalian hosts and a few captive reptiles,[61] but no report of coccidioidomycosis in an avian species exists to date. Animals of virtually any age may be susceptible. It is not understood why some animals have no clinical signs of coccidioidal infection, whereas others develop disease that progresses even in the face of antifungal treatment.[62]

Coccidioidomycosis has been reported in armadillos,[56] cattle, sheep, dogs,[63] swine, horses,[64] burros, rodents, chinchillas,[65] coyotes,[66] cats,[67] and mountain lions (*Felis concolor*).[68] In addition, the disease has been reported in the following captive free-living wild animals: llamas (*Lama* spp.),[69] Bengal tigers (*Leo tigris*) maintained in a Davis, California, compound,[70] in a giant red kangaroo (*Macropus rufus*) shipped from Australia to the El Paso Zoo, Texas,[71] a tapir (*Tapirtis terrestris*),[72] a mountain gorilla (*Gorilla beringeri*) exhibited at the San Diego Zoo, California,[73] a sooty mangabey (*Cercocebus atys*) transported to Davis, California, from Sierra Leone,[74] and a gelada baboon (*Theropithecus gelada*) imported to Canada from Southern California.[75] Even marine species can acquire the infection and develop disease, including the Pacific bottlenose dolphin,[76] the California sea lion,[77,78] and the southern sea otter.[79]

CONCLUSIONS

Because of its apparent regional confinement, coccidioidomycosis is not perceived to have a substantial impact outside the areas classically considered as endemic. This view should be reconsidered, however. An estimated 150,000 new infections of coccidioidomycosis occur annually in areas of the southwestern United States. In recent years the incidence of clinically apparent disease has increased in California and Arizona, which may be partially due to the massive migration of Americans to the Sunbelt states; cases have recently been discovered outside the traditional areas, suggesting the endemic area may be wider than originally described.

Coccidioidomycosis is often diagnosed in nonendemic regions; diagnosis in that case is often delayed because the infection is not considered initially. Travelers visiting regions where *Coccidioides* spp. are endemic should be made aware of the risk of acquiring coccidioidomycosis, and health care providers should be familiar with the presenting signs and symptoms of this disease.

REFERENCES

1. EINSTEIN, H.E. & R.H. JOHNSON. 1993. Coccidioidomycosis: new aspects of epidemiology and therapy. Clin. Infect. Dis. **16:** 349–356.
2. POSADA, A. 1892. Un nuevo caso de micosis fungoidea con psorospermias. Ann. Circulo Medico Argentino **15:** 585–597.
3. HECTOR, R.F. & R. LANIADO-LABORIN. 2005. Coccidioidomycosis: A fungal disease of the Americas. PLoS Med. **2:** 2e.
4. CATANZARO, A. 2004. Coccidioidomycosis. Sem. Respir. Care Med. **25:** 123–128.
5. GALGIANI, J.N. 1993. Coccidioidomycosis. West. J. Med. **159:** 153–171.
6. PAPPAGIANIS, D.1988. Epidemiology of coccidioidomycosis. Curr. Top. Med. Mycol. **2:** 199–238.
7. DESAI, S.A., O.A. MINAI, S.M. GORDON, *et al.* Coccidioidomycosis in non endemic areas: a case series. Respir. Med. **95:** 305–309.
8. KOLIVRAS, K.N., P. JOHNSON, A.C. COMRIE & S.R. YOOL. 2001. Environmental variability and coccidioidomycosis (valley fever). Aerobiologia **17:** 31–42.
9. COMRIE, A.C. 2005. Climate factors influencing coccidioidomycosis seasonality and outbreaks. Environ. Health Perspect. **113:** 688–692.
10. MADDY, K.T. 1958. The geographic distribution of *Coccidioides immitis* and possible ecologic implications. Ariz. Med. **15:** 178–188.
11. KIRKLAND, T.N. & J. FIERER. 1996. Coccidioidomycosis: a reemerging infectious disease. Emerg. Infect. Dis. **2:** 192–199.
12. KOLIVRAS, K.N. & A.C. COMRIE. 2003. Modeling valley fever (coccidioidomycosis) incidence on the basis of climate conditions. Int. J. Biometeorol. **47:** 87–101.
13. PAPPAGIANIS, D. 1994. Marked increase in cases of coccidioidomycosis in California: 1991, 1992, and 1993. Clin. Infect. Dis. **19:** S14–S18.
14. CENTERS FOR DISEASE CONTROL AND PREVENTION. Update. 1994. Coccidioidomycosis—California, 1991–1993. MMWR Morb. Mortal. Wkly. Rep. **43:** 421–423.
15. PAPPAGIANIS, D. & H. EINSTEIN. 1978. Tempest from Tehachapi takes toll or *Coccidioides immitis* conveyed aloft and afar. West J. Med. **129:** 527–530.
16. CAIRNS, L., D. BLYTHE, A. KAO, *et al.* 2000. Outbreak of coccidioidomycosis in Washington State residents returning from Mexico. Clin. Infect. Dis. **30:** 61–64.
17. SCHNEIDER, E., R.A. HAJJEH, R.A. SPIEGEL, *et al.* 1997. A coccidioidomycosis outbreak following the Northridge, Calif, earthquake. JAMA **277:** 904–908.
18. WANKE, B., M. LAZERA, P.C. MONTEIRO, *et al.* 1999. Investigation of an outbreak of endemic coccidioidomycosis in Brazil's northeastern state of Piaui with a review of the occurrence and distribution of *Coccidioides immitis* in three other Brazilian states. Mycopathologia **148:** 57–67.

19. CRUM, N., C. LAMB, G. UTZ, *et al.* 2002. Coccidioidomycosis outbreak among United States Navy SEALs training in a *Coccidioides immitis*-endemic area-Coalinga, California. J. Infect. Dis. **186:** 865–868.
20. SMITH, C.E., R.R. BEARD, H.G. ROSENBERG & E.G. WHITTING. 1946. Effect of season and dust control on cocccidioidomycosis. JAMA **132:** 833–888.
21. AMPEL, N.M., M.A. WIEDEN & J.N. GALGIANI. 1989. Coccidioidomycosis: clinical update. Rev. Infectious Dis. **11:** 897–911.
22. CRUM, N.F., E.R. LEDERMAN, C.M. STAFFORD, *et al.* 2004. Coccidioidomycosis. A descriptive survey of a reemerging disease. Clinical characteristics and current controversies. Medicine **83:** 149–175.
23. DRUTZ, D.J., M. HUPPERT, S.H. SUN & W.I. MCGUIRE. 1981. Human sex hormones stimulate the growth and maturation of *Coccidioides immitis*. Infect. Immun. **32:** 897–907.
24. DERESINSKI, S.C., D. PAPPAGIANIS & D.A. STEVENS. 1979. Association of ABO blood group and outcome of coccidioidal infection. Sabouradia **17:** 261–264.
25. LOUIE, L., S. NG, R. HAJJEH, *et al.* 1999. Influence of host genetics on the severity of coccidioidomycosis. Emerg. Infect. Dis. **5:** 672–680.
26. AMPEL, N.M. 2005. Coccidioidomycosis in persons infected with HIV Type 1. Clin. Infect. Dis. **41:** 1174–1178.
27. AMPEL, N.M., C.L. DOLS & J.N. GALGIANI. 1993. Coccidioidomycosis during human immunodeficiency virus infection: results of a prospective study in a coccidioidal endemic area. Am. J. Med. **94:** 235–240.
28. JONES, J.L., P.L. FLEMING, C.A. CIESIELSKI, *et al.* 1995. Coccidioidomycosis among persons with AIDS in the United States. J. Infect. Dis. **171:** 961–966.
29. AMPEL, N.M., K.J. RYAN, P.J. CARRY, *et al.* 1986. Fungemia due to *Coccidioides immitis*: an analysis of 16 episodes in 15 patients and a review of the literature. Medicine. **65:** 312–321.
30. CRUM, N.F. & G. LANDA-BALLON. 2006. Coccidioidomycosis in pregnancy: case report and review of the literature. Am. J. Med. (online). **119:** 993.e11–e17.
31. CALDWELL, J.W., E.I. ARSURA, W.B. KILGORE, *et al.* 2000. Coccidioidomycosis in pregnancy during an epidemic in California. Obst. Gynecol. **95:** 236–239.
32. LOGAN, J.L., J.E. BLAIR & J.N. GALGIANI. 2001. Coccidioidomycosis complicating solid organ transplantation. Semin. Respir. Infect. **16:** 251–256.
33. BLAIR, J.E. 2006. Coccidioidomycosis in liver transplantation. Liver. Transpl. **12:** 31–39.
34. ABBOTT, K.C., I. HYPOLITE, D.J. TVEIT, *et al.* 2001. Hospitalizations for fungal infections after initiation of chronic dialysis in the United States. Nephron **89:** 426–432.
35. GALGIANI, J.N. 1999. Coccidioidomycosis: A regional disease of national importance. Rethinking approaches for control. Ann. Intern. Med. **130:** 293–300.
36. http://quickfacts.census.gov/qfd/states(accessed on 4 January 2007).
37. Arizona Department of Health Services Website: http://www.azdhs.gov/.
38. 2001.Coccidioidomycosis in workers at an archeologic site: Dinosaur National Monument, Utah, June–July 2001. MMWR **50:** 1005–1008.
39. CHATURVEDI, V., R. RAMANI, S. GROMADZKI, *et al.* 2000. Coccidioidomycosis in New York State. Emerg. Infect. Dis. **6:** 25–29.
40. LEFLER, E., D. WEILER-RAVELL, D. MERZBACH, *et al.* 1992. Traveller's coccidioidomycosis: case report of pulmonary infection diagnosed in Israel. J. Clin. Microbiol. **30:** 1304–1306.

41. Ajello, L. 1967. Comparative ecology of respiratory mycotic disease agents. Bacteriol. Rev. **31:** 6–24.

42. González-Ochoa, A. 1966. La coccidioidomicosis en México Rev. Invest. Salud Publ. (Mex). **26:** 245–262.

43. Laniado-Laborín, R., R.P. Cárdenas-Moreno & M. Álvarez-Cerro. 1991. Tijuana: zona endémica de infección por *Coccidioides immitis*. Salud Pública Mex. **33:** 235–239.

44. Papua, A., V. Martínez-Ordaz, V.M. Velasco-Rodriguez, *et al*. 1999. Prevalence of skin reactivity to coccidioidin and associated risk factors in subjects living in a northern city of Mexico. Arch. Med. Res. **30:** 388–392.

45. Mondragón-González, R., L.J. Méndez-Tovar, E. Bernal-Vázquez, *et al*. 2005. Detección de infección por *Coccidioides immitis* en zonas del estado de Coahuila, México. Rev. Argentina Microbiol. **37:** 135–138.

46. Bialek, R., J. Kern, T. Herrmann, *et al*. 2004. PCR assays for identification of *Coccidioides posadasii* based on the nucleotide sequence of the antigen 2/proline-rich antigen. J. Clin. Microbiol. **42:** 778–783.

47. Fisher, M.C., G.L. Koenig, T.J. White & J. W. Taylor. 2002. Molecular and phenotype description of *Coccidioides posadasii* spp. nov., previously recognized as the non-Californian population of *Coccidioides immitis*. Mycologia **94:** 73–84.

48. Castañón-Olivares, L.R., A. Aroch-Calderón, E. Bazán-Mora & E. Córdova-Martínez. 2004. Coccidioidomicosis y su escaso conocimiento en nuestro país. Rev. Fac. Med. UNAM **47:** 145–148.

49. Castañeda-Godoy, R. & R. Laniado-Laborín. 2002. Coexistencia de tuberculosis y coccidioidomicosis. Presentación de dos casos clínicos. Rev. Inst. Nal. Enf. Resp. Mex. **15:** 98–101.

50. Rios-Olivares, E.O. 1979. 1st human case of coccidioidomycosis in Nicaragua. Rev. Latinoam. Microbiol. **21:** 215–218.

51. Negroni, P., C.R. Bravo, R. Negroni, *et al*. 1978. Estudios sobre el *Coccidoides immitis*. Encuesta epidemiológica efectuada en la Provincia de Catamarca. Bol. Acad. Nal. Med. **56:** 327–339.

52. Masih, D.T., B.E. Marticorena, N. Borletto, *et al*. 1987. Epidemiologic study of bronchopulmonary mycosis in the Province of Cordova, Argentina. Rev. Inst. Med. Trop. São Paulo. **29:** 59–62.

53. Bonardello, N.M. & C.G. de Gagliardi. 1979. Intradermal reactions with coccidioidins in different towns of San Luis Province. Sabouraudia **17:** 371–376.

54. Gomes, O.M., R.R.P. Serrano, H.O.V. Prado, *et al*. 1978. Coccidioidomicose pulmonar: primeiro caso nacional. Rev. Assoc. Med. Bras. **24:** 167–168.

55. Bezerra, C., R, de Lima, M. Lazera,et al. 2006. Viability and molecular authentication of *Coccidioides immitis* strains from Culture Collection of the Instituto Oswaldo Cruz, Rio de Janeiro, Brazil. Rev. Soc. Brasileira Med. Trop. **39:** 241–244.

56. Eulalio, K.D., R.L. Macedo, M.A.S. Cavalcanti, *et al*. 2001. *Coccidioides immitis* isolated from armadillos (*Dasypus novemcinctus*) in the state of Piaui, northeast Brazil. Mycopathologia **149:** 57–61.

57. Nobre-Veras, K., B.C. de Souza-Figueirêdo, L.M. Soares-Martins, *et al*. 2003. Coccidioidomycosis: an unusual cause of acute respiratory distress syndrome. J. Pneumol. **29:** 45–48.

58. Gomez, R. F. 1950. Endemism of coccidioidomycosis in the Paraguayan Chaco. California Med. **73:** 35–38.

59. ROBLEDO, M.V. 1965. Coccidioidomicosis. Antioquia Med. **15:** 361–362.
60. CAMPINS, H. 1961. Coccidioidomicosis. Comentarios sobre la casuistica Venezolana. Mycopathol. Mycol. Appli. **15:** 306–316.
61. TIMM, K.I., R.J. SONN & B.D. HULTGREN. 1988. Coccidioidomycosis in a Sonoran gopher snake, *Pituophis melanoleucus affinis*. J. Med. Vet. Mycol. **26:** 101–104.
62. SHUBITZ, L.F. & S.M. DIAL. 2005. Coccidioidomycosis: a diagnostic challenge. Clin. Tech. Small Anim. Pract. **20:** 220–226.
63. SHUBITZ, L.F., C.D. BUTKIEWICZ, S.M. DIAL & C.P. LINDAN. 2005. Incidence of *Coccidioides* spp. infection among dogs residing in an endemic region. J. Am. Vet. Med. Assoc. **226:** 1846–1850,
64. ZIEMER, E.L., D. PAPPAGIANIS, J.E. MADIGAN, *et al.* 1992. Coccidioidomycosis in horses: 15 cases (1975–1984). J. Am. Vet. Med. Assoc. **201:** 910–916.
65. ASHBURN, L.L. & C.W. EMMONS. 1942. Spontaneous coccidioidal granuloma in the lungs of wild rodents. Archs. Path. **34:** 791–800.
66. STRAUB, M., R.J. TRAUTMAN & J.W. GREENE. 1961. Coccidioidomycosis in 3 coyotes. Am. J. Vet. Res. **22:** 811–812.
67. GREENE, R.T. & G.C. TROY. 1995. Coccidioidomycoisis in 48 cats: a retrospective study (1984–1993). J. Vet. Int. Med. **9:** 86–91.
68. ADASKA, J.M. 1999. Peritoneal coccidioidomycosis in a mountain lion in California. J. Wildlife Dis. **35:** 75–77.
69. FOWLER, M.E., D. PAPPAGIANIS & I. INGRAM. 1992. Coccidioidmycosis in llamas in the United States: 19 cases (1981–1989). J. Am. Vet. Med. Assoc. **201:** 1609–1614.
70. HENRIKSON, R.V. & E.L. BIBERSTEIN. 1972. Coccidioidomycosis accompanying hepatic disease in two Bengal tigers. J. Am. Vet. Med. Ass. **161:** 674–677.
71. HUTCHINSON, L.R., F. DURAN, C.D. LANE, *et al.* 1973. Coccidioidomycosis in a giant red kangaroo (*Macropus rufus*). J. Zoo Anim. Med. **4:** 22–24.
72. DILLEHAY, D.L., T.R. BOOSINGER & S. MACKENZIE. 1985. Coccidioidomycosis in a tapir. J. Am. Vet. Med. Assoc. **187:** 1233–1234.
73. MCKENNEY, F.D., J. TRAUM & A.E. BONESTELL. 1944. Acute coccidioidomycosis in a mountain gorilla (*Gorilla beringeri*) with anatomical notes. J. Am. Vet. Med. Assoc. **104:** 136–140.
74. PAPPAGIANIS, D., J. VANDERLIP & B. MAY. 1973. Coccidioidomycosis naturally acquired by a monkey, *Cercocebus atys*, in Davis, California. Sabouraudia **11:** 52–55.
75. RAPLEY, W.A. & J.R. LONG. 1974. Coccidioidomycosis in a baboon recently imported from California. Can. Vet. J. **15:** 39–41.
76. REIDARSON, T.H., L.A. GRINER, D. PAPPAGIANIS & J. MCBAIN. 1988. Coccidioidomycosis in a bottlenose dolphin. J. Wildlife Dis. **34:** 629–631.
77. REED, R.E., G. MIGAKI & J.A. CUMMINGS. 1976. Coccidioidomycosis in a California sea lion. J. Wildlife Dis. **12:** 372–375.
78. FAUQUIER, D.A., F.M.D. GULLAND, J.G. TRUPKIEWICZ, *et al.* 1996. Coccidioidomycosis in free-living California sea lions (*Zalophus californianus*) in central California. J. Wildl. Dis. **32:** 707–710.
79. CORNELL, L.H., K.G. OSBORN, J.E. ANTRIM JR. & J.G. SIMPSON. 1979. Coccidioidomycosis in California sea otter (*Enhydra lutris*). J. Wildl. Dis. **15:** 373–378.
80. HECTOR, R.F. & R. LANIADO-LABORÍN. 2005. Coccidioidomycosis: a fungal dease of the Americas. PLoS Med. **2(1):** e2.

Ecological Niche Modeling of *Coccidioides* spp. in Western North American Deserts

RAÚL C. BAPTISTA-ROSAS,[a,b] ALEJANDRO HINOJOSA,[c] AND MERITXELL RIQUELME[d]

[a]*Arid Land Ecosystem Management, School of Sciences, Autonomous University of Baja California (UABC), Ensenada, Baja California 22800, Mexico*

[b]*Military Regional Hospital El Cipres, Ensenada, Baja California 22780, Mexico*

[c]*Geographic Information Systems and Remote Sensing Laboratory, Department of Geology, Center for Scientific Research and Higher Education of Ensenada CICESE, Ensenada, Baja California 22860, Mexico*

[d]*Department of Microbiology, Center for Scientific Research and Higher Education of Ensenada CICESE, Ensenada, Baja California 22860, Mexico*

ABSTRACT: Coccidioidomycosis is an endemic infectious disease in western North American deserts caused by the dimorphic ascomycete *Coccidioides* spp. Even though there has been an increase in the number of reported cases in the last years, few positive isolations have been obtained from soil samples in endemic areas for the disease. This low correlation between epidemiological and environmental data prompted us to better characterize the fundamental ecological niche of this important fungal pathogen. By using a combination of environmental variables and geospatially referenced points, where positive isolations had been obtained in southern California and Arizona (USA) and Sonora (Mexico), we have applied Genetic Algorithm for Rule Set Production (GARP) and Geographical Information Systems (GIS) to characterize the most likely ecological conditions favorable for the presence of the fungus. This model, based on environmental variables, allowed us to identify hotspots for the presence of the fungus in areas of southern California, Arizona, Texas, Baja California, and northern Mexico, whereas an alternative model based on bioclimatic variables gave us much broader probable distribution areas. We have overlapped the hotspots obtained with the environmental model with the available epidemiological information and have found a high match. Our model suggests that the most probable fundamental ecological niche for *Coccidioides* spp. is found in the arid lands of the North American deserts and provides the methodological basis to further characterize the realized ecological niche of *Coccidioides*

Address for correspondence: Meritxell Riquelme, Department of Microbiology. Center for Scientific Research and Higher Education of Ensenada CICESE., P.O. Box 430222. San Ysidro, CA 92143-0222. Voice: +52-646-1750500; ext.: 27061; fax: +52- 646-1750595.
riquelme@cicese.mx

Ann. N.Y. Acad. Sci. 1111: 35–46 (2007). © 2007 New York Academy of Sciences.
doi: 10.1196/annals.1406.003

spp., which would ultimately contribute to design smart field-sampling strategies.

KEYWORDS: coccidioidomycosis; fungal ecology; endemic mycosis

INTRODUCTION

Coccidioidomycosis, also known as San Joaquin Valley Fever, is an endemic infectious fungal disease in western North American deserts caused by the dimorphic ascomycete *Coccidioides* spp. The fungus infects mainly mammals, including humans, but also reptiles, and does not get transmitted person to person or by a vector. The host acquires the disease via respiratory inhalation of arthroconidia disseminated in their natural habitat. The primary disease is auto-limited, although in fewer than 1% of the cases do develop complications that result in high morbidity and mortality.

The most important endemic areas in the United States are found in southern California and southern Arizona, and in Mexico, in the states of Sonora, Nuevo Leon, Coahuila, and Baja California. However, this distribution was established according to the epidemiological data gathered more than 30 years ago.[1] Currently, the prevalence of coccidioidomycosis in Mexico is unknown,[2] which presents a serious public health hazard.

Only a few positive isolations from environmental sampling have been obtained in the highly endemic zones in North America. By means of classical culturing methods, 23 positive isolations out of 723 soil samples were reported in California (13 sites),[3] 1 in Arizona,[3] and 1 in northern Mexico near Hermosillo, Sonora.[4] Similarly, in more recent studies, using microbiological selective isolation techniques followed by polymerase chain reaction (PCR) amplification, only 4 positive isolations out of 720 soil samples were obtained in California (3 sites),[5] whereas no positive detection was found from more than 150 soil samples in Arizona.[6]

We believe that a possible reason for such low number of positive isolations from those randomly obtained environmental samples could be due to the poorly defined sampling areas, rather than the methods used to process the collected samples. There is a long history of laboratory research aimed at characterizing the requirements and conditions to grow *Coccidioides,* but there is not a complete characterization of the *Coccidioides* ecological niche, which would include the total range of environmental conditions that are suitable for the existence of the fungus. Collectively, all these factors and the associated problems one encounters when sampling at a microscale limit the effectiveness of soil sampling. Bearing this in mind, we have used several either bioclimatic or environmental variables in combination with geo-referenced positive isolation points to generate a predictive model with the most likely fundamental niche(s) for *Coccidioides* in North American deserts, and to establish the basis to further conduct smart field-sampling strategies.

EXPERIMENTAL METHODS

Epidemiological Distribution of Coccidioidomycosis

As a spatial reference, we used the epidemiological distribution of coccid-ioidomycosis in North America reported by Gonzalez-Ochoa.[1] The published map was scanned, geo-referenced, and incorporated into a geographic informa-tion system (GIS)[2] to compare it with the predicted *Coccidioides* fundamental ecological niche generated from the model. To generate the maps and manage the spatial data, we employed the GIS software ArcView version 3.2.

Modeling

To model the most probable fundamental ecological niche of *Coccidioides* spp. we used the Genetic Algorithm for Rule Set Production (GARP) software version 1.1.6 developed by the University of Kansas Biodiversity Research Center (http://www.lifemapper.org/desktopgarp/). This modeling tool works in an iterative process of rule selection and evaluation, choosing a method from a set of possibilities.[7]

GARP includes several distinct algorithms, such as logistic regression, range, negated range, and atomic rules, in an iterative artificial intelligence–based approach.[8] Individual algorithms are used to produce component "rules" in a broader rule set, and hence portions of the species distribution may be determined within or without its niche, based on these different rules. This method searches for common variable values associated with the spatial oc-currence of a species and looks for other geographic locations where similar environmental variable values are found, while filtering out potential sources of errors. We compared two different models, one based on an environmental database comprising climatic and topographic variables and another based on a bioclimatic database. In both instances we used as reference points 18 sites with previous registries of positive isolations[3–6] dating from 1960 to 2002, which were referred to geographic coordinates.

We conducted 20 experiments and each of them was run either with 1,000 iterations or until convergence. The GARP method was trained and tested using 50, 75, 95, and 100% of the reference points. The output models obtained using 50% of the points were analyzed for consistency by the overlap method with GIS Arc View 3.2.

Bioclimatic Database

A set of 19 climate layers (climate grids) with a spatial resolution of a square kilometer (30 arcsec grid spacing) were used. These data sets, also known as bioclimatic variables, are derived from the monthly temperature and rainfall values in order to generate more biologically meaningful variables. They represent annual trends, seasonality, and extreme or limiting environmental factors. The data grids include annual mean temperature, mean diurnal range, isothermality, temperature seasonality, maximum temperature of warmest month, minimum temperature of coldest month, temperature annual range, mean temperature of wettest quarter, mean temperature of driest quarter, mean temperature of warmest quarter, mean temperature of coldest quarter, annual precipitation, precipitation of wettest month, precipitation of driest month, precipitation seasonality, precipitation of wettest quarter, precipitation of driest quarter, precipitation of warmest quarter, and precipitation of coldest quarter. The global bioclimatic variables can de downloaded from the Worldclim database[9] (http://www.worldclim.org/).

Environmental Database

In addition to modeling with the 19 bioclimatic variables, other GARP runs were tested with a combined set of climatic and topographic spatial variables referred to in the document as environmental variables. We considered 11 surface climatic variables obtained from the Climate Research Unit (CRU) Global Climate Dataset available through the Data Distribution Center (DDC) of the Intergovernmental Panel on Climate Change (IPCC) for the period 1961–1990 for western North America (http://ipcc-ddc.cru. uea.ac.uk/obs/cru_climatologies.html). These variables included mean, maximum and minimum temperature, precipitation, cloud cover, relative humidity, ground-frost frequency, sunshine percent, wet-day frequency, vapor pressure, and wind speed. These grid data sets have a spatial resolution of 30 arcmin or half of a degree.

For topography, we used 30 arcsec grids of digital elevation model and derived sets of slope, aspect, and compound topographic index. These data sets were obtained from the Center for the Earth Resources Observation and Science (EROS) of the U.S. Geological Survey (USGS) (http://edc.usgs. gov/products/elevation/gtopo30/hydro/namerica.html).

RESULTS

Both models identified areas of high response or elevated clusters for potential presence of *Coccidioides,* which we referred to as hotspots.[8] GARP modeling, when including the bioclimatic database and the positive isolation registries, predicted as probable *Coccidioides*-inhabiting areas practically all of what is known as the biome area of the Lower Sonoran desert habitat (FIG. 1). In contrast, GARP modeling using the environmental database and the positive

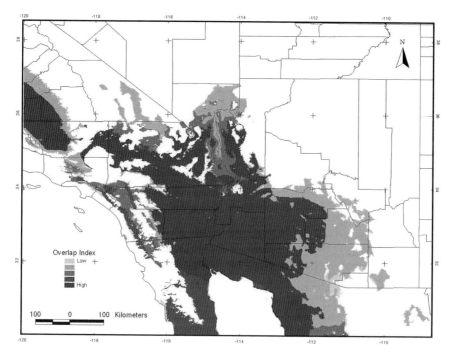

FIGURE 1. GARP modeling of the ecological niche of *Coccidioides* spp. using only bioclimatic variables in western North America. For reference, 18 registries of the fungus in soil samples were used: 16 in California and 1 in Arizona, United States, and 1 in Sonora, Mexico. Twenty experiments were analyzed with GIS ArcView 3.2 to obtain the overlap index, where high equals 100% coincidence and low equals less than 75%, shown in grayscale gradient. The pixel area is equivalent to 1 km^2.

isolation registries allowed us to identify more delimited areas for the presence of *Coccidioides* in California, Arizona, and Texas in the United States and in Baja California, Sonora, Chihuahua, Coahuila, Nuevo Leon, and San Luis Potosi in Mexico (Fig. 2). According to this modeling, in California the most probable hotspots included the southern part of San Joaquin Valley, the western region of Kern County, large extensions of Tulare, Madera, and Mariposa counties in the slope of Sierra Nevada, a great part of San Diego, Ventura, and Orange counties, and the western Mojave Desert in the San Bernardino and Inyo Counties. In Arizona there were important hotspots in the Mohave, Maricopa, Pinal, and Pima counties. Since we did not have fungal isolation data for Texas we could not include reference points of that state for our modeling purposes. However, our analysis identified important hotspots in the counties of Reeves, Loving, Ward, and Winkler, including an important probable hotspot in a region previously reported with a high prevalence near Midland-Odessa cities (Fig. 2).

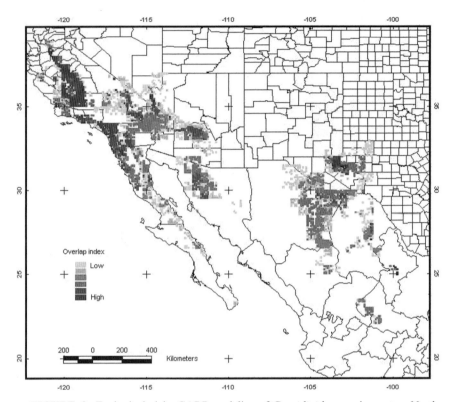

FIGURE 2. Ecological niche GARP modeling of *Coccidioides* spp. in western North America using as environmental data 11 climate and 4 topographic variables. For reference, 18 registries of the fungus in soil samples were used: 16 in California and 1 in Arizona, United States, and 1 in Sonora, Mexico. Twenty experiments were analyzed with GIS ArcView 3.2 to obtain the overlap index, where high equals 100% coincidence and low equals less than 75%, shown in grayscale gradient. The pixel area is equivalent to 10 km^2.

In Baja California we found the most important areas for the development of *Coccidioides* in Mexicali valley. The analysis found important hotspots in the intercoastal and inland valleys across the southern Baja California peninsula: Las Palmas, Santo Tomas, El Rosario, and El Socorro valleys, in the Pacific coast, and Puertecitos, located south of San Felipe, in the Baja Californian gulf coast. In Sonora the model showed probable areas in northern Hermosillo, from Rayon to Magdalena, and involving the municipalities of San Miguel de Horcasitas, Carbó, Opodeme, Benjamin Hill, and Santa Ana. We observed also important probable areas in the northern Mexico region from Sabinas to Monclova in the states of Chihuahua and Coahuila, and a few hotspots in the state of Nuevo Leon between Cadereyta, General Teran, and Montemorelos.

Most of the hotspots predicted by GARP using the environmental database and the positive registries (FIG. 2) fell within the well-known endemic areas

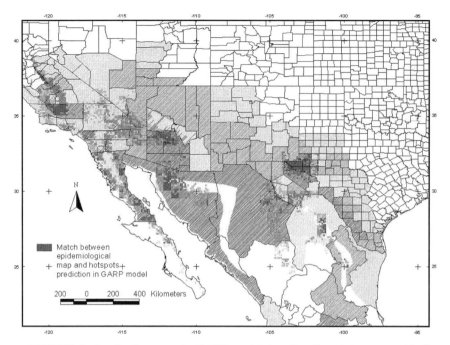

FIGURE 3. Overlap between the GARP model (see FIG. 2) and the geographically referred epidemiological data.[2]

(FIG. 3) identified previously on the basis of epidemiological data,[1] which included in the United States, the San Joaquin Valley (California), southern Arizona and southern Texas, and in Mexico, the northern Baja California peninsula, Sonora, Chihuahua, and Sinaloa. In addition, the model also predicted important hotspots in areas not previously described as endemic, such as the highlands of Coahuila in the Oriental Sierra Madre and the southern region of Baja California.

DISCUSSION AND CONCLUSIONS

Ecological Niche Modeling

Most previous studies of coccidioidomycosis spatial distribution have focused on epidemiological aspects of the disease. In this study, by using a combination of environmental variables and geospatially referenced points from the United States and Mexico, where positive isolations had been obtained, we have generated a possible distribution model for *Coccidioides* spp.

This study provides a better understanding of some of the factors that may determine the distribution of *Coccidioides* spp. in natural environments and suggests clues toward other unexplored factors that may be involved in this distribution. A better knowledge not only of the epidemiology but also of the fundamental ecological niche of this fungal pathogen will help us to understand its typical spatial variations. GARP modeling approaches have been applied earlier to predict the occurrence of animal species with high levels of statistical significance for each species tested.[10] GARP has also been used to model some infectious diseases and their related animal reservoirs.[11–14] The distribution of *Coccidioides* spp. obtained by GARP modeling using environmental variables showed that the predicted hotspots were confined to desert lands of North America, with important coincidence with the areas of highest incidence of the disease. However, these hotspots covered much smaller areas than those delimited by epidemiological testing. In some instances, few predicted hotspots were found out of the known endemic areas in Coahuila and southern Baja California. There were no earlier records of the disease in Coahuila, but more recently epidemiological reports have shown important prevalence rates there, which would support our modeling data.[15,16] Southern Baja California, is an unexplored region for this disease, which is why there are no registries in most of the epidemiological distribution references.

Although we modeled with the best available environmental and climatic data for North America, caution must be taken when interpreting the hotspots' spatial distribution obtained with GARP since they are likely to be biased toward the scarce positive soil samples and influenced by the incomplete specification of the various bioclimatic and environmental limiting variables. Altogether this suggests that other important components contributing to this apparent biased distribution of the disease need to be taken into account in future studies, such as more positive samplings, more detailed information on environmental and bioclimatic variables at different scales, spatial distribution of related potential hosts, and demographic variables, such as migration and urban development in high-risk areas of the desert environments.

In Baja California the predicted hotspots were found mainly in the northern inland valleys, while much less potential activity was predicted in the south. Most of the models indicated low probability of *Coccidioides* presence in the driest zones of the Colorado delta and the southern Baja California peninsula, probably because of the scarce seasonal rain over there. Recent bioclimatic analyses in Arizona have indicated that the weather is apparently one of the main factors involved in the increase of cases reported, although this relationship has proven statistically less important in California's endemic areas, which are typically characterized by a pattern of drought followed by rain.[17–20]

Several hypotheses have proposed that the humidity from the rain contributes to the completion of *Coccidioides'* life cycle.[18–20] This can be inferred from the fact that most members of the fungal kingdom are homoiohydrics, that is, the water level inside the cell depends directly of the surrounding substrate humidity. In addition, we believe that this correlation between pluviometry and coccidioidomycosis incidence rates can also be related to the dynamics of small-mammals' communities.[21,22] Desert blossoms provide more seeds, the main source of nutrients for many small-mammal species, including the ones probably related to the disease. This phenomenon changes the structure and composition of mice populations, which ultimately can result in enrichment of the soil either in form of excretes or in form of organic debris from mice carcasses of successive generations.

This theory is supported by recent bioclimatic findings that showed an increase in the number of cases reported in California during the 2–3 years of unusual rains that followed a drought period.[20]

The probable association of *Coccidioides* spp. ecological niche with small mammals and temporal occurrence of rains after drought can focus the sampling strategy at the microscale, since the hotspots we identified by GARP are 1 km^2 geographic areas in the better case. The search could be oriented to the small mammals' dens after the desert blossoms.

Ecology of Coccidioides *in Soils*

In the arid lands, as in other ecosystems, the microbial biomass is usually dominated by fungi. Of the 1.5 million fungal species that are estimated to exist worldwide, more than 288 fungal species belonging to 87 taxa have been described in North American arid lands,[23] and more recently, the highest microbial diversity has been demonstrated in desert environments.[24–27] There is not much current information on the prevalence of *Coccidioides* spp. in grounds of the affected areas. From the first description of the saprotrophic soil habitat for *Coccidioides* spp.,[28] diverse hypotheses have been proposed to explain their limited spatial distribution. The paradigm generally accepted is that the soil is a natural habitat for the fungus.[3] In general, alkaline soils in the endemic areas have rich levels of essential nutrients like iron, calcium, and magnesium. Nevertheless, there is evidence that this dimorphic fungus is able to grow in most desert soil types and has high tolerance to low pH and extreme temperatures in the barren zones.[29,30] *Coccidioides* tolerance to environmental stresses, such as alkaline grounds and high temperatures, along with the fact that it is a poor competitor among other microorganisms commonly found in soil samples outside the endemic zones, could explain why *Coccidioides* has specialized in such specific substrates. In addition, the high salinity of the soils seems to play an important role in the suppression of the development of *Coccidioides* antagonists.[29–31]

Coccidioides: Fundamental Niche versus Realized Niche

To further focus on the search for the niche of this important pathogen, we should also aim to characterize its realized ecological niche, which in addition to the characteristics of the fundamental niche considered in our study includes the associated biotic interactions and competitive exclusion.[6] *C. immitis* soil isolations have been associated with vestiges of human presence, mainly in Paleo-Indian villages, suggesting a correlation between the fungus and the increased content of organic debris in the ground at these sites.[32] Nevertheless, the only common variables in different grounds were the sandy and alkaline nature of the soil.

Some findings support the probable relationship between the disease and desert wild rodents, since the pathogen has been isolated from heteromyid species and their burrows.[33] There is also evidence of isolations in areas close to bats' caves.[3,29] The dispersion of the disease by necrophagia has been refuted and it has been concluded that the predators of the rodents do not disperse the disease through excretes.[34,35]

Many researchers have suggested that organic matter–enriched soils provide a suitable medium for the development of different fungal pathogenic species[33] and recently, some studies provided evidence of microclimatic conditions for fungal growth associated with heteromyids' dens.[36,37] Many studies have confirmed that spherules and endospores protected in biological fluids survive for long periods of time to extreme environmental conditions.[38] Nevertheless, until now, this theory could not be demonstrated because of the poor isolation efficiency in extensive samplings.[4–6,31–33,38,39]

This analysis will contribute to concentrate our efforts toward soil sampling in the most likely *Coccidioides*-occurring areas of northern deserts of Baja California, which are practically uninhabited and present a good state of conservation.[40,41]

REFERENCES

1. GONZÁLEZ-OCHOA, A. 1966. La Coccidioidomicosis en México. Rev. Invest. Salud. Publ. (Mex). **26:** 245–262.
2. BAPTISTA-ROSAS, R.C. 2006. Coccidioidomicosis en México: bioclimatología y modelación de probable nicho ecológico. MSc thesis. Autonomous University of Baja California. México.
3. SWATEK, F.E. 1970. Ecology of *Coccidioides immitis*. Mycopathol. Mycol. App. **40:** 3–12.
4. SOTOMAYOR, C., G.S. MADRID & E.A. TORRES. 1960. Aislamiento de *Coccidioides immitis* del suelo de Hermosillo, Sonora México. Rev. Latinoam. Microbiol. **3-4:** 237–238.
5. GREENE, D.R., G. KOENIG, M.C. FISHER, *et al.* 2000. Soil isolation and molecular identification of *Coccidioides immitis*. Mycologia **92:** 406–410.
6. TABOR, J., M. ORBACH, L. SHUBITZ, *et al.* 2002. Final Report: Disease Prevention through Detection of *Coccidioides immitis* in the Environment (Project TS 325). Final report to Association of Teachers of Preventive Medicine & Centers for Disease Control and Prevention, October 2002. Arizona College of Public Health,

College of Agriculture and Life Sciences, & Valley Fever Center for Excellence, University of Arizona, Tucson. http://arsc.arid.arizona.edu/dogfever/index.htm.

7. GUISAN, A.A. & N.E. ZIMMERMANN. 2000. Predictive habitat distribution models in ecology. Ecol. Model. **135**: 147–186.

8. TIWARI, N., C.M.S. ADHIKARI, A. TEWARI, *et al.* 2006. Investigation of geo-spatial hotspots for the occurrence of tuberculosis in Almora district, India, using GIS and spatial scan statistic. Int. J. Health Geograph. **5**: 33 (http://www.ijhealthgeographics.com/content/5/1/33).

9. HIJMANS, R.J., S.E. CAMERON, J.L. PARRA, *et al.* 2005. Very high resolution interpolated climate surfaces for global land areas. Int. J. Climatol. **25**: 1965–1978.

10. PETERSON, A.T. 2001. Predicting species' geographic distributions based on ecological niche modeling. The Condor **103**: 599–605.

11. PETERSON, A.T., V. SÁNCHEZ-CORDERO, J. RAMSEY, *et al.* 2002. Identifying mammal reservoirs for Chagas disease in Mexico via ecological niche modeling of primary point occurrence data of parasites and hosts. Emerg. Infect. Dis. **8**: 662–667.

12. FISHER, M.C., W.P. HANAGE, S. DE HOOG, *et al.* 2005. Low effective dispersal of asexual genotypes in heterogeneous landscapes by the endemic pathogen *Penicillium marneffei*. PLoS Pathog. **1**: 159–165.

13. FANG, L., L. YAN, S. LIANG, *et al.* 2006. Spatial analysis of hemorrhagic fever with renal syndrome in China. BMC Infect. Dis. **6**: 77 (http://www.biomedcentral.com/1471-2334/6/77).

14. ROGERS, J.D. & S.E. RANDOLPH. 2003. Studying the global distribution of infectious diseases using GIS and RS. Nat. Rev. Microbiol. **1**: 231–236.

15. PADUA Y GABRIEL, A., V.A. MARTÍNEZ-ORDAZ, V.M. VELASCO-RODRÍGUEZ, *et al.* 1999. Prevalence of skin reactivity to coccidioidin and associated risk factors in subjects living in a Northern City of México. Arch. Med. Res. **30**: 388–392.

16. MONDRAGÓN-GONZÁLEZ, R., L.J. MÉNDEZ-TOVAR, E. BERNAL-VÁZQUEZ, *et al.* 2005. Detección de infección por *Coccidioides immitis* en zonas del estado de Coahuila, México. Rev. Arg. Microbiol. **37**: 135–138.

17. MADDY, K.T. & J. COCCOZZA. 1964. The probable geographic distribution of *C. immitis* in México. Bol. Oficina Sanit. Panam. **57**: 44–54.

18. KOLIVRAS, K.N. & A.C. COMRIE. 2003. Modeling valley fever (Coccidioidomycosis) incidence on the basis of climate conditions. Int. J. Biometeorol. **47**: 87–101.

19. COMRIE, A.C. 2005. Climate factors influencing coccidioidomycosis seasonality and outbreaks. Environ. Health Perspect. **113**: 688–692.

20. ZENDER, C.S. & J. TALAMANTES. 2006. Climate control in valley fever in Kern County California. J. Biometeorol. **50**: 174–182.

21. ILLOLDI-RANGEL, P., M.A. LINAJE & V. SÁNCHEZ-CORDERO. 2002. Distribución de los mamíferos terrestres en la región del Golfo de California, México Anales del Instituto de Biología, Universidad Nacional Autónoma de México. Serie Zoología; **73**: 213–224.

22. CORTÉS-CALVA, P. & S.T. ÁLVAREZ-CASTAÑEDA. 2003. Rodent density anomalies in scrub vegetation areas as a response to ENSO 1997-98 in Baja California Sur, Mexico. Geofísica Internacional **42**: 547–551.

23. BILLS, G.F., M. CHRISTENSEN, M. POWELL & G. THORN. 2004. Saprobic soil fungi. *In* Biodiversity of Fungi. G.M. Mueller *et al.*, Eds.: 271–302. Elsevier Academic Press. Amsterdam.

24. GUNDE-CIMERMAN, N., J.C. FRISVAD, P. ZALAR, *et al.* 2005. Halotolerant and halophilic fungi. *In* Biodiversity of Fungi. Their Role in Human Life. S. K. Deshmukh & M. K. Rai, Eds.: 69–127. Science Publishers Inc.

25. SURYANARAYANAN, T.S. & D.L. HAWKSWORTH. 2005. Fungi from little explored and extreme habitats. *In* Biodiversity of Fungi. Their Role in Human Life. S. K. Deshmukh & M. K. Rai, Eds.:38–48. Science Publishers Inc.

26. GREEN, J.L., A.J. HOLMES, M. WESTOBY, *et al.* 2004. Spatial scaling of microbial eukaryote diversity. Nature **432:** 747–750.

27. FIERER, N. & R.B. JACKSON. 2006. The diversity and biogeography of soil bacterial communities. PNAS **103:** 626–631.

28. BECK, M.D. 1931. Diagnostic laboratory procedure, coccidioidal granuloma; epidemiology. California State Dept. of Public Health **57:** 16–25.

29. KRUTZSCH, P.H. & R.H. WATSON. 1978. Isolation of *Coccidioides immitis* from bat guano and preliminary findings on laboratory infectivity of bats with *Coccidioides immitis*. Life Sci. **22:** 679–684.

30. SORENSEN, R.H. 1967. Survival characteristics of biphasic *Coccidioides immitis* exposed to the rigors of a simulated natural environment. *In* Coccidioidomycosis: Papers from the Second Symposium on Coccidioidomycosis. L. Ajello, Ed.: 313–317. University of Arizona Press. Tucson, AZ.

31. SWATEK, F.E., D.T. OMIECZYNSKI & O.A. PLUNKETT. 1967. *Coccidiodes immitis* in California. *In* Coccidioidomycosis: Papers from the Second Symposium on Coccidioidimycosis. L. Ajello, Ed.: University of Arizona Press. 255–265. Tueson, AZ.

32. MADDY, K.T. & H.G. CRECELIUS. 1967. Establishment of *Coccidioides immitis* in negative soil following burial of infected animals and animal tissues. *In* Coccidioidomycosis: Papers from the Second Symposium on Coccidioidimycosis. L. Ajello Ed.:309–312. University of Arizona Press. Tucson, AZ.

33. LACY, G.H. & F.E. SWATEK. 1974. Soil ecology of *Coccidoides immitis* at Amerindian middens in California. Appl. Microbiol. **27:** 379–388.

34. MANTOVANI, A. 1978. The role of animals in the epidemiology of the mycoses. Mycopathology **65:** 61–66.

35. BORRELLI, D. 1977. *Coccidioides immitis*: transmission by necrophagia among animals. *In* Coccidioidomycosis: Current Clinical and Diagnostic Status. L. Ajello, Ed.:115–123. Miami, FL. Symposia Specialists.

36. HAWKINS, L.K. 1996. Burrows of kangaroo rats are hotspots for desert soil fungi. J. Arid. Environ. **32:** 239–249.

37. AYARBE, J.P. & T.I. KIEFT. 2000. Mammal mounds stimulate microbial activity in a semiarid shrubland. Ecology **81:** 1150–1154.

38. MADDY, K.T. 1965. Observations on *Coccidioides immitis* found growing naturally in soil. Ariz. Med. **22:** 281–288.

39. EGEBERG, R.O. & A.F. ELY. 1956. *Coccidioides immitis* in the soil of the southern San Joaquin Valley. Am. J. Med. Sci. **23:** 151–154.

40. Comisión Nacional de Áreas Protegidas (CONANP). 1995. Programa de manejo Reserva de la Biosfera Alto Golfo de California y Delta del río Colorado.

41. Comisión Nacional de Áreas Protegidas (CONANP). 2002. Valle de los Cirios, tesoro natural de Baja California. Ensenada, Baja California, México.

Coccidioides Niches and Habitat Parameters in the Southwestern United States

A Matter of Scale

FREDERICK S. FISHER,[a] MARK W. BULTMAN,[b] SUZANNE M. JOHNSON,[c] DEMOSTHENES PAPPAGIANIS,[c] AND ERIK ZABORSKY[d]

[a]*Department of Geosciences, University of Arizona, Tucson, Arizona, USA*

[b]*U.S. Geological Survey, Tucson, Arizona, USA*

[c]*Department of Medical Microbiology and Immunology, University of California, Davis, California, USA*

[d]*U.S. Bureau of Land Management, Hollister, California, USA*

ABSTRACT: To determine habitat attributes and processes suitable for the growth of *Coccidioides,* soils were collected from sites in Arizona, California, and Utah where *Coccidioides* is known to have been present. Humans or animals or both have been infected by *Coccidioides* at all of the sites. Soil variables considered in the upper 20 cm of the soil profile included pH, electrical conductivity, salinity, selected anions, texture, mineralogy, vegetation types and density, and the overall geomorphologic and ecological settings. Thermometers were buried to determine the temperature range in the upper part of the soil where *Coccidioides* is often found. With the exception of temperature regimes and soil textures, it is striking that none of the other variables or group of variables that might be definitive are indicative of the presence of *Coccidioides.* Vegetation ranges from sparse to relatively thick cover in lower Sonoran deserts, Chaparral-upper Sonoran brush and grasslands, and Mediterranean savannas and forested foothills. No particular grass, shrub, or forb is definitive. Material classified as very fine sand and silt is abundant in all of the *Coccidioides*-bearing soils and may be their most common shared feature. Clays are not abundant (less than 10%). All of the examined soil locations are noteworthy as generally 50% of the individuals who were exposed to the dust or were excavating dirt at the sites were infected. *Coccidioides* has persisted in the soil at a site in Dinosaur National Monument, Utah for 37 years and at a Tucson, Arizona site for 41 years.

Address for correspondence: Frederick S. Fisher, 520 North Park Ave., Suite 355, Tucson, AZ 85719. Voice: 520-670-5578 or 970-596-2734 (cell); fax: 520-670-5571.
ffisher@swfo.arizona.edu

Ann. N.Y. Acad. Sci. 1111: 47–72 (2007). © 2007 New York Academy of Sciences.
doi: 10.1196/annals.1406.031

KEYWORDS: habitat; niche; soil; temperature; texture; endemic zones; sand; silt; clay; porosity; salinity; microorganisms

INTRODUCTION

Coccidioides is a saprophytic fungus occurring in arid and semiarid regions of the New World (FIG. 1). Soils in these areas are the natural reservoir of the organism and are critical for its survival. In the United States it is endemic in parts of Arizona, California, Nevada, New Mexico, Texas, and Utah. It is also endemic in northern Mexico, Central America, and scattered areas in South America. The endemic zones are, with some exceptions, generally arid to semiarid with mild winters, and long hot seasons (FIG. 2). Two species of *Coccidioides* have been described[1]: *Coccidioides immitis*, found in the central valley of California, southern California, and northern Mexico; and *Coccidioides posadasii*, found in the parts of the endemic area outside of California.

Areas where *Coccidioides* is present in soils may be divided into two major types: growth sites and accumulation sites. Growth sites are where physical, chemical, and biological conditions are suitable for completion of the

FIGURE 1. *Coccidioides* endemic zones in North, Central, and South America. Endemic areas are shown in gray. Blowup shows endemic areas in the United States, including the 2001 outbreak site in Utah.

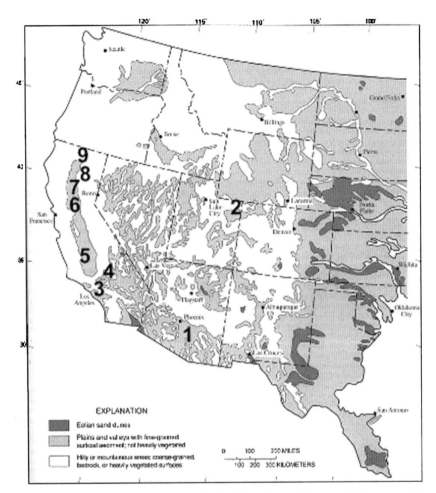

FIGURE 2. *Coccidioides* sites sampled by the authors. 1: Tucson, Arizona area, four sites, (3-2B, MEG, MWB, and DP); 2: SS, Swelter Shelter, Dinosaur National Monument, Utah; 3: Santa Susana Mts., near Simi Valley, California, sampled general area, not a specific site; 4: STH, Sharktooth Hill, near Bakersfield, California; 5: WC, White Creek, Diablo Range, western Fresno County, California; 6: CV, Capay Valley, near Brooks, California; 7: WH, Whiskey Hill, 20 miles west of Williams, California; 8: RS, Richardson Springs, near Chico, California; 9: DC, Dye Creek, near Red Bluff, California. Dark gray color = Eolian sand dunes; medium gray color = plains and valleys with fine-grained surficial sediment, not heavily vegetated; light gray = hilly or mountainous areas, coarse-grained sediments, bedrock, or heavily vegetated surfaces. (Map modified from Breed.[2])

entire growth cycle required by the organism. These growth sites are the focus of this study. Accumulation sites are where arthroconidia (spores) of *Coccidioides* may be deposited on or near the soil surface after being transported from growth sites by wind, water, organisms, or anthropogenic means.

Accumulation sites may, under the right circumstances, evolve into growth sites if the necessary environmental conditions are present for growth and if the arthroconidia are protected from adverse environmental or biological conditions. But in many situations the viability of the arthroconidia (especially on or near the soil surface) may be destroyed by exposure to excessive heat, ultraviolet light, or lack of moisture. The time required, *in situ,* for the viability to be lost is unknown, but undoubtedly varies with differing environmental circumstances. It is important to note that soils in both sites may be infectious to animals and humans at any time of the year under the right conditions.

Field and laboratory studies by numerous authors have demonstrated that many physical, chemical, biological, and temporal factors influence the growth of *Coccidioides.*[3–8] Key factors include: amount and timing of rainfall and available moisture, soil humidity, soil temperature, soil texture, alkalinity, salinity, organic content of soils, degree of exposure to sunlight and ultraviolet light, and competition with other microorganisms or plant species or both. Other factors that may be important are the presence of soils derived from marine sedimentary rocks, presence of borates, soil chemistry, presence of Indian middens, and presence of rodent burrows.[7–10]

Trying to identify, rank as to importance, and describe *Coccidioides* soil niches is a daunting task. Soil is one of the most complex natural systems known, involving a wide range of physical, chemical, and biological processes operating and interacting with each other on a vast range of temporal and spatial scales. Soils are continually changing in response to water supply, temperature, organic activity, erosion and deposition, disturbances (both natural and anthropogenic), and time. These changes can take place slowly, over years, decades, or millennia (e.g., continental erosion and deposition, weathering of parent rocks and minerals, and global climate change), or rapidly over months, days, or minutes (e.g., movement of salts and other soluble compounds on account of capillary forces, microbial growth, flooding, and localized erosion due to wind and/or water). Spatially, the endemic zones for *Coccidioides* are measured in hundreds of kilometers, while individual growth sites within endemic areas are measured in meters and centimeters, and niches within the soil profile (defined by physical, chemical, and biological phenomena and associated processes) are measured on scales of millimeters (e.g., soil pore space) or even nanometers (e.g., water films on clays).

ENDEMIC ZONES

Zones endemic for *Coccidioides* were defined by observations and cataloging the occurrence of coccidioidomycosis in humans and other animals and by skin testing of human populations and cattle.[11–13] Maddy[14] pointed out the concurrence of the areas defined by coccidioidomycosis cases and skin testing with the Lower Sonoran Life Zone (LSLZ) as defined and described by

Merriman.[15] Maddy[14] further suggested (p. 154) that *Coccidioides* can only propagate within the LSLZ. Other researchers have pointed out that there are known *Coccidioides* sites within areas characterized by a Mediterranean climate with different types of flora and fauna from those found in the LSLZ.[7] However, sites examined where *Coccidioides* populations exist in Utah, and in central and northern California, do not share many characteristics with the LSLZ. Most notable is the difference in the types and densities of vegetative ground cover (FIGS. 4–11). In addition, throughout the entire endemic zone, water and temperature are arguably the two main factors, operating at all scales that control and limit the growth and propagation of *Coccidioides*.

Water

Precipitation in the endemic zones of the southwestern United States is characterized by two main regional patterns. The first is a winter rainy season, when most of the annual precipitation occurs that is followed by a prolonged dry season. The second pattern is a rainy winter season followed by a spring dry period and then a second, monsoonal rainy period, during the summer months, which is in turn followed by a drier fall season. The first pattern is characteristic of California, while the second is more typical of Arizona, New Mexico, Texas, and Utah. Annual precipitation within endemic zones generally ranges between 5 cm and 50 cm; however, in northern California near known (FIG. 2) *Coccidioides* sites, annual amounts as great as 66 cm in the Richardson Springs (Mud Creek) area, 63 cm in the Williams (Whiskey Hill) area, and 61 cm in the Red bluff (Dye Creek) area have been recorded (TABLE 1). Rainwater that reaches the ground may evaporate, run off, or infiltrate into the soil. In most cases it will be dispersed by all three methods to varying degrees. More important than the total amount of precipitation is the amount of water that infiltrates the soil profile to depths where *Coccidioides* may exist.

Infiltration of the soil is governed by numerous factors. Topography of the land surface directly affects infiltration as steeper slopes enhance runoff, allowing less time for infiltration, while in flat or depressed areas water will pond or drain off more slowly, allowing more time for infiltration. Infiltration is also affected by the intensity and duration of a precipitation event. Heavy rainfalls of short duration may exceed the infiltration capacity of the soil, resulting in greater runoff than would occur if the same amount of rain fell over a longer period of time. Different types and densities of vegetation can retard runoff, thereby increasing infiltration rates. Plant roots can also enhance infiltration by providing pathways in the soil for water movement. A rough soil surface can slow the runoff and increase the infiltration rate. Infiltration will not occur if the soil is saturated (pore spaces filled with water) prior to a precipitation event.

TABLE 1. Climate factors (data from the nearest government weather station to each site)

Sites	Ambient air temperature (°C)				Annual precipitation (cm)
	Average high	Average low	Extreme high	Extreme low	
3-2B[1]	27.7	12.7	47.2	18.8	30.5
MEG-s-06[1]	22.7	12.7	47.2	18.8	30.5
MWB-05[1]	22.7	12.7	47.2	18.8	30.5
DP-06[1]	22.7	12.7	47.2	18.8	30.5
SS2-05[2]	17.7	−0.5	43.3	−40	21.6
SV[3]	21.1	12.7	43	−5	30.5
STH-06[4]	25	11.6	46.1	−7.2	15.2
WC1-06[5]	22.2	8.3	43.3	−12.2	38.1
CV[6]	23.8	8.8	44.4	−6.6	57.9
WH[7]	24.4	8	46.1	−6	63.5
RS[8]	23.8	8.3	47.7	−12.2	66.0
DC[9]	23.8	10	48.8	−7.7	60.9

[1] Tucson, Arizona.
[2] Dinosaur National Monument, Utah.
[3] Santa Susana Mts., California.
[4] Sharktooth Hill near Bakersfield, California.
[5] White Creek, Diablo Range, western Fresno County, California.
[6] Capay Valley, near Brooks, California.
[7] Whiskey Hill, west of Williams, California.
[8] Richardson Springs, near Chico, California.
[9] Dye Creek, near Red Bluff, California.

Porosity becomes the most important controlling factor as water, driven by gravity, penetrates the soil profile. Soil pores are the open space (voids) between individual soil particles not occupied by solids. Soil pores are always filled with air or water or both in varying proportions. Porosity is determined by the texture, structure, and organic material content of the soil. Permeability is the ability of a soil to transmit water and is also dependent on grain size and shape. Soils with high amounts of same-sized sand particles will have larger continuous pores and will rapidly transmit water and air. In comparison, clay-rich soils, because of the small size of individual particles, will have low permeability and transmit water slowly because of poor connectivity between soil pores and swelling, when wet, of individual clay particles.

Soil texture is the proportion of individual soil particles in different size groups (FIGS. 3, 12). The size most important for microorganisms ranges from clay (<0.002 mm) to sand (0.5 to 2.0 mm). Soil structure is the arrangement of masses of individual soil particles into larger aggregates (sometimes called peds) that are held together by clay minerals, organic material, iron oxides, fungi hyphae, and bacterial polysaccharides.[16] Aggregates are further classified by their size, shape, and cohesion. Between aggregates the pore spaces

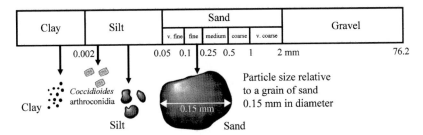

FIGURE 3. Size classification of soil particles. Note the relative size of *Coccidioides* arthroconidia compared to sand, silt, and clay particles (U.S. Department of Agriculture).

and their interconnections are larger than the spaces between individual soil particles and, as such, water infiltration is enhanced in well-aggregated soils. When the soil pores are completely filled with water, the soil is said to be saturated and the pore water will eventually drain to lower levels in the profile in response to gravity. Soils that are frequently or continually saturated with water (e.g., swamps, wetlands, and poorly drained areas), that is, areas where the water table is at or very near the surface, are not considered favorable habitats for *Coccidioides*. As water drains from the profile, it is replaced with air. The soil atmosphere in well-aerated soils is similar to air, composed mainly of nitrogen, oxygen, and carbon dioxide. However, respiration by organisms consumes oxygen and releases carbon dioxide; thus in poorly aerated soil, soil

FIGURE 4. 3-2B *Coccidioides* site on the alluvial slope of the Santa Catalina Mountains, Tucson, Arizona. Dust from excavation of a ditch infected three humans and three dogs. One dog and one adult human male had severe cases of coccidioidomycosis.

FIGURE 5. Landscape adjacent to Swelter Shelter *Coccidioides* site at the edge of an ancient flood plain of the Green River, Dinosaur National Monument, Utah.

with low permeability, and soil within aggregates, carbon dioxide levels are elevated and oxygen levels may fall, even to anaerobic levels.

Available water in the pore spaces, after all of the gravitational water has drained, is held by capillary forces and surface tension between water and solid soil materials. The soil is then described as being at field capacity. Evapotranspiration by plants depletes capillary water and it is also used for nutrient diffusion by microorganisms. As soils dry, water films around soil particles become thinner and bacterial activity and nutrient availability becomes limiting for many microorganisms. Ultimately with continued drying, microorganism activity is confined to filamentous organisms that can use hyphae growth to reach water unavailable to bacteria. It can be speculated that this indeed is a favorable situation for *Coccidioides* to prevail, in drier soils with temperatures too high in the upper soil profiles for robust bacterial activity.

In addition to precipitation, water for *Coccidioides* growth can be supplied by permanent or intermittent streams within, or originating outside of, but flowing through, endemic areas. Numerous *Coccidioides* growth sites are within or on the margins of riparian areas adjacent to these streams. This is the case for the Capay Valley site (FIG. 2, site number 6) within the riparian area along Cache Creek (FIG. 11), Red Bluff (FIG. 2, site number 9) about 100 m from Dye Creek, and Richardson Springs (FIG. 2, site number 8) within 100 m of Mud Creek. It is also reported that sites positive for *Coccidioides* are not uncommonly found next to mostly dry arroyos that carry water only during times of sporadic cloudbursts.[7,8]

FIGURE 6. Landscape surrounding Sharktooth Hill *Coccidioides* site near Bakersfield, California. Note the extensive grasslands. The density of ground cover varies with the amount of winter precipitation.

At the Swelter Shelter *Coccidioides* site (FIG. 2, site number 2) in Dinosaur National Monument, Utah annual rainfall is 21 cm. The soils have a very high percentage of fine sandy material with a low water-holding capacity and thus the soils are frequently exceedingly dry. However, directly above the midden area associated with the archeology site is an extensive steeply dipping rock face whereby during rainstorms water runs off the rock face directly onto the Swelter Shelter midden, forming a localized area in the soil favorable for microorganisms including *Coccidioides*.

Swatek[7] suggested that water requirements for *Coccidioides* growth in natural soils can be supplied by relative (soil?) humidities between 56% and 90% for several weeks. Freidman *et al.*[17] pointed out that dry arthroconidia (*Coccidioides* spores) survived for 6 months at temperatures between 15°C and 37°C and at relative humidities (laboratory controlled) greater than 10% and as high as 95%. Also at a humidity of 10% and a temperature of 37°C there was a significant loss of viability. Maddy[14] suggests that elevated soil humidities (humidity of the air within the pore spaces not filled with water) following rainstorms may reach the optimum for *Coccidioides* growth. It is important to point out that because of restrictive connections between soil pore spaces, soil humidities often cannot equilibrate with atmospheric humidities. Under very dry conditions in soils at or even below the wilting point of plants, water is still present as tightly bound films on soil particles and soil aggregates. Thus, the relative humidity within soil pore space is commonly high, often near 100%. In desert environments soil humidities (even near the surface) have been

FIGURE 7. White Creek *Coccidioides* site in the Diablo Range, western Fresno County, California. The site is at the base of the large rock outcrop, (upper left corner of photograph); note the dense chaparral and grass understory.

frequently measured above 85%. At the same time the relative humidity of the atmosphere above the soil was at 15%.[18]

Temperature

Temperature, second to water, is the most important factor affecting the physical, chemical, and biological processes within the known endemic zones for *Coccidioides*. Temperature affects microorganisms in many ways, including controlling the rates of respiration, rates of growth for which there are different limiting and optimal temperatures for different organisms, and microbial rhythms, such as dispersal and sporulation. At the microscale, in soil niches, temperature affects oxidation–reduction potentials, pressure, volume, diffusion of soil gases, capillary action, Brownian movement, viscosity, and surface tension.

Soil temperatures are a reflection of the ambient air temperatures and vary accordingly, with a lag time that increases with depth in the soil profile. If the air temperature is greater than the soil temperature, heat will be transferred to the soil. If the soil is warmer than the air, heat will be transferred to the atmosphere above the soil. Large fluctuations in the air temperature result in large fluctuations in soil temperatures. The amplitude of diurnal and seasonal temperature fluctuations decreases with depth in the soil profile. At 50 cm the change in temperatures due to varying surface conditions is not pronounced

FIGURE 8. Richardson Springs (Mud Creek) *Coccidioides* site at the extreme southern end of the Cascade Mountains about 8 miles northeast of Chico, California. The site is on a low terrace of Mud Creek; the soils are alluvial material derived from basaltic volcanic sediments, lava flows, breccias, and tuffs, and the climate is Mediterranean.

and at 10 m the soil temperature is nearly constant. Numerous factors affect soil temperature including latitude, altitude, insolation, cloud cover, slope aspect, soil color, organic content, vegetation cover, snow cover, and moisture content. In the endemic zone for *Coccidioides* in the southwestern United States mean annual air temperatures can range from –0.5°C to 24.4°C with extreme lows and highs of –40°C to 48.8°C (TABLE 1). Surface temperatures of soils in the endemic zone range from well below freezing to over 80°C (a temperature lethal for many microorganisms). Spot (isolated individual temperature determinations at selected sites at any given date) temperature readings in our study on July 21, 2006 from a soil in western Fresno County, California near White Creek were taken with an ambient air temperature of 42°C. The soil was dark gray and had a surface temperature of 63°C. At a depth of 2 cm the temperature was considerably less at 35°C, and at a depth of 20 cm the temperature was 30°C. Spot temperatures from a soil at Richardson Springs, California on July 22, 2006 were made with an ambient air temperature of 46°C. Again, the temperature of the dark brown soil at the surface was considerably higher at 70°C. At a 2-cm depth, the temperature had fallen to 40°C, and at 20-cm depth it was 29°C. Comparable measurements in July 2006 at Dye Creek near Red Bluff, California had an ambient air temperature of 47°C. The dark brown soil had a temperature of 82°C at the surface and was 46°C at a 2-cm depth. At 20-cm depth the temperature had dropped to 27°C. These spot temperatures demonstrate that even with high ambient air and higher yet (possible lethal)

FIGURE 9. Dye Creek *Coccidioides* site is located 14 miles southeast of Red Bluff, California in the foothills of the Cascade Mountains. Seventeen of 39 participants of an archaeological dig at this site were infected. The soils are a mixture of alluvial and residual material derived from underlying volcanic lava flows, tuffs, and lahars (mudflows). The climate is Mediterranean; note oak and pine trees and the ground cover of yellow star thistle and needle grass.

soil surface temperatures, conditions at a depth of 20 cm are well within the temperature range for optimal growth of *Coccidioides*.

Most, perhaps all, known *Coccidioides* growth sites in Arizona are within soils classified as hyperthermic arid or thermic arid and semiarid (FIG. 13). It is postulated that this is also true for most soils positive for *Coccidioides* throughout the endemic zones outside of Arizona. Hyperthermic soils have an average annual temperature that is greater than 22°C at 50-cm depth. Thermic soils have an average annual temperature between 15°C and 22°C at 50-cm depth. In both instances, if the interface between the soil and the underlying bedrock is less than 50 cm, the temperature is taken at the rock–soil interface. In addition, both hyperthermic and thermic soils have a greater than 5°C difference between the mean annual summer and the mean annual winter temperature measured at 50-cm depth.

Coccidioides has frequently been isolated from soils at depths from 2 to 20 cm. A study of soil temperatures that included these depths was made for soil profiles from Arizona, California, and Utah (TABLE 2), which is a statistical summary of the results collected by single self-recording thermometers buried at the indicated depth for about 1 year. Most thermometers were set to record a value every hour; approximately 8,760 individual readings from each thermometer. At three sites (MEG-nw, MEG-s, and MWB) the thermometers

FIGURE 10. Landscape adjacent to the Whiskey Hill *Coccidioides* site 20 miles west of the town of Williams, California. The mountainous area is within the California Coast range. The soils are residual material derived from Cretaceous and Jurassic marine sandstone and shale and the climate is Mediterranean.

were set to record a value every 16 min for a year's duration. So for those sites each row of statistical values in TABLE 2 is derived from a population of approximately 32,850 readings from each thermometer. Perhaps, most notable, are the data from the Swelter Shelter site and two nearby locations (most likely negative for *Coccidioides*) in Dinosaur National Monument. Temperatures at these three locations are significantly colder than all of the other sites in Arizona and California at all levels in the soil profiles. Also notable for most sites are the maximum temperatures at the 2-cm depth, which range between 54°C and 60°C with the exception being the OF site in Dinosaur National Monument at 49°C. Temperatures in the high 50s and greater may be lethal for *Coccidioides*, depending on the humidity and duration of exposure.[7,14,17] At the 20-cm depth maximum temperatures were mostly in the 30s and 40s, temperatures possibly close to the optimal for *Coccidioides* growth.

At Dinosaur National Monument three thermometers were placed in soil profiles (2, 10, and 20 cm) at several locations in and near the Swelter Shelter *Coccidioides* site. One profile (SS2-05) was directly within the midden area at the base of a steeply dipping rock outcrop with a shallow overhang, the floor of which was excavated during an archeological dig in 1961. The second profile (OF-05) was located in an open area approximately 100 m east of SS2-05 and away from the bottom edge of the rock face. The third profile (LC-05) was located at the base of the rock outcrop approximately 100 m south of SS2-05. This overall arrangement was designed to test the hypothesis that the southern exposure of SS2-05 and its location at the base of the rock was an

FIGURE 11. Capay Valley *Coccidioides* site located near the town of Brooks, California and within the riparian area of Cache creek. Eleven of 23 individuals were infected after excavating this archaeological site. The photograph is looking from the site across Cache Creek. The landscape adjacent to the riparian area is open grasslands with scattered oak trees and the climate is Mediterranean.

environment in which *Coccidioides* could grow; in large part because of the additional moisture washing off the steeply dipping rock face and the warmer soils associated with the southern exposure, lack of shade throughout the entire year, a thin vegetation cover, and heat reflected off the rock face. Temperatures from SS2-05 and LC-05 are somewhat higher than those recorded at OF-05, with median temperatures of 14.2 to 14.8°C versus 11.7 to 12.3° C, respectively. While these temperature differences are not great, they are satisfactory for the growth and propagation of *Coccidioides*, as human infections occurred at Swelter Shelter in 1964 during an archeological dig and again 37 years later, in June 2001, during archeological work associated with construction of a wall at the site, and yet again in September 2001, when work was restarted at the site and additional infections occurred.[20,21]

The overall colder temperatures associated with the Swelter Shelter area and in particular the SS2-05 *Coccidioides* site are clearly shown by the median values in FIGURE 14. For Swelter Shelter area samples, 50% of the values lie above the median temperature of 14.9°C, which is within the range considered acceptable for the growth of *Coccidioides*.[8,14] Sites MEG-s and MEG-nw are approximately 200 m apart with similar vegetation density, but MEG-s is on a slightly steeper slope that faces directly south whereas MEG-nw faces northeast. The higher median temperatures at MEG-s versus MEG-nw are the result of the sun aspect and slope gradient on soil temperatures. The overall

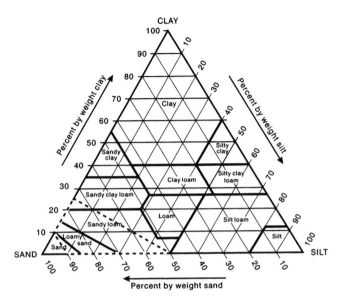

FIGURE 12. Textural triangle of sand, silt, and clay. Soils from all *Coccidioides* sites examined to date have textures that lie within the area bounded by the *heavy dashed lines* (U.S. Department of Agriculture).

distributions for all sites are similar at both depths but at the 20-cm depth the maximum temperatures are 10 to 20 degrees colder than the 2-cm maximums. There is less difference between the minimum temperatures at 20 cm, which are mostly 2 to 10 degrees warmer than the 2-cm depth (FIG. 14).

In summary, the growth of *Coccidioides* may be described by six temperature regimes (TABLE 3). The temperature regimes were determined on the basis of laboratory data and field observations and are by no means meant to be definitive. They are at best speculations that require refinement as additional data are obtained. The temperature regimes are: (1) >55° C: lethal in a relatively short time (hours or days, shorter with lower humidities); (2) 40–55° C: somewhat limiting at higher temperatures (especially under drier conditions); (3) 20–40°C: optimal growth temperature range; (4) 5–20°C: growth becoming marginal at lower temperatures in this range; (5) 0–5° C: mostly dormant but arthroconidia still viable; and (6) <0°C: dormant, but viable and capable of growth with increasing temperatures; possibly lethal if exposed to repeated freezing and thawing.

Most remarkable is the large amount of time the soils at both the 2-cm and the 20-cm depths are within temperature regimes favorable for the growth and propagation of *Coccidioides*. Some regional differences are noteworthy. At the 20-cm depth, soils at the Swelter Shelter site are within the 20°C to 40°C range an average of 32% of the year (~ 2,800 h/year). In comparison, the two California sites have soil temperatures in the 20°C to 40°C range

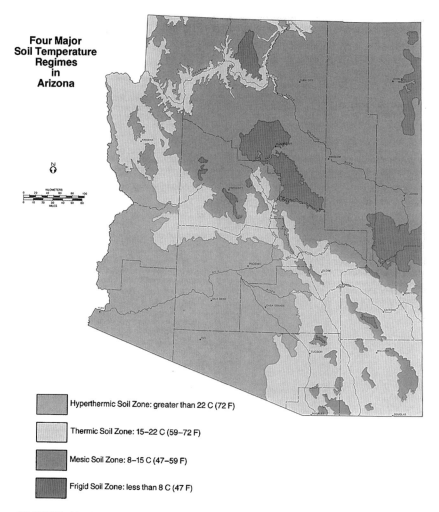

Four Major Soil Temperature Regimes in Arizona

Hyperthermic Soil Zone: greater than 22 C (72 F)

Thermic Soil Zone: 15–22 C (59–72 F)

Mesic Soil Zone: 8–15 C (47–59 F)

Frigid Soil Zone: less than 8 C (47 F)

FIGURE 13. Soil temperature regimes in Arizona (modified from Hendricks[19]).

an average of 56% of the year (~ 4,900 h/year), and the four Arizona sites with soils in the 20°C to 40°C range, an average of 66% of the year (~ 5,700 h/year). One possible conclusion requiring additional study is that *Coccidioides* may be more prevalent in Arizona soils than California soils. Because of the longer duration of time, Arizona soils are at temperatures most favorable for *coccidioides* growth. If this hypothesis is valid then it may suggest that, all things being equal, Arizona soils may be more infective than soils in California. Also noteworthy is the relatively little amount of time soils at the 2-cm depth are warmer than 55°C. However, soils at White Creek were over 55°C for

TABLE 2. Soil temperature profiles at *Coccidioides* **sites**

Collection site	Depth (cm)	Soil temperature profile (°C)					
		Maxi-mum	Mini-mum	Mean	Median	25 Quartile	75 Quartile
DP-06[1]	2	No data					
DP-06[1]	10	45.3	5.4	27.4	27.9	20.9	33.2
DP-06[1]	20	40.3	6.8	26.3	27.2	20.2	31.7
MEG-s-06[1]	2	60.5	−0.04	25.4	24.9	15.8	33.4
MEG-s-06[1]	6	48.5	3.75	24.7	24.8	17.1	31.5
MEG-s-06[1]	12	43.5	6.9	25.3	25.4	18.2	31.8
MEG-s-06[1]	27	39.4	13.7	26.0	26.2	19.3	32.4
MEG-nw-06[1]	2	54.5	0.37	22.5	22.5	13.4	29.6
MEG-nw-06[1]	9	48.2	3.9	23.0	22.7	14.6	30.3
MEG-nw-06[1]	15	41.4	7.2	23.4	23.2	15.3	30.8
MEG-nw-06[1]	31	38.1	12.5	24.1	23.6	16.5	30.9
MWB-05[1]	2	54.0	1.97	24.7	24.6	15.7	32.6
MWB-05[1]	10	43.3	7.4	24.6	24.9	15.5	32.8
MWB-05[1]	20	40.0	9.3	24.9	25.1	15.9	32.6
SS2-05[2]	2	57.4	−3.4	16.1	14.8	5.8	24.2
SS2-05[2]	10	43.0	−1.0	15.7	14.9	5.7	24.7
SS2-05[2]	20	35.6	−0.9	14.6	14.2	5.0	23.5
LC-05[2]	2	56.2	−1.6	16.4	14.4	6.1	24.2
LC-05[2]	10	43.8	−0.52	15.4	14.3	5.6	24.0
LC-05[2]	20	35.7	0.69	15.1	14.3	5.9	23.9
OF-05[2]	2	49.8	−6.5	14.4	12.3	3.1	22.5
OF-05[2]	10	39.1	−5.0	12.9	11.8	2.2	22.5
OF-05[2]	20	31.7	−2.9	12.1	11.7	2.4	22.4
STH-06[3]	2	55.7	2.7	24.5	22.5	14.1	33.8
STH-06[3]	10	47.7	5.7	24.7	23.0	15.2	34.3
STH-06[3]	20	45.8	9.8	24.5	24.1	15.9	33.3
WC1-06[4]	2	59.8	2.7	24.1	21.8	13.0	32.6
WC1-06[4]	10	44.3	6.3	23.2	22.5	13.8	32.1
WC1-06[4]	20	38.7	7.4	22.0	21.8	13.2	30.0

All values measured over a year's duration with individual readings taken mostly at 1-h intervals throughout the year.
[1] Tucson, Arizona.
[2] Dinosaur National Monument, Utah.
[3] Sharktooth Hill near Bakersfield, California.
[4] White Creek, Diablo Range, western Fresno County, California.

1.7% of the year ($\sim$ 149 h). Detailed examination of the temperature data from White Creek shows that temperatures greater than 55°C occurred for five continuous hours on 14 days, for four continuous hours on 9 days, and for three continuous hours on 12 days during July and August of 2005 and July of 2006.

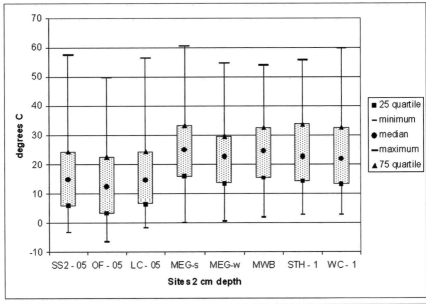

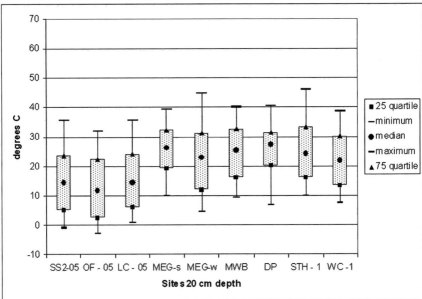

FIGURE 14. Box plots of soil temperatures at *Coccidioides* sites. Utah (SS2-05, OF-05, LC-05), Arizona (MEG-s, MEG-w, MWB, DP), and California (STH-1, WC-1). Data are not available for the 2-cm depth at the DP site. These box plots use TABLE 2 data and give the data maximum and minimum values, the median as the measure of central tendency, and the box containing the middle 50% of the data delineated by the 25th and 75th quartiles.

Texture

As previously mentioned texture is the relative proportions of sand, silt, and clay in a given soil (FIGS. 3, 12). Soils from all of the *Coccidioides* sites examined in this study so far fall within the domain shown by the heavy dashed line on FIGURE 12. These soils are best described as loamy sand or sandy loams. In general, soils with a high amount of sand-sized material have relatively low water-holding capacity, are more susceptible to wind erosion, and most often are poorly aggregated. Mineralogically, sand-sized material in most soils is composed primarily of quartz with much smaller (1 or 2%) amounts of other silicate minerals (e.g., feldspar, pyroxene, hornblende, and mica). Quartz (SiO_2) is chemically quite inert, and as such, its predominance in sandy soils may make it somewhat less favorable for robust permanent microbial colonies, given its poor nutritional status.

Silt-sized material is, in most cases, still dominated by the mineral quartz. The much smaller overall size of the individual silt particles (FIG. 3) and consequently greater surface areas and larger volume of pore spaces give silt a greater water-holding capacity than sand. Silts generally have a lower permeability than sand because the connections between pore spaces in silt are narrower and more restrictive to the movement of water. Because of the greater surface areas in silts, silicate minerals (other than quartz) will chemically degrade faster, providing nutrients for utilization by microorganisms.

The clay-sized fraction of soils is unquestionably of great importance to all soil microorganisms. Some of the clay-sized material in soils is composed of primary rock-forming minerals (quartz, feldspar, and mica). However, most of the clay-sized particles are composed of secondary minerals formed from the weathering and chemical breakdown of the primary minerals. These secondary clay-sized minerals are the most important determinants of the chemical characteristics of most soils. Clay minerals are hydrous aluminum phyllosilicates with mineralogical structures built from layers of SiO_4 tetrahedral bound to layers of (Al, Mg, Fe)(O,OH) octahedral cations. Cations of potassium, sodium, and calcium may also often be bound to the negatively charged clay particles. Clays have a much greater surface area, higher cation exchange capacity, and greater water-holding capacity than sand or silt. Clays are chemically active and their negative charge allows for considerable interaction with organic materials and microorganisms (TABLE 2).[22] The ionic charges on clays provide for the slow release of nutrients to the pore environment. Clays may adhere to the surface of larger silt and sand particles, thereby providing additional areas for microbial colonization.

The distribution of fine-grained surficial sediments in the western United States is shown in FIGURE 2. Comparison of FIGURE 1 with FIGURE 2 shows a close correspondence between the areas mapped as endemic for *Coccidioides* and the distribution of fine-grained material in the southwestern United States. Texture analysis of soils from all of the *Coccidioides* sites examined in this

TABLE 3. Percentage time on an annual basis that soil temperatures were within ranges favorable or unfavorable for the growth of *Coccidioides*

Sites[1]	Collection dates	Reading frequency	Total hours	Percentage time in each temperature range					
				<0°	0–5°	5–20°	20–40°	40–55°	>55°
2-cm depth									
MEG-s[2]	3/18/05–3/14/06	16 min	8,672	0.003	2.1	32.9	51.1	12.9	0.8
MEG-nw[2]	3/18/05–3/14/06	16 min	8,672	0	3.3	39.5	48.8	8.2	0
MWB[2] (5 cm)	2/4/05–1/29/06	16 min	8,663	0	1.3	37.7	49.0	11.6	0
DP-06[2]	no data	–	–	–	–	–	–	–	–
SS2-05[3]	6/20/04–6/20/05	1 h	8,753	5.0	16.7	41.2	32.5	4.0	0.1
OF-05[3]	6/20/04–6/20/05	1 h	8,753	2.8	26.0	38.9	28.5	3.2	0
LC-05[3]	6/20/04–6/20/05	1 h	8,752	1.1	19.5	42.9	31.3	4.6	0.07
STH-1[4]	7/08/05–7/20/06	1 h	9,001	0	0.9	42.6	42.6	13.7	0.07
WC-1[5]	7/10/05–7/21/06	1 h	9,034	0	1.1	44.9	41.1	11.8	1.7
20-cm depth									
MEG-s[2]	3/18/05–3/14/06	16 min	8,672	0	0	29.6	70.3	0	0
MEG-nw[2] (15cm)	3/18/05–3/14/06	16 min	8,672	0	0	39.9	59.0	1.0	0
MWB[2] (17cm)	2/4/05–1/29/06	16 min	8,662	0	0	34.7	64.7	0.17	0
DP-06[2]	2/16/06–11/17/06	1 h	6,577	0	0	25.4	74.3	<1	0
SS2-05[3]	6/20/04–6/20/05	1 h	8,753	0.8	23.1	43.3	32.4	0	0
OF-05[3]	6/20/04–6/20/05	1 h	8,753	17.0	16.0	37.0	29.5	0	0
LC-05[3]	6/20/04–6/20/05	1 h	8,752	0	20.8	45.3	33.6	0	0
STH-1[4]	7/08/05–7/20/06	1 h	9,001	0	0	40.9	58.9	0	0
WC-1[5]	7/10/05–7/21/06	1 h	9,034	0	0	45.2	54.7	0	0

[1] Unless noted, thermometers were placed at either 2- or 20-cm depths.
[2] Tucson, Arizona.
[3] Dinosaur National Monument, Utah.
[4] Sharktooth Hill near Bakersfield, California.
[5] White Creek, Diablo Range, western Fresno County, California.

study was done by dry-sieving samples of all soil material less than 2 mm in size. The total percentage of soil material between 0.05 mm and 0.25 mm in size at different *Coccidioides* sites is shown in TABLE 4. This very fine-to-fine sand-sized soil fraction is notably present at all sites visited during this study and may be a characteristic common to all *Coccidioides* sites in the southwestern United States. In some sites (SS2-05, STH-06, WC1-06, DP) this very fine sandy material, comprising mostly quartz (SiO_2), makes up the major proportion of the soil. All of these sites were associated with notably high rates of infection (often >70%).

Chemical Characteristics

The role of oxygen, carbon, and nitrogen was not assessed in this study, but these elements are all necessary for the growth of *Coccidioides*. Furthermore, these elements must be available in an environment that provides a physical and chemical setting whereby *Coccidioides* can complete its specific biological functions required for life. At the centimeter scale, in the upper parts of sandy well-aerated soils common to the endemic areas of the southwestern United States, the availability of oxygen is unlikely to be limiting to the growth of *Coccidioides*. However, deeper within the soil profile (at a millimeter or smaller scale), if the permeability is low, anaerobic conditions may exist and become a limiting factor.

Carbon and nitrogen required for fungal growth are derived from organic material contained in the soils. The distribution of organic material (controlled largely by vegetation) within the endemic areas of the southwestern United States ranges from sparse in desert areas to abundant in Mediterranean savannas and forested foothills. Visual estimates of organic material in soil samples collected in this study suggest that soils from the deserts of Arizona contain significantly less organic material than the soils collected from California.

One goal of this study was to search for characteristics that might be common to all of the soils from the 13 different known *Coccidioides* sites. The variables investigated are summarized in TABLE 4 and included pH, electrical conductivity, and concentration of fluoride, nitrates, and sulfate. A paste derived from a one-to-one extract with distilled water was used to determine the pH. Arizona soils are slightly more basic and have a smaller range in values than soils from the California sites. The overall range in pH from 6.1 to 8 suggests that, in the natural environment, pH is not a limiting factor for the growth of *Coccidioides* (at least in mid range values). Salinity, derived from the electrical conductivity, is high in three sites and relatively low in all the rest. High sulfate values at Swelter Shelter, Sharktooth Hill, and soils from near Simi Valley (Santa Susana Mts.) are due to the presence of gypsum (calcium sulfate), which was not identified at any of the other sites. The high nitrate

TABLE 4. Physical and chemical parameters of *Coccidioides* sites

Sites	% fines[1]	pH 1:1 extract	Elect. conduct.[2] (µS/cm)	Salinity (mg/L)	Fluoride (µg/g)	Chloride (µg/g)	Nitrate (µg/g)	Sulfate (µg/g)
32B[3]	36	7.76	490	247	<2.0	4.23	68.6	16.7
MEG-s[3]	35	7.96	631	332	<2.0	9.90	97.5	44.9
MEG-nw[3]	28	7.86	349	176	no data	no data	no data	no data
MWB-05[3]	25	7.61	448	225	no data	no data	no data	no data
DP[3]	58	8.02	300	153	<2.0	9.68	61.8	19.1
SS2-05[4]	83	7.38	3200	1600	3.91	169	993	2870
SV[5]	46	7.36	3900	1370	2.12	19.5	63.5	13,000
STH-06[6]	69	6.15	3700	1910	7.5	32.2	91.6	15,000
WC1-06[7]	72	7.86	343	174	<2.0	3.82	<10.0	9.71
CV[8]	19	7.46	301	151	3.31	10.6	12.7	<15.0
WH[9]	40	7.81	430	216	<2.0	8.18	<10.0	15.1
RS[10]	32	6.82	251	126	no data	no data	no data	no data
DC1[11]	20	6.8	3 83	191	<2.0	13.3	51.0	23.7

[1] Percentage of soil material between 0.05 mm and 0.25 mm in size.
[2] Electrical conductivity.
[3] Tucson, Arizona.
[4] Dinosaur National Monument, Utah.
[5] Santa Susana Mts., California.
[6] Sharktooth Hill near Bakersfield, California.
[7] White Creek, Diablo Range, western Fresno County, California.
[8] Capay Valley, near Brooks, California.
[9] Whiskey Hill, west of Williams, California.
[10] Richardson Springs, near Chico, California.
[11] Dye Creek, near Red Bluff, California.

value from the Swelter Shelter soils is probably due to the frequent utilization of the area by birds. Fluoride and chloride are not notable at any of the sites. No one chemical parameter or group of parameters addressed in this study displays a consistent pattern throughout all the sites. This conclusion is in concurrence with the observations of Swatek[6,7] from his studies of midden sites and adjacent soils in California.

Several factors not considered in this study have been suggested as influencing the growth of *Coccidioides*. Elconin *et al.*[23] showed in an 8-year study that high levels of salinity in surface soils could be correlated with the growth of *Coccidioides*. Egeberg *et al.*[5] demonstrated that high soil temperatures and salinity suppressed the growth of bacterial antagonists while enhancing the growth of *Coccidioides*. Pappagianis[8] and Egeberg[10] noted that anomalously high amounts of boron were associated with soils positive for *Coccidioides*. However, these and other factors are not universally observed (this article and Swatek[7]) at all sites known to be positive. One conclusion is that these and other associations are not necessary for the growth of *Coccidioides*, but instead their presence in a given soil may enhance that site by creating an environment more favorable for *Coccidioides*. The improved favorability may be due to any number of reasons; for example, higher organic content may increase the available nutrients, high salinities may reduce competition from other microorganisms, high temperatures may also reduce bacterial numbers, thereby also reducing competition, and high concentrations of sodium borate may be antiseptic for some soil microorganisms but not for *Coccidioides*.

DISCUSSION

During the 20th century *in situ* field studies of the habitat of *Coccidioides* began with the first separation[24] of the organism from the soil in 1932 and reached a peak[25] during the 1950s through the 1970s. With the advent of major advances in the medical field, attention shifted from the soil habitat to the dimorphic parasitic habitat in the lung and the development of new therapies, medicines, and the search for vaccines. While isolated outbreaks were investigated in the last decades of the century, *in situ* field studies waned. However, in the late 1990s and early in the 21st century, investigations[26–29] of soils using polymerase chain reaction (PCR) techniques that allow microbial detection and isolation of DNA directly from soils opened a new era of interest in the soil habitat. In 2001 the outbreak of coccidioidomycosis in Dinosaur National Monument sparked new studies by multidisciplinary teams of earth scientists, microbiologists, and medical doctors of the soil habitat there and at other sites in Arizona and California.[20,21]

Major technological advances in the last two decades provide powerful tools for examining the soil environments favorable for the growth of *Coccidioides*. Examples are PCR techniques as mentioned above; microelectronics that enable physical properties to be determined at the micrometer scale; more powerful computers capable of simulating and modeling complex natural systems; biosensors that are sensitive to a wide range of chemical compounds; and the discoveries of the properties and behavior of nanometer-sized soil particles. Research opportunities abound in the study of the ecology of the saprophytic habitat of *Coccidioides*. These opportunities are best undertaken by a multidisciplinary approach, with the knowledge of the vast spatial and temporal scales involved, and an understanding that soil is a complex system characterized by the interaction of numerous physical, biological, and chemical processes with diverse parameters. These interactions are most typically nonlinear, nonreversible, and difficult to quantify. In such systems the behavior of the system as a whole cannot be predicted by the behavior of its individual agents or processes. New methods needed to help with future research on *Coccidioides* include development of standardized PCR techniques capable of detecting *Coccidioides* routinely from minimally processed soil samples and development of noninvasive *in situ* techniques to characterize the utilization of pore space by *Coccidioides* in the soil profile. Also needed are fractal models of soil structures found in sites positive for *Coccidioides*. Important questions yet to be answered include the discovery of the microbial antagonists of *Coccidioides* and whether there are temporal cycles of competition with *Coccidioides*. Other questions are: What *in situ* substrates are used by *Coccidioides* and what is the activity (growth/death/dormancy) of *Coccidioides* in response to nutrient fluxes in the vadose zone? Also are these nutrient fluxes affected by clay mineralogy and abundance? And most importantly we need a description of the life cycle of *Coccidioides* in the soil from the initiation of mycelium growth to the development of arthroconidia to dormancy, as well as the knowledge that the role that physical, chemical, and biological processes play in the different chases of the cycle. Answering these and other detailed questions would bring us closer to a more complete understanding of the integral life cycle of this organism and help us predict how the infectious areas might change with global shifts in climate and resulting patterns of desertification.

REFERENCES

1. FISHER, M.C., G.L. KOENIG, T.J. WHITE, *et al.* 2002. Molecular and phenotypic description of *Coccidioides posadasii* sp. nov., previously recognized as the non-California population of *Coccidioides immitis*. Mycologia **94:** 73–84.
2. BREED, C.S.1999. Monitoring surface changes in desert areas. *In* Desert Winds, Monitoring Wind-Related Surface Processes in Arizona, New Mexico, and California. C.S. Breed & M.C. Reheis, Eds.:1–27. U.S. Geol. Survey Prof. Paper 1598.
3. FIESE, M.J. 1958. Coccidioidomycosis. Charles C Thomas. Springfield, IL.
4. SORENSEN, R.H. 1964. Survival characteristics of mycelia and spherules of *Coccidioides immitis* in a simulated natural environment. Am. J. HYG. **80:** 275–285.

5. EGEBERG, R.O., A.E. ELCONIN & M.C. EGEBERG. 1964. Effect of salinity and temperature on *Coccidioides immitis* and three antagonistic soil saprophytes. J. Bact. **88:** 473–476.

6. LACY, G.H. & F.E. SWATEK. 1974. Soil ecology of *Coccidioides immitis* at Amerindian middens in California. Appl. Microbiol. **27:** 379–388.

7. SWATEK, F.E. 1975. The epidemiology of coccidioidomycosis. *In* The Epidemiology of Human Mycotic Diseases. Y. Al-Doory, Ed.: 75–102. Charles C Thomas. Springfield, IL.

8. PAPPAGIANIS, D. 1988. Epidemiology of coccidioidomycosis. Curr. Top. Med. Mycol. **2:** 199–238.

9. MADDY, K.T. 1957. Ecological factors of the geographic distribution of *Coccidioides immitis*. J. Am. Vet. Med. Assoc. **130:** 475–476.

10. EGEBERG, R.O. 1962. Factors influencing the distribution of *Coccidioides immitis* in soil. *In* Recent Progress in Microbiology, 8th meeting.: 652–655.

11. SMITH, C.E. 1951. Diagnosis of pulmonary coccidioidal infections. Calif. Med. **75:** 385–391.

12. MADDY, K.T., H.G. CRECELIUS & R.G. CORNELL. 1961. Where can coccidioidomycosis be acquired in Arizona. Ariz. Med. **18:** 184–194.

13. PAPPAGIANIS, D. 1980. Epidemiology of coccidioidomycosis. *In* Coccidioidomycosis. D.A. Stevens, Ed.: 63–85. Plenum Medical Book Company. New York.

14. MADDY, K.T. 1957. Ecological factors possibly relating to the geographic distribution of *Coccidioides immitis*. Proceedings of the Symposium on Coccidioidomycosis.U.S. Public Health Service, Pub. No. **575:** 144–157. CDC, Atlanta, GA.

15. MERRIAM, C.H. 1898. Life zones and crop zones of the United States. Div. Biol. Survey Bull. **10:** USDA, Washington, DC.

16. SINGER, M.J. 1991. Physical properties of arid region soils. *In* Semi-Arid Lands and Deserts. J. Skujins, Ed.: 81-109. Marcel Dekker. New York, NY.

17. BERMAN, R.J., L. FREIDMAN, D. PAPPAGIANIS, *et al.* 1956. Survival of *Coccidioides immitis* under controlled conditions of temperatures and humidity. Am. J. Pub. Health **46:** 1317–1324.

18. CAMERON, R.E. 1966. Properties of desert soils. *In* Biology and the Exploration of Mars. C.S. Pittendrigh, W. Vishniac, J.P.T. Pearman, Eds. National Academy of Sciences–Natural Research Council, NASA. Washington, DC.

19. HENDRICKS, D.M. 1985. Arizona Soils. College of Agriculture Centennial Publication, The University of Arizona, Tucson, Arizona.

20. FISHER, F.S., M.W. BULTMAN, D. PAPPAGIANIS, *et al.* 2002. The ecology of the Swelter Shelter *C. immitis* site, Dinosaur Nat. Monument, UT. *In* Proceedings of the Annual Coccidioidomycosis Study Group Meeting, Meeting number **46:** 3. J. Galgiani, Ed. The Valley Fever Center for Excellence, Tucson, AZ.

21. PETERSEN, L.R., S.L. MARSHAL, C. BARTON-DICKSON, *et al.* 2004. Coccidioidomycosis among workers at an archeological site, northeastern Utah. Emerg. Infect. Dis. **10:** 637–642.

22. CHENU, C. & G. STOTZKY. 2002. Interaction between microorganism and soil particles: an overview. *In* Interactions Between Microorganisms and Soil Particles. P.M. Huang, J.M. Bollag & N. Senesi, Eds.: 3–40. John Wiley & Sons. Hoboken, NJ.

23. ELCONIN, A.E., R.O. EGEBERG & M.C. EGEBERG. 1964. Significance of soil salinity on the ecology of *Coccidioides immitis*. J. Bact. **87:** 500–503.

24. STEWART, R.A. & K.F. MEYER. 1932. Isolation of *Coccidioides immitis* (Stiles) from the soil. Proc. Soc. Exp. Biol. Med. **29:** 937–938.

25. STEVENS, D.A. 1980. Coccidioidomycosis. Plenum Medical Book Company. New York, NY.
26. BURT, A., B.M. DECHAIRO, G.L. KOENIG, et al. 1997. Molecular marker reveals differentiation among isolates of Coccidioides immitis from California, Arizona, and Texas. Mol. Ecol. 6: S781–S786.
27. DANIELS, J.I., W.J. WILSON, T.Z., et al. 2002. Development of a quantitative Taqman-PCR assay and feasibility of atmospheric collection for Coccidioides immitis for ecological studies: final report. UCRL-ID-146977. Lawrence Livermore National Laboratory, Livermore, CA.
28. FISHER, M.C., T.J. WHITE & J.W. TAYLOR.1999. Primers for genotyping single nucleotide polymophisms and microsatellites in the pathogenic fungus Coccidioides immitis. Mol. Ecol. 8: 1075–1092.
29. GREENE, D.R., G.L. KOENIG, M.C. FISHER, et al. 2000. Soil isolation and molecular identification of Coccidioides immitis. Mycologia 92: 406–410.

Fluctuations in Climate and Incidence of Coccidioidomycosis in Kern County, California

A Review

JORGE TALAMANTES,[a] SAM BEHSETA,[b] AND CHARLES S. ZENDER[c]

[a]Department of Physics and Geology, California State University, Bakersfield, California, USA

[b]Department of Mathematics, California State University, Bakersfield, California, USA

[c]Department of Earth System Science, University of California, Irvine, Irvine, California, USA

ABSTRACT: Coccidioidomycosis (Valley Fever) is a fungal infection found in the southwestern United States, northern Mexico, and some places in Central and South America. The fungi that cause it (*Coccidioides immitis* and *Coccidioides posadasii*) are normally soil dwelling, but, if disturbed, become airborne and infect the host when their spores are inhaled. It is thus natural to surmise that weather conditions, which foster the growth and dispersal of *Coccidioides*, must have an effect on the number of cases in the endemic areas. This article reviews our attempts to date at quantifying this relationship in Kern County, California (where *C. immitis* is endemic). We have examined the effect on incidence resulting from precipitation, surface temperature, and wind speed. We have performed our studies by means of a simple linear correlation analysis, and by a generalized autoregressive moving average model. Our first analysis suggests that linear correlations between climatic parameters and incidence are weak; our second analysis indicates that incidence can be predicted largely by considering only the previous history of incidence in the county—the inclusion of climate- or weather-related time sequences improves the model only to a relatively minor extent. Our work therefore suggests that incidence fluctuations (about a seasonally varying background value) are related to biological and/or anthropogenic reasons, and not so much to weather or climate anomalies.

Address for correspondence: Jorge Talamantes, Department of Physics and Geology, 62 SCI, California State University, Bakersfield, 9001 Stockdale Highway, Bakersfield, CA 93311. Voice: 661-654-2335; fax: 661-654-2040.

jtalamantes@csub.edu

Ann. N.Y. Acad. Sci. 1111: 73–82 (2007). © 2007 New York Academy of Sciences.
doi: 10.1196/annals.1406.028

KEYWORDS: Valley Fever; coccidioidomycosis; *Coccidioides immitis*; *Coccidioides posadasii*; disease statistical modeling; GARMA modeling; climate and health

INTRODUCTION

Much is known about the biological, medical, and indeed the epidemiologic aspects of *Coccidiodes immitis* (*C. immitis*) and *Coccidiodes posadasii* (*C. posadasii*), the fungi that cause Valley Fever (coccidioidomycosis) (see, for example, Ref 1 and references therein). *Coccidioides* have a complete life cycle as soil-dwelling organisms, but if they are disturbed and become airborne, they are able to infect a host via the respiratory tract when the fungi spores are inhaled.

Given its wide geographic distribution, it is evident that *Coccidioides* are able to flourish within somewhat varied climatic environments. Endemic areas include[2] the southern part of the San Joaquin Valley in California, southern California, the southern part of Arizona, New Mexico, and Texas, most of northern Mexico, and some areas in Guatemala, Honduras, Venezuela, north-eastern Brazil, Argentina, and Paraguay. There are some variations in climate in these areas: for example, the southern San Joaquin Valley gets most of its precipitation in the winter, whereas the southern part of Arizona gets late summer monsoon rains as well as frontal systems in winter.

Conventional wisdom would suggest that climatic fluctuations might affect the rate at which humans become infected.[1] For example, a wetter-than-normal rainy season might help *Coccidioides* bloom; windy spells might facilitate the dispersal of the fungus; hot summers could be anticipated to suppress competing organisms, thus enhancing the survival of *Coccidioides*.[3] Indeed, anecdotal evidence to this effect is well documented in the literature.[4–13]

There have been a number of attempts at demonstrating this connection quantitatively. They can be divided into two groups: (i) in Arizona, where *C. posadasii* is endemic, a strong connection has been reported[13–15] between climatic patterns and coccidioidomycosis incidence, whereas (ii) in Kern County, California, where *C. immitis* is endemic, only a weak connection has been found.[16,17] In this article, we endeavor to give an overview of our work with incidence and weather data in Kern County. For the purpose of comparison, in this article we discuss some salient points in connection to the works pertaining to Arizona and provide a summary of our own work.

ARIZONA: INCIDENCE–WEATHER CORRELATIONS AND MODELS

Kolivras and Comrie[14] found that antecedent precipitation and temperature are moderate climate risk factors for valley fever in Pima County (which includes Tucson). They developed a multivariate model to account for Valley

Fever incidence in a given month based on climate conditions and anomalies in the antecedent 2 years. Their model uses and predicts a metric called the transformed incidence anomaly. This is the monthly incidence anomaly relative to the annual (rather than the climatological, or climatological monthly) mean. The maximum transformed incidence anomalies they reported in Pima County are about 10% and their model predicts up to half of some anomalies. The transformed incidence is insensitive to uniform increases in monthly incidence, which results in an absolute annual increase (e.g., an epidemic), but which does not change the relative contribution of each month to the annual incidence. (In contrast, the 1991–1995 epidemic in Kern County increased interannual and intra-annual variations in incidence by about 10-fold. This appears to be the largest well-documented epidemic on record.)

Komatsu *et al.*[13] performed a Poisson regression in an effort to model monthly (1998–2001) Valley Fever incidence in Maricopa County, which includes Phoenix. They found a large correlation ($R^2 = 0.75$) between incidence and cumulative rain in the preceding 7 months, average temperature during the preceding 3 months, dust during the preceding month, and precipitation in the preceding 2 months in proportion to the preceding 7 months.

Comrie[15] explored the climate–incidence connection (for 1992–2003 Pima County data) using PM_{10} (particulate matter of size less than 10 μm) concentrations as a proxy for *C. posadasii* abundance in the atmosphere, while at the same time accounting for precipitation time series. The rationale for using PM_{10} is clear: soil dust emitted into the atmosphere from endemic regions may carry a proportional concentration of *Coccidioides* spores (size $\sim 2 \times 5$ μm). Comrie devised an exposure-day methodology, which allowed him to estimate the date of exposure adaptively depending on whether the onset, diagnosis, or report date was available for each patient. Using this approach, he identified a bimodal distribution of the monthly incidence data—a pattern that had not been clearly seen in previous analyses. Comrie then grouped the data into seasons and was able to produce a model that predicted closely the observed disease incidence. This model combined the lagged seasonal precipitation, and concurrent seasonal dust and precipitation, and was able to explain 80% of the variance in coccidioidomycosis seasonal incidence data.

KERN COUNTY: INCIDENCE–WEATHER CORRELATIONS AND MODELS

The Kern County seasonal climate was described in detail in Ref. 16. The area receives in the neighborhood of 16.5 cm of rain a year, largely in the winter. Summers are hot and dry—usually reaching 43°C in July, with virtually no precipitation. May and June experience the largest wind speeds--and average of the order of 3.5 m/sec, with maxima usually less than 15 m/sec, and very rarely exceeding 20 m/sec. Coccidioidomycosis incidence (number of cases per 100,000 population) also has a corresponding yearly cycle, with the number generally increasing toward the late fall (incidence near 17 per month per 100,000 population), decreasing in the winter, and reaching a minimum in the spring and summer (incidence near 4.7 per month per 100,000 population).

The work on Kern County data so far has searched for a connection between the fluctuations of Valley Fever incidence (about a seasonally varying background) and climate anomalies, that is, we do not endeavor to explain the seasonal behavior of incidence in terms of climate parameters; instead, we try to explain *incidence fluctuations* in terms of *weather anomalies*. Furthermore, we do not seek to explain or model effects of relatively infrequent events, such as the December 1977 dust storm in Kern County,[7] or the Northridge Earthquake of 1994—both of which produced large outbreaks of the disease.[18]

The climatic variables we have investigated in connection with coccidioidomycosis incidence are precipitation, surface temperature, and wind speed (which we take as a proxy for spore abundance in the atmosphere). We present in FIGURE 1 the annual cycle of coccidioidomycosis incidence in Kern County, and some potential climate risk factors. FIGURE 1 was graphed from monthly

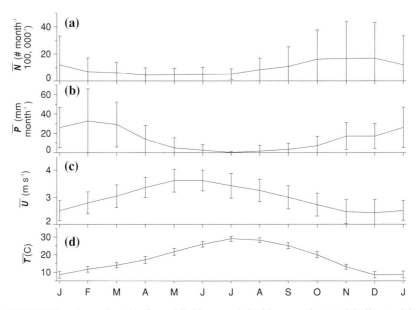

FIGURE 1. Annual cycle of coccidioidomycosis incidence and potential climate risk factors from 1980 to 2002. (**a**) Monthly mean incidence $\bar{N}$ (# per month per 100,000 population), (**b**) precipitation $\bar{P}$ (mm per month), (**c**) wind speed $\bar{U}$ (m per sec), (**d**) surface temperature $\bar{T}$ (C). *Bars.* Two standard deviations of the interannual variability computed separately for each month. Standard deviations computed using 1980–2002 data for incidence, and 1961–2002 for climate variables.

1980–2002 data for incidence and 1961–2002 for climate variables. The bars indicate intramonth variability. Their size was computed as twice the standard deviation for the interannual variability computed separately for each month. We performed a (univariate) lag-correlation analysis between incidence seasonal anomalies on the one hand, and climatic anomalies in each of precipitation, surface temperature, and wind speed on the other hand.[16] We found that only precipitation 8 months prior to incidence had a statistically highly significant correlation with incidence ($P = 0.0020$), but whereas this is consistent with the expectation than an unusually wet winter can lead to larger incidence later in the *Coccidioides* season, the magnitude of the correlation was quite small ($R^2 = 0.04$). Further analysis of bivariate correlations led to inconclusive results. To try to capture nonlinear correlations, we also investigated correlations of incidence anomalies and wind speed anomalies squared and cubed, as well as with wind speeds only within a certain window (from some minimum to some maximum). These efforts also led to no statistically significant correlations.

We used 9 years (Jan 1995–Dec 2003) of weekly mean atmospheric conditions and incidence data to construct a statistical model which allows a degree of prediction.[17] We presented a generalized autoregressive moving average (GARMA) model that maximizes the accuracy of prediction while minimizing the number of input variables.[19] In other words, we maximize the predictive skill of the model while keeping only the most important input parameters, and removing from the input those variables that do not contribute (or contribute minimally) to the prediction task. The method starts from putting into the model parameters that might help predict incidence. We used prior incidence history, as well as precipitation, wind speed, and temperature. As expected (since the number of input parameters is very large), this first step leads to an excellent model. We interpreted this to mean that at least some of the input parameters contain information highly relevant to our prediction task. We called this result the full model (FIG. 2a). The second step entails the minimization procedure mentioned above. We present this result (the final model) in FIGURE 2b. Our most important finding[17] was that the final model does not require any weather parameters—it turned out that in order to predict weekly incidence at some time t, the most important input parameters were weekly incidence at times t-k, where $k = 1, 2, 4, 26$ weeks. We point out that the incidence surges at the end of the period under consideration are predicted by the final model. For comparison, we forced the model to consider the best way to use weather parameters to predict incidence (while not considering information about the history of prior incidence). We called this the environmental model. The result is given in FIGURE 2c. This model predicts only the seasonal variations of incidence (as expected, since weather and incidence have a yearly cycle), but is quite unable to predict the surges in the early 2000s.

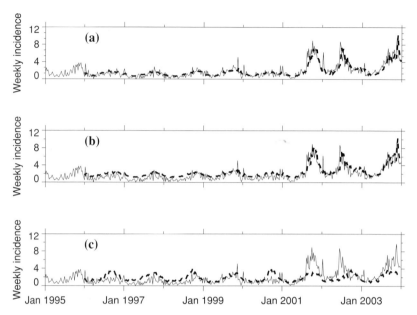

FIGURE 2. Reported Valley Fever incidence in Kern County, California (cases per 100,000 population) and three models for the period Jan 1996–Dec 2003. In all three panels, thin solid lines indicate incidence, and dashed heavy lines correspond to model results: (**a**) full model, (**b**) final model, and (**c**) environmental model. The year 1995 is missing from the model results because at least 1 year of Valley Fever incidence data is required to start predicting future values.

DISCUSSION

At this point, our work suggests that fluctuations in incidence are probably due more to human activities (such as construction on previously undeveloped land) or biological processes taking place in the field (see complex systems discussion below), rather than to climatic fluctuations. In contrast with the results reported in Ref. 14, we found[16] only a weak correlation between anomalies in weather variables and incidence in Kern County. The discrepancy might have been due to the fact that climate exhibits important differences in Pima and Kern counties. The mathematical treatments were also different. The differences in the two approaches are mainly in the processing of raw incidence and weather data, although both papers analyzed correlations in a linear model.

Studies to date have demonstrated a strong incidence–climate connection in Arizona,[15] but not in California.[17] Perhaps this is because the latter studies use wind and not PM_{10} as a proxy for spore abundance in the atmosphere. Unfortunately, there are no actual measurements of *Coccidioides* spore concentrations in the atmosphere as a function of PM_{10} concentrations, wind speed, or any other parameter. In the end, we can do no better than give plausibility arguments as to why a particular parameter should be a good proxy without any hard data to support our surmise. However, given the explanatory power of precipitation and PM_{10} for coccidioidomycosis incidence in Arizona,[15] these predictors should be fully evaluated in Kern County. PM_{10} may be a better proxy for spore abundance in the atmosphere than wind speed, although this may be more so in Arizona than in Kern County, which has one of the worst air pollution problems in the country. Kern County gets large PM_{10} contributions from paved and unpaved road dust, as well as from farming operations (dust that presumably does not contribute spores since the fungus does not grow in cultivated soils). Indeed, as much as 81% of PM_{10} levels is estimated to be of anthropogenic origin in the summer, and 89% in the winter.[20] Kern County is located at the southern end of California's Central Valley and the prevailing winds are northerly, which tends to accumulate pollutants from throughout the valley up against the southern mountains. In Kern County, PM_{10} levels can thus be expected to have a significant contribution from sources not directly related to *C. immitis*. Nevertheless, PM_{10} might be a better proxy to the extent that one is only trying to index dusty days. In phenomenological approaches, such as those in Refs. 13–17, it does not matter what the actual connection is between PM_{10} and spore abundance. It only matters that a good incidence predictor be found.

Interestingly, according to preliminary studies conducted in Arizona,[21] there seems to be a link between smoking and risk of coccidioidomycosis infection: apparently, smokers have a higher risk of infection than nonsmokers; however, smoking cessation restores the risk back to the level of nonsmokers. Clearly, one has to wonder whether this smoking–coccidioidomycosis link results from a suppression of the immune system. And if so, is there a similar effect due to other environmental contaminants (such as NO_x, ozone, and PM_{10})? The effect reported in Ref. 15 might be one instance of a more general effect—perhaps PM_{10} concentrations point to coccidioidomycosis infection not necessarily (or not only) because PM_{10} is a good proxy for *Coccidioides* concentrations in the atmosphere, but because PM_{10} pollution might adversely affect the immune system, thus increasing the risk of infection. Indeed, there is one study[22] in which repeated exposure to diesel exhaust particles showed a sustained pattern of downregulation of T cell–mediated immune responses. This suppression in cell-mediated immunity could be considered a risk for coccidioidomyco-sis. Furthermore, as pointed out by Terashita and Capone–Newton,[23] PM_{10} is possibly also a marker for other air pollutants. Van Loveren *et al.*[24] found that ozone can inhibit resistance to an intrathecal challenge with *Listeria* monocy-togenes, indicating suppression of Th1 immune responses. Again, there may be a similar effect with Valley Fever, turning PM_{10} into a good incidence pre-dictor even if it arises largely from anthropogenic sources, which are unrelated to *Coccidioides* spore concentrations in the atmosphere.

Disentangling the differences between our results and those in Arizona is one of our main endeavors now. Indeed, to that end, we are presently (i) an-alyzing Arizona data using our GARMA method, and (ii) including PM_{10}

measurements in our Kern County GARMA model. One must keep in mind that comparisons between California and Arizona results are less straightforward than might be realized initially. Not only (i) is *C. immitis* endemic in California, whereas *C. posadasii* is endemic in Arizona, but also (ii) climate is different and it is possible that climatic variations are too small in Kern County to explain incidence anomalies, but not so in Arizona (i.e., there is no compelling reason why an incidence–weather connection should be the same in all endemic areas since *Coccidioides* may react differently to different environments); (iii) the extent to which PM_{10} mirrors spore abundance in the atmosphere is probably different in Pima and Kern counties, owing to the high pollution levels in the latter location; and (iv) generally, different types of soils in the various endemic areas and the extent to which they absorb precipitation may also play a role[21]—for example, if rain goes mostly into run-off, then presumably its effect on fungal growth should be less. These effects, individually and in their interplay, need to be examined in detail to sort out the differences alluded to.

Our GARMA model may lead to different conclusions when PM_{10} concentrations in Kern County are used as additional input parameters. This is because the method relies on a multidimensional minimization search. Therefore, it would not be surprising (at least from a modeling perspective) if this algorithm leads, for example, to both precipitation and PM_{10} being important, even though precipitation did not come up in our original search (when PM_{10} was not considered).

The *Coccidioides*–environment system is interesting not only because of its obvious practical implications--economic costs[25] and human suffering,[1] but from a theoretical standpoint as well. It seems very likely that the input–response relationships are nonlinear (with the input being the environmental parameters and incidence time series, and the output being the incidence time series); since incidence depends strongly on its own history, there must be feedback loops, and the system clearly has memory; the system boundaries are difficult to determine; it is an open system far from any sort of equilibrium. In other words, the *Coccidioides*–environment system has many characteristics of what is generally known as a complex system: it might be exhibiting an emergent global behavior (large *Coccidioides* population booms) not imposed by a central controller (climatic fluctuations), but resulting from the interactions between the agents (*Coccidioides* and other organisms in its environment).[26] Perhaps methods and techniques developed for totally unrelated complex systems might prove useful in understanding the behavior of Valley Fever time series, especially surges and epidemics.

ACKNOWLEDGMENTS

We are grateful to Dawn Terashita, M.D. and Peter Capone-Newton, M.D. of the Los Angeles County Department of Health Services, as well as to Brenda

Turner of the San Joaquin Valley Air Pollution Control District for valuable communications regarding this work. We are also thankful to Dr. Karl Clemons and the anonymous referee for useful comments regarding this article.

REFERENCES

1. PAPPAGIANIS, D. 1988. Epidemiology of coccidioidomycosis. *In* Current Topics in Mycology, Vol 2. M. McGinnins, Ed.: 199–238. New York–Berlin–Heidelberg. Springer.
2. PAPPAGIANIS, D. 1994. Marked increase in cases of coccidioidomycosis in California: 1992, 1992, and 1993. Clin. Infect. Dis. **19**(Suppl 1): S14–S18.
3. KOLIVRAS, K.M., P.S. JOHNSON, A.C. COMRIE & S.R. YOOL. 2001. Environmental variability and coccidioidomycosis (valley fever). Aerobiologia **17**: 31–42.
4. SMITH, C.E., R.R. BEARD, H.G. ROSENBERGER & E.G. WHITING. 1946. Effect of season and dust control on coccidioidomycosis. JAMA **132**: 833–838.
5. HUGENHOLTZ, P.G. 1957. Climate and coccidioidomycosis. *In* Proceedings of the Symposium on Coccidioidomycosis, Phoenix, Arizona: 136–143. Public Health Service. Washington, DC.
6. MADDY, K.T. 1957. Ecological factors possibly relating to the geographic distribution of *Coccidiodes immitis*. *In* Proceedings of the Symposium on Coccidioidomycosis, Phoenix, Arizona: 144–157. Public Health Service. Washington, DC.
7. PAPPAGIANIS, D. & H. EINSTEIN. 1978. Tempest from Tehachapi takes toll or *Coccidioides* conveyed aloft and afar. West J. Med. **129**: 527–530.
8. PAPPAGIANIS, D., R.K. SUN, S.B. WERNER, *et al.* 1993. Coccidioidomycosis—United States, 1991–1992. Morb. Mortal Wkly. Rep. **42**: 21–24.
9. JINADU, B.A., G. WELCH, R. TALBOT, *et al.* 1994. Update: coccidioidomycosis—California 1991–1993. Morb. Mortal Wkly. Rep. **43**: 421–423.
10. MOSLEY, D., K. KOMATSU, V. VAZ, *et al.* 1996. Coccidioidomycosis–Arizona 1990–1995. Morb. Mortal Wkly. Rep. **45**: 1069–1073.
11. KIRKLAND, T.N. & J. FIERER. 1996. Coccidioidomycosis: a reemerging infectious disease. Emerg. Infect. Dis. **3**: 192–199.
12. SCHNEIDER, E., R.A. HAJJEH, R.A. SPIEGEL, *et al.* 1997. A coccidioidomycosis outbreak following the Northridge, Calif, earthquake. JAMA **277**: 904–908.
13. KOMATSU, K., V. VAZ, C. MCRILL, *et al.* 2003. Increase in coccidioidomycosis—Arizona 1998–2001. Morb. Mortal Wkly. Rep. **52**: 109–112.
14. KOLIVRAS, K.M. & A.C. COMRIE. 2003. Modeling valley fever (coccidioidomycosis) incidence on the basis of climate conditions. Int. J. Biometeorol. **47**: 87–101.
15. COMRIE, A.C. 2005. Climate factors influencing Coccidioidomycosis seasonality and outbreaks. Environ. Health Perspect. **113**: 688–692.
16. ZENDER, C.S. & J. TALAMANTES. 2006. Climate controls on valley fever incidence in Kern County, California. Int. J. Biometeorol. **50**: 174–182, DOI: 10.1007/s00484-005-0007-6.
17. TALAMANTES, J., S.S. BEHSETA & C.S. ZENDER. 2007. Statistical modeling of valley fever data in Kern County, California. *In* Int. J. Biometeorol. [DOI: 10.1007/s00484-006-0065-4].
18. PAPPAGIANIS, D., G. FELDMAN, K. BILLIMEK, *et al.* 1994. Coccidioidomycosis following the Northridge Earthquake—California 1994. Morb. Mortal Wkly. Rep. **43**: 194–195.

19. BENJAMIN, M.A., R.A. RIGBY & D.M. STASINOPOULOS. 2003. Generalized autoregressive moving average models. J. Am. Stat. Assoc. **98:** 214–223.
20. CALIFORNIA AIR RESOURCES BOARD. 2006. Reports available at http://www.arb.ca.gov.
21. TABOR, J. 2006. Private communication.
22. YIN, X.J., C.C. DONG, J.Y.C. MA, *et al.* 2004. Suppression of cell-mediated immune responses to *Listeria* infection by repeated exposure to diesel exhaust particles in brown Norway Rats. Toxicol. Sci. **77:** 263–271.
23. TERASHITA, D. & P. CAPONE-NEWTON. 2006. Private communication.
24. VAN LOVEREN, H., P.A. STEERENBERG, J. GARSSEN & L. VAN BREE. 1996. Interaction of environmental chemicals with respiratory sensitization. Toxicol. Lett. **86** (2–3): 163–167.
25. JINADU, B.A. 1995. Valley Fever Task Force report on the control of *Coccidioides immitis*. Technical Report, Kern County Health Department, Bakersfield, California.
26. BOCCARA, N. 2004. Modeling Complex Systems. Springer-Verlag. New York.

Assessment of Climate–Coccidioidomycosis Model

Model Sensitivity for Assessing Climatologic Effects on the Risk of Acquiring Coccidioidomycosis

ANDREW C. COMRIE AND MARY F. GLUECK

Department of Geography and Regional Development, University of Arizona, Tucson, Arizona, USA

ABSTRACT: Understanding the predictive relationships between climate variability and coccidioidomycosis is of great importance for the development of an effective public health decision-support system. Preliminary regression-based climate modeling studies have shown that about 80% of the variance in seasonal coccidioidomycosis incidence for southern Arizona can be explained by precipitation and dust-related climate scenarios prior to and concurrent with outbreaks. In earlier studies, precipitation during the normally arid foresummer 1.5–2 years prior to the season of exposure was found to be the dominant predictor. Here, the sensitivity of the seasonal modeling approach is examined as it relates to data quality control (QC), data trends, and exposure adjustment methodologies. Sensitivity analysis is based on both the original period of record, 1992–2003, and updated coccidioidomycosis incidence and climate data extending the period of record through 2005. Results indicate that models using case-level data exposure adjustment do not suffer significantly if individual case report data are used "as is." Results also show that the overall increasing trend in incidence is beyond explanation through climate variability alone. However, results also confirm that climate accounts for much of the coccidioidomycosis incidence variability about the trend from 1992 to 2005. These strongly significant relationships between climate conditions and coccidioidomycosis incidence obtained through regression modeling further support the dual "grow and blow" hypothesis for climate-related coccidioidomycosis incidence risk.

KEYWORDS: coccidioidomycosis; climate and health; Valley Fever; southwestern United States; Arizona; prediction; model sensitivity; incidence

Address for correspondence: Dr. Andrew C. Comrie, Administration 601, P.O. Box 210066, University of Arizona, Tucson, AZ 85721, USA. Voice: 520-621-3512; fax: 520-621-7507.
comrie@arizona.edu

Ann. N.Y. Acad. Sci. 1111: 83–95 (2007). © 2007 New York Academy of Sciences.
doi: 10.1196/annals.1406.024

INTRODUCTION

Coccidioidomycosis (Valley Fever) is undergoing an epidemic in Arizona. Since the disease became reportable to the State in 1997, the number of cases has risen from ~1,000 to almost 5,500 in 2006. The causes of this trend are poorly understood, but leading hypotheses include better reporting, in-migration of susceptible individuals to Arizona, and climate. Year-to-year fluctuations in coccidioidomycosis incidence are observed, with similar hypothetical causes. This article examines the role of climate in the trend and variability of coccidioidomycosis incidence in Arizona.

As soil fungi in semiarid regions, *Coccidioides* spp. require suitable temperatures and soil moisture for growth.[1–3] *Coccidioides* and other soil biota are presumed not to survive extremely high summer temperatures (>60°C) near the soil surface but instead remain viable below the surface. A moist period favorable for growth will be followed by subsequent dry conditions that lead to formation of arthroconidia. These spores remain in the soil until an increase in moisture enables further growth or until they are disturbed and become airborne. Inhalation of the spores by humans, other mammals, and reptiles leads to infected cases, which in turn fluctuate with climate variability.

Initial links between coccidioidomycosis and climate were established in the 1940s[4,5] with additional pioneering work in the 1950s and 1960s.[6,7] Many anecdotal relationships between climate and incidence have been mentioned in the coccidioidomycosis research literature since that time, and it has been suggested that the early 1990s outbreak in California may have been connected to drought.[8] Until recently, there were no new quantitative studies explicitly dealing with the role of climate. Kolivras *et al.*[3] reviewed the available literature dealing with *Coccidioides* in the environment, including climate. Kolivras and Comrie[9] developed the first multivariate analyses linking climate conditions to coccidioidomycosis incidence for the Tucson region. Park *et al.*[10] examined climate and other potential variables for the Phoenix area (also reported as Komatsu *et al.*[11]). Zender and Talamantes[12] investigated statistical links to climate and weather conditions for Kern County, California. Comrie[13] developed seasonal models of coccidioidomycosis incidence based on antecedent precipitation and airborne dust at the time of exposure. Those models were based on data for Pima County, Arizona (effectively the Tucson region) from 1992 to 2003, and they employed case-level exposure adjustment of incidence data.

The models developed by Comrie[13] provide the strongest and most parsimonious results to date, but the sensitivity of the models to the exposure adjustment methodology is unknown, which has implications for translation of this research to public health agencies. Further, although those models were cross-validated on independent subsamples of the original data, they have not been validated on recent incidence data since 2003. This article examines the sensitivity of the seasonal modeling approach to data from different quality control (QC) and adjustment procedures and also evaluates model performance on coccidioidomycosis incidence data updated through 2005.

DATA AND METHODS

Data for this study were acquired for 2004 and 2005 to update the prior data set[13] containing incidence reports for 1992–2003. Monthly precipitation data for five long-term stations in the Tucson area were obtained from the National Climatic Data Center via the Western Regional Climate Center. Dust data (PM_{10}, particulate matter < 10 μm) were obtained for the five sites with suitable PM_{10} records (1991–2006) from Pima County Department of Environmental Quality. Monthly coccidioidomycosis summary case counts for Pima County were obtained from the Arizona Department of Health Services (ADHS). Disease data were aggregated to the seasonal level based on exposure, onset, and report lag computations. Results from application of updated case-level data were not available for this study (research currently in progress). Compiled annual summaries of the seasonal study data are presented in FIGURE 1.

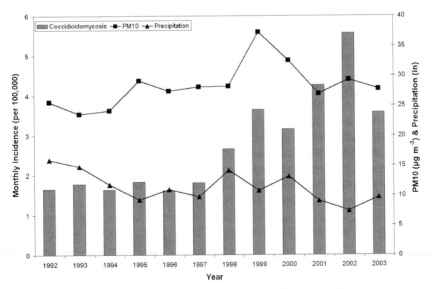

FIGURE 1. Annual values of coccidioidomycosis incidence and climate variables for Pima County. Arizona. The strong upward trend in incidence does not have a counterpart in the annual precipitation or dust (PM_{10}) data. (Note: Forthcoming results from analysis of finer temporal scale incidence data should provide a clearer understanding of these relationships on subseasonal scales.)

Details of the model being evaluated can be found in Comrie.[13] Essentially, a stepwise multiple linear regression model incorporating independent current and antecedent precipitation and dust variables was used to elucidate climate–coccidioidomycosis relationships fundamental to the "blow and grow" hypothesis of coccidioidomycosis incidence. Such seasonal analyses have significant advantages over shorter periods for broad climate studies, as they tend to produce results that are easier to interpret and they allow one to examine the phenomenon at more meaningful scales. Variables were included only if tolerance values (that specifically test for collinearity) were acceptable. Previous studies[5,9] have also indicated seasonal differences in climate–coccidioidomycosis relationships, linking climate and other components to seasonal patterns in disease incidence and to longer trends. Following Comrie,[13] seasons here were defined on the basis of most apparent minima and maxima for each variable in the models. Predictions of coccidioidomycosis incidence were obtained independently for each season. An "annual prediction" series was compiled from these individual predicted series by placing each year's seasonal predictions in calendar order.

Three data analysis experiments were carried out to evaluate model sensitivity, as follows:

(1) How "clean" do the case report data need to be? Case-level data contain duplicates, typographical errors, bad dates, etc. and these data were quality-controlled prior to use in our previous study.[13] For operational purposes, is this time-consuming step necessary? The identical multivariate regression modeling approach was followed here, but instead using raw case-level data from 1992 to 2003 without data QC to generate results for comparison. Data preparation and assignment of exposure dates followed procedures documented in the earlier Comrie[13] study. Mean onset-to-diagnosis and onset-to-report lags were calculated for each individual month's cases in the Pima County incidence time series. Lags were smoothed with a 3-month running mean and used to estimate date of exposure. An offset of 14 days was then included to account for exposure to incubation lag. Monthly case counts were converted to incidence rates per 100,000 using monthly population estimated for Pima County.[14] As in prior experiments,[13] data were then grouped into seasons defined by predominant seasonal maximum and minimum climate variables and incidence occurrence.

(2) In the interests of operational simplicity, how well do the models perform if we skip the exposure adjustment of incidence data and simply use aggregate monthly case report data? Again, the same data preparation and modeling approaches were used, but this time, the model was run using the "raw" aggregate monthly coccidioidomycosis totals without case-level exposure adjustment. Instead, a simple 1-month offset was used to allow for incubation time and diagnosis/reporting lags (14 days, incubation + 15 days, average onset to diagnosis lag = ~30 days) observed in the original case data obtained from ADHS, and noted in our prior study.[13] Models with original and new predictors

were evaluated, using the non-QC data for 1992–2003 and updated to 2005, separately. Attempts were also made to project 2004 and 2005 incidence using the 1992–2003 model results.

(3) It is clear that in data updated through 2005, the strong trends in coccidioidomycosis incidence are not mirrored by corresponding trends in climate variables (FIG. 1). However, seasonal and annual variability is present in all these series and there is strong evidence from prior studies linking climate to coccidioidomycosis variability. If the underlying, possibly nonclimatic, trend in coccidioidomycosis is removed from the data series, how well does climate variability describe the remaining variations in coccidioidomycosis? Aggregate monthly incidence data were linearly detrended, and the residual series was modeled using multivariate regression as before.

RESULTS

Case-Level QC versus Non-QC

TABLE 1 shows the model results from the raw case-level data (non-QC), including model explained variance (R^2) and standardized coefficients (β). FIGURE 2 illustrates the resulting modeled incidence time series. These results can be directly compared to those obtained from quality-controlled data in the prior study.[9] The results are remarkably similar, with many of the same predictors and only slightly diminished explained variance ($R^2 = 0.77$ vs. 0.80 in the prior study[13]). Therefore, duplicates and other data quality issues at the case level lead to only minor differences in model details for 1992–2003.

Omitting Exposure Adjustment

TABLES 2 and 3 show model results based on "raw" aggregate monthly report data to 2003. No case-level exposure adjustment was performed, but an overall 1-month offset was used to approximate the delay between exposure and report date (e.g., November reports are assigned to October exposure). For the first of these two sets of seasonal models in TABLE 2, the same predictor variables as used previously were specified [see also (1) above]. While many of the original predictor variables remained in the seasonal models, several became insignificant with the use of aggregate data and were therefore omitted in the final prediction equation. As indicated in FIGURE 2 as well as TABLE 2, the explained variances of the initial seasonal models (1992–2003) by individual season and by the combined series are strong.

For the second of these two sets of models in TABLE 3, all variables were available for selection via stepwise multiple linear regression rather than being specified a priori. The resulting predictor variables included antecedent foresummer precipitation as in previous models, adding a few other new predictors

TABLE 1. Experiment 1, case-level QC versus non-QC, results in predictors similar to those of prior findings[13] with same role of foresummer precipitation and dust, leading to almost identical results

	Foresummer	Monsoon	Fall	Winter
R^2 (observed–predicted)	0.89	0.82	0.91	0.97
	0.98	*0.60*	*0.61*	*0.95*
Predictors				
Dust PM10	0.72 ($\leq$0.001)			0.65 ($\leq$ 0.001)
	0.75($\leq$0.001)			*0.44 ($\leq$ 0.001)*
Precipitation				
Winter0	NA	NA	NA	
Fall0	NA	NA	−0.51 (0.003)	−0.36 (0.004)
			−0.49 (0.029)	
Monsoon0	NA			
Foresummer0	0.47 ($\leq$0.001)			0.60 ($\leq$0.001)
				0.49 ($\leq$0.001)
Winter1	*0.20 (0.023)*		−0.67 (0.003)	−0.33 (0.004)
Fall1	*−0.26 (0.030)*			0.19 ($\leq$0.001)
Monsoon1				
Foresummer1		0.56 (0.003)	0.82 ($\leq$0.001)	*0.56 ($\leq$0.001)*
		0.45 (0.044)	*0.73 (0.004)*	
Winter2				
Fall2				
Monsoon2				
Foresummer2	1.01($\leq$0.001)	0.66 ($\leq$0.001)		
	1.36 ($\leq$0.001)	*0.64 (0.008)*		
Winter3				
Fall3				
Monsoon3				0.23 (0.010)
Foresummer3				
Winter4				
Fall4			0.44 (0.02)	NA
Monsoon4	0.71 (0.006)		NA	NA
	−0.93 ($\leq$0.001)			
Foresummer4		NA	NA	NA

NOTE: Record period is 1992–2003; *P*-values given in parentheses. Season numbers reflect lag. Italics indicate prior results.[13]

as well. As no precipitation variables were indicated for the winter model, only dust was used in the prediction equation. Again, the explained variances of these seasonal models are strong by individual season in the original model.

Results from the same experiments (non-QC, 1-month adjustment, specified and stepwise determination of predictors) based on an updated database, 1992–2005, were less conclusive. Inclusion of the 2004–2005 data greatly reduced the seasonal and overall series R^2 values, in both the specified and stepwise experiments. FIGURE 3 illustrates the time series of the combined seasonal

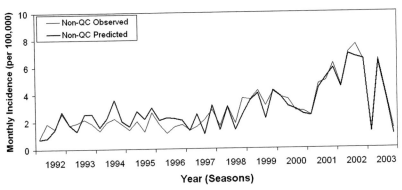

FIGURE 2. Non-quality control (QC) observed and non-QC modeled coccidioidomycosis incidence (exposure adjustment applied) for Pima County, $R^2 = 0.89$. For QC observed and non-QC modeled, $R^2 = 0.77$.

models for both 1992–2005 ($R^2 = 0.06$ specified, $R^2 = 0.10$ stepwise) and a 1992–2003 subperiod extracted from the new results ($R^2 = 0.26$ specified, $R^2 = 0.57$ stepwise).

The most striking result of this latest set of experiments is the inability of these newest models to capture the increasing trend in incidence in 2004 and 2005. However, the models do capture the seasonal variability quite well. Therefore, while important elements of the changes in coccidioidomycosis incidence are explained by climatic variability, the underlying trend in incidence does not have a parallel in the climate data (FIG. 1) and the models cannot be expected to reproduce it. Presumably, the long-term trend has some other cause unrelated to seasonal climate variability.

Climate and Detrended Incidence Data

The final set of analyses examines the role of climate when the underlying, and possibly nonclimatic, trend is removed from the incidence data series. FIGURE 4A shows the fit of the linear trend to the series. Data values were expressed as deviations from this line (FIG. 4B) and then modeled as before for both the 1992–2003 and 1992–2005 periods. The incidence data for these analyses are the same aggregate monthly reports (non-QC) without case-level exposure adjustment, offset by 1 month for incubation and reporting lags as before.

TABLE 4 shows the results of stepwise regression modeling of the detrended data. Explained variance of the seasonal models ranged from 0.29 for the monsoon to 0.99 for the winter model. Overall R^2 for the combined models was 0.66. PM_{10} was a useful predictor in the winter and arid foresummer, as in prior models.[9] However, the antecedent precipitation variables were quite different

TABLE 2. Results obtained using aggregate level, raw monthly incidence data (1-month offset for onset/exposure; no other exposure adjustment) and *fixed* predictors (same as in TABLE 1 but dropping insignificant predictors)

	Foresummer	Monsoon	Fall	Winter
R^2 (observed–predicted)	0.87	0.78	0.78	0.72
Predictors				
Dust PM10	0.81 ($\leq$0.001) *0.75($\leq$0.001*			0.34 ($\leq$ 0.11) *0.44 ($\leq$ 0.001)*
Precipitation				
Winter0	NA	NA	NA	
Fall0	NA	NA	−0.49 (0.25) *−0.49(0.029)*	*−0.36 (0.004)*
Monsoon0	NA			
Foresummer0	*0.47($\leq$0.001)*			0.63 (0.013) *0.49($\leq$0.001)*
Winter1	*0.20 (0.023)*		−0.50(0.022)	*−0.33(0.004)*
Fall1	*−0.26 (0.030)*			
Monsoon1				
Foresummer1		0.58 (0.006) *0.45 (0.044)*	0.68(0.005) *0.73(0.004)*	*0.56 ($\leq$0.001)*
Winter2				
Fall2				
Monsoon2				
Foresummer2	0.92($\leq$0.001) *1.36 ($\leq$0.001)*	0.60 ($\leq$0.005) *0.64 (0.008)*		
Winter3				
Fall3				
Monsoon3				0.58 (0.020)
Foresummer3				
Winter4				
Fall4				NA
Monsoon4	−0.74 (0.007) *−0.93 ($\leq$0.001)*	NA	NA	NA
Foresummer4		NA	NA	NA

NOTE: Record period is 1992–2003; italics indicate prior results.[13]

from those identified in prior models. Most notably, the arid foresummer is no longer a major predictor across all models. Instead, there is a range of predictor variables spread across numerous antecedent seasons, and with few if any consistent variables from model to model. Because these data are necessarily the aggregate monthly report totals, it may be that a clear antecedent climate signal is not well-captured in these "raw" data and case-level exposure adjusted data are needed instead (research currently in progress).

Despite the wide range of predictor variables identified in these detrended models, the time series of model predictions matches the detrended incidence

TABLE 3. Results obtained using aggregate level, raw monthly incidence data (non-QC, 1-month offset for onset/exposure; no other exposure adjustment) and *new* regression model with *new* predictors

	Foresummer	Monsoon	Fall	Winter
R^2 (observed-Predicted)	0.88	0.72	0.73	No new model possible; entries below from Comrie[9]
Predictors				
Dust PM10	0.48 (0.004) *0.75(≤0.001)*		0.44 (0.049)	*0.44 (≤ 0.001)*
Precipitation				
Winter0	NA	NA	NA	
Fall0	NA	NA	*−0.49 (0.029)*	*−0.36 (0.004)*
Monsoon0	NA			
Foresummer0	*0.47(≤0.001)*			*0.49 (≤0.001)*
Winter1	0.20 (0.023)		*−0.67 (0.003)*	*−0.33 (0.004)*
Fall1	*−0.26 (0.030)*			
Monsoon1				
Foresummer1		0.56 (0.003) *0.45 (0.044)*	0.55 (0.017) *0.73 (0.004)*	*0.56 (≤0.001)*
Winter2				
Fall2				
Monsoon2				
Foresummer2	0.46 (0.005) *1.36 (≤0.001)*	0.66 (≤0.001) *0.64 (0.008)*	0.57 (0.16)	
Winter3	−0.58 (0.002)			
Fall3				
Monsoon3				
Foresummer3				
Winter4				NA
Fall4				NA
Monsoon4	*−0.93 (≤0.001)*		NA	NA
Foresummer4	−0.41	NA	NA	NA

NOTE: No significant predictors were obtained here for winter. Record period is 1992–2003; italics indicate prior results.[13]

curve quite well. Once the initial trend is added back into the data series to reconstruct the original pattern (FIG. 4C), the explained variance of linear trend plus the seasonal models is 0.90 over the whole 1992–2005 period. This is dramatically better than the performance of the earlier models when 2004–2005 data are included. FIGURE 4D illustrates those periods when climate variability acts to increase or decrease coccidioidomycosis incidence relative to the linear trend. The most notable aspect of this curve is that the apparent importance of climate variability is perhaps exceeded by the broader trend underlying the data.

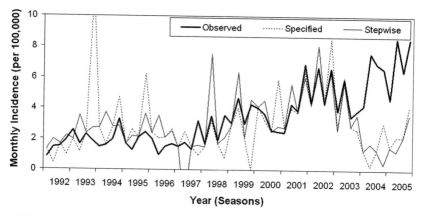

FIGURE 3. Pima County valley fever raw (non-QC, 1-month exposure adjustment) aggregate monthly models. $R^2 = 0.06$ (specified to observed) and $R^2 = 0.10$ (stepwise to observed) for 1992–2005. Through 2003, for the same model, $R^2 = 0.26$ (specified to observed) and $R^2 = 0.57$ (stepwise to observed).

CONCLUDING DISCUSSION

The aims of this study were to analyze the sensitivity of the seasonal modeling approach of Comrie[13] using different QC and adjustment procedures, and to evaluate model performance on coccidioidomycosis incidence data updated through 2005. Three specific issues were addressed: (i) the role of quality-controlled case report data versus the data as supplied; (ii) model performance using simple aggregate monthly incidence instead of exposure adjusted incidence data; and (iii) the relationship of climate variability to incidence after the underlying trend is removed.

It appears that models using case-level data adjustment do not suffer significantly if individual case report data are used "as is." Processing these data to remove occasional duplicates and other inconsistencies (estimated as a few percent of all reports) is time-consuming, and thus omitting this step leads to only a small cost to model performance and can actually speed data processing. An important finding of our prior work[13] was that case-level exposure adjustment led to far greater clarity in the seasonal patterns of estimated coccidioidomycosis exposure dates. Given that the seasonal modeling approach in Comrie[13] was also useful, we evaluated this approach using the aggregate monthly reports (research on case-level data in progress). The result of using the unadjusted "raw" monthly reports was a partial reduction in model explained variance. Some important explanatory power is nonetheless retained with this approach, and a number of the predictor variables so identified were also identified in the full method of Comrie.[13]

When recent data through 2005 are included, it is clear that these models are unable to capture the recent overall upward trend in coccidioidomycosis

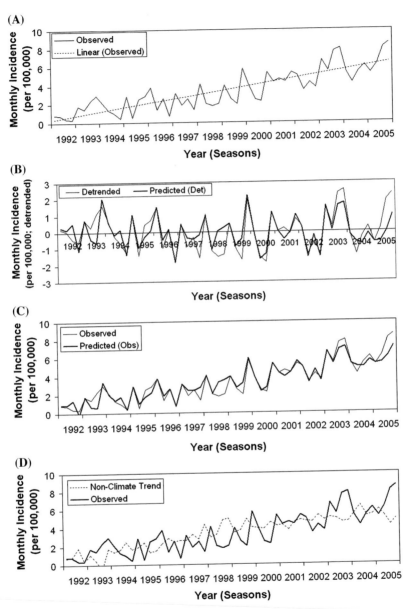

FIGURE 4. (**A–D**) Pima County aggregate detrended coccidioidomycosis incidence model, showing (**A**) best-fit line to observed aggregate incidence data; (**B**) predicted and observed values for the model of detrended incidence; (**C**) predicted and observed values reconstructed to include the original linear trend; (**D**) observed coccidioidomycosis incidence and modeled incidence with climate variability removed (Note: observed curve above nonclimate trend indicates that climate increases incidence; observed below nonclimate trend indicates that climate decreases incidence.)

TABLE 4. Results obtained using detrended aggregate monthly incidence records (non-QC) with 1-month offset for onset/exposure (no other exposure adjustment)

	Foresummer	Monsoon	Fall	Winter
R^2 (observed–predicted)	0.55	0.29	0.74	0.99
Predictors				
Dust PM10	−0.61 (0.016)			0.13 (≤ 0.001)
Precipitation				
Winter0	NA	NA	NA	
Fall0	NA	NA		−0.47 (≤ 0.001)
Monsoon0	NA			−0.54 (≤ 0.001)
Foresummer0				
Winter1				
Fall1			−0.58 (0.018)	
Monsoon1		−0.37 (0.180)		−0.18 (0.011)
Foresummer1				
Winter2			−0.46 (0.034)	
Fall2			−1.22 (≤ 0.001)	0.83 (≤ 0.001)
Monsoon2				
Foresummer2				
Winter3				0.73 (≤ 0.001)
Fall3	0.68 (0.009)	0.47 (0.094)		0.29 (≤ 0.001)
Monsoon3			1.05 (0.002)	
Foresummer3				−0.55 (≤ 0.001)
Winter4				
Fall4				NA
Monsoon4		NA		NA
Foresummer4	NA	NA		NA

NOTE: Detrended raw monthly data are included in the model to account for climate-related variability (not trend), as a new model with new predictors. Record period is updated 1992–2005.

incidence, despite being able to capture much of the seasonal and annual variability. We therefore removed the long-term trend in incidence to investigate this further. Applying an otherwise identical modeling approach (seasonal models with aggregate monthly reports) to the detrended data resulted in models that explained a large portion of the remaining seasonal and annual variability. Antecedent fall precipitation appears to have some long-term influence on detrended incidence, and dust remains important for dispersion in winter and foresummer. The caveat is that these models are based on aggregate monthly data, and that the identified relationships are therefore somewhat more tentative until they can be confirmed using exposure-adjusted data.

While climate explains a large proportion of variability in coccidioidomycosis incidence, it appears that the overall trend exceeds the magnitude of the variability. It is unclear from these models whether the longer trend is due to a nonclimate factor (e.g., urban expansion, improved reporting, etc.) or perhaps to some decadal-scale climate response not resolved in these short periods of data. Still, there are close relationships identified between climate conditions and coccidioidomycosis incidence that further support the broader "grow and blow" hypothesis of moist conditions being followed by drier conditions leading to spore formation, then followed by spore dispersion under dusty conditions.

ACKNOWLEDGMENTS

The Arizona Department of Health Services provided coccidioidomycosis data. J. Tabor processed the incidence data to appropriate population, spatial, and temporal scales for our initial studies. Preliminary model development was funded in part by the Arizona Disease Control Research Commission (now Arizona Biomedical Research Commission). Additional research and development was funded in part by the United States Environmental Protection Agency Science to Achieve Results Program, under Grant R8327540.

REFERENCES

1. SWATEK, F.E. 1970. Ecology of *Coccidioides immitis*. Mycopathol. Mycol. App. **40:** 3–12.
2. FIESE, M.J. 1958. Geographic distribution of *Coccidipides immitis*. *In* Coccidioidomycosis. 53–76. Charles Thomas. Springfield, IL.
3. KOLIVRAS, K.N., P. JOHNSON, A.C. COMRIE & S.R. YOOL. 2001. Environmental variability and coccidioidomycosis (Valley Fever). Aerobiology **17:** 31–42.
4. SMITH, C.E. 1940. Epidemiology of acute coccidioidomycosis with erythema nodosum ("San Joaquin" or "Valley Fever"). Am J. Pub. Health **30:** 600–611.
5. SMITH, C.E., R.R. BEAR, H.G. ROSENBERG & E.G. WHITING. 1946. Effect of season and dust control on coccidioidomycosis. J. Amer. Med. Assoc. **132:** 833–838.
6. HUGENHOLTZ, P. 1957. Climate and coccidioidomycosis. *In* Proceedings of the Symposium on Coccidioidomycosis, Phoenix, Arizona. Publ. **575:** 136–143. U.S. Public Health Services. Washington, DC.
7. MADDY, K. 1965. Observations on *Coccidioides immitis* found growing naturally in soil. Ariz. Med. **22:** 281–288.
8. JINADU, B.A. 1995. Valley Fever Task Force Report on the Control of *Coccidioides immitis*, Kern County. Kern County Health Department. Bakersfield, CA.
9. KOLIVRAS, K.N. & A.C. COMRIE. 2003. Modeling valley fever incidence based on climate conditions in Pima County, Arizona. Int. J. Biometeor. **47:** 87–101.
10. PARK, B.J. *et al.* 2005. An epidemic of coccidioidomycosis in Arizona associated with climate changes, 1998–2001. J. Infect. Dis. Available at http://www. journals.uchicago.edu/JID/journal/rapid.html.
11. KOMATSU, K. *et al.* 2003. Increase in coccidioidomycosis—Arizona, 1998–2001. Morb. Mortal. Wkly. Rep. **50:** 109–112.
12. ZENDER, C.S. & J. TALAMANTES. 2006. Climate controls on valley fever incidence in Kern County, California. Int. J. Biometeor. **50:** 174–182. [DOI 10.1007/s00484-005-0007-6]
13. COMRIE, A.C. 2005. Climate factors influencing coccidioidomycosis seasonality and outbreaks. Env. Health Pers. **113:** 688–692.

Public Health Surveillance for Coccidioidomycosis in Arizona

REBECCA H. SUNENSHINE,[a,b] SHOANA ANDERSON,[a]
LAURA ERHART,[a] ANNE VOSSBRINK,[a] PETER C. KELLY,[a]
DAVID ENGELTHALER,[a] AND KENNETH KOMATSU[a]

[a]Arizona Department of Health Services, Phoenix, Arizona, USA

[b]Centers for Disease Control and Prevention, Atlanta, Georgia, USA

ABSTRACT: Coccidioidomycosis or Valley Fever is a fungal disease that occurs primarily in the southwestern United States. Of the estimated 150,000 U. S. coccidioidomycosis infections per year, approximately 60% occur in Arizona, making this state the focal point for investigation of the disease. In this manuscript, we describe the epidemiology of coccidioidomycosis reported in Arizona over the last decade, hypotheses for the findings, and Arizona's response to the rising epidemic. Coccidioidomycosis surveillance data in Arizona consist of basic demographics of all laboratory and physician-diagnosed cases, the reporting of which has been mandated by law since 1997. The rate of reported coccidioidomycosis has more than quadrupled over the last decade from 21 cases per 100,000 population in 1997 to 91 cases per 100,000 in 2006 ($P < 0.001$). Case rates in older age groups (≥ 65 years old) have more than doubled since 2000 ($P < 0.001$). These data demonstrate the rising coccidioidomycosis epidemic in Arizona, especially among the elderly. The increase in the numbers of reported cases can be partially explained by the institution of mandatory laboratory reporting in 1997, but the cause of the persistent rise after 1999 is unknown. Further investigation of coccidioidomycosis will not only assist with the development of public health interventions to control this disease in Arizona and the southwestern United States, but will also provide important information to prepare for a bioterrorism event caused by this select agent.

KEYWORDS: coccidioidomycosis; lung diseases; fungal; epidemiology

INTRODUCTION

Of the estimated 150,000 infections with *Coccidioides* spp. per year in the United States,[1] approximately 60% occur in people who live in Arizona.[2]

Address for correspondence: Rebecca Sunenshine, M.D., Arizona Department of Health Services, 150 N. 18th Ave, Suite 150, Phoenix, AZ 85007. Voice: 602-768-1682; fax: 602-542-2722. Sunensr@azdhs.gov

Ann. N.Y. Acad. Sci. 1111: 96–102 (2007). © 2007 New York Academy of Sciences. doi: 10.1196/annals.1406.045

Prior to 1997, the Arizona Department of Health Services (ADHS) required only physicians to report coccidioidomycosis cases. Because of the increasing numbers of reported coccidioidomycosis cases,[3] ADHS instituted mandatory reporting of coccidioidomycosis by both laboratories and physicians in 1997. This led to a sharp increase in reported cases of coccidioidomycosis in addition to improving the timeliness and completeness of coccidioidomycosis reporting. In this manuscript, we describe the epidemiology of reported coccidioidomycosis in Arizona over the last decade, hypotheses for the findings, and Arizona's response to the rising epidemic.

METHODS

Clinical laboratories in Arizona routinely submit coccidioidomycosis blood serology results to ADHS using the U.S. postal service, fax, or by courier, depending on the laboratory. Additionally, providers are required to report patients they diagnose, treat, or detect with coccidioidomycosis to the local health agency, who subsequently report to ADHS.

All surveillance data from both laboratories and providers were collected on standard forms, entered into Arizona's Medical Electronic Disease Surveillance and Intelligence System on Microsoft Access or Excel, and analyzed using SAS (SAS Institute, Cary, NC, USA) software. Rates are calculated using Arizona census population data from the respective year as the denominator, except for 2006, which uses the 2005 census data as the denominator. The Mantel–Haenszel chi-square test was used to determine odds ratios for categorical variables. All statistical tests were two-tailed; a *P*-value of 0.05 or less was considered significant.

The case definition used by ADHS for a confirmed coccidioidomycosis case includes at least one of the following laboratory confirmatory tests:

- Cultural, histopathologic, or molecular evidence of the presence of *C. immitis*, or *C. posadasii*
- Immunologic evidence of infection (All titers must be ≥ 1:4)
  1. Serologic (testing of serum, cerebrospinal fluid, or other body fluid):
     (a). detection of coccidioidal IgM by immunodiffusion, enzyme immunoassay (EIA), latex agglutination, or tube precipitin, or
     (b). detection of any titer of coccidioidal IgG by immunodiffusion, enzyme immunoassay (EIA), or complement fixation
  2. Coccidioidal skin test conversion from negative to positive after the onset of clinical signs and symptoms (reagent not currently available)

Data submitted from the laboratories include the patient's name and birth date; date of report; laboratory result, specimen type and site, if applicable; and

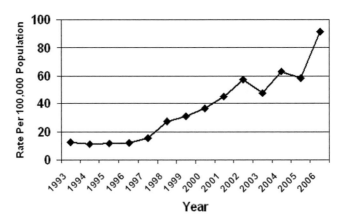

FIGURE 1. Rates of reported coccidioidomycosis by year in Arizona, 1993–2006.

sometimes race/ethnicity. Epidemiologists at ADHS input, review, and analyze
these data weekly.

Mortality due to coccidioidomycosis is obtained from death certificate data,
which originates directly from the ADHS Office of Vital Records. If a death
is due to coccidioidomycosis and is not recorded as either the primary or a
contributing cause of death on the death certificate, the information would not
be captured by our surveillance system.

RESULTS

The rate of reported coccidioidomycosis cases in Arizona increased from a
base line of approximately 21 cases per 100,000 population in 1997 (958 total
cases) to 37 cases per 100,000 population (1,812 total cases) in 1999 ($P <
0.001$) after the institution of mandatory laboratory reporting in 1997 (FIG. 1).
Thereafter, the rate of reported coccidioidomycosis trended upward annually
to a peak of 91 cases per 100,000 in 2006 ($P < 0.001$) for a total 5,535 cases
in 2006.

The sex distribution of reported coccidioidomycosis cases in Arizona has
consistently demonstrated a slight male preponderance and has not changed
over time as is described in the literature[4] (55% males in 2005); however, the
number of cases by age group has changed considerably over time (FIG. 2). The
case rates in older age groups (≥ 65 years old) progressively increased from 83
in 2000 to 206 per 100,000 population in 2006 ($P < 0.001$). This means that
the coccidioidomycosis incidence in the elderly has more than doubled over a
7-year period.

The average rate of reported coccidioidomycosis cases in Arizona in 2005
was 58 per 100,000, with the majority consistently occurring in Maricopa,
Pima, and Pinal counties, Arizona's most densely populated urban counties

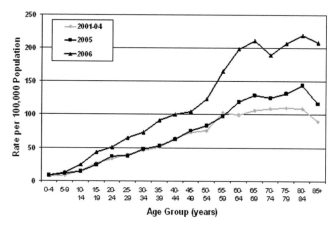

FIGURE 2. Coccidioidomycosis rates per 100,000 by age and year, Arizona, 2001–2006.

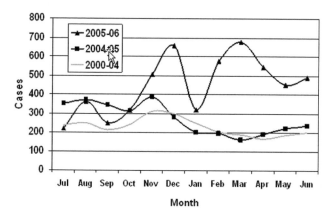

FIGURE 3. Reported coccidioidomycosis cases by month, Arizona, 2000–2006.

in the Mediterranean climatic zone (67.5, 78.3, and 63.7 cases per 100,000, respectively). Maricopa and Pima Counties contain the two largest cities in Arizona, Phoenix and Tucson, respectively.

Coccidioidomycosis has demonstrated a seasonal pattern in Arizona, which typically presents as a pronounced early winter peak from October to January and a smaller summer peak from May to August. The surveillance data from 2000 to 2004 reflect this pattern (FIG. 3). In the year from July 2004 to June 2005, we observed a more pronounced, yet shorter winter peak and a very small summer peak. In contrast, 2005–2006 data exhibit remarkably high summer and winter peaks, well beyond those in the previous 5 years.

The total number of deaths due to coccidioidomycosis has remained essentially unchanged over the last decade, with approximately 30 reported each year (mean and median 28.5, range 23–41). The mortality associated with coccidioidomycosis has actually decreased because of the rising population in Arizona (from 0.9 in 1996 to 0.5 in 2005 per 100,000 population).

DISCUSSION

The ADHS surveillance data demonstrate a steady increase in reported coccidioidomycosis cases, with pronounced demographic and seasonal changes over the last decade. These data prompt a number of questions, the majority of which can only be answered with further investigation.

First, does the increase in reported cases represent an actual increase in coccidioidomycosis in Arizona or is it an artifact of increased reporting or testing? Of note, the population of Arizona has risen steadily from 4.6 million in 1997 to 6.0 million in 2005, whereas the reported rates of coccidioidomycosis have increased even when taking the population into account. The initiation of mandatory laboratory reporting could certainly explain the initial increase in cases after 1997; however, it is unlikely to explain increases after 1999. Our rules for laboratory reporting have not changed since 1997 and although completeness of physician reporting has been historically questionable, no evidence suggests improvement in physician reporting over the last decade.

Regarding the possibility of increased testing by physicians, an investigation performed by Chang and colleagues in two large Phoenix outpatient clinics indicate that <15% of patients with community-acquired pneumonia (CAP), a common presentation of coccidioidomycosis,[5] are tested for the disease.[6] Unfortunately, no statewide data are available regarding current coccidioidomycosis testing practices.

Assuming the increase in coccidioidomycosis reporting represents an actual increase in disease incidence, what is the cause? Numerous factors have been hypothesized, including soil disturbance due to construction of new homes for the rapidly rising population in Arizona;[2,7] climate variations, including changes in temperature and precipitation, in different seasons, which have been shown to be important predictors of coccidioidomycosis incidence;[7–10] and windstorms leading to increased dust exposure.[11] Unfortunately, further research needs to be done to determine why coccidioidomycosis has reached epidemic proportions in Arizona.

Why has the incidence of disease specifically increased in the elderly? Although the population of Arizona has increased in age over the last decade, rates among persons aged $\geq$ 65 years have continued to increase, despite age adjustment. One hypothesis is that older individuals are more likely to have health insurance and therefore seek medical care, which may increase their likelihood of being diagnosed with coccidioidomycosis. A second is that persons aged 65 years and older are more susceptible to developing symptomatic disease. A third is that the increased migration of individuals at retirement age to Arizona from nonendemic areas may be leading to a more susceptible elderly population. Again, further study is required to elucidate whether any of these hypotheses has merit.

An additional question raised is why the mortality due to coccidioidomycosis has actually decreased when reported rates of the disease have increased. One possible explanation is that mortality data obtained from death certificates often underestimate actual disease mortality since physicians often fill out death certificates incorrectly.[12,13] Another possibility, however, is that treatment for coccidioidomycosis and medical care in general has improved substantially, allowing patients to survive longer. Alternatively, there could be a disproportionate increase in the diagnosis of primary pulmonary coccidioidomycosis, filling in a portion of the "iceberg" of total cases previously undiagnosed, as physician and patient awareness increase with the recent media coverage of Arizona's coccidioidomycosis epidemic.

Finally, what is ADHS doing about this important public health issue? First, an ongoing physician-education program has been established to raise awareness among health-care providers about the increase in coccidioidomycosis and its tendency to present as CAP.[5] Secondly, in May 2006, ADHS issued the recommendation to test all patients with CAP in endemic areas of Arizona for coccidioidomycosis. ADHS is also working with partners at the Centers for Disease Control and Prevention to better determine the percentage of patients diagnosed with CAP who actually have coccidioidomycosis. This investigation will provide more information about testing practices, empiric treatment practices, and follow-up of patients with coccidioidomycosis. Lastly, ADHS will begin enhanced surveillance for coccidioidomycosis to further characterize risk factors and clinical outcomes so that public health interventions can be developed. These efforts will enable ADHS to develop public health interventions to control coccidioidomycosis in Arizona and to increase the knowledge about the natural history of the disease to prepare for a potential bioterrorism event involving this select agent.

[*Disclaimer:* The findings and conclusions in this report are those of the authors and do not necessarily represent the views of the Centers for Disease Control and Prevention.]

ACKNOWLEDGMENT

The authors acknowledge Jamie Kokko for her technical assistance with manuscript preparation.

REFERENCES

1. GALGIANI, J.N., N.M. AMPEL, J.E. BLAIR, *et al.* 2005. Coccidioidomycosis. Clin. Infect. Dis. **41:** 1217–1223.

2. CENTERS FOR DISEASE CONTROL AND PREVENTION. 2004. Summary of Notifiable Diseases—United States—2004. Published June 16, 2006, for Morb. Mortal. Wkly. Rep. 53 (No 53).
3. AMPEL, N.M., D.G. MOSLEY, B. ENGLAND, et al. 1998. Coccidioidomycosis in Arizona: increase in incidence from 1990 to 1995. Clin. Infect. Dis. 27: 1528–1530.
4. EINSTEIN, H.E. & R.H. JOHNSON. 1993. Coccidioidomycosis: new aspects of epidemiology and therapy. Clin. Infect. Dis. 16: 349–356.
5. VALDIVIA, L., D. NIX, M. WRIGHT, et al. 2006. Coccidioidomycosis as a common cause of community acquired pneumonia. Emerg. Infect. Dis. 12: 958–962.
6. CHANG, D.C., B.J. PARK, L.A. BURWELL, et al. 2006. Disparities in testing practices for Coccidioides among patients with community-acquired pneumonia—Metropolitan Phoenix, 2003–2004. Poster [M-1608]; 46th Interscience Conference on Antimicrobial Agents and Chemotherapy; Sep 27–30, San Francisco, CA.
7. PAPPAGIANIS, D. 1994. Marked increase in cases of coccidioidomycosis in California: 1991, 1992, and 1993. Clin. Infect. Dis. 19(Suppl 1):S14–S18.
8. SMITH, C.E., R.R. BEARD, H.G. ROSENBERGER & E.G. WHITING. 1946. Effect of season and dust control on coccidioidomycosis. JAMA 132: 833–338.
9. PAPPAGIANIS, D. 1988. Epidemiology of coccidioidomycosis. Curr. Top. Med. Mycol. 2: 199–238.
10. KOLIVRAS, K.N. & A.C. COMRIE. 2003. Modeling valley fever (coccidioidomycosis) incidence on the basis of climate conditions. Int. J. Biometeorol. 47: 87–101.
11. PAPPAGIANIS, D. & H. EINSTEIN. 1978. Tempest from Tehachapi takes toll or Coccidioides conveyed aloft and afar. West J. Med. 129: 527–530.
12. MANT, J., S. WILSON, J. PARRY, et al. 2006. Clinicians didn't reliably distinguish between different causes of cardiac death using case histories. J. Clin. Epidemiol. 59: 862–867.
13. RODRIGUEZ, S.R., S. MALLONEE, P. ARCHER & J. GOFTON. 2006. Evaluation of death certificate-based surveillance for traumatic brain injury–Oklahoma 2002. Public Health Rep. 121: 282–289.

Coccidioidomycosis in California State Correctional Institutions

DEMOSTHENES PAPPAGIANIS AND THE COCCIDIOIDOMYCOSIS
SEROLOGY LABORATORY

*Department of Medical Microbiology and Immunology, University of California,
Davis, Davis, California 95616, USA*

ABSTRACT: Coccidioidomycosis (CM) has been recognized in inmates
of California State prisons since 1919, where it has been diagnosed in
inmates of various correctional facilities inside and outside the known
endemic areas. In recent years construction of new prisons within endemic areas has led to an increase in the number of cases of CM. In
2005 and 2006, the Pleasant Valley State Prison (PVSP) near Coalinga
and Avenal State Prison (ASP) near Avenal on the western side of the
San Joaquin Valley have been particularly affected. In 2005, our serologic testing yielded 150 new cases from PVSP and 30 from ASP. The
incidence rate in 2005 for PVSP (population approximately 5,000) was
at least 3,000 per 100,000, and this will be exceeded in 2006. Some cases
diagnosed in early 2006 likely were infections that were acquired in 2005.
Some cases are medically managed on site, but very ill inmates have had
care in nonprison facilities. Precise numbers of patients who were hospitalized were not made available to the author. Estimates of the cost per
patient have varied from $8,000 in the 1990s to $30,000 more recently.
Thus, this disease has important medical, demographic, and financial
implications for the state.

KEYWORDS: Coccidioidomycosis; prisoners; prisons

INTRODUCTION

For many years, coccidioidomycosis (CM) has been encountered in inmates
of prisons in the endemic areas of the southwestern United States. In recent
years, new prisons have been constructed in coccidioidal endemic areas of California and this has resulted in an expanded number of persons with this disease
among inmates and employees, a problem that has attracted our attention.

CM apparently was first recognized in an inmate of Folsom Prison near
Sacramento, California in 1919.[1] This prison was not in the area(s) to which

Address for correspondence: Demosthenes Pappagianis, M.D., Ph.D., Room 3146 Tupper Hall,
School of Medicine, University of California, Davis, Davis, CA 95616. Voice: 530-752-3391; fax:
530-752-6813.
  savaden@ucdavis.edu

Ann. N.Y. Acad. Sci. 1111: 103–111 (2007). © 2007 New York Academy of Sciences.
doi: 10.1196/annals.1406.011

this disease is endemic, exemplifying how the disease may be encountered outside the endemic zones. This also is exemplified by a prisoner in Boise, Idaho whom we have followed serologically for some years after his acquisition of CM in California.

Other instances of CM in incarcerated persons occurred during World War II among Japanese Americans forced into a camp near Casa Grande, Arizona, and among German prisoners of war in Florence, Arizona.[2] Some German prisoners of war complained of mistreatment as a result of lethal and other coccidioidal infections under the Geneva Convention Rules. This led to discontinuation of the use of the Florence, AZ facility for foreign prisoners; but it has continued to house civilian prisoners. In the 1950s and later, CM was described among young men prisoners who were sent to fight fires in endemic areas of Los Angeles County and elsewhere.[3,4]

For many years, our UC Davis Coccidioidomycosis Serology Laboratory has received serum specimens from incarcerated individuals who have or are suspected of having coccidioidomycosis. For example, sera had been submitted by physicians (Dr. D. Smilovitz and colleagues) from infected inmates in the California Men's Colony in San Luis Obispo County. In the year 2000, our attention was called to an outbreak of CM among inmates of the California Youth Authority, Paso Robles, who had been assigned to fight grass fires in McKittrick in the highly endemic area of Kern County. This led us into compilation of seropositive cases from other state prisons.[5,6] As yet, we have received too little information to assess the broad clinical implications of the disease in these inmates. Nevertheless, the occurrence of CM in inmates has important implications—to the state and its citizens: medical, demographic, and financial.

MATERIALS AND METHODS

The cases of coccidioidomycosis were detected by positive serologic tests on serum or other body fluids. Testing was carried out at our UC Davis Coccidioidomycosis Serology Laboratory. Initial testing was carried out by immunodiffusion of specimens after being concentrated approximately eightfold by evaporation under reduced pressure.[7,8] Patients were identified by name, date of birth, and inmate (California Department of Corrections) identification number. This precluded duplication in our enumeration of cases. In many instances, for logistical reasons, specimens from several inmates were drawn on the same date rather than in relationship to clinical indications. As a result of this, it was not possible to know the date of onset of illness, thus usually precluding recording cases by month. Moreover, on some occasions it was evident that sera from some inmates came by way of some intermediate laboratory, obscuring the provenance of the specimen. In FIGURE 1 we have presented a map of California indicating the location of prisons (name

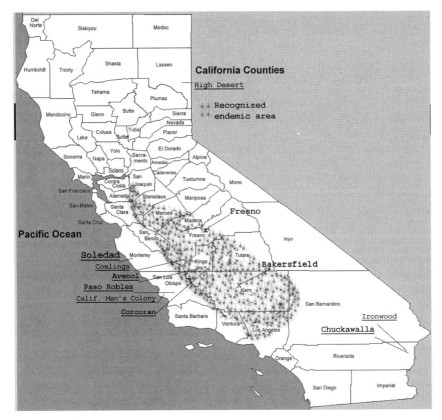

FIGURE 1. Map of California and recognized endemic area with prison locations (names underlined).

underlined) with respect to recognized areas in which CM is endemic. The more detailed map in FIGURE 2 indicates the relative positions of three prisons significantly represented among the cases we have tabulated: Coalinga (Pleasant Valley State Prison), Avenal, and Corcoran.

RESULTS

Simply expressed are the numbers of cases recognized serologically (TABLES 1–4). Note that the data of TABLE 1 were obtained before Pleasant Valley State Prison (PVSP) was completed. Following its inclusion, PVSP became the largest contributor of cases. FIGURE 3 illustrates the influence of "new construction" (including excavation) for a mental health hospital near (perhaps 200 yards from) PVSP. Construction began in late summer to early fall and soon the number of cases increased. As noted in the MATERIALS AND METHODS

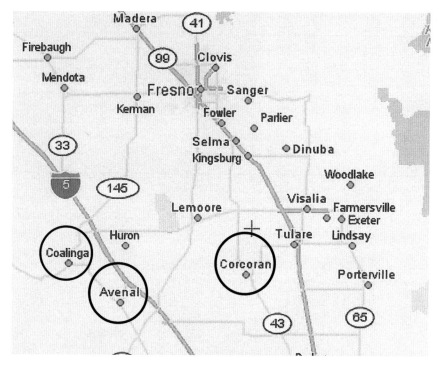

FIGURE 2. Three state prisons in the western San Joaquin Valley.

section some cases recorded for a given month were based on the date of the positive serum, but might have been drawn in an adjacent month. It was evident that PVSP had a higher rate of infections than other institutions, some of which had comparable numbers of inmates. By mid August 2006, PVSP had 300 new cases recognized, far exceeding those recognized (51) in Avenal, the next highest represented. We calculated an incidence of 3,000/100,000 for PVSP in 2005, and in 2006 though mid August the rate was 6,000/100,000. For comparison, the highest incidence rate of CM was 572/100,000 for Kern County during the epidemic year 1998. By mid August, the total number of reported cases of CM in California was approximately 1,300. Thus, the total number of cases, 388, in state prisons (TABLE 4) represented approximately 30% of the cases reported to the California State Department of Health Services. In 2005, the state prison cases (244) represented 15% of the total reported cases (approximately 1,600) in California.

Based on studies in Kern County during the epidemic years of the 1990s, the cost of care per patient was $8,000.[9] The cost of care of the 388 patients detected in state prisons, on basis of the figure of Caldwell *et al.*, would have been $3,104,000.00. Others have calculated that the cost per hospitalized patient (in 1998–2001) was approximately $34,000.[10] Specific numbers of hospitalized

TABLE 1. Coccidioidomycosis in California correctional institution, 2000—2001[a]

Facility	Number of Cases
Avenal	36[b]
Corcoran	14
Chuckawalla	1
Ironwood	1
Vacaville	8
CMC—San Luis Obispo	16
Miscellaneous	10
CA Youth Authority (Paso Robles)	23
Total	109

[a]In some instances onset of disease may have been earlier than 200.
[b]One case in prison employee.

inmates were not provided to the author. Inasmuch as approximately 5% of patients with clinical evidence of coccidioidomycosis undergo metapulmonary dissemination of their disease, at least 20 of the 388 patients would have required hospitalization at a cost of $680,000. Therefore, the fiscal impact to the state is substantial.

DISCUSSION

Incarcerated individuals and employees of correctional institutions in endemic areas may acquire coccidioidomycosis. Because incarcerated individ-

TABLE 2. Coccidioidomycosis in California correctional institutions from 3/2003 through 2/2004

Facility	Number of Cases
Avenal	22 + 1[a]
Pleasant Valley—Coalinga	127 + 1[a]
Corcoran	21
Chowchilla (Women's)	1
CMC–San Luis Obispo	7
Chuckawalla	1
Ironwood—Blythe	3
Vacaville	1
Wasco	1
Miscellaneous	18
Susanville, San Quentin, Salinas Valley, Soledad	
CSAT, Lancaster, CA Youth Authority	5
Total	207

[a]Case in state employees.

TABLE 3. Coccidioidomycosis in California correctional institutions in 2005

Facility	Number of Cases
Pleasant Valley—Coalinga	150
Avenal	47
Corcoran (SATF)	2
Corcoran	23
CMC–San Luis Obispo	3
Ironwood—Blythe	1
North Kern—Delano	1
Solano—Vacaville	4
Soladad	5
Unknown State Prison	8
Total	244

uals have centralized medical care, some compilation of cases is possible. Enumeration of cases among employees is more difficult because they do not have a unified source for medical care.

Coccidioidal infections can be acquired by inmates within the institutions to which they have been confined, or, as illustrated above, by inmates who have been confined in institutions outside the endemic areas, but have been assigned to fight fires in those areas.

Occasionally prisoners are transferred from one California state prison to another. In some instances, an individual already afflicted with coccidioidomycosis may baffle an unsuspecting medical staff because of the mimicry of coccidioidomycosis for other diseases or because the medical staff does not appreciate that the patient/inmate had previously been in the endemic area. One striking example of this is a male prisoner in an institution in nonendemic Idaho who acquired his coccidioidal infection in California.

TABLE 4. Coccidioidomycosis in California correctional institutions from January through mid August 2006

Facility	Number of Cases
Pleasant Valley–Coalinga	300[a]
Avenal	51
Camarillo	1
Corcoran (SATF)	6
Corcoran	13
CMC–San Luis Obispo	7
Ironwood–Blythe	2
Mule Creek	1
Soladad	1
Vacaville	1
Ventura Youth Authority	5
Total	244

[a]Incidence 6,000/100,000.

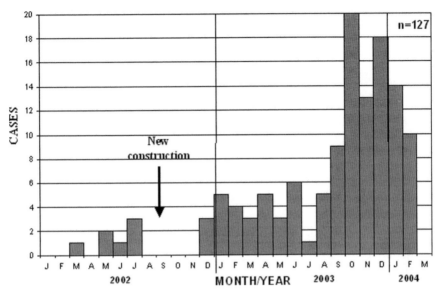

FIGURE 3. Cases of CM at Pleasant Valley State Prison between January 2002 and March 2004. Approximate time of construction of a mental health hospital near the prison is designated with an *arrow*.

Drastic consequences followed in a former prisoner who acquired his coccidioidal disease in prison in Arizona, but who then moved to Alabama, where he became moribund following a lethal cerebrovascular episode. His physicians in Alabama were not initially aware of his prior coccidioidomycosis, and donated his kidneys and liver to awaiting recipients. At 19 and 17 days after transplantation, respectively, the recipient of one kidney and the recipient of the liver died with fulminant coccidioidomycosis.[11] Because of the varying severity of coccidioidomycosis, the intensity and strategy in the treatment of inmate/prisoners can pose a challenge to prison physicians and their often limited resources. As a result there are occasions when the inmate/patient's illness requires more complex means of treatment that are available at nonprison referral hospitals. An example, treatment of spinal coccidioidomycosis among prisoners, but in a nonprison site, provided valuable information of management of severe coccidioidomycosis.[12]

An additional problem pertains to patient/inmates who acquire coccidioidomycosis and are subsequently discharged after they have completed their sentence. Uncertainty about their clinical status and about how and where to seek medical attention may result in or be followed by recrudescence of coccidioidal disease. At least one such individual died following his belated routing into a medical care center.

One aspect of coccidioidomycosis that could be defined is the influence of certain intercurrent diseases present among inmates (e.g., hepatitis C) on the course of coccidioidomycosis. Additional valuable information may also accrue from the medical and surgical attention provided to inmates as mentioned above relative to spinal surgery.[12]

Some cases of CM can be anthropogenic, as in the construction of a mental health facility adjacent to Pleasant Valley State Prison, or can result from expected seasonal/climatic associations, which influence the rise and fall of incidence. However, the incarceration of individuals from nonendemic areas, in federal [13] as well as state prisons within the endemic areas will continue to provide a stream of challenging and costly cases of coccidioidomycosis.

ACKNOWLEDGMENTS

I am very grateful to Dr. Suzanne Johnson for support in the preparation of this report, and to the following members of the UCD Coccidioidomycosis Serology Laboratory for the performance of serologic tests: L. Fortis, D. Griffin, C. MacVean, M. Tamashiro, W. Tsang, and C. Miller.

REFERENCES

1. HELSLEY, G.F. 1919. Coccidioidal granuloma: report of a case. J. Amer. Med. Assoc. **73:** 1697–1699.
2. SMITH, C.E. 1958. Coccidioidomycosis. Chapter XVI. *In* Preventive Medicine in World War II; Vol. IV. Communicable Diseases Transmitted Chiefly through Respiratory and Alimentary Tracts. 285–316. Office of the Surgeon General, Dept. of the Army, Washington, DC.
3. KRITZER, M.D., M. BIDDLE & J.F. KESSEL. 1950. An outbreak of primary pulmonary coccidioidomycosis in Los Angeles County, California. Ann. Intern. Med. **33:** 960–990.
4. RAO, S., M. BIDDLE, O. BALCKU, *et al.* 1972. Focal endemic coccidioidomycosis in Los Angles County. Amer. Rev. Resp. Dis. **105:** 410–416.
5. PAPPAGIANIS, D. & V. VAN KEKERIX. 2002. Resurgent coccidioidomycosis (coccy) in California—emphasis on Tulare County and prison inmates. *In* Proceedings of the Annual Coccidioidomycosis Study Group Meeting. No. 46. Valley Fever Center for Excellence, Tucson, AZ.
6. PAPPAGIANIS, D. & N.D. SACKS. 2004. Outbreak of coccidioidomycosis in California State Prisons 2003–2004. *In* Proceedings of the Annual Coccidioidomycosis Study Group Meeting. No. 48. Valley Fever Center for Excellence, Tucson, AZ.
7. PAPPAGIANIS, D. & B.L. ZIMMER. 1990. Serology of coccidioidomycosis. Clin. Micro. Rev. **3:** 242–268.
8. PAPPAGIANIS, D. 1994. Marked increase in cases of coccidioidomycosis in California: 1991, 1992, and 1993. Clin. Infect. Dis. **19**(Suppl. 1): S14–S18.
9. CALDWELL, I. W., G. WELCH, R. H JOHNSON, *et al.* 1996. The economic impact of coccidioidomycosis in Kern County, California, 1991 to 1993.

In Coccidioidomycosis. H.E. Einstein & A. Catanzaro, Eds.: 88–97. National Foundation for Infectious Disease, Washington, DC.

10. PARK, B.J., K. SIGEL, V. VAZ, *et al.* 2005. An epidemic of coccidioidomycosis in Arizona associated with climatic changes 1998–2001. J. Infect. Dis. **191:** 1981–1987.

11. WRIGHT, P.W., D. PAPPAGIANIS, M. WILSON, *et al.* 2003. Donor-related coccidioidomycosis in organ transplant recipients. Clin. Infect. Dis. **37:** 1265–1269.

12. HERRON, L.D., P. KISSEL & D. SMILOVITZ. 1997. Treatment of coccidioidal spinal infection: Experience in 16 cases. J. Spinal Disorder **10:** 215–222.

13. BURWELL, L.A., B.J. PARK, K. WANNEMUEHLER, *et al.* 2005. Evaluation of an enhanced diagnosis and treatment program for coccidioidomycosis, Kern County, CA. 279. *In* Abstracts of the Infectious Diseases Society of America, 43rd Annual Meeting, San Francisco, CA.

Coccidioidomycosis in the U.S. Military

A Review

NANCY F. CRUM-CIANFLONE

Infectious Diseases Staff, Naval Medical Center San Diego, San Diego, California, USA

ABSTRACT: Coccidioidomycosis has had an impact on the military since the discovery of the causative agent, *Coccidioides immitis,* over a century ago. The first reports of *Coccidioides* outbreaks affecting U.S. military personnel were reported by Smith and others during World War II. Since that time, numerous outbreaks and sporadic cases have occurred, affecting both the health and readiness of our armed forces. This article summarizes the impact of *Coccidioides* sp. on our military troops with a review of the literature, a description of the experience at a tertiary referral hospital treating the disease, and a synopsis of incidence studies conducted at bases in the southwestern United States. The substantial effect that coccidioidomycosis has had on the military supports the development of both newer and more effective treatments as well as a Valley Fever vaccine for prevention of this disease.

KEYWORDS: coccidioidomycosis; military; epidemiology; review

BACKGROUND

Since its initial description over 100 years ago, coccidioidomycosis has been a notable disease, particularly among military personnel. The first human case was identified in an Argentinian soldier in the 1890s.[1] The U.S. military has been affected by this mycosis, which is endemic to the southwestern United States and locations in Mexico and Central and South America. Early large-scale reports of *Coccidioides* infections in U.S. military personnel were reported by C.E. Smith in the 1940s (during World War II), when numerous airmen and other military members training in desert regions became infected.[2] Over the past 50 years, many outbreaks and sporadic cases have

Address for correspondence: Dr. Nancy Crum-Cianflone, c/o Clinical Investigation Department (KCA), Naval Medical Center San Diego, 34800 Bob Wilson Drive, Ste. 5, San Diego, CA 92134-1005. Voice: 619/532-8134/40; fax: 619-532-8137.

nfcrum@nmcsd.med.navy.mil

Ann. N.Y. Acad. Sci. 1111: 112–121 (2007). © 2007 New York Academy of Sciences.
doi: 10.1196/annals.1406.001

occurred, affecting both the health and readiness of the armed forces. Cases are most often present within endemic areas; however, given troop mobility, hundreds of cases have been diagnosed in nonendemic areas in this country as well as overseas,[2,3] emphasizing the need for clinicians around the globe to be able to recognize and treat this disease.

Coccidioidomycosis is an occupational hazard to military members in terms of acute illness, leading to lost workdays (an estimated 35 lost days per case),[4] severe disseminated disease with early/forced retirements, and the need for close follow-up of personnel with clinical disease, which may have a negative impact on their deployment status.[5-7] Currently, more than 350,000 military members are stationed in endemic areas, with thousands more training in these locales annually.[8] Military bases situated near or in endemic areas in California and other endemic states (Arizona, Nevada, New Mexico, Utah, and Texas) are shown in FIGURES 1 and 2, respectively. This article reviews the effect of coccidioidomycosis on U.S. military troops over the past century.

OUTBREAKS AMONG MILITARY PERSONNEL

The first large outbreaks of coccidioidomycosis in the U.S. military occurred in association with training exercises during and shortly after World War II. The generally good weather and semiarid conditions in desert areas of southwestern United States have proven ideal for many military training activities, particularly for airfield operations. C.E. Smith documented annual rates of 25–50 infections per 100 susceptible persons at three Army air fields in the San Joaquin Valley.[9,10] Smith's high rates were related to "newcomer" military enlistees with no prior immunity, the excessive exposure due to dusty training operations, and the elevated endemicity of the area. In addition to reporting asymptomatic and pulmonary cases, Smith and others demonstrated that severe, life-threatening cases occurred among otherwise healthy military men, with a propensity for these cases appearing among African American individuals.[2,9-11]

Troops were also infected, albeit typically at lower rates, in many other locations, including bases in California, Arizona, and Texas.[9,10,12-14] Between 1942 and 1945, nearly 4,000 army men were infected, leading to 39 deaths.[2] In addition to the impact on the U.S. troops during the 1940s and 1950s, prisoners of war in these locations (especially in Florence, Arizona) were also affected, many suffering from both tuberculosis and coccidioidomycosis. During a 70-day period at Minter Field in Bakersfield, California, 10% of POWs were hospitalized with coccidioidomycosis for a total of 150 infections among prisoners.[2]

During and shortly after WW II, the Commission on Epidemiological Survey of the Army Epidemiological Board, Preventive Medicine Division, Office of the Surgeon General, studied the disease and developed preventive strategies

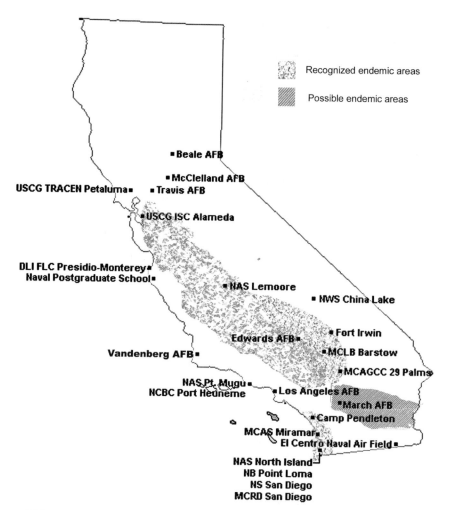

FIGURE 1. U.S. military bases in California located in or near endemic areas for coccidioidomycosis. AFB = Air Force Base; DFI FLC = Defense Logistic Institute Foreign Language Center; ISC = Integrated Support Command; MCAGCC = Marine Corps Air Ground Combat Center; MCAS = Marine Corps Air Station; MCLB= Marine Corps Logistic Base; NAS = Naval Air Station; NB = Naval Base; NCBC = Naval Construction Battalion Center; NS = Naval Station; TRACEN = Training Center; USCG = U.S. Coast Guard.

in an attempt to reduce the impact of *Coccidioides* infections on military personnel.[2] Much current knowledge regarding the rate of asymptomatic disease, diagnostic methods (coccidioidin and complement fixation tests), the seasonality of the disease, locations of endemicity, and control measures were defined by this Board's efforts.[2,10]

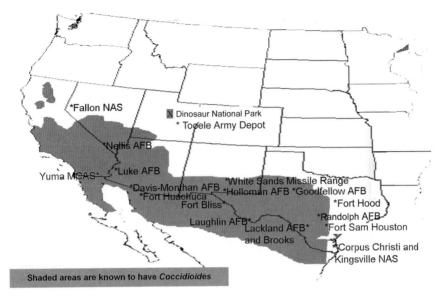

FIGURE 2. Additional U.S. military bases in the southwest United States located in or near areas endemic for coccidioidomycosis.

While cases continued to occur in the military environment over the next 30 years and coccidioidomycosis remained a leading cause of lost work days among troops,[4] the next outbreak associated with military troops was not reported in the literature until 1977.[7] After a natural dust storm at the Lemoore Naval Air Station (NAS), Lemoore, California, 18 persons stationed at the base developed symptomatic infection; four cases were disseminated (all among non-Caucasians), and one person died of the disease.

Standaert *et al.* described an outbreak that occurred in 1992 among a U.S. Marine reserve unit from Tennessee, which had attended a 3-week training exercise in the mountains of San Luis Obispo County, California. The outbreak was recognized after one member required hospital admission for pneumonia in Tennessee, where an astute physician obtained serologic testing for *Coccidioides* in view of the patient's recent travel history. This led to an epidemiologic investigation, wherein 8 of 27 men (30%) were found to have a positive serologic test; of these, 7 (88%) had symptomatic respiratory disease, and 1 had dissemination to the skin.[15]

The most recent reported coccidioidomycosis outbreak occurred among a U.S. Navy SEALs unit during sniper training in Coalinga, California, in 2001. Ten (45%) of 22 men developed the infection during a 6-week training exercise, the highest attack rate of clinical disease ever reported in a military unit. Members had been sleeping in tents on the soil, driving in open military vehicles on convoys, and occasionally, concealing themselves in handmade holes in the desert soil. The high attack rate was likely related to the high in-oculum exposures created by these activities. All patients were symptomatic, with fever, cough, and chills as the most frequent manifestations; 50% had abnormal chest radiographs, despite the fact that X-rays were obtained weeks after the initial symptoms. Fortunately, there were no cases of disseminated disease or death associated with this outbreak.[16]

THE NAVAL MEDICAL CENTER SAN DIEGO (NMCSD) EXPERIENCE

A recent descriptive review of coccidioidomycosis cases evaluated at the NMCSD provided an overview of the impact of coccidioidomycosis on military forces in southern California.[17] The NMCSD is a tertiary referral center for infectious diseases; cases seen at this facility represent the "tip of the iceberg" in terms of the overall burden of *Coccidioides* infections, as only complicated/disseminated cases are generally referred from military bases in the region to this medical facility. A recent published retrospective study of *Coccidioides* infections treated at NMCSD during the years 1994–2002 described 223 confirmed cases of which 130 (58%) were pulmonary, 50 (22%) were disseminated disease, and 32 (14%) had an elevated complement fixation titer ($>1:16$) without evidence of dissemination; 11 (5%) could not be classified because of the unavailability of the medical records.[17]

The mean age of patients was 40.3 years; 73% were males. Race was not a risk factor for the development of pulmonary disease, but it was predictive of dissemination. Rates of dissemination were recalculated from the study data for active duty members. Filipinos were at highest risk (RR 15.4, $P <$ 0.05), followed by African Americans (RR 7.0, $P < 0.05$), when compared to white/non-Hispanic military personnel, confirming earlier rates showing that certain races have a propensity for disseminated disease.[2,10–12,18] Hispanics had a fourfold higher rate of dissemination than Caucasians, but this did not reach statistical significance, likely due to the small sample size of this racial group in the study.

This study also looked at time from symptom onset until the diagnosis of coccidioidomycosis was established. For those with pulmonary disease, it was a mean of 48 days (range 7–270 days) and for disseminated disease, a mean of 116 days (5–730 days).[17] These data highlight the continued efforts needed to reduce the delay in diagnosis of coccidioidomycosis; an earlier diagnosis may assist in reducing unnecessary antibiotics, relieve patient and command concerns, and allow for early recognition of complicated disease. Enhancing the education of both military members and physicians regarding this infectious disease is a priority to aide in the earlier presentation of patients for evaluation and the earlier diagnosis of the disease by clinicians. Given the protean manifestations of chronic or disseminated disease and the nonspecific nature of acute respiratory infections, a high index of suspicion is needed, and travel and exposure histories should be obtained. As noted earlier, the diagnostic challenge is worldwide, not just regional, on account of rapid deployments and mobility of troops.[5]

Since disseminated disease has the most profound impact on health status and early medical retirements, the NMCSD data were evaluated to determine the rate of progression of coccidioidal infections to dissemination disease.[17] Of patients initially presenting with only pulmonary disease who had a complement fixation (CF) titer ≤1:16, 9 of 130 (7%) subsequently developed disseminated disease over a 4–66-month time period; all of these patients initially received antifungal therapy for pulmonary involvement, but then discontinued therapy and developed disseminated disease. Of patients initially with pulmonary disease who had a CF titer >1:16, 4 of 32 (13%) developed progressive disease. Of patients initially presenting with pulmonary disease who went on to disseminated disease, 69% were Filipino. When all Filipinos in this study cohort who initially presented with only pulmonary disease were examined, it was seen that, 30% went on to develop disseminated disease. These data emphasize the need for close continued follow-up of all coccidioidal cases, especially among Filipino military members. In such cases, the member's ability to be deployed may be restricted on account of the need for regular visits at a medical center with reliable CF testing and infectious disease specialists. A total of 50 patients with disseminated cases were treated at NMCSD during the 9-year study period, with 74% of these cases occurring among African Americans and Filipinos. The site of dissemination was skeletal system/joints in 56% and the skin in 36% of cases, but a variety of locations were involved.[17]

The impact on the military embraces both personnel health issues and costs. Twenty-two percent of those with pulmonary disease and 40% with disseminated disease were hospitalized at NMCSD. A prior study by Gray *et al.* examined data from 1981–1994 among active duty Navy and Marine Corps personnel and found the risk factors for hospitalization among those with pulmonary disease included those at junior pay grades, older age, and race other than white/non-Hispanic; these factors may be in part explained by difference in military duties (e.g., administrative vs. field work) and recreational activities among these groups.[19]

Regarding the NMCSD data, 11% of patients had progressive disease or relapse, and 4% had significant disabilities. Such events often led to medical retirement boards. In this series, there were only two deaths, none occurring among active duty personnel. Although the mortality rate for this disease is low overall, the morbidity associated with clinical disease is oftentimes significant; for instance, even among patients with uncomplicated disease, the illness may be protracted, with weeks to months of fatigue affecting the member's readiness to serve.

INCIDENCE RATES OF COCCIDIOIDOMYCOSIS AMONG MILITARY PERSONNEL

The rate of infections among military personnel stationed in endemic areas has varied widely between studies; differences may be due to factors such as location, number of newcomers to the base, activities of the personnel, climactic conditions, and seasonality. During World War II, C.E. Smith reported annual incidence rates of 25–50% among susceptible persons.[9,10] Drips and Smith evaluated skin test conversion rates per year at Lemoore Naval Air Station in the 1960s and found a rate of only 1.6%; over half of the converters noted a history of a respiratory illness.[20] Hopper *et al.* reported up to a 35% skin test conversion rate among tank drivers during a 7-month period in the 1970s at the Marine Corps Base at Twenty-Nine Palms, California; infantry members visiting the base for a 1-month training exercise had an 11% rate. Of those permanently stationed at Lemoore NAS, a conversion rate of 25% over 6 to 8 months was reported.[21]

More recent studies evaluated the incidence rates at the National Training Center in Fort Irwin, California. In August 2002, military units arriving from nonendemic areas were assessed during a 5-week training period; activities included 3 weeks of intense hiking, artillery use, and tank driving. Serologic testing was performed at baseline and then at 5-week and 11–13-week (subset of patients) time periods using enzyme immunoassay and CF testing.[22] The demographic characteristics of the 387 participants were: age, a mean of 24 years; gender, 95% male; and race, 67% Caucasian, 15% African American, 14% Hispanic, and 4% other. Two percent of the personnel had evidence of prior coccidioidal infection. Only one patient (0.6% of the study population with testing at all three time periods) developed confirmed coccidioidomycosis, which was manifested as a flu-like illness; four others had possible infection based on an isolated IgM titer. The calculated annual incidence rate based on this study was 6–32%. The limitations of this study included the lack of an available skin test and the evaluation of only a single time period.

A study at Fort Irwin documented 9 positive skin test results from 101 evaluable military subjects over a mean of 20 months of residence time at the base; one positive test represented a prior exposure (personal communication, Richard Hector). On the basis of these data, the annual incidence for skin test conversion was 4.8% per year, which is higher than the general population's rate, but overall low compared to other military studies. Additional incidence data regarding the rate of coccidioidal infections among military members are needed. This will require further research since data relying on clinical cases are insufficient because of the large number of asymptomatic infections (60% of cases) and pulmonary coccidioidal infections often self-resolve without a specific diagnosis.

CASE SERIES AT A NAVAL BASE

A retrospective review of *Coccidioides* cases at Lemoore NAS was conducted in 2006 after an increasing number of cases were noted among military members at the base. Laboratory data and medical records from January of 2003 to May of 2006 were reviewed for epidemiologic purposes (personal communication, Dr. Rachel Lee). The patient population served at the Lemoore NAS Medical Clinic and Hospital included 1,830 on active duty, 8,756 dependents, and 793 retirees, for a total population of 11,379. The base is located in King and Fresno Counties in the San Joaquin Valley. Approximately 68 cases were noted over the 41-month period, with an average of 175 cases per 100,000 person-year, a rate higher than reported by most studies (43–86/100,000 person-years), and from prior rates found at Lemoore NAS in the 1970s (11 cases/26,000 persons).[7,23]

Data over the 41-month period suggest that the number of cases of coccidioidomycosis is increasing. Data from the Kings County Health Department also demonstrated increasing rates among the civilian population during this time period. In addition, reports from other areas in California and Arizona suggested a rising number of cases; hence the findings at Lemoore appear consistent with temporal trends in a variety of endemic areas. In addition to the rising number of cases at Lemoore NAS, it is noteworthy that many (15%) patients presented with disseminated disease. A similar cluster of disseminated cases was previously observed among cases evaluated at the NMCSD.[24] These reports are remarkable in that most cases occurred in otherwise healthy men without comorbidities. Further investigation into the recent trends at the base is under way.

CONCLUSIONS

Coccidioidomycosis remains a threat to the health and readiness of military personnel, as numerous bases and training sites are located within endemic areas. Physicians should suspect the disease worldwide given the mobility of troops and the current frequent deployments of our personnel stationed in the southwestern United States. Both outbreaks and sporadic cases of coccidioidomycosis have been documented in the U.S. military over the past century. The relevance of this disease to the military is not "going away," but may be increasing as we continue to see rising rates of infections and disseminated forms of the disease at some military bases.

Future endeavors should include active surveillance of coccidioidal infections at military installations to more accurately estimate the current impact of this disease on the U.S. service members and to evaluate temporal trends. Incidence data via research studies using skin tests and serologic studies would be also helpful. Continued education programs for military personnel and clinicians are advocated to more rapidly identify cases. Preventive strategies are needed given that scores of military bases are located within endemic areas, plus the potential for exposure to *Coccidioides* during military training exercises, the continual influx of "newcomers" without immunity to bases in endemic areas, and ethnic diversity of military members. Such prevention efforts could include use of closed-cab vehicles, pavement of the desert soil, avoidance of exercises that create dust formation, and use of facemasks during dusty conditions. However, these strategies may often be difficult to implement, and hence prevention would be best accomplished by development of a vaccine for Valley Fever to provide both military members and civilians residing or training in endemic areas with a means to avoid acquiring the disease.

ACKNOWLEDGMENT

The views expressed in this article are those of the authors and do not reflect the official policy or position of the Department of the Navy, Department of Defense, or the United States Government.

REFERENCES

1. POSADA, A. 1892. Uno nuevo caso de micosis fungoidea con psorospermias. Ann. Circ. Med. Argentino **15:** 585–596.
2. SMITH, C.E. 1958. Coccidioidomycosis. *In* Preventive Medicine in World War II, vol IV: Communicable Diseases Transmitted Chiefly through Respiratory and Alimentary Tracts. Colonel John Boyd, Jr. *et al.*, Eds.: 285–316.
3. CRUM-CIANFLONE, N.F., A.A. TRUETT, N. TENEZA-MORA, *et al.* 2006. Unusual presentations of coccidioiodomycosis: a case series and review of the literature. Medicine. **85:** 263–277.
4. SCOGINS, J.T. 1957. Comparative study of time loss in coccidioidomycosis and other respiratory diseases. *In* Proceedings of symposium on Coccidioidomycosis. Public Health Service Publ. 515.: 132–135.
5. OLSON, P.E., W.D. BONE, R.C. LABARRE, *et al.* 1995. Coccidioidomycosis in California: regional outbreak, global diagnostic challenge. Mil. Med. **160:** 304–308.
6. RUSH, W.L., D.P. DOOLEY, S.P. BLATT, *et al.* 1993. Coccidioidomycosis: a persistent threat to deployed populations. Aviat. Space Environ. Med. **64:** 653–657.
7. WILLIAMS, P.L., D.L. SABLE, P. MENDEZ, *et al.* 1979. Symptomatic coccidioiodomycosis following a severe natural dust storm. An outbreak at the Naval Air Station, Lemoore, Calif. Chest **76:** 566–570.
8. OLIVERE, J.W., P.A. MEIER, S.L. FRASER, *et al.* 1999. Coccidioidomycosis—the airborne assault continues: an unusual presentation with a review of the history, epidemiology, and military relevance. Aviat. Space. Environ. Med. **70:** 790–796.
9. SMITH, C.E., R.R. BEARD, H.G. ROSENBERGER, *et al.* 1946. Effect of season and dust control on coccidioidomycosis. JAMA **132:** 833–838.

10. SMITH, C.E., R.R. BEARD, E.G. WHITING, *et al.* 1946. Varieties of coccidioidal infection in relation to epidemiology and control of diseases. Am. J. Public Health **36:** 1394–1402.

11. WILLET, F.M. & A. WEISS 1945. Coccidioidomycosis in Southern California: report of a new endemic area with a review of 100 cases. Ann. Intern. Med. **23:** 349–375.

12. PAPPAGIANIS, D., S. LINDSAY, S. BEALL, *et al.* 1979. Ethnic background and the clinical course of coccidioidomycosis. Am. Rev. Respir. Dis. **120:** 959–961.

13. SHELTON, R.M. 1942. A survey of coccidioidomycosis at Camp Roberts, California. JAMA **118:** 1186–1190.

14. GOLDSTEIN, D.M. & S. LOUIE 1943. Primary pulmonary coccidioidomycosis: report of epidemic of 75 cases. War. Med. **4:** 299–317.

15. STANDAERT, S.M., W. SCHAFFNER, J.N. GALGIANI, *et al.* 1995. Coccidioidomycosis among visitors to a *Coccidioides immitis*-endemic area: an outbreak in a military reserve unit. J. Infect. Dis. **171:** 1672–1675.

16. CRUM, N., C. LAMB, G. UTZ, *et al.* 2002. Coccidioidomycosis outbreak among United States Navy SEALs training in a *Coccidioides immitis*-endemic area— Coalinga, California. J. Infect. Dis. **186:** 865–868. Epub 2002 Aug 16.

17. CRUM, N.F., E.R. LEDERMAN, C.M. STAFFORD, *et al.* 2004. Coccidioidomycosis: a descriptive survey of a reemerging disease. Clinical characteristics and current controversies. Medicine (Baltimore). **83:** 149–175.

18. MCCRACKEN, B.E. 1953. Final report of coccidioidomycosis research project at Camp Roberts, Ca., 1 September 1952–15 October 1953, Surgeon's Office 6th Army.

19. GRAY, G.C., E.F. FOGLE & K.L. ALBRIGHT 1998. Risk factors for primary pulmonary coccidioidomycosis hospitalizations among United States Navy and Marine Corps personnel, 1981–1994. Am. J. Trop. Med. Hyg. **58:** 309–312.

20. DRIPS, W. Jr. & C.E. SMITH 1964. Epidemiology of coccidioidomycosis. A contemporary military experience. JAMA **190:** 1010–1012.

21. HOOPER, R., G. POPPELL, R. CURLEY, *et al.* 1980. Coccidioidomycosis among military personnel in Southern California. Mil. Med. **145:** 620–623.

22. CRUM, N.F., M. POTTER & D. PAPPAGIANIS 2004. Seroincidence of coccidioidomycosis during military desert training exercises. J. Clin. Microbiol. **42:** 4552–4555.

23. ROSENSTEIN, N.E., K.W. EMERY, B.S. WERNER, *et al.* 2001. Risk factors for severe pulmonary and disseminated coccidioidomycosis: Kern County, California, 1995–1996. Clin. Infect. Dis. **32:** 708–715.

24. CRUM, N.F., E.R. LEDERMAN, B.R. HALE, *et al.* 2003. A cluster of disseminated coccidioidomycosis cases at a US military hospital. Mil. Med. **168:** 460–464.

Coccidioidomycosis Infection in a Predominantly Hispanic Population

MIGUEL ANGEL PENA-RUIZ,[a] ARMANDO D. MEZA,[a] AND ZUBER D. MULLA[b]

[a]Department of Internal Medicine, Texas Tech University, El Paso, Texas, USA

[b]Department of Obstetrics and Gynecology, Texas Tech University, El Paso, Texas, USA

ABSTRACT: In Texas there are limited data on the epidemiology of coccidioidomycosis. Our goal is to determine the prevalence of coccidioidomycosis in a county hospital in El Paso, Texas. The charts of all patients with coccidioidomycosis admitted to the hospital in the past 9 years was retrospectively reviewed statistical analysis performed. Forty-one cases were identified, giving a prevalence of 3.2 cases per 10,000 discharges. Pneumonic consolidation occurred in 14 (44%), miliary pattern in 6 (19%) and cavitation in 6 (19%) cases. Pulmonary involvement occurred in 32 patients (78%) and meningeal involvement in 3 patients. Six patients had disseminated disease. The mortality rate was higher with disseminated disease (50% compared to 3.6%, $P = 0.04$). Four had concomitant pulmonary tuberculosis. Diabetes mellitus was found in 17 patients (41.4%), followed by HIV infection in 15 (36.5%). Patients with HIV had a higher incidence of disseminated disease (36.4% vs. 0%, $P = 0.01$). Four patients died, and the risk of death was increased in disseminated disease ($P < 0.05$). Coccidioimycosis is not as frequent in El Paso, and for that matter in Texas, as in other states, but still has to be taken into consideration both in HIV and in diabetic patients.

KEYWORDS: epidemiology; coccidioidomycosis; Hispanics

BACKGROUND AND SIGNIFICANCE

The area of endemicity of coccidioidomycosis (CM) in the United States is localized in the southwestern region of the country, including California, Arizona, Nevada, Utah, New Mexico, and Texas. Within Texas, the southwestern section of the state is mainly involved.[1] Often, epidemiologic studies are mostly reported from states with high prevalence, such as California and Arizona.[2-7] Because of its large population, El Paso, Texas is an expected high

Address for correspondence: Dr. Miguel Angel Pena-Ruiz, Department of Internal Medicine, Texas Tech University Health Sciences Center, 4801 Alberta Ave., El Paso, TX 79905, USA.

Ann. N.Y. Acad. Sci. 1111: 122–128 (2007). © 2007 New York Academy of Sciences.
doi: 10.1196/annals.1406.038

target area, but the data are limited on the real epidemiology of CM since almost all information comes from individual case reports. In 1966 an outbreak occurred in the most eastern part of the state, where 10 people were affected.[1] Other outbreaks had been reported previously in 1940, 1954, 1961, 1965, and 1966. Only the 1965 outbreak occurred in El Paso and it involved 10 cases.[8] Recent data from this location include case reports related to renal transplant patients.[9–11]

The goal of this study was to obtain information about the prevalence and the general characteristics of the patients with coccidioidomycosis that were admitted to Thomason Hospital, a 400-bed county hospital in El Paso, Texas, with a predominantly Hispanic patient population.

METHODS

A retrospective chart review was performed of medical records corresponding to codes 114.0 through 114.9 of the ICD-9, which includes all patients with CM admitted to the hospital in the past 9 years. We reviewed the logs of the pathology and microbiology laboratories. Once we had the list of patients, we searched our electronic medical record system and the hard copies of the chart for the discharge summaries and laboratory records.

The "de-identified" data were retrieved on a collection sheet, which included the following: date of admission, sex, age, number of admissions before the diagnosis was made, number of days hospitalized when the diagnosis was made, type of CM infection, comorbid conditions, treatment received for the CM and treatment received for the comorbid conditions, whether the patient died during hospitalization, the clinical presentation at admission, and the method of diagnosis of the coccidioidomycosis. We included all patients with an established diagnosis of CM confirmed by biopsy, culture, direct visualization, or antibody titer. The chest radiographs at initial presentation were reviewed (when available) and the patients were categorized according to the type of disease indicated.

Statistical Analysis

The SAS System for Windows, Release 9.1 (SAS Institute, Inc., Cary, NC, USA) was used to analyze the data. The following potential risk factors for disseminated disease were evaluated: age of 39 years or older, sex, HIV infection, and diabetes. We also sought to determine whether there were associations between mortality and the following clinical and demographic variables: age $\geq$ 39 years, sex, HIV infection, diabetes, and disseminated disease. Fisher's exact tests (two-tailed) were performed. A P-value of 0.05 or less indicated a statistically significant result. We attempted to control for confounders using

multivariate analyses, but failed because of the small sample size. We calculated the prevalence of CM in our hospital according to the total hospital
discharges per year.

RESULTS

A total of 41 patients with CM were identified in a period of 9 years (January
1997 to December 2005). The average number of cases per year was 4.55, and
the average number of discharges in our hospital per year was 17,967 patients.
The prevalence of CM was 3.25 cases per 10,000 discharges, and was stable
through the 9-year period. Twenty-seven patients were male and 14 female,
40 patients were Hispanic; the mean age was 39 years (15–72). Thirty-two
patients were from El Paso, 3 from New Mexico, 2 from Mexico, and 4 from
other parts of the United States. The annual incidence of CM was five cases
per year, with the highest incidence in 2001 and 2002. All patients, except one,
were of Hispanic ethnicity.

Presenting clinical Features

The presenting symptoms were similar to those in other published studies.
The most frequent symptom was cough in 18 (44%), and fever and chills in
18 (44%), followed by shortness of breath in 9 (23%), hemoptysis in 5 (16%),
and chest pain in 5 (10%). Other symptoms were altered mental status, leg
swelling/pain, weight loss, and nausea (5% each).

Radiographic Presentation

A total of 35 chest films were obtained from the 41 patients. The most
common presentation was a pneumonic consolidation in 14 patients (44%),
followed by cavitation and miliary forms at 6 each (19%). Other presentations were lung mass, empyema, and effusion. All six patients with miliary
presentation had human immunodeficiency virus (HIV) infection.

There were patients that presented with a combination of consolidation and
cavitation, or consolidation and effusion, and each presentation was registered
separately.

Extent of The Disease

The lung was the most commonly involved organ: 32 patients (78%) had
pulmonary involvement. Three patients had meningeal involvement, of whom

one had active pulmonary CM. Six patients had disseminated disease (miliary pattern on chest X-ray film). The hospital mortality rate was significantly higher in patients with disseminated disease (50% compared to 3.6%, $P = 0.04$). Univariate analysis of risk factors for disseminated disease revealed that only HIV infection was statistically related to disseminated disease ($P = 0.01$).

Two patients had coccidioidal osteomyelitis, both of whom had a history of pneumonia, one at 6 months and the other at 7 months prior to the diagnosis. Of the patients with concomitant infections, one had *Pneumocystis jiroveci* (formerly *Pneumocystis carinii*) infection and four had pulmonary tuberculosis. One of the two patients with neck mass secondary to adenopathy had HIV infection. One HIV-positive patient had a psoas abscess at presentation with concomitant miliary disease.

OTHER COMORBIDITIES

The most frequently associated comorbidity was diabetes mellitus, which was found in 17 patients (41.4%). The second most common comorbidity was HIV infection, present in 15 patients (36.5%), of whom only 8 were on antiretroviral treatment at the time of presentation. Patients with HIV had a higher incidence of disseminated disease than any other patients (36.4% vs. 0%, $P = 0.01$). One patient had both diabetes mellitus and HIV infection. Concomitant pulmonary tuberculosis was present in four patients. Three patients had *Varicella zoster* infection, all of whom were HIV positive. Two patients had leukemia: one had acute lymphocytic leukemia and the other had chronic myelogenous leukemia. Only four patients had only CM infection without comorbidities.

DIAGNOSTIC METHODS

The most common method of diagnosis was culture of bronchoalveolar lavage, which was carried out in 18 patients (43.9%). Biopsy from a body site was performed in 16 patients (39%) and included fine-needle aspiration, pleura decortication, and open resection. Other biopsy specimens were taken from bone, leg, and cervical vertebrae. In four patients the sputum culture was positive. One patient had a positive blood culture. Antibody detection tests were ordered in 8 patients: 5 had pulmonary involvement, 2 had meningeal involvement (the sample was from cerebrospinal fluid), and one had cervical vertebral osteomyelitis. Antibodies were positive in all eight patients.

THERAPY

The most common treatment was high-dose fluconazole in 15 patients (36.5%), followed by combined treatment of in-patient amphotericin B and outpatient high-dose fluconazole in 13 patients (31.7%). In five patients the treatment was not known and two patients received only amphotericin B. Of the three patients with meningeal CM, only one was treated with systemic amphotericin B. Of the two patients with osteomyelitis, only one received amphotericin B. The majority of patients who received amphotericin B alone or in combination with fluconazole had cavitating pulmonary CM.

Outcome

Four patients died during hospitalization from respiratory failure secondary to adult respiratory distress syndrome. Three of them had AIDS and were on no antiretroviral treatment and the other one was diabetic. Of these three HIV-positive patients, the diagnosis of HIV had been made at 5 months, 6 months, and 4 years prior to admission. All of them had miliary CM, one with a concomitant abdominal (psoas) abscess and the other patient was pregnant. Two patients were born in Mexico.

DISCUSSION

The endemic areas of coccidioidomycosis in the United States are mainly limited to the states of California, Arizona, Nevada, Utah, New Mexico, and Texas. The state with the most number of cases is Arizona. Texas is considered an endemic state, but no institution has published the number of yearly cases. The city of El Paso is considered to be one of the most endemic regions for coccidioidomycosis infection in Texas. Consequently, we studied all the cases of CM occurring in the previous 9 years, from 1997 to 2005. There was a steady incidence of 3 to 5 cases per year, with an increase in incidence from 2001 to 2002, during which more than one-third of the cases occurred. Our incidence is far lower than that in other states: for example, as a single institution Maricopa Medical Center in Phoenix had approximately six times more cases of CM and HIV co-infection, although their study period was from 1982 to 1993, before the highly active antiretroviral therapy (HAART) era.[12] In Sonora, Mexico, which share a border with Arizona, the coccidiodin reactivity was found to be 52.5%.[13] In other border regions coccidiodin sensitivity has also been evaluated. For example, in San Diego, California, the coccidiodin sensitivity in 1,027 adolescents was 9.7%[14] and it correlates with a study done at their border neighbor, Tijuana, Mexico, where coccidioidin sensitivity was 10% in 1,128 individuals of all ages.[15] The study by Catanzaro found 50 cases per year in San Diego Hospitals.[14] We do not have data to compare with our border town, the city of Juarez, Chihuahua, Mexico, and coccidioidin surveys have not been done in El Paso, Texas.

Males were affected twice as often as females, comparable with the male predominance in other series, most probably because men tend to work outdoors more often and are exposed to wind and dust. Only one patient was not

Hispanic, which makes this one of the largest series of coccidioidal infection ever reported in the Hispanic population in the United States outside of Arizona. This high predominance is likely due to the demographic distribution of the city of El Paso and not because of a Hispanic's predisposition. Additionally, socioeconomic factors may come into play.

The age range is similar to that of other series. People in their productive years are the ones that get infected, again most probably on account of increased exposure to wind and dust of the outdoors.

By far the most common symptoms were cough, fever, and chills, which can easily be confounded with bacterial community-acquired pneumonia (CAP). Any pneumonia that fails to respond to standard medical therapy should raise the suspicion of coccidioidomycosis in these endemic areas. Tuberculosis should also be taken into account since it is also common in this patient population and as seen in this series both entities can coexist in the same patient. A few patients in our series were treated for CAP in a first admission and were later found to have CM infection in a subsequent admission.

The radiological appearance of CM is very diverse, from a simple focal pulmonary infiltrate to disseminated disease with a miliary pattern on a chest X-ray film. An interesting observation is that although it is impossible to distinguish the etiology of a miliary pattern by radiological means, CM seems to have a "thicker" appearance than the micronodular infiltrate observed in tuberculosis, which observation may help in suspecting this fungal infection.

Of the 15 patients with HIV infection, we saw only six cases of miliary dissemination and one of them was in a patient with concomitant diabetes. The lung was the most affected anatomical site, and three patients had meningeal involvement. It would be expected that in the HAART therapy era, we may be seeing disseminated disease less frequently in HIV patients. Only 4 patients of a total of 41 developed the infection without any underlying disease. The most frequent associated comorbidity was diabetes mellitus, which should be considered as important a risk factor for symptomatic CM as HIV infection.

All of our patients had microbiological or pathological confirmation of CM. Serological testing was not performed systematically in our patients. Of note, at the Maricopa Medical Center study, where most of the HIV patients had serologic testing, results were negative in 21% of their cases, mainly in diffuse pulmonary disease.[12] In our series serology was positive on all tested patients with culture/biopsy–confirmed CM; none of them had diffuse pulmonary disease and there were two cases of extrapulmonary disease.

Our patients were treated with amphotericin and/or high-dose fluconazole. Fluconazole alone was the most frequent therapy. None of our patients with meningeal involvement was treated with intrathecal amphotericin B.

In this series the four patients who died during hospitalization had miliary CM, three of them with HIV who were not on HAART therapy; hence this represents the biggest risk factor for death in this series.

This study finds that CM infection is not as frequent in El Paso as in other states, but still there is an endemicity of the disease that has to be considered both in HIV and in diabetic patients who have an apparent CAP or tuberculosis infection. Further studies are needed to more appropriately determine the extent of burden of disease that CM infection has in El Paso and in the state of Texas.

REFERENCES

1. TEEL, K.W., M.D. YOW & T.W. WILLIAMS JR. 1970. A localized outbreak of coccidioidomycosis in southern Texas. J. Pedia. **77:** 65–73.
2. RUTHERFORD, G.W. & M.F. BARRET. 1996. Epidemiology and control of coccidioidomycosis in California. Prevent. Med. Public Health **165:** 221–222.
3. CENTERS OF DISEASE CONTROL. 2006. Coccidioidomycosis—Arizona, 1990–1995. MMWR Morbid. Mortal Wkly. Rep. **45:** 1069–1073.
4. CENTERS OF DISEASE CONTROL. 1994. Update: Coccidioidomycosis—California, 1991–1993. MMWR Morbid. Mortal. Wkly. Rep. **43:** 421–423.
5. PAPPAGIANIS, D. 1994. Marked increase in cases of coccidioidomycosis in California: 1991, 1992, and 1993. Clin. Infect. Dis. **19**(Suppl 1): S14–S18.
6. AMPEL, N.M., D.G. MOSLEY, B. ENGLAND, et al. 1998. Coccidioidomycosis in Arizona: increase in incidence from 1990 to 1995. Clin. Infect. Dis. **27:** 1528–1530.
7. CENTERS OF DISEASE CONTROL. 2003. Increase in Coccidioidomycosis-Arizona 1998-2001. MMWR Morbid. Mortal. Wkly. Rep. **52:**109–112; JAMA **289:** 1500–1501.
8. ROBERTS, P.L. & R.C. LISCIONDRO. 1967. A community epidemic of coccidioidomycosis. Am. Rev. Resp. Dis. **96:** 766.
9. ANTONY, S., D.C. DOMINGUEZ & E. SOTELO. 2003. Use of liposomal amphotericin B in the treatment of disseminated coccidioidomycosis. J. Natl. Med. Assoc. **95:** 982–985.
10. ANTHONY, S. 2004. Use of the echinocandins (caspofungin) in the treatment of disseminated coccidiodiomycosis in a renal transplant recipient. Clin. Infect. Dis. **39:** 879–880.
11. ANTONY, S., S.J. DUMMER MCNEIL & I. SALAS. 2005. Coccidioidomycosis in renal transplant recipients. Infect. Dis. Clin. Pract. **13:** 250–254.
12. SIGH, V.R., D.K. SMITH, J. LAWERENCE, et al. 1996. Coccidiodomycosis in patients infected with human immunodeficiency virus: review of 91 cases at a single institution. Clin. Infect. Dis. **23:** 563–568.
13. MADRID, G.S. & R. GARCIA. 1974. Coccidioidomicosis. Impresora y Editorial. Hermosillo, México.
14. CATANZARO, A. 1979. Coccidioidin sensitivity in San Diego Schools. Sabouraudia **17:** 85–89.
15. LANIADO-LABORIN, R. & M. ALVAREZ-CERRO. 1991. Tijuana: zona endémica de infección por coccidioidomicosis immitis. Salud Publica Mex. **33:** 235–239.

Compensability of, and Legal Issues Related to, Coccidioidomycosis

LUANN HALEY

Industrial Commission, Tucson, Arizona, USA

ABSTRACT: Legal issues that may develop when treating patients with coccidioidomycosis include allegations of medical malpractice, claims for workers' compensation benefits, and civil actions against business owners. In states where the disease is most prevalent, California recognizes cocci diodomycosis as a compensable condition, although Arizona does not. In civil actions, the state courts have not imposed liability on any business or institution for those that claim to have developed cocci diodomycosis on or near the premises of the business.

KEYWORDS: negligence; workers' compensation; legal liability; civil actions; litigation; coccidioidomycosis

INTRODUCTION

As litigation has become commonplace in American culture, medical professionals must be prepared to address legal liability issues in their clinical and consultation practices. Clinicians who are unfamiliar with coccidioidomycosis may be faced with allegations of negligence for failure to recognize and diagnose coccidioidomycosis. Further, physicians treating employees who may have contracted coccidioidomycosis at the workplace should be aware of the workers' compensation statutes that have an impact on the benefits their patients may be entitled to receive. Finally, consultants or those who act as forensic experts may be called on to address safety issues and allegations of negligence from businesses located in areas where coccidioidomycosis is prevalent. Accordingly, it is important for all medical practitioners who treat patients with coccidioidomycosis to have a general understanding of the liability issues that can develop with this unique disease.

For physicians who are not familiar with the complications that can arise with a patient who contracts coccidioidomycosis there may be some difficulty in the proper and timely diagnosis of the condition. If there are difficulties with making the diagnosis, there may be allegations of negligence involving

Address for correspondence: LuAnn Haley, Industrial Commission, 2675 E. Broadway Blvd., Tucson, AZ 85716. Voice: 520-320-4209; fax: 520-628-5182.
  lhaley7859@aol.com

Ann. N.Y. Acad. Sci. 1111: 129–132 (2007). © 2007 New York Academy of Sciences.
doi: 10.1196/annals.1406.002

the treating doctor. Failure to diagnose is the number one reason that gives rise to medical negligence claims. In order to avoid claims of negligence, practitioners should be sure to have sufficient documentation to show that they had a reasonable basis for their diagnosis. Further, under Arizona law, a medical practitioner must keep legible notes of the exam, or a presumption of sloppy or incomplete exam will be raised. Similarly, after any claim is made, it is also important not to change or attempt to recreate patient records as it will detract from the credibility of the evidence and the testifying medical expert.

WORKERS' COMPENSATION STATUTES

Most physicians today understand the general principles of Workers' Compensation Statutes and that these laws will have an impact on what benefits their patients may receive for compensable injuries that they have sustained while at work. In general, all state statutes provide that injured workers may receive medical and wage-loss benefits for injuries that arise in the course and scope of their employment. Of special interest to medical practitioners dealing with coccidioidomycosis are the occupational disease provisions of the statutes that allow for compensation for cumulative conditions not resulting from a specific accident. Medical professionals who treat injured workers should be familiar with standards for proof of medical causation, the link between the work-related activity and the injury, that will form the foundation for compensable claims in each state. Despite the similarities in the industrial laws of California and Arizona, the evolving case law in each state has reached contrary results in the determination of whether coccidioidomycosis is a compensable condition.

The Arizona courts addressed whether coccidioidomycosis, or what is commonly referred to as Valley Fever, was compensable in several cases in the 1970s. The Arizona Court of Appeals, in two separate industrial claims, held that it is not possible for employees to sustain their burden of proving that Valley Fever resulted from their working conditions, and therefore, both claims were denied (see *O'Conner v. ICA*, 19 Ariz.App.43 (1972) and *Crawford v. ICA*, 23 Ariz. App. 578 [1975]). The Court held that testimony that the work exposure only statistically increased the probability of contracting Valley Fever did not meet the required standards for proof to show a causal link between the claimant's work and the injury. Both decisions further provide that it is hard to prove that a disease that is endemic to Arizona can have a sufficient nexus to the workplace activity to fall under Arizona's worker compensation statutes. In a 1983 decision, the Arizona Court of Appeals further held that coccidioidomycosis cases would not be considered compensable under the occupational disease provisions of the Workers' Compensation Statute, despite finding that a worker who contracted pulmonary cryptococcosis could be compensated for his disease.

In contrast, the Workmen's Compensation Appeals Board in California authored several decisions finding coccidioidomycosis a compensable condition where medical experts similarly opined that coccidioidomycosis is endemic to the southern California area (see *Cypress Insurance Company v. WCAB & Delaney*, 71 Cal. Rptr. 915 [1968]). The California courts had no trouble finding compensable coccidioidomycosis claims despite the inability of the medical experts to pinpoint the actual exposure that caused the workers to develop the disease. With the continuing acceptance of Valley Fever as a compensable condition in subsequent court determinations, it appears that California workers will not have the same proof-of-causation problems as employees who contract coccidioidomycosis in Arizona.

SAFETY AND CIVIL ACTIONS

Of perhaps more interest to practitioners who act as forensic or industry consultants are the decisions that resolve whether civil damages may be awarded to individuals who contract coccidioidomycosis while in the Southwest for business, educational, recreational, or athletic pursuits. The theory of these cases involves allegations that the defendants, identified as landowners, have a duty to warn plaintiffs, identified as invitees, of the possibility that Valley Fever can be contracted while on the landowners' premises. In 1973, the Arizona Court of Appeals rejected a claim by an African American student who contracted Valley Fever while enrolled at the University of Arizona and held that a finding of liability against the University would result in a "stream of endless cases" that may be filed by visitors who contract coccidioidomycosis (see *Randolph v. Arizona Board of Regents*, 19 Ariz. App. 121 [1973]).

Most recently, the Supreme Court of Arizona considered a similar claim by a PGA professional, Gregory Kraft, who contracted Valley Fever after participating in a PGA golf tournament in Tucson. In February of 2006, the Arizona Court of Appeals denied the golfer's negligence claim and cited the same public policy reasons for refusing to hold the PGA and the resort sponsor liable for his disease. In the appeal of the case, Kraft's attorneys cited legal precedent from a Federal Court determination on the issue of negligence for failure to warn. In *Crim v. International Harvester*, 646 F2 d 161 (1981), the Federal Court, applying Arizona law, found a vehicle manufacturer liable for a Texas vehicle dealer's developing Valley Fever while test-driving a vehicle in Arizona. The Court in *Crim* distinguished the *Randolph* case by finding that the dealer was able to prove that he contracted the disease at the defendant's proving grounds, since that was the only area he visited in Arizona. Kraft's attorneys, therefore, relied on the same rationale that Mr. Kraft was invited to Arizona from a nonendemic region (Florida) for the purpose of engaging in an activity (golf) that guaranteed the participant's exposure to soil containing fungus spores. The Supreme Court of Arizona in August of 2006 refused to accept Kraft's petition and confirmed the lower Court's holding that business landowners in Arizona should not incur liability for a condition that is endemic to the region. In September of the same year, a jury in the Superior Court of Kern County in California returned a similar verdict of no liability against an energy company that had constructed a power plant in Bakersfield, California. Several of the workers alleged that they contracted Valley Fever on the jobsite; however, the court would not impose liability on the company as the fungus is also endemic to the Kern County area in California. With these recent decisions, it appears the Courts have closed the door to negligence law suits in both Arizona and California for those who are afflicted with Valley Fever.

CONCLUSION

In addition to dealing with the complicated issues of medical diagnosis and treatment of Valley Fever, it is evident from the above court decisions that medical practitioners must also understand legal liability issues that may arise when treating coccidioidomycosis patients. Whether assisting a patient in pursuing industrial benefits or advising institutions regarding limiting their risk in liability claims, physicians and lawyers need to be aware of the current legal trends in cases involving patients who contract Valley Fever.

The Application of Proteomic Techniques to Fungal Protein Identification and Quantification

JAMES G. ROHRBOUGH,[a,b] JOHN N. GALGIANI,[c,d] AND VICKI H. WYSOCKI[a]

[a] *Department of Biochemistry and Molecular Biophysics, University of Arizona, Tucson, Arizona, USA*

[b] *Civilian Institution Programs, Air Force Institute of Technology, Wright-Patterson Air Force Base, Ohio, USA*

[c] *Medical and Research Services, Southern Arizona VA Healthcare System, Tucson, Arizona, USA*

[d] *Valley Fever Center for Excellence and Department of Internal Medicine, University of Arizona, Tucson, Arizona, USA*

ABSTRACT: The number of sequenced genomes has increased rapidly in the last few years, supporting a revolution in bioinformatics that has been leveraged by scientists seeking to analyze the proteomes of numerous biological systems. The primary technique employed for the identification of peptides and proteins from biological sources is mass spectrometry (MS). This analytical process is usually in the form of whole-protein analysis (termed "top-down" proteomics) or analysis of enzymatically produced peptides (known as the "bottom-up" approach). This article will focus primarily on the more common bottom-up proteomics to include topics such as sample preparation, separation strategies, MS instrumentation, data analysis, and techniques for protein quantification. Strategies for preparation of samples for proteomic analysis, as well as tools for protein and peptide separation will be discussed. A general description of common MS instruments along with tandem mass spectrometry (MS/MS) will be given. Different methodologies of sample ionization including matrix-assisted laser desorption ionization (MALDI) and electrospray ionization (ESI) will be discussed. Data analysis methods including database search algorithms and tools for protein sequence analysis will be introduced. We will also discuss experimental strategies for MS protein quantification using stable isotope labeling techniques and fluorescent labeling. We will introduce several fungal proteomic studies to illustrate the use of these methods. This article will allow investigators to gain a working knowledge of proteomics along with some strengths and weaknesses associated with the techniques presented.

Address for correspondence: Dr. Vicki H. Wysocki, 1306 E. University Blvd., University of Arizona, Tucson, AZ 85721-0041. Voice: 520-621-2628; fax: 520-621-8407.
vwysocki@email.arizona.edu

Ann. N.Y. Acad. Sci. 1111: 133–146 (2007). © 2007 New York Academy of Sciences.
doi: 10.1196/annals.1406.034

KEYWORDS: proteomics; mass spectrometry; protein identification; peptide analysis

INTRODUCTION

With the increasing number of sequenced genomes, proteomics is a field of study that has expanded rapidly in the last decade to encompass a wide variety of techniques and technologies. While protein analysis is not new, many recent advances in bioinformatics as well as in mass spectrometry (MS) have increased the speed and breadth of samples that can be efficiently analyzed. Older methodologies, such as protein sequence determination by Edman degradation, amino acid composition analysis, or gel-based analytical methods like Western blots still have their place in specific research endeavors, but are somewhat less applicable to the high-throughput nature of modern proteomic analyses. This article is intended to be a brief overview of modern proteomics as it relates to the characterization and identification of fungal proteins. Readers desiring more comprehensive information can look to articles such as the 2003 review by Aebersold and Mann,[1] or more recent articles by Swanson and Washburn,[2] and Wysocki et al.[3]

A proteomic analysis of a system involves the collection, separation, identification, and functional determination of the expressed proteins of a sample,[4] which can lead to a better understanding of protein function and regulation of a system. A detailed analysis of the proteome can then lead to protein targets for disease identification, treatment, or vaccine development. The general steps of a proteomic analysis after the collection of a sample of interest are extraction of the proteins from the sample mixture, proteolytic digestion to produce peptides, peptide separation, peptide identification, and determination of identity and function of proteins present in the original sample. A brief overview of a typical proteomic analysis strategy is shown in FIGURE 1. The methods (sample preparation, ionization type, mass analyzer, etc.) employed for a proteomic analysis may vary depending on the starting material and the goal of the analysis.

There are many difficulties encountered in the analysis of proteins in a biological sample. The most obvious is the inherent complexity of many samples. Many proteins are present in locations that prohibit easy analysis, such as membrane-bound proteins which are difficult to solubilize. Very large and very small proteins can also be difficult to analyze and detect. Not all proteins are present in equal abundance, a concept known as dynamic range.[5] This leads to a difficulty in detection of low-abundance proteins in the presence of highly abundant ones. Unlike RNA-based methods of transcript amplification, sample protein levels cannot be increased to facilitate analysis of low-abundance proteins. Any undertaking of a proteomic analysis will likely require addressing at least one of these difficulties.

Typical Proteomic Analysis Strategy

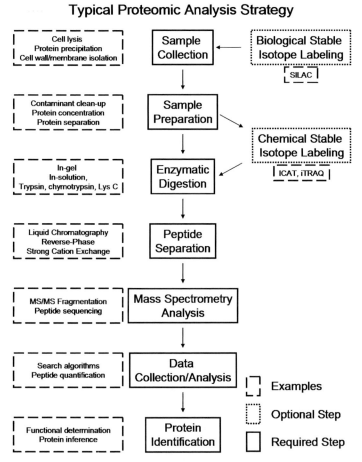

FIGURE 1. Typical proteomic analysis strategy. The typical strategy for a bottom-up proteomic analysis, including optional stable isotope labeling for protein quantification and examples of some of the techniques employed in each step.

Recent advances in transcript identification may lead investigators to use mRNA analysis to try to infer protein presence. It is important to note, however, that while analysis of the mRNA levels of a system provides insights into gene expression, those levels may not correlate with protein abundance. Protein and corresponding mRNA levels can vary dramatically. It is common for protein or mRNA levels to vary by as much as 30-fold with respect to each other, or in some cases for either to be absent.[6]

The most common methods of analysis use MS for protein and peptide identification. Sample proteins can then be analyzed whole, in what is known as a top-down proteomic analysis,[7] or analyzed as peptides from protein digestion in a bottom-up approach. Top-down proteomics is less popular, primarily because of the need for expensive and complex high-resolution mass spectrometers. Bottom-up methods can include peptide identification by high-resolution mass determination from a single round of MS, known as peptide mass fingerprinting,[8] or more commonly, peptides are identified by peptide fragmentation in a tandem mass spectrometry experiment (MS/MS) to facilitate amino acid sequence correlation.[9] In MS/MS, the peptides are separated by the mass analyzer and then subjected to fragmentation. The masses of the fragment ions are determined by a second mass analyzer (or in a second round of MS in a trapping-type instrument). Since different peptides will fragment differently, this technique allows not only for the identification of peptides with different masses, but also those with the same or similar masses.[10,11]

PROTEIN/PEPTIDE SEPARATION

Protein complexity can be reduced with some relatively simple methods of protein separation, such as one-[12] and two-dimensional[13] electrophoresis (1DE and 2DE). These gels are run under denaturing conditions, including heat, detergent (such as SDS), and a reductant (such as dithiothreitol [DTT] or β-mercaptoethanol) for disulfide bond cleavage. In addition to reducing the sample complexity, gel electrophoresis can also be used for sample cleanup (removal of salts, detergents, etc.). One-dimensional electrophoresis involves the separation of denatured proteins based on size, while 2DE starts with a separation of proteins by isoelectric point, followed by separation by size. A variation of the 2DE method using fluorescent dyes for protein quantification is known as difference gel electrophoresis (DIGE), which will be discussed later. After protein separation using electrophoresis, in-gel digestion is often used to produce and extract peptides prior to MS analysis.[14]

The complex peptide mixtures from a 1DE gel or solution digest in a bottom-up experiment can be further separated by using high-performance liquid chromatography (HPLC) online peptide separation methods. The most common peptide separation is known as reverse-phase (RP) LC. Using this method, the peptides are separated by hydrophobicity by eluting peptides bound to the RP packing material by means of an organic solvent (such as acetonitrile or methanol) gradient flow by HPLC.[15] Another LC separation method used is strong cation exchange (SCX), which separates peptides by charge, where the peptides bound to the SCX material are eluted by a salt gradient. SCX is often used in conjunction with RP separation in a method known as,[16] MultiDimensional Protein Identification Technology (MudPIT). MudPIT uses a column containing both RP and SCX chromatography phases described above, which allows for easier and automated analysis of biological mixtures, by reducing the complexity of peptides with a method that does not require protein separation by electrophoresis. While MudPIT is often used to analyze solution-digested proteins, 1DE has been used as a sample clean-up step, followed by MudPIT analysis.

IONIZATION METHODS

There are several types of ionization methods used for MS analysis, but there are two primary ones used in proteomics. Electrospray ionization (ESI),[17] and its closely-related small-volume cousin nanoelectrospray (nano-ESI),[18] involve injection of analyte (peptide or protein) molecules exiting the LC in solution into the mass spectrometer. A major advantage of ESI is the ease of coupling online separation methods, such as RP and SCX prior to ionization and MS analysis. Also, ESI produces multiply-charged ions, allowing for identification of larger ions in an instrument with a low mass-to-charge (m/z) ratio limit.

The second major ionization method used in proteomics is matrix-assisted laser desorption/ionization (MALDI).[19,20] The sample of interest is mixed with a matrix, spotted onto a sample plate, and then excited by a laser beam, ionizing and transferring analyte molecules into the gas phase for analysis. Advantages of MALDI ionization include a larger analyte mass range and higher tolerance of salts than ESI.

MASS ANALYZERS

After the protein or peptide molecules have been ionized and put into the gas phase, they enter the mass spectrometer for analysis. There are several methods of mass analysis in MS, but there are four major types used in proteomic analysis. All mass analyzers operate on the same basic principle of separation of ions by their m/z ratio. The first type is the quadrupole mass analyzer,[21] which is one of the oldest and best defined and has been a mainstay of MS analysis for decades. A major disadvantage associated with quadrupoles is the necessity of multiple mass analyzers to accomplish MS/MS.

There are two common ion-trap mass analyzers based on the physics of the quadrupole mass analyzer. The first is the quadrupole ion trap (QIT),[22] the second is the newer linear ion trap (LIT).[23] Another ion-trap mass analyzer is the Fourier transform ion cyclotron resonance (FTICR, or just FT)[24] MS that uses a superconducting magnet for ion control. The benefits of ion-trapping mass analyzers include the ability to perform multiple rounds of MS and prevent ion fragmentation within the same mass analyzer, as well as an increase in signal-to-noise ratio. Disadvantages of the QIT/LIT analyzers include limited resolution and mass accuracy. While the FT has excellent resolution and mass accuracy appropriate for top-down sequencing, it is relatively large and expensive and is more difficult to couple to LC-based sources.

The final common mass analyzer is the time of flight (TOF),[25] which is a much simpler system than either a quadrupole or FT-based mass analyzer. In the TOF, ions are separated by the time it takes them to travel the length of the analyzer, with the smaller m/z ratio ions having an impact before larger ones. In addition to its simplicity of operation, the TOF mass analyzer is also valued for its enhanced mass accuracy and resolution over the quadrupole-based instruments. Disadvantages of TOF include the difficulty of coupling the pulsed analysis with continuous-ionization LC-based peptide separation, and the need for an additional mass analyzer to accomplish MS/MS. TOF instruments that are commonly used in proteomic analysis include the quadrupole coupled to a TOF (QTOF) or two TOF analyzers in sequence (TOF–TOF).

DATA ANALYSIS

MS/MS spectra typically do not directly provide a peptide sequence. Instead, spectral information is matched to known peptide sequences from protein sequence databases. Some of the more popular database collections include the National Center for Biotechnology Information (NCBI), which includes most of the public domain sequence databases, included in the nonredundant (NR) database. Another collection of sequences is the Swiss Prot database, which has many sequences, but also has a large amount of annotation included with the sequences to allow for easier identification of functionality of listed proteins.[11] If the genome of the organism being analyzed has not been sequenced, the best strategy is to build a database of closely related species, or search against the NR database, with the realization that the larger the database, the longer the search process will take. There are several different sources for sequenced fungal genomes. Among these are the Broad Institute (http://www.broad.mit.edu), the Sanger Institute (http://www.sanger.ac.uk), The Institute for Genomic Research (TIGR) (http://www.tigr.org), and Génolevures (http://cbi.labri.fr/Genolevures). Peptide sequences are matched to spectral information using a database search algorithm. Two of the most common licensed programs are SEQUEST[26] and Mascot.[27] Both algorithms use the fragmentation patterns of MS/MS analysis to compare to the predicted fragmentation patterns of database protein sequences. A third algorithm that functions in a similar manner and is becoming popular is the open-source XTandem program.[28]

After the identified peptides are matched to protein sequences, there may be a need to further analyze the protein sequence to elucidate function, modification, or cellular location. One of the tools available for this is the basic local assignment search tool (BLAST).[29] BLAST can search both nucleotide and protein databases to identify protein homology, which is useful when the database used for peptide identification is insufficiently annotated. Additional information, such as subcellular localization can be found using the TargetP[30] localization predictor. Another useful program is the Gene Ontology Tool[31] to infer functional classification of proteins. More helpful sequence analysis programs can be found at the Expert Protein Analysis System (ExPASy) proteomics server (http://www.expasy.ch). Additional tools for analysis of MS data can be found at the Protein Prospector (http://prospector.ucsf.edu), as well as helpful proteomics software at the Proteome Commons (http://www.proteomecommons.org).

PROTEIN QUANTIFICATION

The area of differential proteomics has seen several advances in recent years. Three of the newer techniques can be used for differential quantitative analysis using MS (a technology that does not lend itself well to quantitative measurement unless internal standards are used). Each of the techniques involves differential labeling of proteins from different samples using stable isotopes, such as deuterium, ^{15}N, or ^{13}C. There are two ways to label proteins with these isotopes: biological incorporation, where cells are grown on media enriched with the isotope being used. The second method is known as chemical incorporation, where the isotopically labeled tag is added to proteins after extraction. Regardless of label incorporation method, corresponding labeled and unlabeled peptides will be detected in the mass spectrometer at the same time. Quantitative data are derived by comparing the ratio of areas of the MS peaks for labeled and unlabeled peptides.[32]

The first technique is stable isotope labeling by amino acids in cell culture (SILAC),[33] which is a biological incorporation method. Next is the isotope coded affinity tag (ICAT),[34] which involves a chemical incorporation of a deuterium-labeled (and unlabeled) reagent to cysteine residues. Last is a derivative of the ICAT method known as isotope tagging for relative and absolute protein quantitation (iTRAQ).[35]

The primary advantage of ICAT and iTRAQ labeling is the ability to label proteins that cannot be labeled using biological incorporation, such as human serum samples. ICAT also allows for decreasing the complexity of a complex biological mixture by the use of biotin affinity separation. The main disadvantage of ICAT compared to other labeling methods is the fact that only cysteine-containing peptides can be identified. The iTRAQ has the advantage over ICAT in being able to label all peptides in a mixture. It also allows for the analysis of four different samples at once, rather than just two with ICAT. The biggest disadvantage of iTRAQ is the more complex MS identification, as well as its increased cost.

Another widely used quantification technique is the two-dimensional difference gel electrophoresis (DIGE).[36] DIGE uses a system of fluorescent markers that bind to proteins in a sample, allowing for quantification of labeled proteins in a 2D gel upon excitation of the marker by a laser. Samples of interest can then be in-gel digested with a protease and analyzed by MS/MS.

APPLICATION OF PROTEOMIC ANALYSIS
TO FUNGAL SYSTEMS

There are several examples of proteomic analyses of fungal organisms to be found in the current literature. Many of these, such as an analysis of the obligate plant pathogen *Uromyces appendiculatus*, use some of the techniques described above.[37] In this analysis, proteins were extracted from uredospores, digested, and then separated by MudPIT. The analysis identified over 400 proteins, many of which are associated with protein production, such as translation factors, ribosomal proteins, and amino acid synthetases. These results led the authors to hypothesize that the uredospores exist in a suspended state of translation that allows the spore to begin protein production rapidly upon germination.

An analysis of the human pathogen *Candida albicans* incorporated 2DE separation of cell wall proteins prior to MS analysis.[38] In this study, the cell walls of the yeast and hyphae morphologies were subjected to protein extraction by SDS and DTT or cyanogen bromide (CNBr)/trypsin digestion. This study produced a total of 82 SDS/DTT-extractable cell wall proteins from both yeast and hyphal samples. Seven of these proteins were shown to be upregulated in the yeast–hyphae transition, and two were downregulated. There were an additional 29 proteins identified from the CNBr/trypsin digestion of both cell types, 12 of which are hyphae-specific, and 6 are yeast-specific. These protein identifications have not only increased the understanding *C. albicans* biology, but also identified a heat-shock protein that is upregulated in the yeast–hyphae transition, but not at the mRNA level. These results suggest that this protein is regulated at a posttranslational level in the fungal cell wall.

Another analysis also focused on *C. albicans*, illustrating the applicability of proteomics to vaccine development.[39] In this study, an extract of yeast cell wall proteins was shown to be effective in protecting mice from infection. This study identified and characterized 20 proteins that reacted with antibodies from the serum of immunized animals. Many of the identified proteins were determined to play important roles in adhesion, cell-surface hydrophobicity, and immunogenic activity. These protein identifications have produced target antigens to be used in the development of a subcellular vaccine against *C. albicans* infection.

There are other examples of fungal proteomics, such as the analysis of proteins secreted by the phytopathogen *Sclerotinia sclerotiorum*.[40] In this study, both mycelial and secreted proteins were separated by 2DE. This analysis identified 18 secreted proteins, along with 95 mycelial proteins that provide insight into the fungal life cycle and pathogenicity. One protein had not been previously identified in the analysis of mRNA levels, highlighting the value of direct protein identifications, rather than protein presence inferred from transcript analysis.

The quantitative technique SILAC was used in a study of the complete proteome of *Saccharomyces cerevisiae*.[41] In this analysis, yeast cells were grown in normal media or media containing labeled lysine. The proteins collected from the cells were digested and the resulting peptides were analyzed on a linear ion trap-FT mass spectrometer capable of extremely high peptide mass accuracy. Peptides were identified by MS/MS fragmentation, resulting in identification of over 2,000 *S. cerevisiae* cytoplasmic proteins. These identifications included low-abundance proteins corresponding to about 100 protein copies per cell.

Another recent analysis used SILAC-like stable isotope labeling for protein quantification in *Schizosaccharomyces pombe*.[42] In this study, fungal cells were treated with Cd^{2+} and labeled with deuterated leucine to determine what effect the toxic metal had on protein production. This study identified 106 proteins that were upregulated and 55 that were downregulated in response to Cd^{2+} treatment. In addition, 28 of the upregulated proteins were revealed to be proteins involved in detoxification of reactive oxygen species (ROS) or repair of damaged cellular components. This study serves to highlight the applicability of proteomics to analysis of environmental effects on cellular metabolism.

There are multiple examples of proteins of interest identified from various cellular preparations of *Coccidioides*. Included among these is the coccidioides-specific antigen (CSA), which was first isolated from the soluble wall fraction of infectious arthroconidia[43] in 1989 by acetone extraction followed by electrophoresis separation and protein removal by electroelution. It was not until 1995 that the protein sequence for CSA was determined by sequencing of the N-terminal and proteolytically produced peptides, as well as sequence determination of isolated cDNA.[44] Another protein, Antigen 2 (Ag2), was first identified in 1978 by two-dimensional immunoelectrophoresis (IEP) from the crude antigen preparations coccidioidin and spherulin,[45] and also found in the alkali-soluble, water-soluble mycelial and spherule extracts.[46] Ag 2 was not sequenced until 1996.[47] A parallel antigen discovery of the proline-rich antigen (PRA) was identified from a toluene spherule lysate.[48] PRA was also sequenced in 1996,[49] at which point it was discovered that Ag2 and PRA were the same protein. These studies highlight some of the important protein discoveries made in *Coccidioides*, and provide much of the groundwork for future protein antigen identifications.

While there are numerous examples of protein antigen identifications in *Coccidioides,* more modern proteomic analyses using MS have only recently been reported. These studies have primarily focused on identification of antigenic proteins. The analysis of T–cell reactive antigens associated with the spherule cell wall by 1D and 2D electrophoresis protein separations followed by peptide identification via MS/MS identified a protective aspartyl protease (Pep1).[50] Another analysis of seroreactive spherule cell wall proteins separated by 2DE and analyzed by MS/MS identified two more protective protein antigens, phospholipase B (Plb) and alpha-mannosidase (Amn1), in addition to Pep1, all of which were shown to be protective in mice as a multivalent recombinant protein vaccine.[51] A 2D DIGE analysis of differential protein expression between the mycelial and spherule phases of *C. posadasii*, resulted in the identification of a new vaccine candidate protein, a peroxisomal matrix protein known as Pmp1, also shown to be protective in mice against coccidioidal infection.[52] Immunoblot analysis of a 2D gel of the thimerisol-inactivated spherule vaccine (T27K) was analyzed by MS, resulting in the identification of a putative Cu, Zn superoxide dismutase (SOD)[53] as well as Amn1.[54] A summary of the identifications of Amn1 and SOD is included elsewhere in the proceedings. Another study[55] purified N-glycan containing glycoproteins from T27K by lectin-affinity chromatography followed by SDS-PAGE separation. From this study, a 60-kDa protein component was identified by MS, with homology to a 1,3 glucanosyltransferase from *C. posadasii* and other fungi.

FUTURE DIRECTIONS

The above list of fungal proteomic analyses, while not exhaustive, provides a glimpse into the wide range of experimental questions that can be answered by deliberate application of proteomic techniques. Future proteomic analyses of *Coccidioides* and other fungal species will undoubtedly prove beneficial to vaccine development efforts, as well as increase the understanding of the biology of these organisms. Current efforts are under way at the University of Arizona to analyze the spherule cell wall proteome, and to comprehensively analyze differential protein expression between the mycelial and spherule phases of *Coccidioides posadasii* by stable isotope labeling. These types of studies will likely prove beneficial to *Coccidioides* vaccine development, as to well as to other fungal systems.

Future proteomic studies of the *Coccidioides* species are likely to focus on specific proteins involved in virulence that could lead to development of novel drug treatments. Many of the techniques presented here would also be effective in elucidating protein biomarkers for coccidioidal infection or perhaps provide clues to the nature of sexual reproduction in these fungal species. Regardless of the scope of studies yet to come, modern proteomic techniques are likely to be included in the tools researchers use to answer important questions about this important fungal pathogen.

ACKNOWLEDGMENTS

The authors would like to thank Dr. Amy Hilderbrand and Samanthi Wickramasekara for helpful suggestions regarding manuscript preparation. J.R. would like to thank Dr. Paul Haynes for guidance and support. Funding sources include the Air Force Institute of Technology, the Department of Veteran's Affairs, and the Pacific Southwest Regional Center of Excellence for Biodefense and Emerging Infectious Diseases Research (Grant 1054-A 1065359-01 from the NIAID). The views expressed in this article are those of the author and do not reflect the official policy or position of the Air Force, Department of Defense, or the United States Government.

REFERENCES

1. AEBERSOLD, R. & M. MANN. 2003. Mass spectrometry-based proteomics. Nature **422:** 198–207.
2. SWANSON, S.K. & M.P. WASHBURN. 2005. The continuing evolution of shotgun proteomics. Drug Discov. Today **10:** 719–725.
3. WYSOCKI, V.H., K.A. RESING, Q. ZHANG, *et al.* 2005. Mass spectrometry of peptides and proteins. Methods **35:** 211–222.
4. HUNTER, T.C., N.L. ANDON, A. KOLLER, *et al.* 2002. The functional proteomics toolbox: methods and applications. J. Chromatogr. B Analyt. Technol. Biomed. Life Sci. **782:** 165–181.
5. CORTHALS, G.L., V.C. WASINGER, D.F. HOCHSTRASSER, *et al.* 2000. The dynamic range of protein expression: a challenge for proteomic research. Electrophoresis **21:** 1104–1115.
6. PRADET-BALADE, B., F. BOULME, H. BEUG, *et al.* 2001. Translation control: bridging the gap between genomics and proteomics? Trends Biochem. Sci. **26:** 225–229.
7. KELLEHER, N.L., H.Y. LIN, G.A. VALASKOVIC, *et al.* 1999. Top down versus bottom up protein characterization by tandem high-resolution mass spectrometry. J. Am. Chem. Soc. **121:** 806–812.
8. THIEDE, B., W. HOHENWARTER, A. KRAH, *et al.* 2005. Peptide mass fingerprinting. Methods **35:** 237–247.
9. YATES, J.R., III. 1998. Database searching using mass spectrometry data. Electrophoresis **19:** 893–900.
10. YATES, J., J. ENG, A. MCCORMACK, *et al.* 1995. Method to correlate tandem mass spectra of modified peptides to amino acid sequences in the protein database. Anal. Chem. **67:** 1426–1436.
11. SIUZDAK, G. 2003. The Expanding Role of Mass Spectrometry in Biotechnology. MCC Press. San Diego, CA.
12. HAMES, B.D. 1998. Gel electrophoresis of proteins : a practical approach. Oxford University Press. Oxford and New York.
13. WITTMANN-LIEBOLD, B., H.R. GRAACK & T. POHL. 2006. Two-dimensional gel electrophoresis as tool for proteomics studies in combination with protein identification by mass spectrometry. Proteomics **6:** 4688–4703.
14. SHEVCHENKO, A., M. WILM, O. VORM, *et al.* 1996. Mass spectrometric sequencing of proteins silver-stained polyacrylamide gels. Anal. Chem. **68:** 850–858.
15. YATES, J.R., A. MCCORMACK, A. LINK, *et al.* 1996. Future prospects for the analysis of complex biological systems using micro-column liquid chromatography-electrospray tandem mass spectrometry. Analyst **121:** R65–R76.
16. WASHBURN, M., D. WOLTERS & J.R. YATES, III. 2001. Large-scale analysis of the yeast proteome by multidimensional protein identification technology. Nat. Biotechnol. **19:** 242–247.

17. MANN, M. & M. WILM. 1995. Electrospray mass spectrometry for protein characterization. Trends Biochem. Sci. **20:** 219–224.
18. WILM, M. & M. MANN. 1996. Analytical properties of the nanoelectrospray ion source. Anal. Chem. **68:** 1 8.
19. KARAS, M., D. BACHMANN, U. BAHR, *et al.* 1987. Matrix-assisted ultraviolet laser desorption of non-volatile compounds. Int. J. Mass Spectrom. Ion Process. **78:** 53–68.
20. KARAS, M. & F. HILLENKAMP. 1988. Laser desorption ionization of proteins with molecular masses exceeding 10,000 daltons. Anal. Chem. **60:** 2299–2301.
21. STEEL, C. & M. HENCHMAN. 1998. Understanding the quadrupole mass filter through computer simulation. J. Chem. Ed. **75:** 1049–1054.
22. JONSCHER, K.R. & J.R. YATES, III. 1997. The quadrupole ion trap mass spectrometer—a small solution to a big challenge. Anal. Biochem. **244:** 1–15.
23. SCHWARTZ, J.C., M.W. SENKO & J.E. SYKA. 2002. A two-dimensional quadrupole ion trap mass spectrometer. J. Am. Soc. Mass Spectrom. **13:** 659–669.
24. WILLIAMS, E.R. 1998. Tandem FTMS of large biomolecules. Anal. Chem. **70:** A179–A185.
25. GUILHAUS, M., D. SELBY & V. MLYNSKI. 2000. Orthogonal acceleration time-of-flight mass spectrometry. Mass. Spectrom. Rev. **19:** 65–107.
26. ENG, J., A. MCCORMACK & J.R. YATES, III. 1994. An approach to correlate tandem mass spectral data of peptides with amino acid sequences in a protein database. J. Am. Soc. Mass Spectrom. **5:** 976–989.
27. PERKINS, D.N., D.J. PAPPIN, D.M. CREASY, *et al.* 1999. Probability-based protein identification by searching sequence databases using mass spectrometry data. Electrophoresis **20:** 3551–3567.
28. CRAIG, R., J.P. CORTENS & R.C. BEAVIS. 2004. Open source system for analyzing, validating, and storing protein identification data. J. Proteome Res. **3:** 1234–1242.
29. ALTSCHUL, S., W. GISH, W. MILLER, *et al.* 1990. Basic local alignment search tool. J. Mol. Biol. **215:** 403–410.
30. EMANUELSSON, O., H. NIELSEN, S. BRUNAK, *et al.* 2000. Predicting subcellular localization of proteins based on their N-terminal amino acid sequence. J. Mol. Biol. **300:** 1005–1016.
31. ASHBURNER, M., C.A. BALL, J.A. BLAKE, *et al.* 2000. Gene ontology: tool for the unification of biology. The Gene Ontology Consortium. Nat. Genet. **25:** 25–29.
32. ONG, S.E., L.J. FOSTER & M. MANN. 2003. Mass spectrometric-based approaches in quantitative proteomics. Methods **29:** 124–130.
33. ONG, S.E., B. BLAGOEV, I. KRATCHMAROVA, *et al.* 2002. Stable isotope labeling by amino acids in cell culture, SILAC, as a simple and accurate approach to expression proteomics. Mol. Cell. Proteomics **1:** 376–386.
34. GYGI, S.P., B. RIST, S.A. GERBER, *et al.* 1999. Quantitative analysis of complex protein mixtures using isotope-coded affinity tags. Nat. Biotechnol. **17:** 994–999.
35. DESOUZA, L., G. DIEHL, M.J. RODRIGUES, *et al.* 2005. Search for cancer markers from endometrial tissues using differentially labeled tags iTRAQ and cICAT with multidimensional liquid chromatography and tandem mass spectrometry. J. Proteome Res. **4:** 377–386.
36. LILLEY, K.S. & D.B. FRIEDMAN. 2004. All about DIGE: quantification technology for differential-display 2D-gel proteomics. Expert Rev. Proteomics **1:** 401–409.

37. COOPER, B., W.M. GARRETT & K.B. CAMPBELL. 2006. Shotgun identification of proteins from uredospores of the bean rust *Uromyces appendiculatus*. Proteomics **6:** 2477–2484.

38. EBANKS, R.O., K. CHISHOLM, S. MCKINNON, *et al.* 2006. Proteomic analysis of *Candida albicans* yeast and hyphal cell wall and associated proteins. Proteomics **6:** 2147–2156.

39. THOMAS, D.P., A. VIUDES, C. MONTEAGUDO, *et al.* 2006. A proteomic-based approach for the identification of *Candida albicans* protein components present in a subunit vaccine that protects against disseminated candidiasis. Proteomics **6:** 6033–6041.

40. YAJIMA, W. & N.N. KAV. 2006. The proteome of the phytopathogenic fungus *Sclerotinia sclerotiorum*. Proteomics **6:** 5995–6007.

41. DE GODOY, L.M., J.V. OLSEN, G.A. DE SOUZA, *et al.* 2006. Status of complete proteome analysis by mass spectrometry: SILAC labeled yeast as a model system. Genome Biol. **7:** R50.01–R50.15.

42. BAE, W. & X. CHEN. 2004. Proteomic study for the cellular responses to Cd2+ in *Schizosaccharomyces pombe* through amino acid-coded mass tagging and liquid chromatography tandem mass spectrometry. Mol. Cell. Proteomics **3:** 596–607.

43. COLE, G.T., S.W. ZHU, S.C. PAN, *et al.* 1989. Isolation of antigens with proteolytic activity from *Coccidioides immitis*. Infect. Immun. **57:** 1524–1534.

44. PAN, S. & G.T. COLE. 1995. Molecular and biochemical characterization of a *Coccidioides immitis*-specific antigen. Infect. Immun. **63:** 3994–4002.

45. HUPPERT, M., N.S. SPRATT, K.R. VUKOVICH, *et al.* 1978. Antigenic analysis of coccidioidin and spherulin determined by two-dimensional immunoelectrophoresis. Infect. Immun. **20:** 541–551.

46. COX, R.A., M. HUPPERT, P. STARR, *et al.* 1984. Reactivity of alkali-soluble, water-soluble cell wall antigen of *Coccidioides immitis* with anti-*Coccidioides immunoglobulin* M precipitin antibody. Infect. Immun. **43:** 502–507.

47. ZHU, Y., C. YANG, D.M. MAGEE, *et al.* 1996. Molecular cloning and characterization of *Coccidioides immitis* antigen 2 cDNA. Infect. Immun. **64:** 2695–2699.

48. DUGGER, K.O., J.N. GALGIANI, N.M. AMPEL, *et al.* 1991. An immunoreactive apoglycoprotein purified from *Coccidioides immitis*. Infect. Immun. **59:** 2245–2251.

49. DUGGER, K.O., K.M. VILLAREAL, A. NGYUEN, *et al.* 1996. Cloning and sequence analysis of the cDNA for a protein from *Coccidioides immitis* with immunogenic potential. Biochem. Biophys. Res. Commun. **218:** 485–489.

50. TARCHA, E.J., V. BASRUR, C.Y. HUNG, *et al.* 2006. A recombinant aspartyl protease of *Coccidioides posadasii* induces protection against pulmonary coccidioidomycosis in mice. Infect. Immun. **74:** 516–527.

51. TARCHA, E.J., V. BASRUR, C.Y. HUNG, *et al.* 2006. Multivalent recombinant protein vaccine against coccidioidomycosis. Infect. Immun. **74:** 5802–5813.

52. ORSBORN, K.I., L.F. SHUBITZ, T. PENG, *et al.* 2006. Protein expression profiling of *Coccidioides posadasii* by two-dimensional differential in-gel electrophoresis and evaluation of a newly recognized peroxisomal matrix protein as a recombinant vaccine candidate. Infect. Immun. **74:** 1865–1872.

53. LUNETTA, J., K.A. SIMMONS, S.M. JOHNSON & D. PAPPAGIANIS. 2007. Molecular cloning and expression of a cDNA encoding a *Coccidioides posadasii* Cu,Zn & superoxide dismutase identified by proteomic analysis of the coccidiodel T27K vaccine. Ann. N.Y. Acad. Sci. This volume.

54. LUNETTA, J., S. JOHNSON & D. PAPPAGIANIS. 2006. Identification of a class I 1,2 alpha-mannosidase protein in the coccidioidal T27K vaccine using immuno-proteomic methods. Presented at the 6th International Symposium on Coccidioidomycosis. Stanford University, CA, Aug 25.

55. JOHNSON, S. & D. PAPPAGIANIS. 2006. Identification and molecular cloning of a pH sensitive protein-like protein present in the coccidioidal vaccine T27K. Presented at the 6th International Symposium on Coccidioidomycosis. Stanford University, CA, Aug 25.

The Population Biology of *Coccidioides*

Epidemiologic Implications for Disease Outbreaks

B. M. BARKER, [a] K. A. JEWELL, [b] S. KROKEN, [c] AND M. J. ORBACH[c]

[a] *Genetics Interdisciplinary Program, University of Arizona, Tucson, Arizona, USA*

[b] *Arizona Department of Health Services, Phoenix, Arizona, USA*

[c] *Division of Plant Pathology and Microbiology, University of Arizona, Tucson, Arizona, USA*

ABSTRACT: Studies of field- and patient-derived isolates conducted over the past 75 years have provided a general picture of the population structure of *Coccidioides*, the cause of coccidioidomycosis. Premolecular studies provided a general outline of the geographical range, epidemiology and distribution of the fungus. Recent studies based on molecular markers have demonstrated that the genus is comprised of two genetically diverse, and genetically isolated, species: *Coccidioides immitis* and *C. posadasii*. Both species are composed of biogeographically distinct populations. Structure for two of these populations (*C. immitis* from central California, and *C. posadasii* from southern Arizona) indicates that frequent genetic recombination occurs within the entire geographic range of each population, even though sex has never been observed in the genus. Outbreaks of coccidioidomycosis are not the result of the spread of a single clonal isolate, but are caused by a diversity of genotypes. Although it is now possible to match patient isolates to populations, the lack of apparent structure within each population and the current paucity of environmental isolates limit map-based epidemiological approaches to understanding outbreaks. Therefore, a comprehensive database comprised of soil-derived isolates from across the biogeographic range of *Coccidioides* will improve the utility of this approach. Appropriate collection of environmental isolates will assist the investigation of remaining questions regarding the population biology of *Coccidioides*. The comparative genomics of representative genotypes from both species and all populations of *Coccidioides* will provide a thorough set of genetic markers in order to resolve the population genetics of this pathogenic fungus.

Address for correspondence: Bridget M. Barker, P.O. Box 210036, Department of Plant Sciences, Division of Plant Pathology and Microbiology, University of Arizona, Tucson, AZ 85721. Voice: 520-626-9942; fax: 520-621-7186.
  bmbarker@email.arizona.edu

Ann. N.Y. Acad. Sci. 1111: 147–163 (2007). © 2007 New York Academy of Sciences.
doi: 10.1196/annals.1406.040

KEYWORDS: coccidioidomycosis; population genetics; RFLPs; SSRs; STRs; SNPs; microsatellites; morphological species concept; phylogenetic species concept

INTRODUCTION

Coccidioides spp. are soil-borne fungi endemic to the southwestern portions of North America, with disjunct populations in Central and South America. The species *Coccidioides immitis* is proposed to be restricted to central and southern California,[1] although it potentially extends south into Baja California and east into Arizona. The species *C. posadasii* is present in southern Arizona extending eastward to Texas and south into Mexico, with disjunct and genetically depauperate populations in South America.[2] Both the U. S. Department of Health and Human Services (HHS) and the U. S. Department of Agriculture (USDA) list the species *C. immitis* as a Select Agent, although *C. posadasii* is listed as a Select Agent only by the HHS.[3]

Coccidioidomycosis is contracted via the inhalation of air-dispersed arthroconidia (clonal single-cell fungal propagules), primarily by nonimmunocompromised hosts living in or visiting endemic areas. On the basis of many years of study, it is generally accepted that 60% of infections are asymptomatic, whereas 40% of cases result in mild-to-severe coccidioidomycosis.[4] Two types of asexual propagules are the only known reproductive structures of *Coccidioides*: soil-borne arthroconidia and host-borne endospores. Sexual structures have never been observed and the presence of functional mating-type genes is as yet undetermined although, recently, typical ascomycete mating-type loci have been identified in both species (M.A. Mandel, B. Barker, S. Rounsley, and M. J. Orbach, unpublished data). However, *Coccidioides'* closest known relative, *Uncinocarpus reseii*,[5] has a known functional sexual life cycle, and population genetic data suggest that *Coccidioides* undergoes recombination.[6,7]

Understanding the distribution of genotypes among populations of a pathogen, as well as its ecology, is important for monitoring outbreaks, determining variance in virulence and predicting disease progression and outbreaks. It is difficult to determine whether an intentional release or a natural outbreak has occurred without monitoring genotypes of the pathogen of interest. Correlating disease severity with pathogen genotype may assist in identifying more virulent strains of a pathogen. Monitoring disease progression of various strains of a pathogen in patients may provide information that could assist with better treatment options. Finally, a better understanding of ecological and environmental factors that influence the growth and reproduction of the organism will assist in predicting and preventing exposure to the pathogen.

DISTRIBUTION AND EPIDEMIOLOGY OF *COCCIDIOIDES*

The understanding that exposure to *Coccidioides* spores could result in morbidity without mortality was not widely recognized until 1937, when E.C.

Dickson presented a paper at a session of the California Medical Association regarding five cases of acute, but nonlethal, pulmonary infections.[8] Prior to this time, most physicians considered the disease to be fatal and rare.[9,10] One of the first studies of a localized epidemic of coccidioidomycosis was concerned with determining the source of the infection.[9] This 1942 study determined that the outbreak was the result of group exposure to the fungus caused by several students digging a rattlesnake out of a ground squirrel hole, and that the majority of infections resulted in illness rather than death. It was known that *Coccidioides* could be isolated from soil,[11] and the distribution of the fungus was sporadic and localized.[12] The determination that the 1942 outbreak was caused by exposure to infected soil, and the realization that coccidioidomycosis was not necessarily a fatal disease, led to an interest in understanding of the patchy distribution of *Coccidioides* in the environment, the geographical range of the fungus, and how many people in the endemic region had been exposed without a resulting serious illness.[13]

Conducting skin test surveys in regions thought to be endemic for the disease was one approach taken to map the distribution of the fungus in the environment.[14] Several studies over many years using the method of spherulin or coccidioidin skin testing indicated that the distribution of *Coccidioides* is limited to the southwestern United States,[15,16] and appears to have regions with higher proportions of skin reactivity. Maddy *et al.*[17] showed that in Arizona, Maricopa County (Phoenix), Pima County (Tucson), and Pinal County had over 70% positive skin test rates, compared to rates of 10–40% in surrounding counties. In California, rates of 50–70% positive were detected in Kern County (Bakersfield), Tulare County and Kings County, whereas the rates in areas a few hundred miles away dropped to 10%.[18]

These results prompted researchers to define ecological factors associated with the presence of *Coccidioides*, as well as determining the distribution of the fungus in the soil at local sites.[19,20] In addition, early work at environmental sites known to be positive for *Coccidioides* included animal trapping and surveys of infection in these animals.[21] Finally, there was confirmation that the distribution of the fungus extended into Mexico,[22] Central America,[23] and South America.[24] All of these studies taken together indicate that specific factors determine the distribution, growth, and persistence of *Coccidioides* in the environment.

VARIATION AMONG STRAINS

Early in the study of *Coccidioides*, researchers observed that there was phenotypic variation among strains of the fungus.[25] Differences included variation in hyphal and colony morphology,[26] growth on different media[27,28] and virulence in mice.[29] Virulence studies were conducted as a result of observations of variation in disease severity in patients.[30] Friedman *et al.*[30] showed that although there was variation in virulence, all strains were infective in mice.

Friedman and Smith[31] showed that there was no correlation between virulence of *Coccidioides* in mice and man. Later work showed that patient immune response, inhalation dosage of the infectious agent, and variance in virulence are all causes of variation in disease and is reviewed elsewhere.[4]

Despite known variation in virulence among strains, it remains to be determined whether there are differences between the species in their pathogenic phase, or in the disease that they cause, and controversy remains over the assignment of species status to *C. posadasii*.[4, 10] Although there are no reported differences between these two species in regards to disease severity or progression, there are quantifiable species-dependent variations in halotolerance[1] and thermotolerance (B. Barker and M. Orbach, unpublished data) in the mycelial phase. These physiological differences may play a role in the saprobic, soil-borne phase of the *Coccidioides* life cycle. This phenotypic difference could also reflect a disjunct distribution of the two species in different environments.

Preliminary analysis of environmental isolates in the Tucson area (B. Barker and M. Orbach, unpublished data), and a large study of clinical isolates by the Arizona Department of Health Services (K. Jewell *et al.*, unpublished data), found no *C. immitis* in soil samples, and only two Arizona patients were infected with *C. immitis*, both of whom had a history of travel to California. These results suggest that only *C. posadasii* is present in southern Arizona. However, it is unknown at this time whether both species occur together in southern California, as no molecular genetic analyses of samples collected from soil surveys have been done south of the Tehachapi Mountains, where *Coccidioides* is known to occur.[13] Although characterization of variation among isolates with respect to drug therapy[32] and vaccine development[33,34] has been addressed, interspecific variation should be considered in the development of treatment strategies, as shown in Johannesson *et al.*[35]

DELINEATION OF TWO SPECIES

There are three main methods for delineating species: application of a morphological species concept (MSC), a biological species concept (BSC), or a phylogenetic species concept (PSC).[36] The MSC is problematic because it can be difficult to apply to microorganisms that have indistinct phenotypic differences among closely related species. The BSC, based on the inability of species to interbreed, requires an understanding of mating requirements for sexual reproduction, which is unknown for a large number of fungi, including *Coccidioides*. The PSC is applicable to all organisms, and has been used to characterize the species structure of many other medically important fungi, including *Histoplasma capsulatum*,[37] *Cryptococcus neoformans*,[38] *Paracoccidioides brasiliensis*,[39] and *Aspergillus fumigatus*.[40]

In general, the PSC is used to recognize species by the distribution of unique molecular markers. Phylogenetic species result from genetic isolation, and

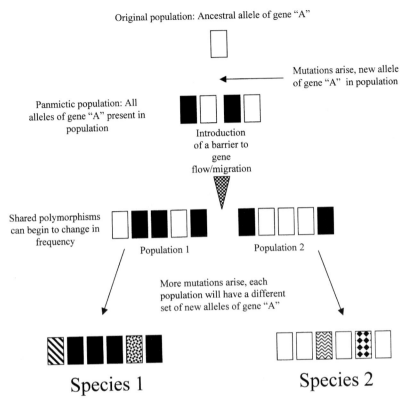

FIGURE 1. A diagrammatic representation of the speciation process in a recombining organism from the point of view of the history of a single gene. Over time, new mutations arise; drift and selection affect genetic diversity. Boxes of different patterns represent individual alleles for a given gene. A species with no breeding barries (panmixia) will have no structure; allelic frequencies across the species will not be significantly different. Limited gene flow (movement of alleles and/or genotypes) results in populations with different allelic frequencies of common alleles, and no rare alleles are shared. A continued and complete lack of gene flow results in genetically isolated species. Over time, no alleles will be shared between phylogenetic species.

therefore accumulate unique alleles for each locus observed (FIG. 1). During the course of speciation, endogenous breeding barriers develop, so that the application of the BSC and PSC eventually result in the same diagnosis of separate species. As species continue to diverge, phenotypic differences accumulate. Therefore, the MSC also will eventually become congruent with the PSC. Phylogenetic species may be initially cryptic.[36,39,41] However, awareness of this species divergence is important so that studies of the ecology of *Coccidioides*, and the etiology, prevention, and treatment of coccidioidomycosis, compare representatives from both species.

TABLE 1. Summary of population studies using nucleotide loci as markers

Study[a]	Marker	No. of isolates	Origin of isolates[b]	Results
Zimmerman 1994[42]	3 restriction length fragment polymorphisms (RFLPs) of whole-genome DNA digests by AvaII, HaeIII, and HincII	15	13 CA (2 soil, 1 sea lion), 1 strain Silveira, 1 Venezuela	For each digest, 1 main RFLP pattern with two subgroups (now known to correspond to 2 species) low diversity for this marker.
Burt 1994[49]	single nucleotide polymorphism (SNP) discovery	6	(not reported)	SNP characterization methods
Burt 1996[6]	14 (SNPs among 12 anonymous loci, internal transcribed spacer (ITS) and chitin synthase (CS)	25	Single hospital in Tucson AZ, 1979–1990	Evidence for sexual recombination in Arizona population, high diversity for these markers.
Burt 1997[51]	14 SNPs among 12 anonymous loci (2 loci with 2 SNPs each)	70	25 AZ (Burt 1996[6]); 25 Bakersfield CA 1993–1994 outbreak, 20 San Antonio TX 1992–1993.	AZ population very different from CA, less different from TX (now known to correspond to 2 species, and 2 populations of 1 of those species). High rates of monomorphism in CA and TX populations for sites that are polymorphic in AZ
Koufopanou 1997,[43] 1998[44]	33 SNPs among 5 known protein-coding loci: CS, dioxygenase3, orotidine decarboxylase, serine protease, and chitinase.	17	Silveira, 2 AZ, 2 TX, 2 MX, 5 SA, 3 CA; 2 CA soil isolates	Recognition of genetically isolated and genetically diverse "California" vs. "non-California" species. High diversity for these markers.
Fisher 1999[52]	13 SNPs & 9 simple tandem repeats (STRs) characterized	7	CA	Characterization of STRs and SNPs that are polymorphic in CA
Fisher 2000[7]	14 SNPs among 7 anonymous loci, 4 protein-coding loci, and ITS; and 2 STRs	39	37 Bakersfield CA 1993–1994 outbreak, 2 non-CA	Unusually large-scale outbreak was due to environmental factors, rather than hypervirulent clonal isolate.

Continued

TABLE 1. *continued*

Study[a]	Marker	No. of isolates	Origin of isolates[b]	Results
Fisher 2000[45]	7 STRs including sequenced flanks 7 SNPs from 6 anonymous loci and ITS	20	Bakersfield (6) San Diego, CA (3) Tucson, AZ (7) San Antonio, TX (4)	STRs useful markers: rapidly evolving, but not hypervariable. Congruent with RFLP and SNP loci. Species separation 10–12 mya
Fisher 2001[2]	9 STRs	163	Entire geographic range, includes 2 soil isolates	"CA" species has 2 populations: central and southern CA. "Non-CA" species has 3 populations: AZ, MX, and TX. South American population was recently introduced, perhaps by humans.
Fisher 2002[1]	9 STRs	167	Entire geographic range, includes 2 soil isolates	Formal description of "non-CA" species as *C. posadasii*
Fisher 2002[58]	9 STRs	179	163 from entire geographic range, and 16 isolates from patients in New York state	Ability to assign 16 isolates to specific population with 97% accuracy for *C. posadasii* and 93% for *C. immitis*

[a] Studies listed by first author and date. Complete references in bibliography.
[b] Isolates are derived from human patients unless otherwise noted.

One of the first molecular analyses of genetic variation in *Coccidioides* was a restriction fragment length polymorphism (RFLP) study (see TABLE 1 for a list of all molecular-based studies discussed in this section).[42] This RFLP-based study looked at fifteen strains and found one RFLP pattern among 13 of 15 isolates, and 2 isolates with a second pattern. These patterns could be consistent with a predominately clonal species; all of the strains were obtained from patients in California. The authors concluded that other DNA typing methods might give a greater resolution of genetic variation that was not be detectable by the methods they employed. For example, we know now that one of the isolates tested, the commonly used strain Silveira, is a member of the species *C. posadasii*.[1] Although obtained from a patient in the San Joaquin Valley of California,[29] the patient could have traveled to areas where *C. posadasii* is commonly present. Alternatively, *C. posadasii* may be present, albeit rarely, in California.

Prior to 2002, *Coccidioides* was described as a genus with the single morphologically defined species: *C. immitis*. The application of a PSC, based on the congruence of five protein-coding gene genealogies, demonstrated that the "California" and "non-California" populations correspond to two phylogenetic species[43,44] that have been genetically isolated for an estimated twelve million years.[45]

The complete lack of observed genetic exchange between these lineages may be consistent with the BSC, if an endogenous breeding barrier exists in nature between the two lineages. Cox and Magee[4] argued that an overlapping distribution of the two lineages would support variety status. However, if the geographic ranges of the two lineages were found to overlap, the argument for two species would be strengthened, and not diminished, as the opportunity for genetic exchange would not be limited by spatial separation.

Fisher *et al.*[1] expanded the initial observation of genetic variation by employing a broad geographic survey of 167 strains. The allelic distribution of nine simple tandem repeat (STR) loci also supported the previous observation of two lineages,[43,44] and resulted in the separation of the two lineages of *Coccidioides* into two separate species defined by a PSC. The species that is endemic to California retains the name *C. immitis*, and the species found primarily in Arizona, Texas, Mexico, and South America has been named *C. posadasii*.[1]

Coccidioides can be considered as one morphological species, or as two phylogenetic species. The medical community tends to prefer definitions of species to reflect phenotypic differences that are relevant to pathogenesis, treatment, and prevention measures.[4,10] Recent studies have shown that candidate vaccines are cross-reactive between *C. immitis* (strain RS) and *C. posadasii* (strain Silveira).[33,34,46] The recognition that *Coccidioides* is composed of two divergent lineages leads to the inclusion of representative isolates in vaccine trials, and therefore the knowledge that the candidate vaccines are universally effective, in spite of genetic variation within and between species. The value of recognizing the two clades as species is to raise awareness of the potential for medically important phenotypic differences that are necessary to consider in candidate drug trials and in studies of the progression of the disease caused by the two lineages.

SNP STUDIES

Single nucleotide polymorphisms (SNPs) in DNA sequences are commonly used as polymorphic genetic markers. The nucleotide change may or may not cause a change in an amino acid, or it may be located in the promoter region of a gene or in intergenic regions.[47] If a SNP has no effect on the amino acid sequence, it is considered to be a neutral mutation, and thus not under selection. These neutral SNPs can go to fixation within a population or species due to the evolutionary process of drift, that is, random chance. Upon fixation, they become markers that indicate genetic isolation among populations or species (FIG. 1). SNPs now can be discovered easily by comparing genomes of multiple sequenced strains, as has been done with the human HAPMAP[48] project; this information has been a crucial tool in the discovery of disease-related genes. For the following early studies on *Coccidioides*, anonymous loci were found by combinations of arbitrary primer pairs. Polymorphisms were found by single-strand conformation polymorphism (SSCP) gels, sequencing of selected polymorphic loci, and scoring selected SNP sites by restriction enzyme digests.[49] These early studies did not use the term "SNP," but, the data gathered and analyses performed are equivalent to those of current SNP studies.[47,50]

The first population genetic study of *Coccidioides* based on SNPs found a population structure consistent with recombination and cryptic sex.[6] This data set consisted of 30 isolates collected from 25 patients from a single hospital in Tucson, Arizona from 1979 to 1990, and subsequently described as "non-California" *C. immitis*.[43,44] Twelve phylogenetically informative random polymorphic SNPs from anonymous loci were used, in addition to two SNPs from known regions of the genome (ITS region of the rRNA gene and a chitin synthase gene). Isolates obtained from the same patients were genetically identical, but there were no identical genotypes found among the 25 different patients. Further analysis of the data also showed a highly statistically significant signature of recombination. If reproduction is clonal, there should be evidence of linkage among alleles of various loci, and the phylogenetic trees for the isolates for any given locus should be similar. However, no linkage or congruity among loci was present in the data.

A subsequent study made use of five SNP-containing loci of known genes: chitin synthase, dioxygenase, orotidine decarboxylase, serine protease, and chitinase,[43,44] and also found that these loci are not linked, consistent with a history of recombination. In addition, reproductive isolation was found between the 5 California and 12 non-California strains collected across the North

American range of *Coccidioides*. For each of the five loci, the two lineages were found to contain unique sets of alleles (FIG. 1).

A study published the same year[51] based on 12 of the 14 previously characterized loci[6] found genetic isolation among *Coccidioides* populations from California, Arizona, and Texas. SNP sites within these loci that were previously determined to be polymorphic in the Arizona population were found to be monomorphic in the California and/or the Texas populations. The California isolates used in this study were all collected during a major outbreak of coccidioidomycosis in the San Joaquin Valley of California, which lasted three years (1991–1994) and resulted in a tenfold increase in cases. One explanation of the data was that the epidemic was the result of a single, and potentially novel, genotype of enhanced virulence.

However, a subsequent genotypic study[7] of this 1991–1994 San Joaquin Valley outbreak made use of markers that were found to be polymorphic in this population. These markers comprised fourteen different SNPs distributed among 12 loci characterized by the methods described in Burt *et al.*,[49] and two STR loci (STR loci are discussed in detail in the following section). The primer sequences and more detailed information are available in Fisher *et al.*[52] By means of these markers, all 37 isolates derived from patients in Bakersfield were found to be "California" *C. immitis*. Among these isolates, there were 34 different genotypes, and their population structure was consistent with a recombining, rather than a clonal, population.

From this population structure data it was hypothesized that the 1991–1994 outbreak, although unusual in its magnitude, was similar in its origin to those of more typical outbreaks. The increased case numbers were hypothesized to be due to an overgrowth in *Coccidioides* as a direct result of favorable weather conditions. It was noted in 1991 and 1993 that there was significantly higher than normal rainfall in Tulare County, California.[53] An extremely dry period followed by heavy rainfall could have resulted in the increased formation of arthroconidia in the soil, thus providing a source of inoculum that was subsequently disturbed and distributed by windstorms during the following dry season.[7,54] The magnitude and temporal aspects of coccidioidomycosis outbreaks are thought to be largely the result of these alternating climatic factors.[55]

STR, SSR, AND MICROSATELLITE STUDIES

Simple tandem repeats (STRs), simple sequence repeats (SSRs), and microsatellites all refer to similar types of molecular markers of genetic variation, and in this document we will refer to all of these motifs as STRs for simplicity. They are generally dinucleotide repeats and are designated as $(NN)_{\#-\#}$ to indicate the nucleotide pair that is repeated and the range of frequency of that repeat (i.e., AT_{10-15}); these repeats can range from 1 to 6 nucleotides, and are repeated in tandem.[56] These repeats can also occur in combination (i.e., $AT_{10-15} CT_{8-12}$). The size variation in STRs between strains can be determined using polymerase chain reaction (PCR) primers designed for conserved regions flanking the repeated regions, and then analyzed by direct sequencing, or by determining the length of the repeat by the use of PCR primers labeled with fluorescent dyes.[1,57]

In 1999, Fisher *et al.*[52] defined nine STR loci in *Coccidioides*. A subsequent paper[45] found that the histories of these STR loci are congruent with the SNP loci from previous work[51] in recovering two genetically isolated lineages. These data were used to support the delineation of species.[41] In addition both species of *Coccidioides* have high levels of intraspecific STR variation. In fact, in a database of 167 isolates, almost all patient isolates are genetically unique, based on the analysis of nine previously published STR loci.[1,52] These STRs are rapidly evolving, such that a larger number of alleles are found per STR locus than in each SNP locus, yet they are not hypervariable, and are therefore not subject to convergent evolution of size class by expansion and contraction of the STR.[41] It was found that STR data gave similar results as phylogenetic trees obtained from sequence analysis. This result supports the use of STRs to determine phylogenetic species, and not simply population structure within a species. Therefore, the STR loci are sufficiently powerful and accurate to resolve distinct genotypes and species.[1]

STR DATABASE OF *COCCIDIOIDES* ISOLATES

The initial database of 167 isolates developed by Fisher *et al.*[1] was based on strains collected mainly from human patients from a wide geographic range. One criticism of the population genetic analyses performed to date is that the number of isolates is relatively small (167 as of 2004).[4] However, the database of isolates whose STR genotypes have been determined is growing rapidly and currently includes an additional ~120 recent (Jewell *et al.*, unpublished data) and ~250 historical collections from patients in Arizona (Barker and Kroken, unpublished data), as well as ~50 Arizona isolates recovered from soil and from domestic and wild animals (Barker and Orbach, unpublished data). The combined published and unpublished data demonstrate that almost all patients from Arizona are infected with *C. posadasii*, as expected on the basis of the biogeographic separation of *C. immitis* and *C. posadasii*. Of approximately 400 patient isolates collected in Arizona, only five are *C. immitis*. These anomalous *C. immitis* infections may not have been acquired locally, as all five patients had travel history to California, where *C. immitis* is endemic. However, long-distance spore dispersal or presence of *C. immitis* in Arizona cannot yet be ruled out. As in previous studies, no genetic linkage was found among the nine loci, and no obvious genetic structure was observed among the southern Arizona isolates. This result is consistent with sexual recombination and a high level of genotype migration within the Arizona population of *C. posadasii*.

The lack of clonality and the current observed lack of any subpopulation structure within populations of *Coccidioides*[6,7,43,44,51] challenge the use of population genetic information in epidemiological studies of *Coccidioides*. In addition, no matching genotypes have been found between soil and patient isolates (Barker and Orbach, unpublished data). These findings severely limit the ability to pinpoint a genotype to a location. Movement of inoculum by long-distance spore dispersal from anthropogenic soil disturbance and windstorms, as well as hosts that travel far from the area of infection, further obscure the source of the inoculum.[58,59] Finally, the possibility of late diagnosis in human hosts,[60] and the potential for reactivation of the disease after a period of dormancy,[61] also makes it difficult to identify when and where the patient was initially exposed.

These observations suggest that the continued addition of patient isolates alone to the STR database will not lead to a significantly improved understanding of the population biology of *Coccidioides*. Rather, the data need to include a larger number of environmental isolates, which should provide an improved resolution of the local structure of *Coccidioides* populations. Such a comprehensive database will improve our ability to match a given genotype of a patient isolate to its most likely place of origin.[50,51,58] In addition, more accurate mapping of when and where *Coccidioides* is present in the soil will improve our understanding of the epidemiology of coccidioidomycosis, and may result in more accurate forecasting of the times and places of increased risk of infection.

COMPARATIVE GENOMICS OF *COCCIDIOIDES*

Interspecies comparisons for all genes are now possible, as a single genotype of both species of *Coccidioides* has been sequenced. *C. immitis* strain RS has been sequenced by the Broad Institute,[62] and *C. posadasii* strain C735 has been sequenced by the Institute for Genomic Research (TIGR).[63] In addition, the Broad Institute is in the process of sequencing 13 strains representing both species. Some strains were selected to represent the populations found within each species. Fisher *et al.*[2] provide evidence, based on nine STR loci, that *C. immitis* is subdivided into central and southern California populations, and that *C. posadasii* is subdivided into three populations, one centered in Arizona, one in Mexico, and one in Texas. In addition, the Texas population is the apparent source of a recent, and perhaps anthropogenic, introduction into South America sometime within the last 9,000–140,000 years.[2]

On the basis of this observed population structure, genotypes selected for sequencing include at least two isolates from each of the five populations. Other strains were selected in order to cover differences in virulence in a murine model, representing both high and low virulence. A final set of isolates, recovered from the soil, was selected in order to compare environmental isolates to genetically similar clinical isolates.[62] The comparative genomic analyses of the resulting data sets will provide detailed information on the level of diversity within and between these species, and has the potential to find genetic differences that correlate with host specificity and virulence.

This genomic database will also provide a resource to develop population- and species-specific SNPs. As SNPs are more slowly evolving than STRs, there is the potential to resolve population structure more finely than the nine STRs used for the current database.[47] With a genomic database including both soil and animal isolates, and with the resolving power of SNPs, the potential exists to accurately place a group of patient isolates to a specific geographic location.

DISCUSSION

Studies of the distribution of *Coccidioides* in the environment and its incidence in humans over the past 75 years established the groundwork for the population biology of *Coccidioides*. Early studies showed that coccidioidomycosis was not necessarily fatal, and that skin reactivity could be used to determine the percentage exposure and rate of conversion to seropositivity within the human population. This information led to surveys of large groups of people throughout the southwestern United States, particularly at military bases. Additional work in Mexico, Central America and South America further expanded the known range of *Coccidioides*. Specific procedures were developed to recover the pathogen from soil, to model the disease in murine models, and to work safely with the pathogen in the laboratory. All of these advances have allowed us to expand our knowledge of the ecology and population biology of *Coccidioides*.

In order to address remaining questions regarding the epidemiology of coccidioidomycosis, molecular-based approaches have been used to estimate the population structure of *Coccidioides*. Outbreaks have been found not to be the result of a single, and perhaps more virulent, genotype, but are more likely the result of soil disturbance of positive soil sites, and/or increased inoculum in the soil due to biotic and/or abiotic factors. The genotypes of *Coccidioides* collected from patients are diverse, and show a history of genetic recombination, in spite of the fact that sex has never been observed in this genus. It appears that the population genetics of each species of *Coccidioides* is that of a sexually reproducing eukaryote, and are not similar to the population genetics of clonal bacterial pathogens. Although each genotype in the soil produces infectious arthroconidia via asexual reproduction, the origin of each of these genotypes appears to be the result of genetic recombination. However, the timing and location of putative sexual recombination in *Coccidioides* spp. remains to be determined.

In spite of the extensive diversity, genetic recombination in *Coccidioides* shows boundaries. Both species have a recombining population structure, with unrestricted recombination and genotype migration within populations, restricted recombination and/or genotype migration among currently defined populations, and no recombination between species. It is worth noting that the putative populations in each species may reflect either sharp boundaries in gene flow or be an artifact of biased sampling, that is, most isolates are derived from patients who live in metropolitan areas, with fewer patient isolates from intervening areas.

We have noted other unanswered questions, including whether *C. immitis* and *C. posadasii* have sympatric (overlapping) ranges or allopatric (separate) ranges, and how individual genotypes are structured spatially and temporally. Other questions that remain to be answered include where and how *Coccidioides* undergoes genetic recombination, and how many additional isolates will be required to find genotype matches between isolates derived from soil and from patients. These questions will only be addressed adequately by the collection of environmental isolates at the appropriate physical and/or temporal scale for each question, as well as additional patient isolates from geographic areas not already exhaustively covered in previous studies. It is hoped that answers to these questions will improve our understanding of the epidemiology of coccidioidomycosis, with an end goal of reducing exposure to the causative agent.

REFERENCES

1. FISHER, M.C., G.L. KOENIG, T.J. WHITE, *et al.* 2002. Molecular and phenotypic description of *Coccidioides posadasii* sp. nov., previously recognized as the non-California population of *Coccidioides immitis*. Mycologia **94:** 73–84.

2. FISHER, M.C., G.L. KOENIG, T.J. WHITE, *et al.* 2001. Biogeographic range expansion into South America by *Coccidioides immitis* mirrors New World patterns of human migration. Proc. Nat. Acad. Sci. U.S.A. **98:** 4558–4562.

3. CDC. HHS and USDA select agents and toxins 7 CFR part 331, 9 CFR part 121, and 42 CFR part 73 [online]. 2006 February [cited 2006 August 8]. Available from http://www.cdc.gov/od/sap/docs/salist.pdf.

4. COX, R.A. & D.M. MAGEE. 2004. Coccidioidomycosis: host response and vaccine development. Clin. Microbiol. Rev. **17:** 804–839.

5. SIGLER, L., A.L. FLIS & J.W. CARMICHEAL. 1998. The genus *Uncinocarpus* (Onygenaceae) and its synonym *Brunneospora*: new concepts, combinations and connections to anamorphs in *Chrysosporium*, and further evidence of relationship with *Coccidioides immitis*. Can. J. Bot. **76:** 1624–1636.

6. BURT, A., D.A. CARTER, G.L. KOENIG, *et al.* 1996. Molecular markers reveal cryptic sex in the human pathogen *Coccidioides immitis*. Proc. Nat. Acad. Sci. U.S.A. **93:** 770–773.

7. FISHER, M.C., G.L. KOENIG, T.J. WHITE, *et al.* 2000. Pathogenic clones versus environmentally driven population increase: analysis of an epidemic of the human fungal pathogen *Coccidioides immitis*. J. Clin. Microbiol. **38:** 807–813.

8. FEISE, M.J. 1958. Coccidioidomycosis. Charles C Thomas. Springfield, IL.
9. DAVIS, B.L., R.T. SMITH & C.E. SMITH. 1942. An epidemic of coccidioidal infection (coccidioidomycosis). JAMA **118:** 1182–1186.
10. DICAUDO, D.J. 2006. Coccidioidomycosis: a review and update. J. Am. Acad. Dermatol. **55:** 929–942.
11. STEWART, R.A. & K.F. MEYER. 1932. Isolation of *Coccidioides immitis* (Stiles) from the soil. Proc. Soc. Exp. Biol. Med. **29:** 937–938.
12. EMMONS, C. 1942. Isolation of *Coccidioides* from soil and rodents. Pub. Health Rep. **57:** 109–111.
13. SWATEK, F.E. 1970. Ecology of *Coccidioides immitis*. Mycopathol. Mycolog. Appl. **41:** 3–12.
14. SMITH, C.E., E.G. WHITING, E.E. BAKER, *et al.* 1948. The use of coccidioidin. Am. Rev. Tubercul. **57:** 330–360.
15. PAPPAGIANIS, D. 1988. Epidemiology of coccidioidomycosis. *In* Current Topics in Medical Mycology, Vol. 2. M. McGinnis, Ed.: 199–238. Springer Verlag. New York.
16. AJELLO, L. 1971. Coccidioidomycosis and histoplasmosis—review of their epidemiology and geographical distribution. Mycopatholog. Mycolog. Appl. **45:** 221–230.
17. MADDY, K.T., I.L. DOTO, M.L. FURCOLOW, *et al.* 1957. Coccidioidin, histoplasmin, and tuberculin sensitivity of students in selected high schools and colleges in Arizona. CDC Proceeding Symposium on Coccidioidomycosis: 121–126.
18. EDWARDS, P.Q. & C.E. PALMER. 1957. Prevalence of sensitivity to coccidioidin with special reference to specific and nonspecific reactions to coccidioidin and to histoplasmin. Dis. Chest. **31:** 35–60.
19. MADDY, K.T. 1958. Ecological factors possibly relating to the geographic distribution of *Coccidioides immitis*. Ariz. Med. **15:** 178–188.
20. ELCONIN, A.F., R.O. EGEBERG & R. LUBARSKY. 1957. Growth patterns of *Coccidioides immitis* in the soil of an endemic area. CDC Proceeding of Symposium on Coccidioidomycosis: 168–170.
21. EMMONS, C.W. & L.L. ASHBURN. 1942. The isolation of *Haplosporangium parvum* nov. sp. and *Coccidioides immitis* from wild rodents—their relationship to coccidioidomycosis. Pub. Health Rep. **57:** 1715–1727.
22. OCHOA, A.G. 1967. Coccidioidomycosis in Mexico. Coccidioidomycosis: Proceedings of the Second Coccidioidomycosis Symposium: 293–299.
23. MAYORGA, R.P. & H. ESPINOZA. 1970. Coccidioidomycosis in Mexico and Central America. Mycopatholog. Mycolog. Appl. **40:** 13–23.
24. CAMPINS, H. 1970. Coccidioidomycosis in South America—a review of its epidemiology and geographic distribution. Mycopatholog. Mycolog. Applicata. **40:** 25–34.
25. FRIEDMAN, L., D. PAPPAGIANIS, R.J. BERMAN, *et al.* 1953. Studies on *Coccidioides immitis*—morphology and sporulation capacity of 47 strains. J. Lab. Clin. Med. **42:** 438–444.
26. HUPPERT, M., S.H. SUN & J.W. BAILEY. 1967. Natural variability in *Coccidioides immitis*. Proceedings of Second Symposium on Coccidioidomycosis.:323–328.
27. BAKER, E.E. & E.M. MRAK. 1943. Taxonomy and distribution of *C. immitis*. Farlowia **1:** 199–244.
28. HUPPERT, M. & L.J. WALKER. 1958. The selective and differential effects of cycloheximide on many strains of *Coccidioides immitis*. Am. J. Clin. Pathol. **29:** 291–295.

29. FRIEDMAN, L., C.E. SMITH, W.G. ROESSLER, *et al.* 1956. The virulence and infectivity of 27 strains of *Coccidioides immitis*. Am. J. Hyg. **64:** 198–210.

30. FRIEDMAN, L., C.E. SMITH & L.E. GORDON. 1955. The assay of virulence of *Coccidioides* in white mice. J. Infect. Dis. **97:** 311–316.

31. FRIEDMAN, L. & C.E. SMITH. 1957. The comparison of four strains of *Coccidioides immitis* with diverse histories. Mycopatholog. Mycolog. Applicata. **8:** 47–53.

32. COLLINS, M.S. & D. PAPPAGIANIS. 1977. Uniform susceptibility of various strains of *Coccidioides immitis* to amphotericin-B. Antimicrob. Agents Chemother. **11:** 1049–1055.

33. PENG, T., K.I. ORSBORN, M.J. ORBACH, *et al.* 1999. Proline-rich vaccine candidate antigen of *Coccidioides immitis*: conservation among isolates and differential expression with spherule maturation. J. Infect. Dis. **179:** 518–521.

34. PAPPAGIANIS, D. 2001. Seeking a vaccine against *Coccidioides immitis* and serologic studies: expectations and realities. Fung. Genet. Biol. **32:** 1–9.

35. JOHANNESSON, H., P. VIDAL, J. GUARRO, *et al.* 2004. Positive directional selection in the proline-rich antigen (PRA) gene among the human pathogenic fungi *Coccidioides immitis*, *C. posadasii* and their closest relatives. Molec. Biol. Evol. **21:** 1134–1145.

36. TAYLOR, J.W., D.J. JACOBSON, S. KROKEN, *et al.* 2000. Phylogenetic species recognition and species concepts in fungi. Fung. Genet. Biol. **31:** 21–32.

37. KASUGA, K., T.J. WHITE, G. KOENIG, *et al.* 2003. Phylogeography of the fungal pathogen *Histoplasma capsulatum*. Mol. Ecol. **12:** 3383–3401.

38. KWON-CHUNG, K. & A. VARMA. Do major species concepts support one, two, or more species within *Cryptococcus neoformans*? FEMS Yeast Res. **6:** 574–587.

39. MATUTE, D.R., J.G. MCEWEN, R. PUCCIA, *et al.* 2006. Cryptic speciation and recombination in the fungus *Paracoccidioides brasiliensis* as revealed by gene genealogies. Mol. Biol. Evol. **23:** 65–73.

40. PRINGLE, A., D.M. BAKER, J.L. PLATT, *et al.* 2005. Cryptic speciation in the cosmopolitan and clonal human pathogenic fungus *Aspergillus fumigatus*. Evolution **59:** 1886–1899.

41. KOUFOPANOU, V., A. BURT, T. SZARO, *et al.* 2001. Gene genealogies, cryptic species, and molecular evolution in the human pathogen *Coccidioides immitis* and relatives (Ascomycota, Onygenales). Mol. Biol. Evol. **18:** 1246–1258.

42. ZIMMERMANN, C.R., C.J. SNEDKER & D. PAPPAGIANIS. 1994. Characterization of *Coccidioides immitis* isolates by restriction fragment length polymorphisms. J. Clin. Microbiol. **32:** 3040–3042.

43. KOUFOPANOU, V., A. BURT & J.W. TAYLOR. 1997. Concordance of gene genealogies reveals reproductive isolation in the pathogenic fungus *Coccidioides immitis*. Proc. Nat. Acad. Sci. U.S.A. **94:** 5478–5482.

44. KOUFOPANOU, V., A. BURT & J.W. TAYLOR. 1998. Concordance of gene genealogies reveals reproductive isolation in the pathogenic fungus *Coccidioides immitis* [correction]. Proc. Nat. Acad. Sci. U.S.A. **95:** 8414.

45. FISHER, M.C., G. KOENIG, T.J. WHITE, *et al.* 2000. A test for concordance between the multilocus genealogies of genes and microsatellites in the pathogenic fungus *Coccidioides immitis*. Mol. Biol. Evol. **17:** 1164–1174.

46. PAPPAGIANIS, D. 1999. *Coccidioides immitis* antigen. J. Infect. Dis. **180:** 243–244.

47. MORIN, P.A., G. LUIKART, R.K. WAYNE, *et al.* 2004. SNPs in ecology, evolution and conservation. Trends Ecol. Evol. **19:** 208–216.

48. International HAPMAP project website. Available from http://www.hapmap.org/. Updated January 17, 2007. Accessed January 21, 2007.

49. BURT, A., D.A. CARTER, T.J. WHITE, *et al.* 1994. DNA sequencing with arbitrary primer pairs. Mol. Ecol. **3:** 523–525.

50. TAYLOR, J.W. & M.C. FISHER. 2003. Fungal multilocus sequence typing—it's not just for bacteria. Curr. Opin. Microbiol. **6:** 351–356.

51. BURT, A., B.M. DECHAIRO, G.L. KOENIG, *et al.* 1997. Molecular markers reveal differentiation among isolates of *Coccidioides immitis* from California, Arizona and Texas. Mol. Ecol. **6:** 781–786.

52. FISHER, M.C., T.J. WHITE & J.W. TAYLOR. 1999. Primers for genotyping single nucleotide polymorphisms and microsatellites in the pathogenic fungus *Coccidioides immitis*. Mol. Ecol. **8:** 1082–1084.

53. PAPPAGIANIS, D. 1994. Marked increase in cases of coccidioidomycosis in California—1991, 1992, and 1993. Clin. Infect. Dis. **19:** S14–S18.

54. DURRY, E., D. PAPPAGIANIS, S.B. WERNER, *et al.* 1997. Coccidioidomycosis in Tulare county, California, 1991: reemergence of an endemic disease. J. Med. Vet. Mycol. **35:** 321–326.

55. COMRIE, A.C. 2005. Climate factors influencing coccidioidomycosis seasonality and outbreaks. Environ. Health Persp. **113:** 688–692.

56. SELKOE, K.A. & R.J. TOONEN. 2006. Microsatellites for ecologists: a practical guide to using and evaluating microsatellite markers. Ecol. Lett. **9:** 615–629.

57. RASSMANN, K., C. SCHLOTTERER & D. TAUTZ. 1991. Isolation of simple-sequence loci for use in polymerase chain-reaction based DNA fingerprinting. Electrophoresis **12:** 113–118.

58. FISHER, M.C., B. RANNALA, V. CHAURVEDI, *et al.* 2002b. Disease surveillance in recombining pathogens: multilocus genotypes identify sources of human *Coccidioides* infections. Proc. Nat. Acad. Sci. U.S.A.. **99:** 9067–9071.

59. DESAI, S.A., O.A. MINAI, S.M. GORDON, *et al.* 2001. Coccidioidomycosis in non-endemic areas: a case series. Resp. Med. **95:** 305–309.

60. STEVENS, D.A. 1995. Coccidioidomycosis. N. Eng. J. Med. **332:** 1077–1082.

61. OLDFIELD, E.C., W.D. BONE, C.R. MARTIN, *et al.* 1997. Prediction of relapse after treatment of coccidioidomycosis. Clin. Infect. Dis. **25:** 1205–1210.

62. Broad Institute. *Coccidioides* Group Database. Available from http://www.broad.mit.edu/annotation/genome/coccidioides_group/-MultiHome.html. Updated November 03, 2006. Accessed December 4, 2006.

63. TIGR Database. *Coccidioides posadasii* Genome Project. Available from http://www.tigr.org/tdb/e2k1/cpa1/. Updated March 2006. Accessed Dec 4, 2006.

Molecular Cloning and Expression of a cDNA Encoding a *Coccidioides posadasii* 1,2-α-Mannosidase Identified in the Coccidioidal T27K Vaccine by Immunoproteomic Methods

JENNINE M. LUNETTA, KEIRA A. SIMMONS, SUZANNE M. JOHNSON, AND DEMOSTHENES PAPPAGIANIS

Department of Medical Microbiology and Immunology, School of Medicine, University of California, Davis, California, USA

ABSTRACT: The coccidioidal T27K vaccine is protective in mice against respiratory challenge with *Coccidioides posadasii* (*C. posadasii*) arthroconidia. The vaccine is a subcellular multicomponent preparation that has not been fully characterized. To identify potential protective antigens in the heterogeneous mixture, the vaccine has been separated by two-dimensional gel electrophoresis and then analyzed for seroreactive proteins using immunoblot analysis with pooled sera from patients with coccidioidomycosis. Two seroreactive spots of identical apparent molecular weight were identified and sequenced using tandem mass spectrometry. Three peptides were generated, two of which matched a tentative consensus sequence in the TIGR *C. posadasii* 2.0 gene index database that is similar to fungal 1,2-α-mannosidases. The $5'$ and $3'$ ends of the mannosidase cDNA were mapped using rapid amplification of cDNA ends (RACE) polymerase chain reaction (PCR), and a full-length cDNA was then obtained using reverse-transcription (RT) PCR. The cDNA was cloned and sequenced and expressed as a recombinant protein. The predicted protein consists of 519 amino acids, has a theoretical molecular weight and pI of 56,918 Da and 4.84, respectively, and is very similar ($>60\%$) to other fungal 1,2-α-mannosidases. Class I 1,2-α-mannosidase enzyme activity was also detected in the T27K vaccine using the substrate, Man-α-1,2-Man-α-OCH$_3$ in a spectrophotometric assay.

KEYWORDS: *Coccidioides*; vaccine; mannosidase; fungus; two-dimensional gel electrophoresis; immunoblot; seroreactive; mass spectrometry; rapid amplification of cDNA ends; reverse-transcription polymerase chain reaction

Address for correspondence: Jennine M. Lunetta, Department of Medical Microbiology and Immunology, School of Medicine, University of California, Davis, CA 95616. Voice: 530-752-7214; fax: 530-752-8692.

jmlunetta@ucdavis.edu

Ann. N.Y. Acad. Sci. 1111: 164–180 (2007). © 2007 New York Academy of Sciences.
doi: 10.1196/annals.1406.015

INTRODUCTION

Previous studies designed to develop a vaccine against coccidioidomycosis have established that protective antigens are present during the spherule-endospore (SE) phase of the fungal life cycle.[1] For instance, it has been demonstrated that formaldehyde-killed whole spherules and mechanically disrupted spherules protect mice against respiratory challenge with the organism.[2–5] The development of a safe and effective human vaccine will most likely require the isolation of the protective component(s) from these crude preparations. The coccidioidal T27K vaccine is a soluble, subcellular fraction from the SE phase of *C. posadasii* consisting mainly of protein and carbohydrate which has been shown to protect mice against intranasal challenge with *C. posadasii*.[5] Recent efforts have focused on the molecular characterization of the crude T27K vaccine so that potential protective antigens can be identified.[6–8]

In this study, we report the identification of a class I 1,2-α-mannosidase protein detected in the coccidioidal T27K vaccine using immunoproteomic methods. The cloning, sequencing, and expression of the cDNA encoding the protein are also described. This work was presented, in part, at the 49th Annual Coccidioidomycosis Study Group Meeting and the 6th International Symposium on Coccidioidomycosis.[7,8]

MATERIALS AND METHODS

Vaccine Preparation

The coccidioidal T27K vaccine was prepared from thimerosal-killed, mechanically disrupted mature endosporulating spherules as previously described.[5] The Silveira strain of *C. posadasii*, formerly known as the non-California *C. immitis*, was used.[9]

Serum Samples

Serum samples from anonymous patients positive for the complement fixation antibody were obtained from the Coccidioidomycosis Serology Laboratory (Davis, CA, USA) in accordance with the policies of the Institutional Review Board at the University of California, Davis.

Two-Dimensional Gel Electrophoresis and Immunoblot Analysis

The T27K vaccine was separated using two-dimensional gel electrophoresis and then immunoblot analysis was performed using pooled sera from patients with coccidioidomycosis. First dimension isoelectric focusing was carried out with a Zoom IPGRunner System as described in the manual (Invitrogen, Carlsbad, CA, USA). The T27K vaccine (800 μg protein) was dissolved in sample rehydration buffer (8 M urea, 2% (w/v) CHAPS, 2% (v/v) Nonidet P40, 0.5% (v/v) Zoom carrier ampholytes, 0.002% (w/v) bromophenol blue, and 100 mM dithiothreitol) and used to rehydrate a Zoom immobilized pH gradient strip. A step voltage protocol was used for the isoelectric focusing of the strip (200 V for 15 min, 450 V for 15 min, 750 V for 15 min, and 2,000 V for 2 h). The proteins were separated in the second dimension in a Ready Gel precast Tris-HCl polyacrylamide gel (BioRad, Hercules, CA, USA) using standard SDS-PAGE methods and then stained with Gelcode Blue (Pierce, Rockford, IL, USA).[10] A replicate gel was transferred overnight at 30 V to an Immunoblot PVDF membrane (BioRad) in transfer buffer (25 mM Tris, pH 8.3, 192 mM glycine, and 20% methanol). The membrane was incubated with 5% nonfat dried milk (NFDM) in phosphate-buffered saline containing 0.05% (v/v) Tween-20 (PBST) for 1 h. The membrane was incubated with pooled sera diluted 1:100 in PBST containing 5% NFDM for 2 h. The membrane was washed three times with PBST for 10 min and then incubated for 1 h with HRP-conjugated goat anti-human IgG (MP Biomedicals, Solon, OH, USA) antibody diluted 1:2,000 in PBST containing 5% NFDM. The membrane was washed three times with PBST for 10 min and then developed with diaminobenzidine (Sigma-Aldrich, St. Louis, MO, USA).

Protein Identification by Tandem Mass Spectrometry (MS)

Two seroreactive spots detected by immunoblot analysis were selected for internal peptide sequencing using MS. The replicate two-dimensional gel was submitted to the UC Davis Molecular Structure Facility (http://msf.ucdavis.edu) for mass spectrometric analysis. In brief, the two spots of interest were excised from the two-dimensional gel and prepared for MS as previously described.[11] Tryptic peptides were analyzed by a hybrid nanospray/ESI-Quadrupole-TOF-MS and MS/MS in a QSTAR mass spectrometer (Applied Biosystems Inc., Foster City, CA, USA). Obtained values (uninterpreted MS/MS fragmentation data) were used for database searches for protein identification using the Mascot search engine. *De novo* sequencing of peptides was carried out using the QSTAR software (Analyst QS) and confirmed via a manual interpretation of MS/MS spectra.

Rapid Amplification of cDNA Ends (RACE)

The 5′ and 3′ ends of the cDNA encoding the mannosidase-related protein identified in the T27K vaccine by mass spectrometric analysis were mapped using a GeneRacer Kit (Invitrogen) for RACE. Endosporulating spherules were grown as previously described and total RNA was isolated from 24-h cultures using an RNeasy Plant Mini Kit (Qiagen, Valencia, CA, USA).[12] RACE-ready first-strand cDNA was prepared from total RNA using SuperScript II reverse transcriptase and the Oligo dT primer exactly as described in the manual provided with GeneRacer Kit. The design of the gene-specific primers for RACE amplification were based on two peptides ("LSDITGDPEYGR" and "VPEAQAEFYK") generated from mass spectrometric analysis of two seroreactive spots detected on the immunoblot of the T27K vaccine. The nucleotide sequence coding for each peptide was obtained from the institute for Genomic Research (TIGR) *C. posadasii* genome database 8x_1285 contig. For 5′ RACE, the cDNA was amplified with the GeneRacer 5′ primer and a reverse gene-specific primer (5′-GTACTCCGGGTCGCCAGTTATGT-3) derived from the ¨LSDITGDPEYGR¨ peptide. The cDNA (2 µL) was amplified in a 50-µL reaction mix containing 20 mM Tris-HCl (pH 8.4 at 25°C), 50 mM KCl, 1.5 mM MgCl$_2$, 400 nM of each primer, 125 µM of each dNTP, and 2.5 U *Taq* DNA polymerase (Promega, Madison, WI, USA). The thermocycler conditions included an initial denaturation at 94°C for 5 min, 35 cycles at 94°C for 30 sec, 68°C for 30 sec, and 72°C for 1 min, followed by a final extension at 72°C for 7 min. The same amplification conditions were used for 3′ RACE, except the cDNA was amplified using the GeneRacer 3′ primer and a forward gene-specific primer (5′-GAGGCCCAAGCAGAATTCTAC-3′) derived from the ¨VPEAQAEFYK¨ peptide. The RACE PCR products were separated by electrophoresis through an agarose gel (0.8%) and the bands visualized with ethidium bromide after UV illumination. The RACE PCR products were cloned using an Original TA Cloning Kit (Invitrogen) as described in the manual. Plasmid DNA was isolated using a QIAprep Spin Miniprep Kit (Qiagen) and the presence and size of insert DNA was determined using restriction enzyme digestion by *Eco*RI followed by agarose gel electrophoresis. The entire insert DNA was sequenced using vector-specific primers. The plasmid DNA was sequenced at the UC Davis Division of Biological Sciences DNA Sequencing Laboratory (http://dnaseq.ucdavis.edu).

Amplification of Full-Length cDNA

The coding region of the mannosidase-related protein was amplified using a Thermoscript system for reverse-transcription polymerase chain reaction (RT-PCR) (Invitrogen). SE phase total RNA was prepared as described earlier and poly (A)$^+$ RNA was isolated using an Oligotex mRNA Mini Kit (Qiagen). First-strand cDNA synthesis was carried out using an oligo (dT)$_{20}$ primer and Thermoscript reverse transcriptase as described in the manual. The gene-specific primers for RT-PCR were prepared using the nucleotide sequences obtained from RACE mapping of the 5′ and 3′ ends of the mannosidase cDNA. First-strand cDNA (2 µL) was amplified in a 50-µL reaction mix containing the following: 20 mM Tris-HCl, (pH 8.8), 2 mM $MgSO_4$, 1 mM $MgCl_2$, 10 mM KCl, 10 mM $NH_4(SO_4)_2$, 0.1% (v/v) Triton X-100, 100 μg/mL bovine serum albumin, 1 μM of each primer (5′-ATGAAGGGATCCCCCGTA-3′, 5′-ACACCCCTGCTGAAGAC-3′), 200 μM of each dNTP and 1.25 U of Platinum *Taq* DNA Polymerase. The thermocycler conditions were an initial denaturation at 94°C for 5 min, followed by 50 cycles at 94°C for 30 sec, 60°C for 30 sec, and 72°C for 1 min, and a final 72°C for 10 min. The RT-PCR reaction was separated by electrophoresis through an agarose gel (0.8%) and the bands visualized with ethidium bromide after UV illumination. The RT-PCR product was cloned and plasmid DNA isolated and evaluated for insert DNA as described earlier for the RACE PCR products. Clones that contained insert DNA were cultured on a large scale and plasmid DNA was isolated using a Plasmid Midi Kit (Qiagen). The entire insert DNA was sequenced as described earlier using vector-specific and gene-specific primers. The gene-specific primers for DNA sequencing were designed using the partial cDNA sequence information obtained from RACE analysis.

Sequence Analysis

C. posadasii sequence data were obtained from the Institute for Genomic Research (TIGR) through the website at http://www.tigr.org. The theoretical isoelectric point and molecular weight were determined using the Compute pI/MW tool at the ExPASy server (http://www.expasy.org). The SignalP program at the Center for Biological Sequence Analysis (CBSA) website (http://www.cbs.dtu.dk) was used for signal peptide and cleavage site prediction and NetNGlyc at the CBSA site was used for the prediction of *N*-glycosylation sites. A hydropathy plot was generated using the Kyte-Doolittle method.[13] The TMpred program at the ch.EMBnet.org website was used to predict the presence of transmembrane α-helices. Multiple sequence alignment was performed using the ClustalW software located at the European Bioinformatics Institute website, http://www.ebi.ac.uk. The National Center for Biotechnology Information GenBank database accession numbers for the *C. posadasii* mannosidase cDNA nucleotide sequence and deduced amino acid sequence are DQ233502 and ABB36773, respectively.

Expression and Purification of Recombinant Fusion Protein

The cDNA encoding the mannosidase was expressed as a recombinant fusion protein using a pEXP5-NT/TOPO TA Expression Kit (Invitrogen). The cDNA was amplified as described earlier for RT-PCR and the product was inserted into a pEXP5-NT/TOPO expression vector containing a bacteriophage

T7 promoter and an N-terminal polyhistidine tag. TOP10 chemically competent *E. coli* cells (Invitrogen) were transformed with the expression plasmid as described in the manual. Several transformants were picked and cultured at 37°C overnight in 5 mL of Luria Bertani (LB) broth containing ampicillin (100 µg/mL). Plasmid DNA was isolated and evaluated for insert DNA as described earlier except *Xba*I and *Eco*RI were used for restriction enzyme digestion. The 5′ end of the plasmid DNA from transformants with inserts of the appropriate size was sequenced using a T7 forward primer. Clones that contained mannosidase cDNA in-frame with the N-terminal tag were cultured on a large scale and plasmid DNA was isolated using a Plasmid Midi Kit (Qiagen). The entire insert DNA was sequenced using gene-specific primers. An expression plasmid (pEXP5-NT-ManI) with the correct cDNA sequence in frame with N-terminal tag was selected for expression experiments. Chemically competent BL21 *E. coli* cells (Invitrogen) were transformed with pEXP5-NT-ManI (10 ng), added to 10 mL of LB medium containing ampicillin (100 µg/mL) and cultured at 37°C overnight with shaking ($OD_{600} = 1.0$). Fresh medium was inoculated with the overnight culture to an $OD_{600} = 0.1$ and incubated at 37°C with shaking for 3 h ($OD_{600} = 0.8$). Expression of the recombinant protein was induced by adding 0.5 mM isopropyl-β-D-thiogalactopyranoside (Fisher Scientific, Pittsburgh, PA, USA) and the cultures were allowed to incubate for an additional 4 h. The cells were harvested and the cell pellet from 100 mL of culture was resuspended in 6 mL of resuspension buffer (50 mM sodium phosphate buffer [pH 8.0] and 500 mM sodium chloride) containing 0.01 mg/mL DNase (Calbiochem, San Diego, CA, USA) and 0.1 mg/mL lysozyme (Sigma-Aldrich) and incubated on ice for 30 min. The cells were then sonicated (Sonic Dismembrator, Model 100, Fisher Scientific) on ice four times at a setting of 5 for 10-sec intervals and centrifuged at 6,000 × g for 15 min at 4°C. The cells were resuspended in 6 mL of cold isolation buffer (50 mM sodium phosphate buffer [pH 8.0], 500 mM sodium chloride, and 2 M urea) containing 2% (v/v) Triton X-100 and sonicated and centrifuged as described earlier and then repeated. The pellet was washed in cold isolation buffer, resuspended in 1 mL of cold isolation buffer and added dropwise to 5 mL of binding buffer (50 mM sodium phosphate buffer [pH 8.0], 500 mM sodium chloride, and 6 M guanidine hydrochloride) and stirred overnight at 4°C. The lysate was centrifuged at high speed for 15 min at 4°C and the supernatant was filtered through a 0.45-µm syringe filter. The recombinant mannosidase fusion protein was then purified from the cleared lysate by fast protein liquid chromatography (FPLC) with a HiTrap Chelating HP 1-mL column (GE Healthcare, Piscataway, NJ, USA). After equilibrating the column with 10 mL of binding buffer, the lysate was applied to the column and washed with 10 mL of binding buffer. The protein was then eluted from the column by means of a step pH gradient consisting of binding buffer (10 mL each) at pH 6.0, pH 5.3, and pH 4.0. The eluate was collected in 1-mL fractions and an aliquot of each was evaluated for the presence of pure recombinant mannosidase using SDS-PAGE and Western blot analysis. SDS-PAGE was performed using standard methods and the gels were stained with Gelcode Blue (Pierce).[10] For Western blot analysis, a replicate gel was preequilibrated in transfer buffer (25 mM Tris, pH 8.3, 192 mM glycine, and 20% methanol) for 15 min and then transferred overnight at 4°C to nitrocellulose at 30 V. The membrane was incubated with blocking buffer (5% (w/v) nonfat dried milk in phosphate-buffered saline containing 0.05% (v/v) Tween-20) for 1 h and then incubated for 1 h with anti-HisG-HRP antibody (Invitrogen) diluted 1:5000 in blocking buffer. The membrane was washed three times with PBST for 10 min each and then developed with tetramethylbenzidine (Pierce).

Mannosidase Enzyme Activity Assay

Class I 1,2-α-mannosidase enzyme activity was measured using the substrate, Man-α-1,2-Man-α-OCH$_3$ in a spectrophotometric assay.[14,15] Jack Bean mannosidase (Sigma-Aldrich) was used as a positive control and deoxymannojirimycin hydrochloride (Axxora, San Diego, CA, USA) was used as an inhibitor at a final concentration of 0.5 mM. Mannosidase activity was quantified as the amount of mannose generated per microgram of protein per hour. Class II mannosidase activity was also assayed using either p-nitrophenyl α-D-mannopyranoside (Sigma-Aldrich) or 4-methylumbelliferyl-α-D-mannopyranoside (Sigma-Aldrich) as substrate.[16,17]

RESULTS

Identification of 1,2-α-Mannosidase in the Coccidioidal T27K Vaccine Using Immunoproteomic Methods

The coccidioidal T27K vaccine was separated using two-dimensional gel electrophoresis, and immunoblot analysis was performed with pooled sera from patients positive for the coccidioidal complement fixation (IgG) antibody. A representative gel and immunoblot are shown in Figure 1. A number of seroreactive spots were detected on the immunoblot including two spots with a migration just above the 50-kDa protein standard. The corresponding region on the replicate gel was selected for internal peptide sequencing using tandem MS and three peptides were generated ("LSDITGDPEYGR," "TIDIETGLFR," and "VPEAQAEFYK"). All three peptides matched the TIGR *C. posadasii* genome database contig 8x_1285 and two peptides ("LSDITGDPEYGR" and "TIDIETGLFR") matched the TIGR *C. posadasii* 2.0 gene index tentative consensus sequence, TC2332. A BLAST search of the National Center for Biotechnology Information (NCBI) GenBank database showed that contig 8x_1285 and TC2332 shared homology with 1,2-α-mannosidases.

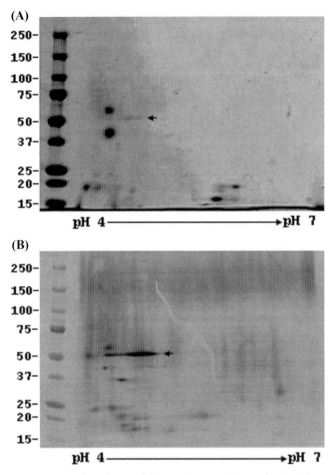

FIGURE 1. Electrophoretic separation and immunoblot analysis of the coccidioidal T27K vaccine. Coomassie blue–stained two-dimensional gel of the coccidioidal T27K vaccine (**A**) and corresponding immunoblot using pooled sera from patients positive for the complement fixation antibody (**B**). The *arrow* indicates the position of the seroreactive region that was sequenced using mass spectrometry (MS). The numbers to the left of the images indicate the molecular weight in kDa of the adjacent protein standards.

Cloning and Sequencing of the cDNA Encoding the 1,2-α-Mannosidase Protein Detected in the T27K Vaccine Using Immunoproteomic Methods

The nucleotide sequence of the cDNA encoding the 1,2-α-mannosidase protein detected in the T27K vaccine was obtained using a combination of RACE and RT-PCR methods. RACE analysis was first performed to map the 5′ and 3′ ends of the mannosidase cDNA. Two amplification reactions were performed with RACE-ready cDNA prepared from 24-h SE phase RNA and either the GeneRacer 5′ primer and a reverse gene-specific primer based on the

"LSDITGDPEYGR" peptide obtained from MS (5′ RACE) or the GeneRacer 3′ primer and a forward gene-specific primer based on the "VPEAQAEFYK" peptide (3′ RACE). A single amplification product was obtained for both the 5′ and 3′ RACE reactions (data not shown) and the two products were cloned and DNA sequenced. The 5′ RACE product consisted of an 803-bp fragment containing 44 bp corresponding to the GeneRacer RNA Oligo and 759 bp of coccidioidal-specific DNA. Translation of the cDNA in an open reading frame that contained the "LSDITGD" peptide sequence revealed that the initiation codon was the first "ATG," 99 nucleotides downstream from the 5′ end. For the 3′ end, the RACE PCR product consisted of a 653-bp DNA sequence containing 54 bp corresponding to the GeneRacer Oligo dT primer and 599 bp of coccidioidal-specific DNA. Translation of the cDNA revealed a partial open reading frame that contained the "EAQAEFYK" peptide sequence and a termination codon. The complete coding region was then determined using RT-PCR with forward and reverse primers designed using the nucleotide sequence from 5′ and 3′ RACE. A single RT-PCR amplification product was obtained (data not shown) and the product was cloned and DNA sequenced. A 1560-bp DNA fragment was obtained, which included sequence that overlapped with the DNA fragments from RACE and represented an open reading frame that included the three peptides from MS. Together the DNA fragments obtained from the RACE mapping and RT-PCR formed an 1866-bp cDNA sequence with a 1560-bp open reading frame flanked by 99 bp of a 5′ untranslated region (UTR) and a 3′ UTR of 207 bp. The deduced amino acid sequence consists of 519 amino acids and has a theoretical molecular weight and isoelectric point of 56,918 Da and 4.84, respectively (FIG. 2).

Analysis of the Deduced Amino Acid Sequence and Comparison to Other Fungal 1,2-α-Mannosidases

A BLAST search of the NCBI GenBank database confirmed that the *C. posadasii* predicted protein was related to the class I 1,2-α-mannosidase gene family. The deduced amino acid sequence was aligned with other fungal mannosidase proteins identified by the BLAST search (FIG. 3). ClustalW alignment analysis showed that the deduced amino acid sequence shares between 59% and 62% identity to other fungal 1,2-α-mannosidases. The highest percentage of identity (62%) was to the *Aspergillas saitoi* 1,2-α-mannosidase (GenBank accession no. BAA08634), followed by 61% identity to both the *Penicillium citrinum* 1,2-α-mannosidase (GenBank accession no. BAA08275) and the *Aspergillus fumigatus* mannosidase I (GenBank accession no. AAS77884). The sequence also shares 59% identity to form IB of the *Aspergillus nidulans* class I α-mannosidase gene family (GenBank accession no. AAG48159). The deduced amino acid sequence contains a number of conserved regions that include a glutamic acid residue

```
  -99           atacataagtatatactctctcctcttttagacgtgcctagggagt
  -54   acggagtccgatcctttgaagaggggggaaaacaaaaaattatagtaataaag
    1   ATGAAGGGATCCCCCGTACTCGCCGTATGCGCTGCAGCGCTGACGCTCATTCCA
    1    M  K  G  S  P  V  L  A  V  C  A  A  A  L  T  L  I  P
   55   TCTGTCGTTGCCCTTCCCATGATTGATAAGGACCTCCCAAGCTCCATCAGTCAA
   19    S  V  V  A  L  P  M  I  D  K  D  L  P  S  S  I  S  Q
  109   TCTTCCGACAAGACGAGTCAAGAACGAGCCGAAGCCGTCAAAGATGCCTTCAGA
   37    S  S  D  K  T  S  Q  E  R  A  E  A  V  K  D  A  F  R
  163   TTTGCCTGGGAAGGGTATTTGGAACATGCATTCCCAAACGACGAGTTACATCCG
   55    F  A  W  E  G  Y  L  E  H  A  F  P  N  D  E  L  H  P
  217   GTGTCCAATACGCCCGGCAATTCTCGGAACGGCTGGGGCGCTTCTGCCGTCGAT
   73    V  S  N  T  P  G  N  S  R  N  G  W  G  A  S  A  V  D
  271   GCGCTCTCGACCGCGATCATCATGGACATGCCAGACGTCGTCGAGAAGATTCTC
   91    A  L  S  T  A  I  I  M  D  M  P  D  V  V  E  K  I  L
  325   GACCATATCTCCAACATCGACTACTCCCAAACAGATACCATGTGTAGCCTGTTT
  109    D  H  I  S  N  I  D  Y  S  Q  T  D  T  M  C  S  L  F
  379   GAAACCACCATTCGCTACCTAGGAGGCATGATATCCGCCTATGATCTGCTGAAG
  127    E  T  T  I  R  Y  L  G  G  M  I  S  A  Y  D  L  L  K
  433   GGACCTGGTTCACATCTTGTGTCGGATCCTGCAAAAGTCGACGTGCTGCTCGCG
  145    G  P  G  S  H  L  V  S  D  P  A  K  V  D  V  L  L  A
  487   CAATCGCTGAAACTGGCCGATGTACTGAAGTTCGCCTTCGATACAAAGACTGGC
  163    Q  S  L  K  L  A  D  V  L  K  F  A  F  D  T  K  T  G
  541   ATACCGGCAAACGAGTTGAATATCACGGATAAGTCTACGGACGGTTCGACAACC
  181    I  P  A  N  E  L  N  I  T  D  K  S  T  D  G  S  T  T
  595   AACGGGCTCGCCACAACTGGCACCCTAGTCTTGGAGTGGACTCGTCTCTCGGAC
  199    N  G  L  A  T  T  G  T  L  V  L  E  W  T  R  L  S  D
  649   ATAACTGGCGACCCGGAGTACGGCAGGCTAGCACAGAAGGGGGAATCGTACCTC
  217    I  T  G  D  P  E  Y  G  R  L  A  Q  K  G  E  S  Y  L
  703   CTCAACCCCACAACCGTCATCGACGGAGCCATTTCCCGGCCTGGTTGGTCGCACT
  235    L  N  P  Q  P  S  S  S  E  P  F  P  G  L  V  G  R  T
  757   ATTGACATTGAGACGGGCCTATTCCGCGATGATTATGTCAGTTGGGGAGGAGGA
  253    I  D  I  E  T  G  L  F  R  D  D  Y  V  S  W  G  G  G
  811   TCGGATTCGTTTTACGAGTATCTTATCAAGATGTACGTCTATGACAAGGGCCGG
  271    S  D  S  F  Y  E  Y  L  I  K  M  Y  V  V  Y  D  K  G  R
  865   TTTGGAAAGTACAAGGACCGGTGGGTGACTGCCGCCGAATCCACCATTGAACAC
  289    F  G  K  Y  K  D  R  W  V  T  A  A  E  S  T  I  E  H
  919   TTAAAATCTTCGCCTTCTACGAGAAAGGACTTGACATTTGTGGCGACGTATTCC
  307    L  K  S  S  P  S  T  R  K  D  L  T  F  V  A  T  Y  S
  973   GGGGGGAGACTCGGTCTCAACTCTGGTCATTGACATGCTTTGATGGCGGTAAT
  325    G  G  R  L  G  L  N  S  G  H  L  T  C  F  D  G  G  N
 1027   TTCCTCCTGGGAGGCCAGATACTTAACCGAGACGACTTCACCAAGTTTGGGCTC
  343    F  L  G  G  Q  I  L  N  R  D  D  F  T  K  F  G  L
 1081   GAGCTTGTTGAAGGCTGCTACGCTACGTATGCTGCGACCGCAACGAAAATCGGC
  361    E  L  V  E  G  C  Y  A  T  Y  A  A  T  A  T  K  I  G
 1135   CCCGAGGGATTCGGCTGGGATGCCACGAAGGTCCCTGAGGCCCAAGCAGAATTC
  379    P  E  G  F  G  W  D  A  T  K  V  P  E  A  Q  A  E  F
 1189   TACAAGGAGGCCGGGTTCTATATCACAACCAGCTACTATAACCTCCGACCGGAG
  397    Y  K  E  A  G  F  Y  I  T  T  S  Y  Y  N  L  R  P  E
 1243   GTCATCGAGAGCATCTATTATGCTTACCGAATGACGAAAGATCCTAAGTATCAA
  415    V  I  E  S  I  Y  Y  A  Y  R  M  T  K  D  P  K  Y  Q
 1297   GAGTGGGCATGGGATGCCTTCGTGCCGATCAACGCAACGACGCGCACGAGCACC
  433    E  W  A  W  D  A  F  V  A  I  N  A  T  T  R  T  S  T
 1351   GGTTTCACCGCCATTGGGGACGTCAATACGCCCAGACGGAGGTCGAAAGTATGAC
  451    G  F  T  A  I  G  D  V  N  T  P  D  G  G  R  K  Y  D
 1405   AACCAAGAGAGCTTCCTCTTTGCTGAAGTGATGAAGTACTCCTATCTCATCCAC
  469    N  Q  E  S  F  L  F  A  E  V  M  K  Y  S  Y  L  I  H
 1459   TCACCAGAAGCAGACTGGCAGGTAGCCGGTCCCGGAGGAACGAATGCATACGTA
  487    S  P  E  A  D  W  Q  V  A  G  P  G  G  T  N  A  Y  V
 1513   TTCAACACCGAAGCACATCCGGTGAAGGTCTTCAGCAGGGGGTGTTGAtcctgc
  505    F  N  T  E  A  H  P  V  K  V  F  S  R  G  C  *
 1567   gttctgtcatagcgcgcagagatgcggaagtttcgtgacgacttgctggacctt
 1621   tttttttttttgcgacggagagtagttaatgaggtcgagcaaattgctctgtagg
 1675   agcttggcgagtgtgttttggacgtagtactccgtagccagcccacagcaaaat
 1729   ccaccctatgaatgaaaacgcgtcagcacatctgtacataa
```

FIGURE 2. The nucleotide sequence of the *C. posadasii* 1,2-α-mannosidase cDNA and deduced amino acid sequence. The numbers to the left of each line indicate either DNA base or predicted amino acid positions. The *asterisk* indicates the termination codon. The 5′ and 3′ untranslated regions are indicated by lowercase letters. The *italicized* sequence is the predicted cleavable signal peptide (amino acids 1–22). The underlined amino acids indicate the peptide sequences generated by MS.

```
Cp MNGSPVLAVCAAALTLIPSVVALPMIDKDLP SSISQS SDKTSQERAEAVKDAF   53
As -MHLPSLSLSLTALAIASPSAAYPH--FGSSQPVLHSSSDTTQSRADAIKAAF     50
Pc MRLPVSFPLT----VLSLLGSTIAHP-YGETEAVLRSEPKSNQAKADAVKEAF     48
Af -MHLPSLSV-----ALALVSSSLAL--PQAVLPENDVSS-----RAAAVKEAF     40
An MRTLLALAA-----FAGFAAARVPA--YAITRPVMRSDS-----RADAVKEAF     41
                                    :.      :* *:* **

Cp RFRWEGYLEHAFPNDELHPVSNTPGNSRNGWGASAVDALSTAIIMDMPDVVEK    106
As SHAWDGYLQYAFPHDELHPVSNGYGDSRNGWGASAVDALSTAVIMRNATIVNQ    103
Pc QHAWNGYMKYAFPHDELTPVSNGHADSRNGWGASAVDALSTAVIMGKADVVNA    101
Af SHAWDGYMKYAFPHDELLPVSNSYGDSRNGWGASAVDALSTAIVMRNATIVSQ     93
An SHAWDGYNYAFPHDELHPISNGYGDSRNHWGASAVDALSTAIMMRNATIVNQ      94
   .**:**   ::  ***:*** *:*:      .:*** ************:*   . :*.

Cp ILDHISNIIDYSQTDTMCSLFETTIRYLGGMISRYDLLKGPGSHLVSDPAKVDV   159
As ILDHVGKIDYSKTNTTVSLFETTIRYLGGMLSGYDLLKGPVSDLVQNSSKIDV    156
Pc ILEHVADIDFSKTSDTVSLFETTIRYLAGMLSCYDLLQGPAKNLVDNQDLIDG    154
Af ILDHIAKIDYSKTSDMVSLFETTIRYLSGMLSGYDLLKGPAADLVEDRTKVDM    146
An ILDHIAAVDYSKTNAMVSLFETTIRYLAGMISGYDLLKGPAAGLVDD-SRVDV    146
   **:.*:   :*:.*:* **********_.**:*_*.****:** **:  .:*

Cp LLAQSLKLADVLKFAFDTKTGIPANELNITDKSTDGSTTNGLATTGTLVLEWT   212
As LLTQSRNLADVLKFAFDTPSGVPYNNLNITSGGNDGAKTNGLAVTGTLALEWT   209
Pc LLDQSRNLADVLKFAFDTPSGVPYNNINITSHGNDGATTNGLAVTGTLVLEWT   207
Af LLQQSKNLGDVLKFAFDTPSGVPYNNINITSHSNDGATTNGLAVTGTLVLEWT   199
An LLEQSQNLAEVLKFAFDTPSGVPYNNMINITSGGNDGATTNGLAVTGTLVLEWT   199
   ** **  :*.:******** :*:* * .***_.   **:.***** .**** ****

Cp RLSDITGDPEYGRLAQKGE SYLLNPQPSSSEPFPGLVGRTIDIETGLFRDIYV   265
As RLSDLTGDTTYADLSQKAESYLLNPQPKSAEPFPGLVGSNINISNGQFTDAQV    262
Pc RLSDLTGDEYAKLSQKAESYLLKPQSSSEPFPGLVGSININDGQFADSRV      260
Af RLSDLTGDQEYAKLSQWAESYLLAPQPSSGEPFPGLVGSAISIQTGQFTNGFV   252
An RLSDLTGNDEYARLSQPAEDYLLHPEPAQYEPFPGLIGSAVNIADGKLANGHI   252
   ****:**:     *.  *.*.*_.***_*:*_. ******:*   :.*   *:: :

Cp SWGGGSDSFYEYLIKMYVYDNGRFGKYKDRWVTAAESTIEHLKSSPSTRKDLT   318
As SWNGGDDSYYEYLIKMYVYDPKRFGLYKDRWVAAAQSTMQHLASHPSSRPDLT   315
Pc SWNGGDDSFYEYLIKMYVYDPKRFETYKDRWVLAAESTIKHLKSHPKSRPDLT   313
Af SWNGGSDSFYEYLMRMYVYDPKRFATYKDRWVATAESSIDHLASNPASRPDLT   305
An SWNGGADSYYEYLIKMYVYDPERFGLYKDRWVAAAESSINHLASHPSTRPDVT   305
   **.** **:****:****** ** *:**** :*:*:.. ** * * :* *:*

Cp FVRTYSGG-RLGLNSGHLTCFDGGNFLLGGQILNRDDFTKFGLELVEGCYRTY   370
As FLASYNNG-TLGLSSQHLTCFDGGSFLLGGTVLNRTDFINFGLDLVSGCHDTY   367
Pc FLSSYSNR-NYDLSSQHLTCFDGGSFLLGGTVLDRQDFIDFGLKLVDGCEATY   365
Af FLATYNKG-SLGLSSQHLACFDGGSYLLGGTVLDRADLIGFGLKLVDGCAETY   357
An FLATYNEEHQLGLTSQHLTCFDGGSFLLGGTLLDRQDFVDFGLDLVAGCHETY   358
   *:::*.    .*.*_**:*****.:****_:*:*.* *:   ***.** ** **

Cp ARTATKIGPEGFCVWDATKVPERQAEFYKEAGFYITTSYYNLRPEVIESIYYAY   423
As NSTLTGIGPESFSWDTSDIPSSQQSLYEKAGFYITSGAYILRPEVIESFYYAW   431
Pc NSTLTKIGPDSWGWDPKRVPSDQKEFYEKAGFYISSGSYVLRPEVIESFYYAH   418
Af HQTLTGIGLESFGWDEKSVPADQKELYERAGFYVQSGAYILRPEVIESFYYAY   410
An NSTLTGIGPEQFSWDPNGVPDSQKELFERAGFYINSGQYILRPEVIESFYYAW   411
   * * **  :: .**  .:* *.:::   ****: :  *  **********.***

Cp RMTKDPKYQEVBWDAF VAINATTRT STGFTAIGDVNTPDGGRKYDNQESFLFA   476
As RVTGQETYRDWIWSAFSAVNDYCRTSSGFSGLTDVNAANGGSRYDNQESFLFA   473
Pc RVTGKEIYRDWVWNAFVAINSTCRTDSGFAAVSDVMKANGGSRYDNQESFLFA   471
Af RVTGKKQYRDWVWNAFKNINKYCRTESGFAGLTNVNAVNGGGRYDNQESFLFA   463
An RVTGDGTYLEWVWNAFTNINKYCRTATGFAGLENVNAANGGGRIDNQESFMFA   464
   *:*. .  *:* *.**  :*    **  :**   *   :** :*** ******:.**

Cp EVMKYSYLIHSPERDWKQVAGPGGTNRYVFNTKAHPVKVESRGC   519
As EVMKRSYMAFAEDAAWQVQ-PGSCNQFVFNTEAHPVRVSST--   513
Pc EVMKRSYLAHSEDAAWQVQ-KGGKNTFVYNTEAHPISVARN--   511
Af EVMKYAYLTHAPEDEWQVQ-RGSGNKFVYNTEAHPVRIHHT--   503
An EVLKYSFLTFAPEDDWQVQ-KGSGNTFVYNTEAHPFKVYTPQ-   505
   **:**:::  .: :   ***   *. * :*:******. :
```

FIGURE 3. Multiple amino acid sequence alignment of fungal mannosidase proteins from *C. posadasii* (Cp), *A. saitoi* (As), *P. citrinum* (Pc), *A. fumigatus* (Af), and *A. nidulans* (An). The numbers to the right of the lines indicate amino acid positions. An *asterisk* indicates an amino acid identity, a *colon* (:), a conserved substitution, and a *dot* (.), a semiconserved substitution.

(Glu-127), that is, the putative carboxylic acid involved in the catalytic mechanism and two cysteine residues (Cys-337 and Cys-366) that form a disulfide bond and contribute to the thermostability of the enzyme.[18,19] A hydropathy plot of the deduced amino acid sequence revealed a region of hydrophobicity at the N-terminal end, and a transmembrane α-helix was indicated by the TMpred program (amino acids 6–23). An overlapping region (amino acids 1–22) was predicted to be a cleavable signal peptide by SignalP software. Analysis using NetNGlyc showed two potential *N*-glycosylation sites at amino acid positions 187 (NITD) and 443 (NATT).

Expression and Purification of Recombinant C. posadasii *1,2-α-Mannosidase*

Recombinant *C. posadasii* 1,2-α-mannosidase was expressed in bacteria as an N-terminal 6xHis tagged fusion protein using a pEXP5-NT/TOPO expression system. Expression of the fusion protein was monitored using SDS-PAGE and Western blot analysis with an anti-HisG antibody, which detects the 6xHis tag followed by glycine. A prominent band of the expected size for the fusion protein was observed in the induced culture between the 41- and 83-kDa standards (FIG. 4A, lane 2). In addition, a positive band of the same size was detected on the Western blot (FIG. 4B, lane 2). A single band of the expected size also reacted with the anti-HisG antibody in the uninduced bacterial cul-

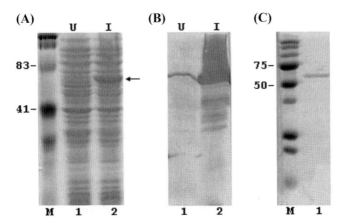

FIGURE 4. Expression and purification of recombinant *C. posadasii* 1,2-α-mannosidase fusion protein. Coomassie blue–stained SDS PAGE gel (**A**) and corresponding Western blot (**B**) of bacterial cell lysates from uninduced and induced cultures at 4 h. Lane 1, uninduced (U); lane 2, induced (I); lane M, kaleidoscope protein standards (BioRad). Coomassie blue–stained SDS PAGE gel of purified recombinant mannosidase (**C**). Lane 1, purified recombinant mannosidase; lane M, Precision Plus protein standards (BioRad). The numbers to the left of the gels indicate the size in kDa of the adjacent protein standard. The *arrow* indicates the position of the recombinant fusion protein.

tures (FIG. 4A, B, lane 1). The pEXP5-NT/TOPO system is a T7-based system and basal expression of T7 RNA polymerase occurs in the absence of inducer. Analysis of the soluble and insoluble proteins in the culture demonstrated that the fusion protein was present in the inclusion bodies of the bacteria (data not shown). Inclusion bodies were isolated from an induced culture and the fusion protein was purified under denaturing conditions. The fusion protein was further purified using a HiTrap chelating column and FPLC. A highly purified fraction was obtained after FPLC (FIG. 4C).

Analysis of Mannosidase Activity in the Coccidioidal T27K Vaccine

Class I 1,2-α-mannosidase activity was measured in the T27K vaccine and culture filtrate using the disaccharide substrate, Man-α-1,2-Man-α-OCH$_3$ in a spectrophotometric assay. Activity was detected in the T27K vaccine (0.58 nmol/μg protein/h) and preincubation of the vaccine with the class I inhibitor deoxymannojirimycin reduced the activity by 80%. Mannosidase activity was not detected in the culture filtrate obtained from the SE phase cultures used to prepare the vaccine. No mannosidase activity was detected in the T27K vaccine or culture filtrate when the synthetic class II substrates, p-nitrophenyl α-D-mannopyranoside or 4-methylumbelliferyl-α-D-mannopyranoside were used in standard assays.

DISCUSSION

A previously uncharacterized seroreactive protein that is related to the class I 1,2-α-mannosidase gene family (glycosylhydrolase family 47) has been identified in the coccidioidal T27K vaccine by immunoproteomic methods. A full-length cDNA encoding the protein was obtained and a predicted protein was generated from the largest open reading frame. The deduced protein sequence consists of 519 amino acids and has a theoretical molecular weight and pI of 56,918 Da and 4.84, respectively. The deduced amino acid sequence confirmed that the protein is homologous to 1,2-α-mannosidases with the highest identity (62%) to the *A. saitoi* protein (GenBank accession no. BAA08634). Class I 1,2-α-mannosidase activity was also detected in the vaccine.

Class I 1,2-α-mannosidase enzymes (glycosylhydrolase family 47) catalyze the cleavage of 1,2-α-linked D-mannose residues from the asparagine-linked oligosaccharide intermediate, Man$_9$GlcNAc$_2$ during the maturation process of glycoprotein N-glycans.[20] The class I 1,2-α-mannosidase genes have been cloned from the filamentous fungi *A. saitoi, A. nidulans, P. citrinum,* and *Trichoderma reesei* and the fungal enzyme has been expressed as a recombinant protein.[15,21–25] The structural and enzymatic properties of the *A. saitoi, T. reesei,* and *P. citrinum* class I 1,2-α-mannosidases have also been studied extensively.[18,19,24,26–28] The filamentous fungal enzymes belong to a functionally distinct subgroup of the class I enzymes that include the mammalian Golgi 1,2-α-mannosidases.[20] The enzymes in this subgroup produce a $Man_5GlcNAc_2$ intermediate that is required for the production of complex and hybrid *N*-glycans.[20]

The deduced *C. posadasii* sequence 1,2-α-mannosidase protein sequence shares about 60% identity with other fungal class I 1,2-α-mannosidases and contains a number of conserved regions including a conserved glutamic acid residue at position 127 and two conserved cysteine residues (Cys-337 and Cys-366).[18,19] The N-terminal region of the deduced amino acid sequence contains a hydrophobic amino acid sequence, a putative transmembrane α-helix, and a predicted cleavable signal peptide. The presence of a cleavable signal peptide seems uncharacteristic since the predicted protein shares homology with the class I 1,2-α-mannosidases family, which are type II transmembrane proteins; however, this is a common feature for the fungal enzymes.[21,22,25] It appears that fungal 1,2-α-mannosidases that share homology with the class I family constitute a unique subgroup of enzymes that are secreted.

Although the fungal class I 1,2-α-mannosidases belong to a group of enzymes that are described as type II membrane proteins, a number of studies have provided evidence that supports the notion that the fungal proteins are secreted. For example, class I 1,2-α-mannosidase activity has been detected in the culture filtrate from *A. saitoi* and *P. citrinum* and the enzyme has been purified from the culture filtrate.[29,30] Furthermore, the soluble form of the enzyme purified from *P. citrinum* lacked the hydrophobic signal sequence.[30] Class I 1,2-α-mannosidase activity was also detected in the culture filtrate from *A. nidulans*, although the majority of activity was found in intracellular protein extracts.[15] Class I 1,2-α-mannosidase activity was detected in the T27K vaccine, suggesting that, at least during the SE phase, *C. posadasii* contains an active intracellular form of the enzyme. Class I mannosidase activity was not detected in the culture filtrate from the SE phase cultures used to prepare the vaccine, although it is possible that the enzyme is not released to the extracellular medium during the 33-h SE phase growth or it is released at levels below the detection limits of the activity assay. Additional studies are required to determine whether or not the *C. posadasii* class I 1,2-α-mannosidase is secreted.

The *C. posadasii* 1,2-α-mannosidase protein was identified by mass spectrometric analysis of two different spots on the two-dimensional gel of the T27K vaccine. The location of the two spots indicated species with identical molecular weights, but slightly different isoelectric points, and suggested the existence whether isoforms of the enzyme. Further analysis is necessary to determine whether isoforms exist, and if so, how the isoforms are generated (e.g., posttranslation modification). The existence of two isoenzymes of 1,2-α-mannosidase has been reported for *P. citrinum*, although the exact mechanism giving rise to the two isoenzymes was not definitively determined.[30]

In summary, a protein with homology to the class I family of 1,2-α-mannosidase enzymes was identified in the coccidioidal T27K vaccine by means of immunoproteomic methods, and the cDNA encoding the protein was cloned, sequenced, and expressed as a recombinant fusion protein. Class I 1,2-α-mannosidase activity was also detected in the vaccine which is prepared from the SE phase of the fungus. Further studies are warranted to evaluate the immunogenicity of the *C. posadasii* 1,2-α-mannosidase protein and to determine whether or not it contributes to the protective effect of the T27K vaccine.

ACKNOWLEDGMENTS

This work was supported by a grant from the California HealthCare Foundation, the Department of Health Services of the State of California and California State University, Bakersfield.

REFERENCES

1. PAPPAGIANIS, D. 2001. Seeking a vaccine against *Coccidioides immitis* and serologic studies: expectations and realities. Fungal Genet. Biol. **32:** 1–9.
2. LEVINE, H.B., J.M. COBB & C.E. SMITH. 1960. Immunity to coccidioidomycosis induced in mice by purified spherule, arthrospore, and mycelial vaccines. Trans. N. Y. Acad. Sci. **22:** 436–449.
3. LEVINE, H.B., Y-C.M. KONG & C.E. SMITH. 1965. Immunization of mice to *Coccidioides immitis*: dose, regimen and spherulation stage of killed spherule vaccine. J. Immunol. **94:** 132–142.
4. PAPPAGIANIS, D., R. HECTOR, H.B. LEVINE & M.S. COLLINS. 1979. Immunization of mice against coccidioidomycosis with a subcellular vaccine. Infect. Immun. **25:** 440–445.
5. ZIMMERMANN, C.R., S.M. JOHNSON, G.W. MARTENS, *et al.* 1998. Protection against lethal murine coccidioidomycosis by a soluble vaccine from spherules. Infect. Immun. **66:** 2342–2345.
6. JOHNSON, S.M., K.M. KEREKES, C.R. ZIMMERMANN, *et al.* 1999. Identification and cloning of an aspartyl proteinase from *Coccidioides immitis*. Gene **241:** 213–222.
7. LUNETTA, J.M., K.A. SIMMONS & D. PAPPAGIANIS. 2005. Cloning and expression of the gene encoding a mannosidase protein detected in the T27K coccidioidal vaccine. Presented at the 49th Annual Meeting of the Coccidioidomycosis Study Group. Bass Lake, CA, April 2.
8. LUNETTA, J., S. JOHNSON & D. PAPPAGIANIS. 2006. Identification of a class I 1,2-α-mannosidase protein in the coccidioidal T27K vaccine using immunoproteomic methods. Presented at the Sixth International Symposium on Coccidioidomycosis. Stanford, CA, August 23–26.
9. FISHER, M.C., G.L. KOENIG, T.J. WHITE & J.W. TAYLOR. 2002. Molecular and phenotypic description of *Coccidioides posadasii* sp. nov., previously recognized as the non-California population of *Coccidioides immitis*. Mycologia **94:** 73–84.
10. LAEMMLI, U.K. 1970. Cleavage of structural proteins during the assembly of the head of bacteriophage T4. Nature **227:** 680–685.

11. SHEVCHENKO, A., M. WILM, O. VORM & M. MANN. 1996. Mass spectrometric sequencing of proteins from silver stained polyacrylamide gels. Anal. Chem. **68:** 850–858.

12. ZIMMERMANN, C.R. & D. PAPPAGIANIS. 1994. Extraction and purification of poly $(A)^+$ RNA from *Coccidioides immitis* spherule/endospore and hyphal phases. *In* Molecular Biology of Pathogenic Fungi. B. Marcesa and G.S. Kobayashi, Eds.: 257–263. Telos Press, New York.

13. KYTE, J. & R.F. DOOLITTLE. 1982. A simple method for displaying the hydropathic character of a protein. J. Mol. Biol. **15:** 105–132.

14. SCAMAN, C.H., F. LIPARI & A. HERSCOVICS. 1996. A spectrophotometric assay for α-mannosidase activity. Glycobiology **6:** 265–270.

15. EADES, C.J. & W.E. HINTZ. 2000. Characterization of the class I α-mannosidase gene family in the filamentous fungus *Aspergillus nidulans*. Gene **255:** 25–34.

16. MATTA, K.L. & O.P. BAHL. 1972. Glycosidases of *Aspergillus niger*. IV. Purification and characterization of α-mannosidase. J. Biol. Chem. **247:** 1780–1787.

17. FOSTER, J.M. & D.B. ROBERTS. 1997. The soluble α-mannosidases of *Drosophila melanogaster*. Insect. Biochem. Molec. Biol. **27:** 657–661.

18. TATARA, Y., B.R. LEE, T. YOSHIDA, *et al.* 2003. Identification of catalytic residues of $Ca2^+$-independent 1,2-α-D-mannosidase from *Aspergillus saitoi* by site-directed mutagenesis. J. Biol. Chem. **278:** 25289–25294.

19. TATARA, Y., T. YOSHIDA & E. ICHISHIMA. 2005. A single free cysteine residue and disulfide bond contribute to the thermostability of *Aspergillus saitoi* 1,2-α-mannosidase. Biosci. Biotechnol. Biochem. **69:** 2101–2108.

20. HERSCOVICS, A. 2001. Structure and function of class I α1,2 mannosidases involved in glycoprotein synthesis and endoplasmic reticulum quality control. Biochimie **83:** 757–762.

21. INOUE, T., T. YOSHIDA & E. ICHISHIMA. 1995. Molecular cloning and nucleotide sequence of the 1,2-α-D-mannosidase gene, *msdS*, from *Aspergillus saitoi* and expression of the gene in yeast cells. Biochim. Biophys. Acta **1253:** 141–145.

22. YOSHIDA, T. & E. ICHISHIMA. 1995. Molecular cloning and nucleotide sequence of the genomic DNA for 1,2-α-D-mannosidase gene, msdC from *Penicillium citrinum*. Biochim. Biophys. Acta **1263:** 159–162.

23. YOSHIDA, T., T. NAKAJIMA & E. ICHISHIMA. 1998. Overproduction of 1,2-α-mannosidase, a glycochain processing enzyme, by *Aspergillus oryzae*. Biosci. Biotechnol. Biochem. **62:** 309–315.

24. ICHISHIMA, E., N. TAYA, M. IKEGUCHI, *et al.* 1999. Molecular and enzymic properties of recombinant 1,2-α-mannosidase from *Aspergillus saitoi* overexpressed in *Aspergillus oryzae* cells. Biochem. J. **339:** 589–597.

25. MARAS, M., N. CALLEWAERT, K. PIENS, *et al.* 2000. Molecular cloning and enzymatic characterization of a *Trichoderma reesei* 1,2-α-D-mannosidase. J. Biotechnol. **77:** 255–263.

26. FUJITA, A., T. YOSHIDA & E. ICHISHIMA. 1997. Five crucial carboxylic residues of 1,2-α-mannosidase from *Aspergillus saitoi* (*A. phoenicis*), a food microorganism, are identified by site-directed mutagenesis. Biochem. Biophys. Res. Commun. **238:** 779–783.

27. VAN PETEGEM, F., H. CONTRERAS, R. CONTRERAS & J. VAN BEEUMEN. 2001. *Trichoderma reesei* α-1,2-mannosidase: structural basis for the cleavage of four consecutive mannose residues. J. Mol. Biol. **312:** 157–165.

28. LOBSANOV, Y.D., F. VALLÉE, A. IMBERTY, *et al.* 2002. Structure of *Penicillium citrinum* α1,2 mannosidase reveals the basis for differences in specificity of the

endoplasmic reticulum and golgi class I enzymes. J. Biol. Chem. **277:** 5620–5630.

29. ICHISHIMA, E., M. ARAI, Y. SHIGEMATSU, H. KUMAGAI & R. SUMIDA-TANAKA. 1981. Purification of an acidic α-D-mannosidase from *Aspergillus saitoi* and specific cleavage of 1,2-α-D-mannosidic linkage in yeast mannan. Biochim. Biophys. Acta **658:** 45–53.

30. YOSHIDA, T., T. INOUE & E. ICHISHIMA. 1993. 1,2-α-D-mannosidase from *Penicillium citrinum*: molecular and enzymic properties of two isoenymes. Biochem. J. **290:** 349–354.

Molecular Cloning and Expression of a cDNA Encoding a *Coccidioides posadasii* Cu,Zn Superoxide Dismutase Identified by Proteomic Analysis of the Coccidioidal T27K Vaccine

JENNINE M. LUNETTA, KEIRA A. SIMMONS, SUZANNE M. JOHNSON, AND DEMOSTHENES PAPPAGIANIS

Department of Medical Microbiology and Immunology, School of Medicine, University of California, Davis, California, USA

ABSTRACT: Previous studies have demonstrated that the coccidioidal T27K vaccine preparation is protective in mice against respiratory challenge using *Coccidioides posadasii* (*C. posadasii*) arthroconidia. Proteomic methods have been employed to define the molecular components within the vaccine. This method has led to the identification of novel and previously uncharacterized coccidioidal proteins including a Cu,Zn superoxide dismutase. A two-dimensional gel of the T27K vaccine was run and spots were excised for mass spectrometric analysis. One peptide was obtained from the T27K gel that matched a TIGR *C. posadasii* 2.0 gene index tentative consensus sequence, TC1072, which is similar to fungal Cu,Zn superoxide dismutase. Activity assays performed with native PAGE gels of the T27K vaccine showed that the vaccine contains superoxide dismutase. The cDNA encoding the enzyme has been cloned and sequenced and expressed as a recombinant protein.

KEYWORDS: *Coccidioides*; vaccine; superoxide dismutase; SOD1; fungus; metalloenzyme; two-dimensional gel electrophoresis; mass spectrometry; rapid amplification of cDNA ends; reverse-transcription polymerase chain reaction

INTRODUCTION

A human vaccine against coccidioidomycosis has been sought for many years and considerable effort has been directed toward this goal.[1-4] The

Address for correspondence: Jennine M. Lunetta, Department of Medical Microbiology and Immunology, School of Medicine, University of California, Davis, California, 95616. Voice: 530-752-7214; fax: 530-752-8692.
  jmlunetta@ucdavis.edu

Ann. N.Y. Acad. Sci. 1111: 181–197 (2007). © 2007 New York Academy of Sciences.
doi: 10.1196/annals.1406.025

coccidioidal T27K vaccine, prepared from mechanically disrupted spherules, has shown promise as a potential vaccine in animal studies; however, the vaccine is a complex mixture of proteins and carbohydrate, which has not been fully characterized.[5] The development of a safe and effective human vaccine from the coccidioidal T27K vaccine will require that the vaccine components be well defined. Proteomic methods can be very useful for rapidly characterizing the molecular components within the vaccine and identifying antigens that can be evaluated as potential vaccine candidates.

In this study, the coccidioidal T27K vaccine has been subjected to a preliminary proteomic analysis. The vaccine was separated using two-dimensional gel electrophoresis and spots were selected for mass spectrometric analysis. A number of proteins were detected in the vaccine using these methods including a Cu,Zn superoxide dismutase. The cDNA encoding the Cu,Zn-superoxide dismutase protein was cloned and sequenced and expressed as a recombinant protein.

MATERIALS AND METHODS

Vaccine Preparation

The coccidioidal T27K vaccine was prepared from *Coccidioides posadasii* (*C. posadasii*) Silveira strain as previously described for the 27K vaccine, except the spherules were inactivated with thimerosal instead of formaldehyde.[5] The *C. posadasii* species was formerly known as the non-California *C. immitis*.[6]

Two-Dimensional Gel Electrophoresis

Isoelectric focusing was performed using a Zoom IPGRunner System (Invitrogen, Carlsbad, CA, USA). The T27K vaccine (800 μg of protein determined by BCA assay [Pierce, Rockford, IL, USA]) was pretreated using a ReadyPrep 2-D Cleanup kit (BioRad, Hercules, CA, USA), dissolved in sample rehydration buffer [8 M urea, 2% (w/v) CHAPS, 2% (v/v) Nonidet P40, 0.5% (v/v) Zoom carrier ampholytes, 0.002% (w/v) bromophenol blue, and 100 mM dithiothreitol] and applied to a Zoom immobilized pH gradient strip (pH4–7) contained within a Zoom IPGRunner Cassette. The following step voltage protocol was used for the isoelectric focusing: 200 V for 15 min, 450 V for 15 min, 750 V for 15 min, and 2,000 V for 2 h. The proteins were separated in the second dimension in a Ready Gel precast 12% resolving Tris-HCl polyacrylamide gel (BioRad) using standard SDS-PAGE methods.[7] A Precision Plus protein standard (BioRad) was run in parallel in order to estimate protein molecular weights. The two-dimensional gel was stained using Gelcode Blue (Pierce). Two-dimensional gel spots were then randomly selected for protein identification using mass spectrometry.

Protein Identification by Mass Spectrometry

The two-dimensional gel of the T27K vaccine was submitted to the UC Davis Genome Center Proteomics Facility (http://proteomics.ucdavis.edu) for mass spectrometry (LC-MS/MS)-based protein identification. Protein spots of interest were excised from the two-dimensional gel and washed thoroughly four times with Milli-Q (Millipore, Billerica, MA, USA) water. The gel pieces were diced into ~1-mm squares and dried in a SpeedVac (Savant, Holbrook, NY, USA). Proteins were reduced and alkylated as previously described and digested with sequencing grade, modified trypsin (Promega, Madison, WI, USA).[8] Protein identification was performed using a NanoLC-2D system (Eksigent, Dublin, CA, USA) coupled to an LTQ linear ion-trap mass spectrometer (Thermo Fisher Scientific, Waltham, MA, USA) through a PicoView nanospray source (New Objective, Woburn, MA, USA). Peptides were loaded on to a nanotrap (Zorbax 300SB-C18, Agilent Technologies, Santa Clara, CA, USA) at a loading flow rate of 5 μL/min. Peptides were then eluted from the trap and separated by a nanoscale PicoFrit column (75 μm × 15 cm, New Objective) packed in-house with Magic C18AQ packing material (Michrom Bioresources, Auburn, CA, USA). Peptides were eluted using a 40-min gradient of 2% to 80% buffer B (buffer A: 0.1% formic acid, buffer B: 95% acetonitrile, 0.1% formic acid). The 10 most abundant ions in each survey scan were subjected to automatic low energy collision-induced dissociation and the resulting uninterpreted MS/MS spectra were searched against an NCBI nonredundant fungal database using Mascot (Matrix Science, London, UK) and x!Tandem2 (Proteome Software, Portland, OR, USA.) software. Scaffold software (Proteome Software) was used to validate MS/MS-based peptide and protein identifications.

RNA Isolation

Spherule-endospore (SE) phase cultures were grown as previously reported and total RNA was isolated from 24-h cultures using an RNeasy Plant Mini kit (Qiagen, Valencia, CA, USA).[9] Contaminating genomic DNA was removed using RNase-free DNase (Qiagen). RNA integrity was evaluated by electrophoresis through a denaturing 1% agarose gel containing formaldehyde (2%). RNA purity and concentration were determined by measuring the absorbance at 260 nm and 280 nm.

Rapid Amplification of cDNA Ends (RACE)

The 5′ and 3′ ends of the cDNA encoding the superoxide dismutase protein identified by MS analysis were mapped using a GeneRacer kit (Invitrogen) for RACE. First-strand cDNA for RACE was prepared from 24-h SE phase total RNA as described in the manual. Briefly, total RNA (500 ng) was treated with calf intestine phosphatase to remove 5′ phosphates. The dephosphorylated RNA was then treated with tobacco acid pyrophosphatase to remove the 5′ cap structure from intact full-length mRNA. A GeneRacer RNA Oligo was ligated to the 5′ end of the mRNA using T4 RNA ligase. The ligated mRNA was reverse-transcribed using SuperScript II reverse transcriptase and a GeneRacer Oligo dT primer. The RACE-ready first-strand cDNA (2 μL) was then amplified in a reaction (50 μL) containing 20 mM Tris-HCl (pH 8.4 at 25°C), 50 mM KCl, 1.5 mM $MgCl_2$, 125 μM of each dNTP, 2.5 U *Taq* DNA polymerase (Promega) and 400 nM of each primer [GeneRacer 5′ primer and a reverse gene-specific primer (5′-GGTGATGTTGCCCAGGTCA-3′) or the GeneRacer 3′ primer and a forward gene-specific primer (5′-TGACCTGGGCAACATCACC-3′)]. The amplification conditions were an initial denaturation at 94°C for 5 min, followed by 35 cycles at 94°C for 30 sec, 60°C for 30 sec, 72°C for 1 min, and a final cycle at 72°C for 7 min. The RACE PCR products were cloned using an Original TA Cloning kit (Invitrogen) and plasmid DNA was isolated using a QIAprep Spin Miniprep kit (Qiagen). The presence and size of the insert DNA were determined using restriction enzyme digestion of the plasmid DNA by *Eco*RI followed by agarose gel electrophoresis. The insert DNA was sequenced using vector-specific primers.

Reverse-Transcription Polymerase Chain Reaction (RT-PCR)

The superoxide dismutase cDNA coding region was amplified using RT-PCR. Poly $(A)^+$ RNA was prepared from 24-h SE phase total RNA using an Oligotex mRNA Mini kit (Qiagen) and RT-PCR was carried out using a Thermoscript RT-PCR kit (Invitrogen). Template poly $(A)^+$ RNA (25 ng) was denatured at 65°C for 5 min in the presence of the oligo $(dT)_{20}$ primer (2.5 μM) and dNTPs (250 μM each). A reaction mix containing 50 mM Tris-acetate (pH 8.4), 75 mM potassium acetate, 8 mM magnesium acetate, 5 mM dithiothreitol, RNase Out (40 U), and Thermoscript reverse transcriptase (15 U) was then added to the RNA to yield a 20-μL reaction volume. The reaction was then incubated at 55°C for 1 h, followed by 85°C for 5 min. RNase H (2 U) was added to the reaction and the sample incubated at 37°C for 20 min. The first-strand cDNA (2 μL) was then amplified in a reaction (50 μL) containing 20 mM Tris-HCl, (pH 8.8), 2 mM $MgSO_4$, 1 mM $MgCl_2$, 10 mM KCl, 10 mM $NH_4(SO_4)_2$, 0.1% (v/v) Triton X-100, 100 μg/mL bovine serum albumin, 1 μM of the forward primer (5′-ATGGTCAGAGCAGTTGCT-3′) and reverse primer (5′-TTATGCAGCAATGCCAATAAC-3′), 200 μM of each dNTP and

1.25 U of Platinum *Taq* DNA Polymerase (Invitrogen). The amplification conditions were an initial denaturation at 94°C for 5 min, followed by 35 cycles at 94°C for 30 sec, 55°C for 30 sec, 72°C for 30 sec, and a final extension at 72°C for 10 min. An aliquot of the RT-PCR reaction was run on an agarose gel (0.8%) and visualized with ethidium bromide. The RT-PCR product was cloned and plasmid DNA was isolated as described earlier. The presence and size of the insert DNA were determined as described earlier for RACE products and insert DNA was sequenced using vector-specific primers.

Sequence Analysis

DNA sequencing was carried out at the UC Davis Division of Biological Sciences DNA Sequencing Laboratory (http://dnaseq.ucdavis.edu). *C. posadasii* sequence data were obtained from the genome and the gene indices database at the Institute for Genomic Research (http://www.tigr.org). The Compute pI/MW tool at the ExPASy server (http://www.expasy.org) was used to calculate theoretical isoelectric point (pI) and theoretical molecular weight. The Center for Biological Sequence Analysis website (http://www.cbs.dtu.dk) was used for signal peptide and cleavage site prediction (SignalP) and for the prediction of *N*-glycosylation sites (NetNGlyc). The ClustalW software located at the European Bioinformatics Institute website (http://www.ebi.ac.uk) was used for multiple sequence alignment. The *C. posadasii* Cu,Zn superoxide dismutase cDNA nucleotide (accession no. DQ530599) and deduced amino acid (accession no. ABF73315) sequences have been deposited in the National Center for Biotechnology Information (NCBI) GenBank database.

Expression and Purification of Recombinant Superoxide Dismutase

The full-length *C. posadasii* Cu,Zn superoxide dismutase cDNA was expressed as a recombinant polyhistidine-tagged fusion protein using a T7-based expression system. Full-length cDNA was amplified as described earlier and cloned into a pEXP5-NT/TOPO expression vector containing a bacteriophage T7 promoter and an N-terminal 6xHis tag (Invitrogen). TOP10 chemically competent *E. coli* cells (Invitrogen) were transformed with the expression plasmid, plated on Luria-Bertani (LB) agar plates containing ampicillin (100 μg/mL) and cultured at 37°C overnight. Several colonies were picked, inoculated into LB medium (5 mL) containing ampicillin (100 μg/mL) and cultured at 37°C overnight. Plasmid DNA was isolated from the cultures using a QIAprep Spin Miniprep kit (Qiagen) and evaluated for the presence of insert DNA using restriction enzyme digestion with *Xba*I and *Eco*RI. Clones with inserts of the correct size were DNA sequenced and a construct with the correct full-length cDNA sequence in frame with the N-terminal tag was selected for expression. BL21 chemically competent *E. coli* cells (Invitrogen) were transformed with the construct (10 ng) and cultured at 37°C overnight. Fresh medium was inoculated with the overnight culture and incubated at 37°C for 3 h. The cells were then cultured for 4 h in the presence or absence of 0.5 mM isopropyl-beta-D-thiogalactopyranoside (Fisher Scientific, Pittsburgh, PA, USA). The bacterial cells were harvested by centrifugation, the cell pellet resuspended in 6 mL of resuspension buffer (50 mM sodium phosphate buffer [pH 8.0] and 500 mM sodium chloride) containing 0.01 mg/mL DNase (Calbiochem, San Diego, CA, USA) and 0.1 mg/mL lysozyme (Sigma, St. Louis, MO, USA) and incubated on ice for 30 min. The cells were sonicated (Sonic Dismembrator, Model 100, Fisher Scientific) on ice four times at a setting of 5 for 10-sec intervals, centrifuged, and the cell pellet was resuspended in 6 mL cold isolation buffer (50 mM sodium phosphate buffer [pH 8.0], 500 mM sodium chloride and 2 M urea) containing 2 % (v/v) Triton X-100. The cells were sonicated twice as described earlier and centrifuged. The pellet was washed with isolation buffer, resuspended in 1 mL of isolation buffer, added dropwise to 5 mL of binding buffer (50 mM sodium phosphate buffer [pH 8.0], 500 mM sodium chloride, and 6 M guanidine hydrochloride) and stirred overnight at 4°C. The lysate was centrifuged at high speed for 15 min at 4°C, filtered through a 0.45-μm syringe filter, and further purified using fast protein liquid chromatography (FPLC; GE Healthcare, Piscataway, NJ, USA). The lysate was applied to a HiTrap Chelating HP 1-mL column (GE Healthcare) preequilibrated with 10 mL of binding buffer and the column was washed with a step pH gradient consisting of binding buffer (10 mL) at pH 6.0, 5.3, and 4.0. The eluate was evaluated for the presence of pure recombinant superoxide dismutase fusion protein using SDS-PAGE and Western blot analysis as described below.

Protein Electrophoresis and Western Blot Analysis of Recombinant Superoxide Dismutase

SDS-PAGE was performed using standard methods.[7] After electrophoresis, the gel was either stained with Gelcode Blue (Pierce) or transferred to nitrocellulose under standard conditions.[10] For Western blot analysis, the membrane was incubated with 5% (w/v) nonfat dried milk (NFDM) in phosphate-buffered saline containing 0.05% (v/v) Tween-20 (PBST) for 1 h and then incubated for 1 h with anti-HisG-HRP antibody (Invitrogen) diluted 1:5,000 in PBST plus 5% (w/v) NFDM. After washing three times with PBST for 10 min each, the membrane was developed with tetramethylbenzidine (Pierce).

Superoxide Dismutase Enzyme Activity Assay

Superoxide dismutase enzyme activity was evaluated in the coccidioidal T27K vaccine using an in-gel negative staining method.[11] The T27K vaccine was dissolved in water (15 μg protein/μL), added to 2× sample buffer

(125 mM Tris-HCl, pH 6.8, 20% glycerol, and 0.1% [w/v] bromophenol blue) and applied to two wells (75 μg protein/well) of a native 6% polyacrylamide gel. The proteins were separated by electrophoresis at 100 V, the gel cut into two portions and either stained for protein with Gelcode Blue (Pierce) or used for superoxide dismutase enzyme activity detection. The gel portion used for superoxide dismutase enzyme activity detection was washed briefly with water and then incubated with an aqueous solution containing 1.23 mM nitroblue tetrozolium (Fisher Scientific) for 15 min at room temperature. The gel was then incubated with 28 mM riboflavin (Eastman Chemical Co., Kingsport, TN, USA) and a 0.4% (v/v) solution of TEMED (BioRad) for 15 min in the dark, washed briefly in water and developed on a light box. The composition of the superoxide dismutase activity band was evaluated by SDS-PAGE analysis and tandem mass spectrometry. The colorless activity band was excised from the native gel, minced and placed in 50 μL of water. After adding 50 μL of 2× sample buffer (125 mM Tris-HCl, pH 6.8, 20% glycerol, 4% SDS, 10% β-mercaptoethanol and 0.1% [w/v] bromophenol blue), the mixture was boiled for 5 min and then applied to a single well of a 12% SDS-PAGE gel. The gel pieces were overlaid with 0.5% agarose and electrophoresis was performed using standard conditions.[7] The gel was stained with Gelcode Blue (Pierce) and the resultant single band was subjected to LC-MS/MS as described earlier.

RESULTS

Identification of Cu,Zn Superoxide Dismutase in the Coccidioidal T27K Vaccine Using Proteomic Methods

The coccidioidal T27K vaccine was separated by two-dimensional gel electrophoresis using a pH gradient of 4 through 7 and stained with Coomassie blue (FIG. 1). A large number of intensely stained spots were revealed in pH range 5 to 6 with apparent molecular weight between 15 and 75 kDa. Very few spots were detected between pH 4 and 5 and in pH range 6 to 7 most of the spots were between the 15- and 37-kDa standards. Twelve spots were randomly selected for sequencing using mass spectrometry and 10 proteins were identified (TABLE 1). One peptide ("HVGDLGNITTDSQGNSTGSVEDK"), obtained from a spot migrating between ∼15 and 20 kDa and pI of ∼7 (FIG. 1, spot #12), matched the TIGR *C. posadasii* 2.0 gene index tentative consensus sequence, TC1072, which is similar to Cu,Zn superoxide dismutase.

Cloning and Sequencing of the Coccidioidal cDNA Encoding the Superoxide Dismutase Protein Detected in the T27K Vaccine

The 5′ and 3′ ends of the cDNA encoding the superoxide dismutase protein detected in the T27K vaccine were mapped using RACE. The nucleotide

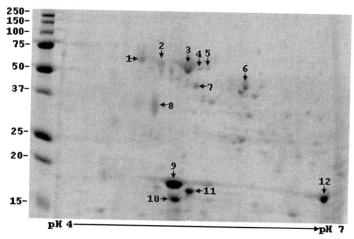

FIGURE 1. Separation of the coccidioidal T27K vaccine by two-dimensional electrophoresis. Coomassie blue–stained two-dimensional electrophoresis gel of the T27K vaccine. The *arrows* indicate the spots that were sequenced using mass spectrometry. The numbers to the left of the gel indicate the molecular weight in kDa of the adjacent Precision Plus protein standard (BioRad).

sequence for the region encoding the "HVGDLGNITTDSQGNSTGSVEDK" peptide was obtained from the TIGR *C. posadasii* 2.0 gene index tentative consensus sequence, TC1072 (Frame +3) and used to design a forward (HVGD-F) and reverse (HVGD-R) primer for RACE. Two RACE PCR reactions were performed, a 5′ RACE reaction with the GeneRacer 5′ primer and the HVGD-R primer and a 3′ RACE reaction with the GeneRacer 3′

TABLE 1. Proteins identified in the T27k vaccine using proteomic methods

Spot No.	Protein Match	TIGR TC No(s)	GenBank Accession No.	Reference[a]
1	pH-sensitive protein-like protein (similar to β-1,3-glucanosyltransferase)	TC4708	ABA38727	
2	Hypothetical protein (similar to Gag protein)	TC4131	EAS33091	
3	Heat shock protein 60	TC4079	AAD00521	Thomas *et al.*, 1997[19]
4,5	Hypothetical protein (similar to repressible alkaline phosphatase)	TC1977/ TC6003	EAS33539	
6	Complement-fixation chitinase	TC1363	AAB48567	Zimmermann *et al.*, 1996[18]
7	1,3-β-Glucanosyltransferase	TC3927	AAL37628	Delgado & Cole, 2003[20]
8	ELI-antigen 1	TC3981	AAO62547	Ivey *et al.*, 2003[21]
9,10	Peroxisomal matrix protein	TC742	AAB42829	Orsborn *et al.*, 2006[22]
11	Heat-stable 19 kDa antigen	TC1485	AAB00101	Pan & Cole, 1995[17]
12	Cu,Zn superoxide dismutase	TC1072	ABF73315	

[a]Published reports for the proteins that have been previously characterized in *Coccidioides* spp.

primer and the HVGD-F primer, and the products cloned and sequenced. The 5′ RACE PCR reaction generated a 383-bp product that consisted of 44 bp corresponding to the GeneRacer RNA Oligo and 339 bp of coccidioidal-specific cDNA and the 3′ RACE PCR reaction yielded a 507-bp product that included 54 bp corresponding to the GeneRacer Oligo dT primer and 453 bp of coccidioidal-specific cDNA. The nucleotide sequence of the 5′ (339 bp) and 3′ (453 bp) end fragments overlapped at the position of the HVGD primer and collectively formed a 773 bp cDNA sequence. Analysis of the resulting sequence revealed a 465-bp open reading frame flanked by a 5′ and 3′ untranslated regions consisting of 72 bp and 236 bp, respectively. The open reading frame encoded a 154–amino acid protein that included the entire sequence of the peptide ("HVGDLGNITTDSQGNSTGSVEDK") obtained from mass spectrometry. The cDNA encoding sequence was also determined using RT-PCR with primers corresponding to the 5′ and 3′ ends obtained by RACE analysis. A single RT-PCR product was obtained and the product cloned and sequenced. A cDNA sequence of 465 bp was obtained that was identical to the encoding region obtained using RACE. The nucleotide and deduced amino acid sequence of the superoxide dismutase cDNA are shown in FIGURE 2. The predicted protein has a molecular mass of 15,997 Da and pI of 6.21.

Comparison of the Deduced Amino Acid Sequence to Cu,Zn Superoxide Dismutase Proteins

A BLAST search of the NCBI GenBank database confirmed that the *C. posadasii*–predicted protein was related to Cu,Zn superoxide dismutase. ClustalW alignment analysis revealed that the deduced amino acid sequence shares between 79% and 83% identity to fungal Cu,Zn superoxide dismutases identified by the BLAST search (FIG. 3). The highest percentage of identity (83%) was to the *Penicillium marneffei* Cu,Zn superoxide dismutase (Genbank accession no. ABD67502), followed by 81, 80, and 79% identity to the Cu,Zn superoxide dismutases from *Paracoccidiodes brasiliensis* (Genbank accession no. AAX13803), *Humicola lutea* (Genbank accession no. P83684), and *Aspergillus fumigatus* (Genbank accession no. AAD42060), respectively. The sequence also shares 72% and 57% identity to the *Candida albicans* (Genbank accession no. AAC12872) and *Homo sapiens* (Genbank accession no. AAB05662) Cu,Zn superoxide dismutases, respectively. The deduced amino acid sequence contains conserved metal binding sites for copper (His-46, His-48, His-64, and His-121) and zinc (His-64, His-72, His-81, and Asp-84) and a conserved arginine residue (Arg-144) that is required for enzyme activity.[12] In addition, two cysteine residues (Cys-58 and Cys-147) that form a disulfide bond and amino acid residues involved in dimer formation are conserved.[12] The sequence lacks a putative signal peptide.

```
 -72                        caaagcagtaaaaagagcatcccaaaaata
 -42      atttaaaaaaaaaggggccaataacaccttccatcagccaaa
   1      ATGGTCAGAGCAGTTGCTGTCCTCCGCGGTGACTCCCTCGTG
   1       M  V  R  A  V  A  V  L  R  G  D  S  L  V
  43      AAGGGCACAGTCACCTTCGAACAGGCTGACGAGAAAAGCCCG
  15       K  G  T  V  T  F  E  Q  A  D  E  K  S  P
  85      ACCACCATCTCCTGGAACATCTCTGGCCACGATGCCAACGCT
  29       T  T  I  S  W  N  I  S  G  H  D  A  N  A
 127      CAGCGTGGCTTCCACATTCATCAATTCGGTGACAACACCAAC
  43       Q  R  G  F  H  I  H  Q  F  G  D  N  T  N
 169      GGCTGCACTTCTGCTGGCCCTCACTACAACCCATTCTCCAAG
  57       G  C  T  S  A  G  P  H  Y  N  P  F  S  K
 211      AATCACGGAGCTCCATCCGACGTAGATCGCCATGTTGGTGAC
  71       N  H  G  A  P  S  D  V  D  R  H  V  G  D
 253      CTGGGCAACATCACCACTGATTCCCAGGGCAACTCCACCGGC
  85       L  G  N  I  T  T  D  S  Q  G  N  S  T  G
 295      AGCGTTGAGGACAAACAGATCAAGCTTATTGGGGAGCACAGT
  99       S  V  E  D  K  Q  I  K  L  I  G  E  H  S
 337      GTTCTCGGTCGCACTGTTGTCGTTCATGCTGGCACTGATGAC
 113       V  L  G  R  T  V  V  V  H  A  G  T  D  D
 379      CTCGGCAAGGGAGGCAACGAGGAATCCAAGAAGACTGGAAAT
 127       L  G  K  G  G  N  E  E  S  K  K  T  G  N
 421      GCTGGGCCCCGTCCTGCCTGCGGTGTTATTGGCATTGCTGCA
 141       A  G  P  R  P  A  C  G  V  I  G  I  A  A
 463      TAAgcttctctcccgagaactctcccctaaggctttcactag
 155       *
 505      attgcgcgaaagcatcaatgaacagttaatgagattgcttag
 547      caattaactagttacctagttagcgatgtaatgatcaaacaa
 589      tccagtttctagaggtgtaatgaaattacatatctcggtgaa
 631      ttagctctacaaaataatctgtacatagttcaatttggagtt
 673      gggagctcatatgaaccagtaacttaatt
```

FIGURE 2. The nucleotide sequence and deduced amino acid sequence of the *Coccid-ioides posadasii* (*C. posadasii*) Cu,Zn superoxide dismutase cDNA. The numbers to the left of each line indicate either nucleotide or amino acid positions. The asterisk (*) indicates the stop codon. The lower case letters indicate the 5′ and 3′ untranslated regions. The underlined amino acid sequence indicates the peptide generated from the protein detected in spot #12 of the T27K vaccine two-dimensional gel.

Expression and Purification of Recombinant C. posadasii *Cu,Zn Superoxide Dismutase*

The full-length cDNA encoding the *C. posadasii* Cu,Zn superoxide dismu-tase was cloned into a pEXP5-NT/TOPO vector and expressed as an N-terminal

```
Cp  MVRAVAVLRGDSLVKGTVTFEQADEKSPTTISWNISGHDA  40
Pm  -MVKAVAVLRGDSNIKGTVTFEQADENSPTTISWNITGHDA  40
Pb  -MVKAVAVLRGDSNVKGTVVFEQASESSTTVITYNLSGNDP  40
Hl  --VKAVAVLRGDSKITGTVTFEQANESAPTTVSWNITGHDP  39
Af  -MVKAVAVLRGDSKITGTVTFEQADENSPTTVSWNIKGNDP  40
Ca  -MVKAVAVVRGDSKVQGTVHFEQESESAPTTISWEIEGNDP  40
Hs  MATKAVCVLKGDGPVQGIINFEQKESNGPVKVWGSIKG-LT  40
    .:**.*::**. : * : *** ...... : .: * .


Cp  NAQRGFHIHQFGDNTNGCTSAGPHYNPFSKNHGAPSDVDR  80
Pm  NAERGIHVHQFGDNTNGCTSAGPHFNPFGKTHGAPTDDER  80
Pb  NALRGFHIHQFGDNTNGCTSAGPHFNPFGKTHGSPSDAER  80
Hl  NAERGMHIHQFGDNTNGCTSAGPHYNPFKKTHGAPTDEVR  79
Af  NAKRGFHVHQFGDNTNGCTSAGPHFNPYGKTHGAPEDSER  80
Ca  NALRGFHIHQFGDNTNGCTSAGPHFNPFGKQHGAPEDDER  80
Hs  EGLHGFHVHEFGDNTAGCTSAGPHFNPLSRKHGGPKDEER  80
    :. :*:*:*:***** ********:** : **.* * *


Cp  HVGDLGNITTDSQGNSTGSVEDKQIKLIGEHSVLGRTVVV  120
Pm  HVGDLGNFKTDAQGNAVGFVEDKLIKLIGAESVLGRTIVV  120
Pb  HVGDLGNITTDAQGNASGTMEDIFIKLIGEHSVLGRTVVV  120
Hl  HVGDLGNIKTDAEGNAVGSVQDKLIKVIGAESILGRTIVV  119
Af  HVGDLGNFETDAEGNAVGSKQDKLIKLIGAESVLGRTLVV  120
Ca  HVGDLGNISTDGNGVAKGTKQDLLIKLIGKDSILGRTIVV  120
Hs  HVGDLGNVTADKDGVADVSIEGSVISLSGDHCIIGRTLVV  120
    *******. :* :* : :. *.: * ..::***:**


Cp  HAGTDDLGKGGNEESKKTGNAGPRPACGVIGIAA  154
Pm  HAGTDDLGRGGNEESKKTGNAGPRPACGVIGISA  154
Pb  HAGTDDLGRGGNEESKKTGNAGPRPACGVIGISA  154
Hl  HAGTDDLGRGGNEESKKTGNAGPRPACGVIGIA-  152
Af  HAGTDDLGRGGNEESKKTGNAGARPACGVIGIAA  154
Ca  HAGTDDYGKGGFEDSKTTGHAGARPACGVIGLTQ  154
Hs  HEKADDLGKGGNEESTKTGNAGSRLACGVIGIAQ  154
    * :** *:** *:*..**:**.* ******::
```

FIGURE 3. Comparison of the *C. posadasii* deduced amino acid sequence with Cu,Zn superoxide dismutase proteins from *Penicillium marneffei* (*Pm*), *Paracoccidioides brasiliensis* (*Pb*), *Humicola lutea* (*Hl*), *Aspergillus fumigatus* (*Af*), *Candida albicans* (*Ca*), and *Homo sapiens* (*Hs*). The numbers to the right of the lines indicate amino acid positions. An asterisk (*) indicates an amino acid identity, a colon (:), a conserved substitution, and a dot (.), a semiconserved substitution.

6xHis tagged fusion protein in *E. coli*. Transformed *E. coli* cells were cultured for 4 h in the presence or absence of inducer and then evaluated for recombinant fusion protein expression using SDS-PAGE and Western blot analysis with an anti-HisG antibody. A prominent Coomassie blue–stained band of apparent molecular weight of ~19 kDa that reacted with the anti-HisG antibody

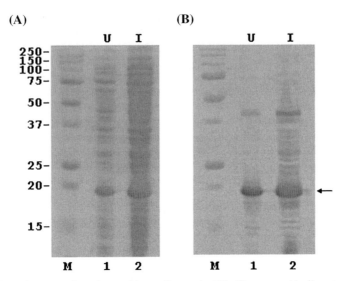

FIGURE 4. Expression of recombinant *C. posadasii* Cu,Zn superoxide dismutase fusion protein. Coomassie blue–stained SDS-PAGE gel (**A**) and corresponding Western blot (**B**) of recombinant Cu,Zn superoxide dismutase bacterial cultures. Lane 1, uninduced (U) culture at 4 h; 2, induced (I) culture at 4 h; M, Precision Plus protein standard (BioRad). The numbers to the left of the gel (**A**) indicate the size in kDa of the adjacent protein standard. The *arrow* indicates the position of the recombinant fusion protein.

was observed in both cultures (FIG. 4). The expression of the fusion protein in the uninduced culture can be explained by the fact that the pEXP5-NT/TOPO expression system is a T7-based system and basal expression of T7 RNA polymerase can occur in the absence of inducer. The transformed bacteria were cultured on a large scale and induced for 4 h for purification of the recombinant fusion protein. Inclusion bodies were isolated from the bacterial cells and solubilized under denaturing conditions. The fusion protein was then purified using a metal-chelating column and fast protein liquid chromatography (FPLC). An SDS-PAGE gel of a highly purified fraction obtained after FPLC is shown in FIGURE 5.

Analysis of Superoxide Dismutase Activity in the Coccidioidal T27K Vaccine

The T27K vaccine was separated on a native polyacrylamide gel and evaluated for superoxide dismutase activity using an in-gel negative staining assay. Multiple achromatic bands were observed on the gel, indicating the presence of superoxide dismutase activity in the vaccine (FIG. 6A,B). Since multiple forms of superoxide dismutase exist, the source of the activity was evaluated

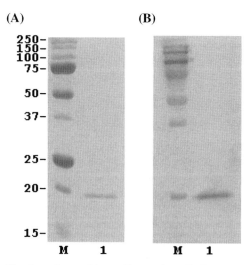

FIGURE 5. Purification of recombinant *C. posadasii* Cu,Zn superoxide dismutase fusion protein. Coomassie blue-stained SDS-PAGE gel (**A**) and corresponding Western blot (**B**) of purified recombinant *C. posadasii* Cu,Zn superoxide dismutase. Lane 1, purified recombinant Cu,Zn superoxide dismutase; M, Precision Plus protein standard (BioRad). The numbers to the left of the gel (**A**) indicate the size in kDa of the adjacent protein standard.

using SDS-PAGE analysis and mass spectrometry. The achromatic band at the top of the activity gel was excised and separated on an SDS-PAGE gel. A single band was observed between the 15- and 20-kDa protein standards, which was the expected size for a Cu,Zn superoxide dismutase monomer (FIG. 6C). The activity band was excised from the SDS-PAGE gel and sequenced using mass spectrometry. The peptide sequences obtained from the excised activity band matched the deduced amino acid sequence for the *C. posadasii* Cu,Zn superoxide dismutase (FIG. 6D).

DISCUSSION

A number of proteins have been identified in the multicomponent coccidioidal T27K vaccine using proteomic methods including one that showed homology to Cu,Zn superoxide dismutase. The cDNA encoding the protein was cloned and sequenced and a predicted protein was generated from the largest open reading frame. The predicted protein consists of 154 amino acids and has a theoretical molecular weight and pI of 15,997 Da and 6.21, respectively. The deduced amino acid sequence confirmed that the protein is homologous to Cu,Zn superoxide dismutase with the highest identity (83%) to the *P. marneffei* protein (Genbank accession no. ABD67502). The *C. posadasii* Cu,Zn

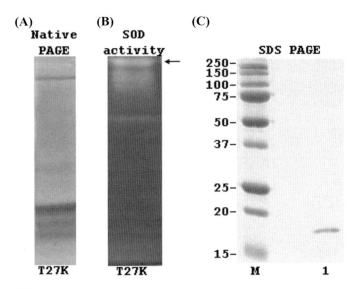

(D)

```
MVRAVAVLRGDSLVKGTVTFEQADEKSPTTISWNISGHD
ANAQRGFHIHQFGDNTNGCTSAGPHYNPFSKNHGAPSDV
DRHVGDLGNITTDSQGNSTGSVEDKQIKLIGEHSVLGRT
VVVHAGTDDLGKGGNEESKKTGNAGPRPACGVIGIAA
```

FIGURE 6. Analysis of superoxide dismutase activity in the T27K vaccine. Coomassie blue–stained native polyacrylamide gel of T27K vaccine (**A**), corresponding superoxide dismutase activity gel (**B**), Coomassie blue–stained SDS-PAGE gel of excised activity band (**C**), and the *C. posadasii* Cu,Zn superoxide dismutase deduced amino acid sequence (**D**). Lane 1, SOD activity band excised from activity gel in (**B**); M, Precision Plus protein standard (BioRad). The *arrow* in panel B indicates the SOD activity band that was excised and applied to the SDS-PAGE gel (**C**). The numbers to the left of the SDS-PAGE gel (**C**) indicate the size in kDa of the adjacent protein standard. The peptides obtained from LC-MS/MS analysis of the excised SOD activity band are underlined (**D**).

superoxide dismutase was expressed as a recombinant His-tagged fusion protein and affinity-purified. Superoxide dismutase activity was detected in the coccidioidal T27K vaccine and it was determined that the activity was generated by the Cu,Zn form of the enzyme.

Superoxide dismutase enzymes catalyze the conversion of superoxide radical anion to hydrogen peroxide and molecular oxygen. The Cu,Zn form is present in all eukaryotes and is located in the cytosol. The protein is a homodimer of 32 kDa. Although the main function of the Cu,Zn form is to eliminate superoxide anion produced in the cytosol, it has been suggested that the enzyme may detoxify host superoxide anion, and therefore play a role in fungal pathogenesis as a virulence factor.[13] An s*od1* (the gene encoding

Cu,Zn superoxide dismutase) mutant strain of *Cryptococcus neoformans* exhibited decreased virulence that was attributed to an increase in susceptibility to oxygen radicals within macrophages.[14] In addition, an *sod1* mutant strain of *C. albicans* has been shown to have attenuated virulence and increased susceptibility to macrophage attack.[15] An *SOD5* mutant strain of *C. albicans* has also been shown to be defective in virulence, although it did not appear to be due to an increased sensitivity to macrophage attack.[16] Similar studies are warranted to evaluate whether *C. posadasii* Cu,Zn superoxide dismutase enzyme plays a role in the virulence of the organism.

It should be noted that an additional protein with homology to Cu,Zn superoxide dismutase has been identified in the T27K vaccine (data not shown). This Cu,Zn superoxide dismutase-related protein matched the TIGR *C. posadasii* 2.0 gene index tentative consensus sequence, TC4698. The cDNA encoding the protein has been cloned and sequenced and the nucleotide and deduced amino acid sequences have been deposited in the NCBI GenBank database (DQ237935 and ABB36775). The *C. posadasii* Cu,Zn superoxide dismutase–related protein showed 8% identity to the *C. posadasii* Cu,Zn superoxide dismutase described here and 27%, 30%, and 20% identity to the *Candida albicans* predicted proteins Sod4p, Sod5p, and Sod6p, respectively (orf 19.2062, orf 19.2060, orf 19.2108 in version 19 of Stanford's *C. albicans* genome sequence assembly, Stanford Genome Technology Center, http://www.sequence.stanford.edu/group/candida).[16]

The main goal of our work is to define the components of the coccidioidal T27K vaccine preparation and to assess their individual protective ability in an attempt to develop a subunit vaccine. In this study, a limited proteomic analysis of the multicomponent coccidioidal T27K vaccine has identified 10 proteins. Six of the ten proteins have been previously characterized in *Coccidioides* spp., heat-shock protein 60 (HSP60), chitinase, β-1,3-glucanosyltransferase (Gel1), ELI-antigen 1 (ELI-Ag1), peroxisomal matrix protein (pmp1), and the heat-stable 19kDa antigen, and three (GEL1, ELI-Ag1, and Pmp1) of the six have been reported to be protective antigens.[17–22] The fact that three of the ten proteins identified here as being present in the T27K vaccine have been shown to be protective antigens is extremely encouraging and demonstrates that the crude coccidioidal T27K vaccine is a good source for potential vaccine candidates. Future experiments will include a more thorough proteomic analysis of the T27K vaccine and molecular studies to clone and sequence the uncharacterized proteins.

In summary, the coccidioidal T27K vaccine has been subjected to a limited proteomic analysis, which identified 10 proteins including a Cu,Zn superoxide dismutase. The cDNA encoding the Cu,Zn superoxide dismutase was cloned and sequenced and expressed as a recombinant fusion protein. The recombinant fusion protein was purified and can now be evaluated for protective ability in animal experiments.

ACKNOWLEDGMENTS

This work was supported by a grant from the California HealthCare Foundation, the Department of Health Services of the State of California, and California State University, Bakersfield.

REFERENCES

1. PAPPAGIANIS, D. 2001. Seeking a vaccine against *Coccidioides immitis* and serologic studies: expectations and realities. Fungal Genet. Biol. **32:** 1–9.
2. KIRKLAND, T. & G.T. COLE. 2002. Coccidioidomycosis: pathogenesis, immune response and vaccine development. *In* Fungal Pathogenesis: Principles and Clinical Applications. R. A. Calderone & R. L. Cihlar, Eds.: 365–400. Marcel Dekker. New York.
3. COLE, G.T., J.-M. XUE, C.N. OKEKE, *et al.* 2004. A vaccine against coccidioidomycosis is justified and attainable. Med. Mycol. **42:** 189–216.
4. MAGEE, D.M. & R.A. COX. 2004. Vaccine development for coccidioidomycosis. *In* The Mycota XII Human Fungal Pathogens. J. E. Domer & G. S. Kobayashi, Eds.: 243–257. Springer-Verlag. Heidelberg, Germany.
5. ZIMMERMANN, C.R., S.M. JOHNSON, G.W. MARTENS, *et al.* 1998. Protection against lethal murine coccidioidomycosis by a soluble vaccine from spherules. Infect. Immun. **66:** 2342–2345.
6. FISHER, M.C., G.L. KOENIG, T.J. WHITE & J.W. TAYLOR. 2002. Molecular and phenotypic description of *Coccidioides posadasii* sp. nov., previously recognized as the non-California population of *Coccidioides immitis*. Mycologia **94:** 73–84.
7. LAEMMLI, U.K. 1970. Cleavage of structural proteins during the assembly of the head of bacteriophage T4. Nature **227:** 680–685.
8. SHEVCHENKO, A., M. WILM, O. VORM & M. MANN. 1996. Mass spectrometric sequencing of proteins from silver stained polyacrylamide gels. Anal. Chem. **68:** 850–858.
9. ZIMMERMANN, C.R. & D. PAPPAGIANIS. 1994. Extraction and purification of poly $(A)^+$ RNA from *Coccidioides immitis* spherule/endospore and hyphal phases. *In* Molecular Biology of Pathogenic Fungi. B. Marcesa & G. S. Kobayashi, Eds.: 257–263. Telos Press. New York.
10. TOWBIN, H., T. STAEHELIN & J. GORDEN. 1979. Electrophoretic transfer of proteins from polyacrylamide gels to nitrocellulose sheets: procedures and some applications. Proc. Natl. Acad. Sci. USA **76:** 4350–4354.
11. BEAUCHAMP, C. & I. FRIDOVICH. 1971. Superoxide dismutase: improved assays and an assay applicable to acrylamide gels. Anal. Biochem. **44:** 276–287.
12. BORDO, D., K. DJINOVIC & M. BOLOGNESI. 1994. Conserved patterns in the Cu,Zn superoxide dismutase family. J. Mol. Biol. **238:** 366–386.
13. HAMILTON, A. J. & M.D. HOLDOM. 1999. Antioxidant systems in the pathogenic fungi of man and their role in virulence. Med. Mycol. **37:** 375–389.
14. COX, G.M., T.S. HARRISON., H.C. MCDADE, *et al.* 2003. Superoxide dismutase influences the virulence of *Cryptococcus neoformans* by affecting growth within macrophages. Infect. Immun. **71:** 173–180.
15. HWANG, C.-S., G.-E. RHIE, J.-H. OH, *et al.* 2002. Copper- and zinc-containing superoxide dismutase (Cu/ZnSOD) is required for the protection of *Candida*

albicans against oxidative stresses and the expression of its full virulence. Microbiology **148**: 3705–3713.

16. MARTCHENKO, M., A.-M. ALARCO, D. HARCUS & M. WHITEWAY. 2004. Superoxide dismutases in *Candida albicans*: transcriptional regulation and functional characterization of the hyphal-induced *SOD5* gene. Mol. Biol. Cell. **15**: 456–467.

17. PAN, S. & G.T. COLE. 1995. Molecular and biochemical characterization of a *Coccidioides immitis*-specific antigen. Infect. Immun. **63**: 3994–4002.

18. ZIMMERMANN, C.R., S.M. JOHNSON, G.W. MARTENS, *et al.* 1996. Cloning and expression of the complement fixation antigen-chitinase of *Coccidioides immitis*. Infect. Immun. **64**: 4967–4975.

19. THOMAS, P.W., E.E. WYCKOFF, E.J. PISHKO, *et al.* 1997. The *hsp60* gene of the human pathogenic fungus *Coccidioides immitis* encodes a T-cell reactive protein. Gene **199**: 83–91.

20. DELGADO, N. & G.T. COLE. 2003. A recombinant β-1,3-glucanosyltransferase homolog of *Coccidioides posadasii* protects mice against coccidiodomycosis. Infect. Immun. **71**: 3010–3019.

21. IVEY, F.D., D.M. MAGEE, M.D. WOITASKE, *et al.* 2003. Identification of a protective antigen of *Coccidioides immitis* by expression library immunization. Vaccine **21**: 4359–4367.

22. ORSBORN, K.I., L.F. SHUBITZ, T. PENG, *et al.* Protein expression profiling of *Coccidioides posadasii* by two-dimensional differential in-gel electrophoresis and evaluation of a newly recognized peroxisomal matrix protein as a recombinant vaccine candidate. Infect. Immun. **74**: 1865–1872.

Nuclear Labeling of *Coccidioides posadasii* with Green Fluorescent Protein

LEI LI,[a,b,d] MONIKA SCHMELZ,[a,d] ELLEN M. KELLNER,[a,b,d]
JOHN N. GALGIANI,[a,c,d] AND MARC J. ORBACH[a,b,d]

[a] *Valley Fever Center for Excellence, Tucson, Arizona, USA*

[b] *Department of Plant Sciences, Tucson, Arizona , USA*

[c] *Department of Internal Medicine, Tucson, Arizona, USA*

[d] *University of Arizona, and Southern Arizona Veteran's Affairs Health Care System, Tucson, Arizona, USA*

ABSTRACT: Coccidioidomycosis is a mild to life-threatening disease in otherwise healthy humans and other mammals caused by the fungus *Coccidioides* spp. Understanding the development of the unique dimorphic life cycle of *Coccidioides* spp. and its role in pathogenesis has been an area of research focus. However, nuclear behavior during the saprobic and parasitic life cycle has not been studied intensively. In this study, green fluorescent protein (GFP) was fused to histone H1 and introduced into *Coccidioides posadasii (C. posadasii)* strain Silveira to monitor the nuclear behavior of the fungus during the saprobic and parasitic stages of the life cycle. We constructed an *Agrobacterium tumefaciens*–mediated transformation (ATMT) vector that had in its T-DNA region a hygromycin-resistance gene as well as the fused *histone H1-GFP* gene under the control of the histone H3 promoter of *C. posadasii*. More than 30 hygromycin-resistant transformants were obtained and 23 were purified to homozygosity through multiple passages of the original transformants on hygromycin-containing media. One strain (VFC1420) transformed with a single copy of the fusion *histone H1-GFP* gene was selected for cytological studies. Strong nuclear-localized GFP signals were observed in arthroconidia, hyphae, as well as in spherules and endospores developed *in vitro*. Thus GFP can be used to study the expression pattern of potential virulence genes identified in serial analysis of gene expression (SAGE) or expressed sequence tags (EST) libraries, and could be a useful tool to monitor disease development in the murine model.

KEYWORDS: *Coccidioides*; coccidioidomycosis; Valley Fever; green fluorescent protein; GFP

Address for correspondence: Marc J. Orbach, Ph.D., Department of Plant Sciences, Forbes Bldg., Room 303, 1140 E. South Campus Dr., P. O. Box 210036, Tucson, AZ 85721. Voice: 520-621-3764; fax: 520-621-7186.
orbachmj@ag.arizona.edu

Ann. N.Y. Acad. Sci. 1111: 198–207 (2007). © 2007 New York Academy of Sciences.
doi: 10.1196/annals.1406.014

INTRODUCTION

Coccidioides spp. are ascomycetous fungi that cause Valley Fever, a respiratory disease, in humans and other mammals. They reside in the soil of the disease-endemic areas including the southwest part of the United States, Mexico, and the South America.[1] Arthroconidia are produced by intercalary fragmentation of mature hyphae and dispersed by wind following soil disturbance. Inhalation of the arthroconidia by humans or other mammals causes mild to severe disease, ranging from influenza-like symptoms to severe pneumonia, to extrapulmonary dissemination, depending on the immune status of the hosts.[2] Typically, upon inhalation through the respiratory route by the host, arthroconidia develop into spherules that enlarge over time with nuclear divisions followed by septation and internal cell development to produce hundreds of endospores capable of reinitiating the parasitic cycle and causing damage to lung tissue.[3] In the laboratory, the spherulation process can be mimicked by incubating the arthroconidia at 37–39°C with greater than 8% CO_2 in Converse media.[4]

Fluorescent proteins, including green fluorescent protein (GFP), are powerful tools in cell biology. Applications of the original GFP from the jellyfish *Aequorea victoria*[5] and a few derivatives of GFP[6,7] include localization of proteins, dynamic gene expression analysis, and protein–protein interactions. In the past decade, various forms of GFPs have been adapted for use in many prokaryotic and eukaryotic systems, including fungi.[8–10]

The purpose of this study was to monitor nuclear behavior during the *in vitro* life cycle of *Coccidioides* with strains whose nuclei were labeled with a histone h1-gfp fusion protein. The application of the study can be extended to monitor disease development in the murine model. The GFP fusion technique is also useful in studying gene expression at all stages of the life cycle of *Coccidioides* spp.

MATERIALS AND METHODS

Strains and Culture Conditions

All the transformants in this study were derived from *Coccidioides posadasii* (*C. posadasii*) strain Silveira (ATCC 28868) using *Agrobacterium tumefaciens*–mediated transformation (ATMT). All strains were maintained as saprobic cultures on 2 × GYE (2% glucose, 1% yeast extract, and 1.5% agar)[11] at room temperature. Arthroconidia were collected using a spinning magnetic bar procedure as previously reported.[12] Mycelial cultures were grown in 2 × GYE liquid media at 37°C with shaking at 125 rpm. Spherules were generated in modified Converse medium[4] at 38°C under 20% CO_2 as described previously.[2] All manipulations of viable *Coccidioides* spp. were performed using biological safety level 3 (BSL3) conditions and practices in a laboratory registered with the Centers for Disease Control for possession of this select agent.

Construction of Histone H1-GFP Fusion Plasmid pAM1420 and ATMT Transformation

A GFP derivative, sGFP, which has a serine to threonine substitution at position 65 (S65T)[13] and results in increased brightness, has been used in this study and is referred to as GFP. To generate the GFP fusion vector pAM1420 for ATMT in *C. posadasii*, a hygromycin-resistance gene (*hph*) and a *histone H1-GFP* fusion gene under the control of the histone H3 promoter of *C. posadasii* were cloned into the *A. tumefaciens* binary vector pAM1145 (FIG. 1).[14] Plasmid pAM1145 contained the *virG^{N54D}* gene, a mutated form of *virG*, to allow inducer-independent T-DNA transformation.[15] First, the *C. posadasii* histone H3 promoter was amplified from Silveira genomic DNA with primers OAM837 (5'-TTTGCGGCCGCTGTTTGCTACTGAAGTGGGTG-3') that had a *Not*I site near the 5' end and OAM838 (5'-GGGTCTAGATTTGAGATGTGTGAAGAATTGAG-3) with an *Xba*I site attached to its 5 end. Plasmid pAM1418 was generated by cloning the histone H3 promoter PCR fragment into pMF280[10] treated with *Not*I and *Xba*I to replace the *grg-1* promoter. Vector pAM1418, containing the *P histone H3-histone H1-GFP* fusion, was then digested with *Not*I, followed by treatment with the DNA polI Klenow fragment in the presence of dNTPs to fill in the cohesive ends. The fragment containing the *P histone H3-histone H1-GFP* fusion was then released with *Eco*RI. The *hph* cassette was isolated from pCB1004[16] by treatment with *Hpa*I, and cloned into pBluescript II SK (+) digested with *Sma*I to generate pAM804. Then the *hph* gene was released from pAM804 with *Bam*HI and *Eco*RV. To create pAM1420, a three-fragment ligation was performed with pAM1145 treated with *Bgl*II and *Eco*RI, the *P_{histionH3}-histone H1-GFP* fusion construct containing an *Eco*RI site and a blunt end, and the *Bam*HI-*Eco*RV *hph* gene fragment

FIGURE 1. A schematic diagram of pAM1420. *Open arrow*: virGN54D gene that enables T-DNA transfer without induction using acetosyringone; *filled box*: left border of T-DNA; *arrow with vertical bar*: hygromycin-resistance gene; *hatched box*: histone H3 promoter from Silveira; *dotted box*: histone H1 gene from *Neurospora crassa*; *filled arrow*: GFP gene; *open box*: right border of T-DNA; arrow with grid: *E. coli* replication origin; arrow with horizontal bar: ampicillin-resistance gene; B/B: ligated *Bam*HI and *Bgl*II compatible end; EV/N: blunted *Not*I site ligated with *Eco*RV; E: *Eco*RI.

(FIG. 1). Plasmid pAM1420 was introduced into *A. tumefaciens* strain AD965 by electroporation to create strain A1420, and Silveira was then transformed using A1420 as described previously.[14] In brief: arthroconidial germlings and A1420 were mixed and dispersed onto plates with sterile 82-mm nitrocellulose filters (Millipore Corporation, Bedford, MA, USA) containing AB induction medium.[14] After 48-h incubation at 28°C, the nitrocellulose filters were transferred to plates containing 2XGYE agar with 50 µg/mL of hygromycin and 100 µg/mL of kanamycin, which selected for *C. posadasii* strains that were transformed, and against further *A. tumefaciens* growth, respectively. Transformants were transferred to fresh plates with 2XGYE agar and 50 µg/mL of hygromycin. Then homokaryotic transformants were isolated by conidial passaging of the primary transformants as described previously.[14]

Culture and Slide Preparations for Imaging of C. posadasii

Mycelia, arthroconidia, or spherules from VFC1420, a transformant that contained a single integration of the T-DNA region from plasmid pAM1420, were spun down in 2-mL screw cap microcentrifuge tubes at 4,000 *g* for 5 min. The cells were then fixed with 3.7% formaldehyde in phosphate-buffered saline (PBS) (137 mM NaCl, 2.7 mM KCl, 10 mM Na_2HPO_4, and 2 mM KH_2PO_4, pH7.4) for 30 min, washed twice with PBS, and mounted.

Fluorescence and Confocal Microscopy

Expression of GFP in *C. posadasii* was examined with an Olympus BX51 epifluorescent microscope (Olympus America Inc., Center Valley, PA, USA) equipped with a 49002 ET GFP filter set (Chroma Technology Group, Rockingham, VT, USA). Images were recorded with a MicroFire™ digital camera (Model S99809) and the PictureFrame software version 1.1 (Olympus America Inc.). For confocal microscopy, GFP was excited at 488 nm with an argon ion laser source and signals emitted between 499 and 520 nm were collected using a Leica TCS SP-Leica DM IRBE confocal laser-scanning microscope (Leica Microsystems AG, Wetzlar, Germany). Images were collected as 8 to 12 serial sections along the Z axis of the cell, and processed with Adobe Photoshop 7.0 software (Adobe Systems, San Jose, CA, USA).

RESULTS

Transformant Selection and Homokaryon Isolation

We constructed a binary vector pAM1420 (FIG. 1), which has in its T-DNA region a hygromycin-resistance gene (*hph*) and a *Neurospora crassa* histone

H1 gene with a carboxy-terminal fusion to *GFP* under the control of the *C. posadasii* histone H3 promoter. This vector was transformed into Silveira by ATMT. More than 30 primary hygromycin-resistant transformants were obtained and 23 transformants were purified to homokaryons by two to three serial conidial isolations. Southern blot analysis of DNA isolated from nine transformants confirmed that three have single integrations of the T-DNA region of pAM1420. The other six transformants contained at least two copies of the T-DNA insertion. One of the three single insertion transformants, VFC1420, was selected for further observations.

Nuclei Observation in Arthroconidia, Mycelia, and Spherules

In VFC1420, strong expression of the GFP fusion protein was achieved with the *C. posadasii* histone H3 promoter in nuclei of arthroconidia (FIG. 2A), vegetative hyphae (FIG. 2B), developing spherules and endospores (FIG. 2C–G). All six transformants that contained more than two copies of the fusion genes also expressed GFP during saprobic and *in vitro* parasitic cycles (data not shown). GFP expression was stable in VFC1420 because the GFP signals were still detectable in arthroconidia after three conidial passages and when the transformant was kept at –20°C for at least 3 months.

Under *in vitro* spherulation conditions (see MATERIALS and METHODS), arthroconidia had initiated isotropic growth by 24 h post inoculation (FIG. 2C). The number of fluorescent nuclei in each arthroconidium ranged from one to four, with two nuclei being the most prevalent (FIG. 2A). The average number of nuclei increased during spherulation (FIG. 2), indicating continuous nuclear division up to 96 h post inoculation. At 48 h, there were two nuclei in each spherule on average (11 spherules observed) (FIG. 2D). At 72 h, the average number of nuclei increased to nine (nine spherules observed) (FIG. 2E). At 96 h, nuclear division resulted in an average of 12 nuclei in each spherule (five spherules observed) (FIG. 2F). After that, the average number of nuclei remained at 12, based on the observation from seven spherules that were lysing at 120 h post inoculation (FIG. 2G). A central vacuole was observed in each of the spherules developed between 72 and 96 h post inoculation (FIG. 2E, F). At 120 h post inoculation, most of the spherules were mature and lysed, releasing endospores (FIG. 2G). However, a few spherules were observed to be lysed as early as 96 h post inoculation (data not shown), which was consistent with an earlier observation.[17]

With the same laser intensity used for confocal microscopy, we detected autofluorescence on the mature spherule cell wall at 120 h post inoculation (FIG. 2G), but not in the spherule walls at earlier stages, suggesting possible alterations in cell wall structure during spherule maturation. However, the composition of the cell wall autofluorescence material was not defined.

DISCUSSION

GFP fusion methodology for examining gene expression patterns in fungi has been used in several other fungal systems.[8–10] A versatile construct containing a *histone H1-GFP* fusion gene developed originally for *N. crassa*,[10] allows localization of GFP to nuclei. Because the fusion protein is produced in the cytoplasm and imported into the nucleus, it allows for uniform labeling of all the nuclei in a common cytoplasm. Cytological studies were carried out in *N. crassa* to monitor nuclear organization and dynamics during vegetative growth and sexual reproduction.[10] During sexual reproduction, all eight ascospores were labeled brightly with GFP.[10] The same *histone H1-GFP* fusion cassette has been used in the rice blast fungus *Magnaporthe grisea* to label nuclei and observe the pattern of nuclear division during infection structure (appressorium) development.[18] In both *N. crassa* and *M. grisea*, fungal strains transformed with the *histone H1-GFP* fusion gene were normal in growth and development and the *M. grisea* strains were also normal in plant infection.[10,18] No toxicity effects were reported in these fungi transformed with the fusion construct.[10,18]

In the study reported here, the nuclear-localized GFP construct developed for *N. crassa* was used to examine nuclear divisions during *in vitro* growth of *C. posadasii*. As in *N. crassa* and *M. grisea*, no defect in saprobic growth was observed following introduction of the *histone H1-GFP* fusion in *C. posadasii*, and nuclei in hyphae and conidia were easily visualized. Nuclear dynamics were also visualized during *in vitro* spherulation. Spherule development was followed from initiation to lysis at 120 h after inoculation. At maturity, spherules contained an average of 12 nuclei expressing GFP, fewer than we expected, although we noted that the mature spherules were between 10 and 15 μm in diameter. These spherules were similar in size to spherules produced *in vitro* from untransformed *C. posadasii*. It is unlikely that the small number of nuclei observed, and thus the low number of endospores that would be produced, were due to a developmental defect, or toxicity from the overexpression of the histone H1-GFP fusion protein in VFC1420. We speculated that the low number of nuclei resulted from the relatively small spherules developed by the arthroconidia *in vitro*. In contrast, upon infection of mice, VFC1420 produced large spherules (100 μm) with more than 100 nuclei expressing GFP in the lung tissue 5 days after inoculation (L. F. Shubitz *et al.*, unpublished data). The normal virulence of VFC1420 also indicated that the expression of the GFP fusion in *C. posadasii* did not cause any apparent toxicity. Further systematic and comparative studies on spherule size and endospore production by different strains during *in vitro* and *in vivo* growth are needed. However, in an earlier study by Huppert and co-workers[19] where clear indication of the size of the spherules produced *in vitro* was available, the diameter of the spherules before rupture was 10–15 μm, similar to what we have observed in this study. In the Pappagianis' lab, the size of spherules produced

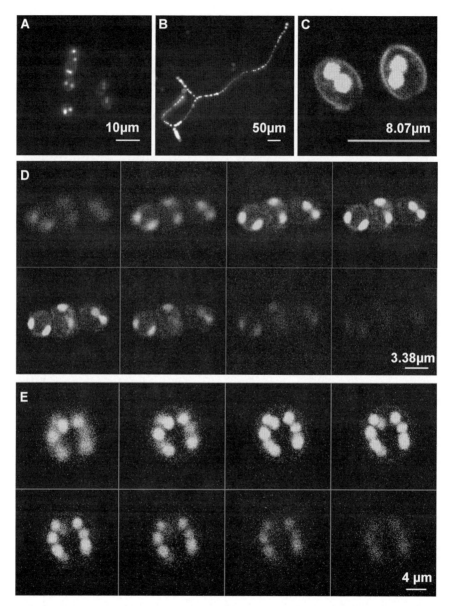

FIGURE 2. VFC1420 expressing GFP in the nuclei of athroconidia, hyphae, spherules, and endospores. GFP expression was detected in the nuclei of arthroconidia (**A**) and hyphae (**B**) by means of an epifluorescent microscope. Spherules of VFC1420 were observed with a confocal microscope. At 24 h (**C**), arthroconidia started to swell and form spherules. Serial sections along the Z axis of the spherules developed at 48 h (**D**), 72 h (**E**), 96 h (**F**), and 120 h (**G**) post inoculation were scanned, and the intervals between each step were 1.17 μm, 1.34 μm, 1.12 μm, 1.03 μm, respectively.

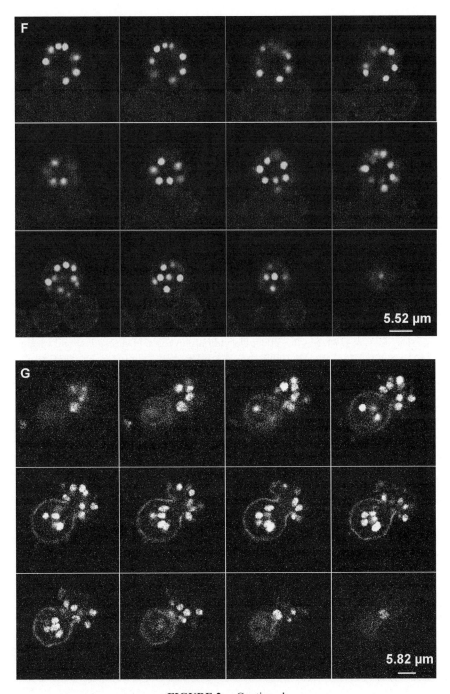

FIGURE 2. Continued.

in vitro from endospores was approximately 20 μm as well (Suzanne Johnson, personal communication).

Our data showed that *C. posadasii* nuclei labeled with histone h1-GFP exhibited bright and localized fluorescence, allowing one to observe nuclei throughout the fungal life cycle. Thus GFP is suitable to be used as a marker gene to study expression patterns of genes identified in serial analysis of gene expression (SAGE) and expressed sequence tags (EST) libraries of *Coccidioides* spp. or others for which expression is of interest. In addition, the GFP-expressing transformant VFC1420 can be used in a murine model to monitor disease development *in vivo*.

ACKNOWLEDGMENTS

This work was supported by the NIH-NIAD Grant 1 PO 1AI061310-01.

REFERENCES

1. PAPPAGIANIS, D. 1980. Epidemiology of coccidioidomycosis. *In* Coccidioidomycosis. D.A. Stevens, Ed. Plenum Medical Book Company. New York.
2. COX, R.A. & D.M. MAGEE. 2004. Coccidioidomycosis: host response and vaccine development. Clin. Microbiol. Rev. **17:** 804–839.
3. COLE, G.T. & S.H. SUN. 1985. Arthroconidium-spherule-endospore transformation in *Coccidioides immitis*. *In* Fungal Dimorphism: with Emphasis on Fungi Pathogenic for Humans. P.J. Szaniszlo & J. L. Harris, Eds.: 281–333. Plenum Press. New York.
4. CONVERSE, J.L. & A.R. BESEMER. 1959. Nutrition of the parasitic phase of *Coccidioides immitis* in a chemically defined liquid medium. J. Bacteriol. **78:** 231–239.
5. CHALFIE, M. *et al.* 1994. Green fluorescent protein as a marker for gene expression. Science **263:** 802–805.
6. TSIEN, R.Y. 1998. The green fluorescent protein. Annu. Rev. Biochem. **67:** 509–544.
7. ZIMMER, M. 2002. Green fluorescent protein (GFP): applications, structure, and related photophysical behavior. Chem. Rev. **102:** 759–781.
8. BOURETT, T.M. *et al.* 2002. Reef coral fluorescent proteins for visualizing fungal pathogens. Fungal Genet. Biol. **37:** 211–220.
9. LORANG, J.M. *et al.* 2001. Green fluorescent protein is lighting up fungal biology. Appl. Environ. Microbiol. **67:** 1987–1994.
10. FREITAG, M. *et al.* 2004. GFP as a tool to analyze the organization, dynamics and function of nuclei and microtubules in *Neurospora crassa*. Fungal Genet. Biol. **41:** 897–910.
11. FISHER, M.C. *et al.* 2002. Molecular and phenotypic description of *Coccidioides posadasii* sp. nov., previously recognized as the non-California population of *Coccidioides immitis*. Mycologia **94:** 73–84.
12. HUPPERT, M., S.H. SUN & A.J. GROSS. 1972. Evaluation of an experimental animal model for testing antifungal substances. Antimicrob. Agents Chemother. **1:** 367–372.

13. HEIM, R., A.B. CUBITT & R.Y. TSIEN. 1995. Improved green fluorescence. Nature **373:** 663.

14. KELLNER, E.M. *et al.* 2005. *Coccidioides posadasii* contains a single 1,3-beta-glucan synthase gene that appears to be essential for growth. Eukaryot. Cell **4:** 111–120.

15. GUBBA, S., Y.H. XIE & A. DAS. 1995. Regulation of *Agrobacterium tumefaciens* virulence gene expression: isolation of a mutation that restores *virG^{D52E}* function. Mol. Plant Microb. Interact. **8:** 788–791.

16. CARROLL, A.M., J.A. SWEIGARD & B. VALENT. 1994. Improved vectors for selecting resistance to hygromycin. Fungal Genet. Newsl. **41:** 22.

17. SUN, S.H. & M. HUPPERT. 1976. Cytological study of morphogenesis in *Coccidioides immitis*. Sabouraudia **14:** 185–198.

18. VENEAULT-FOURREY, C. *et al.* 2006. Autophagic fungal cell death is necessary for infection by the rice blast fungus. Science **312:** 580–583.

19. HUPPERT, M., S.H. SUN & J.L. HARRISON. 1982. Morphogenesis throughout saprobic and parasitic cycles of *Coccidioides immitis*. Mycopathologia **78:** 107–122.

Experimental Animal Models of Coccidioidomycosis

KARL V. CLEMONS,[1-3] JAVIER CAPILLA,[1-3] AND DAVID A. STEVENS[1-3]

[1] *California Institute for Medical Research, San Jose, California, USA*

[2] *Department of Medicine, Division of Infectious Diseases, Santa Clara Valley Medical Center, San Jose, California, USA*

[3] *Department of Medicine, Division of Infectious Diseases and Geographic Medicine, Stanford University, Stanford, California, USA*

ABSTRACT: Experimental models of coccidioidomycosis performed using various laboratory animals have been, and remain, a critical component of elucidation and understanding of the pathogenesis and host resistance to infection with *Coccidioides* spp., as well as to development of more efficacious antifungal therapies. The general availability of genetically defined strains, immunological reagents, ease of handling, and costs all contribute to the use of mice as the primary laboratory animal species for models of this disease. Five types of murine models are studied and include primary pulmonary disease, intraperitoneal with dissemination, intravenous infection emulating systemic disease, and intracranial or intrathecal infection emulating meningeal disease. Each of these models has been used to examine various aspects of host resistance, pathogenesis, or antifungal therapy. Other rodent species, such as rat, have been used much less frequently. A rabbit model of meningeal disease, established by intracisternal infection, has proven to model human meningitis well. This model is useful in studies of host response, as well as in therapy studies. A variety of other animal species including dogs, primates, and guinea pigs have been used to study host response and vaccine efficacy. However, cost and increased needs of animal care and husbandry are limitations that influence the use of the larger animal species.

KEYWORDS: coccidioidomycosis; animal models; *Coccidioides*

INTRODUCTION

The use of laboratory animals is an integral part of the conduct of medical and biological research. For coccidioidomycosis, much of what we know about host

Address for correspondence: Karl V. Clemons, Ph.D., Division of Infectious Diseases, Santa Clara Valley Medical Center, 751 South Bascom Ave., San Jose, CA 95128-2699. Voice: 408-998-4557; fax: 408-998-2723.

clemons@cimr.org

Ann. N.Y. Acad. Sci. 1111: 208–224 (2007). © 2007 New York Academy of Sciences.
doi: 10.1196/annals.1406.029

resistance, pathogenesis, virulence, disease progression, treatment, and even the geographic epidemiology of the disease are results of the use of animal models. Various animals have been used for modeling of coccidioidomycosis and include mice, rats, guinea pigs, primates, and dogs.[1,2] However, the primary animal used for experimental infection with *Coccidioides immitis* or *Coccidioides posadasii* (*C. posadasii)* remains the mouse. The availability of genetically defined strains including specific gene knockout mice, the lower costs of purchase and housing compared to larger animal species, the relative ease of handling, and the ready availability of immunological reagents all contribute to their usefulness. The next most frequently used animal, over the last several years, has become the rabbit.

Prior to discussing the performance of animal models of coccidioidal infection, one must be aware of the biological safety hazard that working with *Coccidioides* spp. presents to the investigator. The formation of arthroconidia when cultured on agar and their ready dispersion into air presents a severe risk for accidental exposure by inhalation. Because of this risk, *Coccidioides* spp. are considered Biosafety Level 3 pathogens and must be worked with in the appropriate physical facility inside of class II or class III biosafety cabinets, as described by the Centers for Disease Control (CDC) in the fourth edition of *Biosafety in Microbiological and Biomedical Laboratories*.[3] Although dry arthroconidia present a great risk, once the organism has converted to the spherule form in tissues there is much less risk because the spherule form is substantially less virulent.[4] Thus, animals inoculated parenterally are maintained under Animal Biosafety Level 2 conditions. In addition, the enactment of the Select Agent Final Rule (49 CFR part 73; 9 CFR part 121) by the United States government has resulted in *Coccidioides* spp. being considered potential agents of bioterrorism. As such, laboratories in the United States working with this organism must be registered with the Select Agent Program under the auspices of the CDC or the United States Department of Agriculture (USDA). All personnel must be cleared by the U.S. Department of Justice, and institutions must adhere to regulations concerning various items including security, training, biosafety, personnel, records, and transfer of listed agents.

In this brief review our aim is to give the reader a concise overview of the animal models of coccidioidomycosis that have been developed and how they were used. Other published reviews have more fully described the details of performing studies in animal models of various fungal infections including coccidioidomycosis, the benefits and drawbacks of animal data, as well as the types of parameters used to evaluate infection.[1,2,5–10] The choice of the model used is critical, as are its reproducibility, how closely it approximates human disease, the specific parameters followed, and the relative ease with which it can be used to obtain statistically valid data. Drawbacks and confounding factors include differences in host susceptibility, differences in fungal strain virulence, the initiation of disease using an unnatural route of infection, and the often acutely severe nature of the induced laboratory infection. However, many of the drawbacks can be overcome by the benefit of the exquisite control the investigator has over many variables, such as infecting fungal strain, route of infection, number of organisms in the inoculum, species and strain of laboratory animal, and duration of disease. With these thoughts in mind, we will review how laboratory animal models of coccidioidomycosis have been used and provide examples derived primarily from studies done in mice and rabbits.

ROUTES AND TYPES OF INFECTION

Five routes of infection have been used to establish disease in mice. The pulmonary route (intranasal, intratracheal, or inhalational; for primary pulmonary disease) and intraperitoneal or intravenous inoculation (for disseminated disease) using arthroconidia are the most commonly used.[1,2,5] Intracranial inoculation using endospores or intrathecal inoculation using arthroconidia are done to model central nervous system disease and meningitis.[11–15] Arthroconidia are highly virulent, with as few as 10 arthroconidia delivered to the lungs causing lethality. By other routes of infection arthroconidia remain virulent, but may require several hundred to cause lethality by the intraperitoneal route; injection of ca. 30 arthroconidia by the intravenous or intrathecal route is also lethal. Endospores or spherules are significantly less virulent than are arthroconidia.[4] Interestingly, high numbers (ca. 10^6) of endospores given intraperitoneally, but not intravenously have been reported to cause a circling syndrome without meningitis.[4] Similar to work in mice, pulmonary, intravenous, and intrathecal routes have been used to establish infection in other species of laboratory animals.[1,2]

One interesting use of mice has been intraperitoneal injection of suspected isolates of *Coccidioides* to confirm identification by conversion *in vivo* into the endosporulating spherule, whereas fungi that closely resemble *Coccidioides in vitro* do not convert into spherules. In addition, mice have been used to isolate the organism from environmental samples of soil.[16] However, direct intraperitoneal injection of soil suspensions can often be lethal on account of bacterial load or to other toxicities of the suspension.

STUDIES OF PATHOGENESIS AND HOST RESPONSE

Pathogenesis and Virulence

Through the years, a variety of investigators have used murine models to examine pathogenesis and virulence. Early studies demonstrated that after intravenous inoculation of large numbers of arthroconidia, cellular changes occurred very early and that spherules could be observed in the tissues as early as 48 h post infection.[17] Because natural infection is acquired via inhalation, there have been studies done following the temporal progression of pulmonary infection. Primary pulmonary infection in either mice or guinea pigs demonstrated that extrapulmonary dissemination appears to occur as early as 7 days post infection.[18,19] Overall, in numerous studies, in mice and other animal species, the course of infection has been shown to be progressive with increases in the fungal burden in the organs, and increases in organ weight.[1, 17–26]

The relative virulence of a pathogen is determined by both the organism and the host. For *Coccidioides,* differences in the virulence of strains have been shown in murine models of infection. Although most are virulent in mice, there is a continuum of virulence among the strains tested.[27–30] What genes are responsible for the virulence of *Coccidioides* is a question under study. Various gene products and processes have been suggested as virulence factors, evidence arising from murine studies where, for example, a strain of *Coccidioides* that is deficient in a process or gene product causes less severe or no disease compared to the parental wild-type strain.[28–36] Among the proposed virulence factors are proteinases, urease, dimorphism from arthroconidia to spherules, heat-shock proteins, and cell wall components. Global examination of differential expression of genes between mycelial and spherule forms has recently been reported and may elucidate additional potential virulence factors.[31]

Host Susceptibility and Response

Also important to the virulence of *Coccidioides* is the susceptibility of the host. Natural infections of various animal species have shown that some animals, like simians, dogs, and rodents are highly susceptible, whereas others, such as cows or cats are very resistant to infection.[1,22] Interestingly, susceptibility difference within a species, similar to that described in humans,[21,22,32–34] is also evident, which is the case for dogs, with some breeds of dogs, like boxers, much more susceptible than other breeds of dogs. Risk factors include amount of time spent outdoors and in areas of endemicity.[21,22,35,36] However, experimental infection studies have not been done to further substantiate differences in canine breed susceptibility.

Laboratory studies of host susceptibility have used murine models.[18,37–40] After pulmonary or intraperitoneal infection, BALB/c mice are highly susceptible compared to DBA/2 mice, which are significantly more resistant [37–40]; Swiss Webster mice were more resistant than C57BL/6 or BALB/c after intravenous inoculation.[18] A genetic basis for resistance to intraperitoneally induced infection has been related to a single locus in the mouse.[38] In contrast to susceptibility differences of mouse strains after pulmonary, intraperitoneal, or intravenous routes of infection, no significant differences in mouse strain resistance were found in studies of murine meningitis,[14] which supports the idea that differences in resistance of mice to coccidioidomycosis may be in part dependent on route of infection, similar to that reported for *Blastomyces dermatitidis.*[41]

Host Response

Differences in mouse strain resistance are also due in large part to differences in host response to coccidioidal infection.[34] The importance of macrophages, lymphocytes, polymorphonuclear leukocytes (PMNs), and various cytokines to resistance against coccidioidomycosis has been examined in experimental murine models, as well as in dogs and primates.[1,2,5,34,42–44] In early studies by Savage and Madin,[45] the cellular response in the lungs of vaccinated mice was examined after intranasal challenge and suggested the involvement of macrophages as cells better able to kill infecting arthroconidia; similar histopathologic studies of vaccinated animals were done by several other investigators, demonstrating the development of the granulomatous response.[20,24,46] T cell-deficient *nu/nu* mice, which are unable to mount an effective adaptive immune response, were highly susceptible to infection, and macrophages were particularly important to resistance of previously immunized mice in comparison with normal control mice, whereas PMN dysfunction in *bg/bg* mice did not appear to play as significant a role in resistance.[18,43,47]

The host response of DBA/2 mice has been shown to be directed toward a protective Th1 response, related to interleukin (IL)-12 and interferon-γ, whereas susceptible BALB/c exhibit a Th2 response, related to IL-4 and IL-10, and less resistance to infection.[34,39,40,48] Additionally, gene knockout mice have proven useful in demonstrating that IL-10 plays a significant role in the susceptibility of the mouse. IL-10 knockouts were more resistant to infection than were normal control mice.[49–52] More recently, dendritic cells and their involvement in the development of adaptive immunity have been examined, further indicating differences in Toll-like receptors (TLR)2 or TLR4 and IL-12 expression between DBA/2 and BALB/c mice and that transfected dendritic cells could be used to immunize naïve mice.[53,54]

A major tenet of fungal immunology is that a protective adaptive response is due to T cells and specifically a Th1 response. Interestingly, gene expression studies have shown that a large number of B cell immunoglobulin-related genes are upregulated in BALB/c mice previously vaccinated with formalin-killed spherules (FKS) and then challenged as compared to sham-vaccinated controls.[55] Furthermore, MuMT mice (these animals lack mature B cells and hence are antibody-deficient mice) could not be protected to the same extent as immunocompetent control mice by vaccination with FKS, indicating that B cells (possibly immunoglobulins) are indeed playing a role in protective immunity against coccidioidomycosis.[55]

Vaccine Studies

A major effort over the last 50 or more years has been the development of a useful protective vaccine against coccidioidomycosis.[34,56–60] Various animal models have been used during this time including mice, dogs, rats, guinea pigs, and simians.[1,34,56,59,61,62] In many early studies mice were vaccinated intraperitoneally, intramuscularly, or subcutaneously using a variety of preparations including killed arthroconidia, mycelia fragments or spherules, or live organisms and then challenged by intraperitoneal or pulmonary routes.[19,23,45,46,60,62] Interestingly, studies done in mice demonstrated that administration of spherules or endospores intravenously rather than intramuscularly or subcutaneously resulted in impaired immunity and lack of protection to challenge.[63–67] Overall, these animal model studies led to the development of the FKS vaccine, which was taken to clinical trial.[68]

With advances in laboratory technology, vaccine preparations for prevention of coccidioidomycosis have focused on subcellular and more recently to specifically cloned antigens, as well as improved ways to present the antigens, the use of engineered chimeric proteins and different adjuvants, many of which have shown some protective activity.[34,56,69–83] These studies have primarily used mouse models of infection and have been performed by pretreatment of the mice with the vaccine preparation, usually multiple doses given subcutaneously, intramuscularly, or intraperitoneally, and challenged by intraperitoneal or pulmonary route 2 to 4 weeks later. Evaluation of protection is then based on prolonged survival and in some studies on the number of CFUs of *Coccidioides* remaining in the target organs (e.g., lungs).

ANTIFUNGAL THERAPY STUDIES

The overall increase in the number of serious fungal infections, including coccidioidomycosis, over the last two decades has stimulated industry to expand the research into new antifungal compounds. We have recently reviewed the utility and methods used for performing animal models of fungal infection in preclinical trial studies.[8,9] Preclinical trial testing of antifungals is integral to their development, and models of coccidioidomycosis have been included during this phase of testing. In general, the design of the studies will depend on the question being asked. Included should be appropriate placebo or untreated controls, a positive comparator control drug that is known to have efficacy and one or more groups that are receiving the compound being tested (often these are dose-escalation evaluations). After infection, the parameters for efficacy are most often prolongation of survival or reduction of fungal burden from the target organs.

For coccidioidomycosis, it has been primarily mice and rabbits (discussed separately below) that have been used, although primates and dogs have also been involved in testing. Both pulmonary and systemic models of infection have been used to test the efficacy of polyenes (e.g., amphotericin B deoxycholate, lipid-based amphotericin B, etc.), polyketides (e.g., ambruticin), azoles (e.g., miconazole, ketoconazole, itraconazole, fluconazole, posaconazole, etc.), sordarins, echinocandins (e.g., cilofungin and caspofungin), and other experimental antifungal compounds or therapies.[11,12,15,84–116] These preclinical trials have contributed to the approval and availability of miconazole, ketoconazole, itraconazole, fluconazole, and posaconazole, as well as the lipid-based formulations of amphotericin B.

Interestingly, a majority of studies have examined the efficacy of monotherapy and few studies have been done testing combination therapy, such as have been done for aspergillosis and other fungal infections.[8,9] The potential role of immunotherapy as an adjunct to conventional antifungal therapy has been examined only to a limited degree, with IL-2, showing no deleterious or beneficial effects in murine model studies.[103] The potential adjunctive use of other cytokines, such as IL-12, interferon-γ, or colony-stimulating factors, have not been reported.

RABBIT MODELS

Although not as often used as mice, rabbits have been used as a model of coccidioidomycosis, particularly for the study of vasculitis and meningitis associated with dissemination of systemic coccidioidomycosis to the central nervous system (CNS).[117] Rabbits have proven to be very successful as a model for the study of coccidioidal meningitis (CM) because the induced disease closely mimics CNS involvement observed in patients with acute infection. To reproduce meningitis the animals are infected with a viable arthroconidial suspension of C. immitis, while under general anesthesia. Inoculation is then done by direct puncture into the cisterna magna. Inocula of 3×10^5 arthroconidia per animal produce a progressive illness of 2 to 7 weeks' duration; use of higher inocula results in an acute course of disease requiring euthanasia of infected animals as early as 7 days post infection. Histologically, infected rabbits show chronic granulomatous cerebrospinal meningitis with cerebritis, arteritis, and infarction, simulating the findings in complicated human coccidioidal disease.[117] The inflammatory process and the progressive pleocytosis and the protein elevation found in the animals reproduce also the findings in humans,[33,118,119] which makes this model useful for understanding the pathologic process of the disease, as well as for assessing antifungal therapy.

Due to the size of the animals, normally within 2–3 kg, it is relatively easy to do multiple sampling of the animals during the infection and obtain adequate volumes of blood and cerebrospinal fluid (CSFs) to perform assays that are not always possible if smaller laboratory animals are used. The temporal course of colony-forming units (CFUs) of Coccidioides, white cells, glucose, protein, and drug levels in CSF and serum can be assayed at different time points during the infection without euthanasia of animals to obtain sufficiently sized samples (0.5 to 1 mL CSF and up to 40 mL whole blood),[117,120–123] as opposed to similar studies performed in mice, which require terminal procedures for collection of only 10 to 20 μL of CSF or 1 mL of whole blood.

The efficacy of therapy or the impact of the disease in animal models is classically measured by comparing survival rate, tissue burdens, but not always by histopathological examination. Using rabbits, we can also get additional information from clinical observation. As demonstrated previously, clinical parameters like posturing, mobility, temperature, and body weight are good indicators of the infection progress.[117,120–123]

Host Response Studies in Rabbits

Using the rabbit model, we demonstrated an increasing level of matrix metalloproteinase-9 (MMP-9) in the CSF with time in infected animals.[117,123,124] MMP-9 is involved in vascular regeneration and during the infectious process may be involved in inflammation and vasculitis, which are the major complications associated with CM.[118,119]

Due to the size of the internal organs of rabbits in comparison to mice, rabbit models allow us to focus on one aspect of the infection more easily than can be done using mice. In a recent publication, Zucker *et al.* used the rabbit model to quantify gene expression of inflammatory mediators in brain basilar artery.[124] The physical size and architecture of the vessels of the rabbit's CNS allow us to dissect and isolate large vessels for later analysis. Using RT-PCR we demonstrated differences between infected and uninfected animals in expression of inflammatory mediators, such as iNOS, interferon-γ, IL-6, and IL-10 among others, through the course of the infection, which furthers the understanding of the infectious process and how the host responds immunologically.

Therapy Studies in Rabbits

The efficacy of different antifungal agents including fluconazole (FCZ), itraconazole (ICZ), terbinafine (TF), conventional amphotericin B (AMB), liposomal AMB (AmBisome; LAMB), and lipid-complexed amphotericin B (Abelcet; ABLC) against CM has been tested using rabbits as model.[120–122,125] These studies showed the efficacy of the drugs as increasing and diminishing, survival and fungal burdens, respectively. FCZ, ICZ, and AMB, LAMB or ABLC showed greater efficacy than TF in prolonging survival and reducing CFU from CNS. Assay of the drugs in CSF showed great penetration of FCZ and lower to not detectable for ICZ, LAMB, ABLC, AMB, or TF. Studies on the comparative pharmacokinetics of amphotericin B colloidal dispersion (ABCD) and conventional AMB, both given intrathecally, were also done in rabbits.[126] Intra-CSF administration of conventional AMB is given to patients who are acutely ill or that do not respond satisfactorily to oral azole treatment. However, problems of toxicity are common and must be carefully monitored.[127–130] Intrathecal delivery of AMB in rabbits resulted in higher levels of drug than administration of ABCD, but was significantly more toxic and suggestive that intrathecal administration of lipid-carried AMB might be considered clinically.

Although using rabbits for models of infection has advantages, rabbits as the model system also have drawbacks. Rabbits require more monitoring and care than small mammals and as a consequence experimental groups often consist of fewer animals. In addition, the lack of genetically defined or modified gene knockout rabbits and specific immunological reagents, such as antibodies and gene primers, makes using the rabbit as a model more difficult.

CONCLUSIONS

In this brief review we have presented a summary of how animal models of coccidioidomycosis have contributed to our understanding of many aspects of this disease. Models have been performed on studies ranging from virulence of the organism to host response to prevention to therapy of the disease. In addition, identification of environmental and clinical isolates is still often confirmed using mice. As laboratory technology continues to become more sophisticated, we will continue to use well defined and reproducible models of infection to further our understanding of the immunological response and mechanisms whereby the organism counters the host or is able to cause primary, as well as associated, disease and to develop new and more effective therapeutics.

It is apparent to us that the use of animal models of coccidioidomycosis has contributed greatly to our overall understanding of this disease with respect to its pathogenesis, how the host responds, its prevention and its treatment. However, models have limitations, which must be taken into consideration. Changes in the apparent virulence of the organism (e.g., pulmonary infection is more lethal than intraperitoneal) based on route of infection, susceptibility of the host (e.g., strain of mouse used) or even morphologic form of the organism (e.g., arthroconidia are virulent whereas spherules or endospores are relatively avirulent) are examples of limitations and pitfalls in interpretation. Our knowledge of the host–parasite interaction has expanded through the use of models. As genetically defined animals have become available, we can examine how a single cytokine alters the host response to infection; studies such as these cannot be accomplished by *in vitro* studies where the multitude of cellular interactions and regulatory processes are not mimicked. Where we once believed that cell-mediated immunity was solely responsible for a protective immunological response to coccidioidomycosis, through the use of models we now understand that the protective response is the result of many immunoregulatory molecules that affect both the cell-mediated (e.g., Th1 or Th2 responses) and antibody-mediated arms of immunity and that both may have a role to play in host resistance. The data reviewed in this article also demonstrate to us that what is efficacious in a model may not prove so in humans. This was the case for the FKS vaccine studies, which showed great promise in mice and monkeys, but did not ultimately prove useful in humans on account of reactogenic side effects. Thus, while animal models will continue to be the mainstay of vaccine development against coccidioidomycosis, determination of efficacy and utility will ultimately require human clinical trials. Lastly, the data reviewed show us how important models are to the development of antifungal therapy and how well the animal results predict drug efficacy in humans, which continues to improve our therapeutic options. Although *in vitro* antifungal activity may be shown by a drug against *Coccidioides*, *in vivo* that same drug may have no efficacy, as was the case, for example, with the echinocandin, cilofungin. Overall, we have been reminded that an *in vitro* study cannot substitute for study of what happens *in vivo*, and thus we do not foresee animal models of coccidioidomycosis losing their utility in the future.

REFERENCES

1. BRUMMER, E. & K.V. CLEMONS. 1987. Animal models of systemic mycoses. *In* Animal Models in Medical Mycology. M. Miyaji, Ed.: 79–95. CRC Press. Boca Raton, FL.
2. SORENSEN, K.N., K.V. CLEMONS & D.A. STEVENS. 1999. Murine models of blastomycosis, coccidioidomycosis, and histoplasmosis. Mycopathologia **146:** 53–65.
3. RICHMOND, J.Y. & R.W. MCKINNEY. 1999. Biosafety in microbiological and biomedical laboratories. 4th Edition, U.S. Government Printing Office, Washington D.C.
4. SCALARONE, G.M. & R.W. HUNTINGTON. 1983. Circling syndrome and inner ear disease in mice infected intraperitoneally or intravenously with *Coccidioides immitis* spherule-endospore phase cultures. Mycopathologia **83:** 75–86.
5. CLEMONS, K.V. & D.A. STEVENS. 2001. Overview of host defense mechanisms in systemic mycoses and the basis for immunotherapy. Semin. Respir. Infect. **16:** 60–66.
6. CLEMONS, K.V. & D.A. STEVENS. 2002. Immunomodulation of fungal infections: do immunomodulators have a role in treating mycoses? EOS Rivista di Immunologia ed Immunofarmacologia **22:** 29–32.
7. CLEMONS, K.V. & D.A. STEVENS. 2005. The contribution of animal models of aspergillosis to understanding pathogenesis, therapy and virulence. Med. Mycol. **43**(Suppl 1): S101–S110.
8. CLEMONS, K.V. & D.A. STEVENS. 2006. Animal models testing monotherapy versus combination antifungal therapy: lessons learned and future directions. Curr. Opin. Infect. Dis. **19:** 360–364.
9. CLEMONS, K.V. & D.A. STEVENS. 2006. Animal models of *Aspergillus* infection in preclinical trials, diagnostics and pharmacodynamics: What can we learn from them? Med. Mycol. **44:** 119–126.
10. STEVENS, D.A. 1996. Animal models in the evaluation of antifungal drugs. J. Mycol. Med. **6**(Suppl. I): 7–10.

11. ALLENDOERFER, R. *et al.* 1992. Comparison of SCH 39304 and its isomers, RR 42427 and SS 42426, for treatment of murine cryptococcal and coccidioidal meningitis. Antimicrob. Agents Chemother. **36:** 217–219.
12. ALLENDOERFER, R. *et al.* 1992. Comparison of amphotericin B lipid complex with amphotericin B and SCH 39304 in the treatment of murine coccidioidal meningitis. J. Med. Vet. Mycol. **30:** 377–384.
13. GRAYBILL, J.R., S.H. SUN & J. AHRENS. 1986. Treatment of murine coccidioidal meningitis with fluconazole (UK 49,858). J. Med. Vet. Mycol. **24:** 113–119.
14. KAMBERI, P. *et al.* 2003. A murine model of coccidioidal meningitis. J. Infect. Dis. **187:** 453–460.
15. DEFAVERI, J., S.H. SUN & J.R. GRAYBILL. 1990. Treatment of murine coccidioidal meningitis with SCH39304. Antimicrob. Agents Chemother. **34:** 663–664.
16. SWATEK, F.E. & D.T. OMIECZYNSKI. 1970. Isolation and identification of *Coccidioides immitis* from natural sources. Mycopathol. Mycol. Appl. **41:** 155–166.
17. SINSKI, J.T. & P.J. SOTO, JR. 1966. Onset of coccidioidomycosis in mouse lung after intravenous injection. Mycopathol. Mycol. Appl. **30:** 41–46.
18. CLEMONS, K.V., C.R. LEATHERS & K.W. LEE. 1985. Systemic *Coccidioides immitis* infection in nude and beige mice. Infect. Immun.. **47:** 814–821.
19. COX, R.A., E.F. PAVEY & C.G. MEAD. 1981. Course of coccidioidomycosis in intratracheally infected guinea pigs. Infect. Immun. **31:** 679–686.
20. CONVERSE, J.L. *et al.* 1962. Pathogenesis of *Coccidioides immitis* in monkeys. J. Bacteriol. **83:** 871–878.
21. DRUTZ, D.J. & A. CATANZARO. 1978. Coccidioidomycosis. Part II. Am. Rev. Respir. Dis. **117:** 727–771.
22. DRUTZ, D.J. & A. CATANZARO. 1978. Coccidioidomycosis. Part I. Am. Rev. Respir. Dis. **117:** 559–585.
23. HUPPERT, M. *et al.* 1976. Lung weight parallels disease severity in experimental coccidioidomycosis. Infect. Immun. **14:** 1356–1368.
24. KONG, Y.C. *et al.* 1964. Fungal multiplication and histopathologic changes in vaccinated mice infected with *Coccidioides immitis.* J. Immunol. **92:** 779–790.
25. SORENSEN, R.H., T. BREEN & V. BENZLER. 1973. Experimental coccidioidal meningitis in rabbits with a procedure for cerebrospinal fluid sampling. Evaluation of the lactic dehydrogenase isoenzymes and coccidioidin latex agglutination results. Am. J. Med. Sci. **265:** 55–61.
26. CONVERSE, J.L. & R.E. REED. 1966. Experimental epidemiology of coccidioidomycosis. Bacteriol. Rev. **30:** 678–695.
27. BERMAN, R.J. *et al.* 1956. The virulence and infectivity of twenty-seven strains of *Coccidioides immitis.* Am. J. Hyg. **64:** 198–210.
28. FRIEDMAN, L., C.E. SMITH & L.E. GORDON. 1955. The assay of virulence of *Coccidioides* in white mice. J. Infect. Dis. **97:** 311–316.
29. KARRER, H.E. 1953. Virulence of *Coccidioides immitis* determined by intracerebral inoculation in mice. Proc. Soc. Exp. Biol. Med. **82:** 766–768.
30. PAPPAGIANIS, D., C.E. SMITH & G.S. KOBAYASHI. 1956. Relationship of the *in vivo* form of *Coccidioides immitis* to virulence. J. Infect. Dis. **98:** 312–319.
31. JOHANNESSON, H. *et al.* 2006. Phase-specific gene expression underlying morphological adaptations of the dimorphic human pathogenic fungus, *Coccidioides posadasii.* Fungal Genet. Biol. **43:** 545–559.
32. GALGIANI, J.N. 1999. Coccidioidomycosis: a regional disease of national importance. Rethinking approaches for control. Ann. Intern. Med. **130:** 293–300.

33. STEVENS, D.A. 1995. Coccidioidomycosis. N. Engl. J. Med. **332:** 1077–1082.

34. COX, R.A. & D.M. MAGEE. 2004. Coccidioidomycosis: host response and vaccine development. Clin. Microbiol. Rev. **17:** 804–39, table of contents.

35. BUTKIEWICZ, C.D., L.E. SHUBITZ & S.M. DIAL. 2005. Risk factors associated with *Coccidioides* infection in dogs. J. Am. Vet. Med. Assoc. **226:** 1851–1854.

36. SHUBITZ, L.E. *et al.* 2005. Incidence of *Coccidioides* infection among dogs residing in a region in which the organism is endemic. J. Am. Vet. Med. Assoc. **226:** 1846–1850.

37. KIRKLAND, T.N. & J. FIERER. 1983. Inbred mouse strains differ in resistance to lethal *Coccidioides immitis* infection. Infect. Immun. **40:** 912–916.

38. KIRKLAND, T.N. & J. FIERER. 1985. Genetic control of resistance to *Coccidioides immitis*: a single gene that is expressed in spleen cells determines resistance. J. Immunol. **135:** 548–552.

39. MAGEE, D.M. & R.A. COX. 1995. Roles of gamma interferon and interleukin-4 in genetically determined resistance to *Coccidioides immitis*. Infect. Immun. **63:** 3514–3519.

40. MAGEE, D.M. & R.A. COX. 1996. Interleukin-12 regulation of host defenses against *Coccidioides immitis*. Infect. Immun. **64:** 3609–3613.

41. MOROZUMI, P.A., E. BRUMMER & D.A. STEVENS. 1981. Strain differences in resistance to infection reversed by route of challenge: studies in blastomycosis. Infect. Immun. **34:** 623–625.

42. BEAMAN, L. & C.A. HOLMBERG. 1980. *In vitro* response of alveolar macrophages to infection with *Coccidioides immitis*. Infect. Immun. **28:** 594–600.

43. BEAMAN, L., D. PAPPAGIANIS & E. BENJAMINI. 1977. Significance of T cells in resistance to experimental murine coccidioidomycosis. Infect. Immun. **17:** 580–585.

44. WEGNER, T.N. *et al.* 1972. Some evidence for the development of a phagocytic response by polymorphonuclear leukocytes recovered from the venous blood of dogs inoculated with *Coccidioides immitis* or vaccinated with an irradiated spherule vaccine. Am. Rev. Respir. Dis. **105:** 845–849.

45. SAVAGE, D.C. & S.H. MADIN. 1968. Cellular responses in lungs of immunized mice to intranasal infection with *Coccidioides immitis*. Sabouraudia **6:** 94–102.

46. PAPPAGIANIS, D. 1967. Histopathologic response of mice to killed vaccines of *Coccidioides immitis*. J. Invest. Dermatol. **49:** 71–77.

47. CLEMONS, K.V., C.R. LEATHERS & K.W. LEE. 1985. Role of activated macrophages in resistance to experimental *Coccidioides immitis* infections. *In* Coccidioidomycosis: Proceedings of the Fourth International Conference. H.E. Einstein & A. Cantanzaro, Eds.: 149–159. The National Foundation for Infectious Diseases. Washington D.C.

48. COX, R.A. & D.M. MAGEE. 1998. Protective immunity in coccidioidomycosis. Res. Immunol. **149:** 417–428; discussion: 506-7.

49. FIERER, J. 2006. IL-10 and susceptibility to *Coccidioides immitis* infection. Trends Microbiol. **14:** 426–427.

50. FIERER, J. *et al.* 1998. Importance of interleukin-10 in genetic susceptibility of mice to *Coccidioides immitis*. Infect. Immun. **66:** 4397–4402.

51. FIERER, J., L. WALLS & T.N. KIRKLAND. 2000. Genetic evidence for the role of the Lv locus in early susceptibility but not IL-10 synthesis in experimental coccidioidomycosis in C57BL mice. J. Infect. Dis. **181:** 681–685.

52. FIERER, J. *et al.* 1999. Genes influencing resistance to *Coccidioides immitis* and the interleukin-10 response map to chromosomes 4 and 6 in mice. Infect. Immun. **67:** 2916–2919.

53. AWASTHI, S. *et al.* 2005. Efficacy of antigen 2/proline-rich antigen cDNA-transfected dendritic cells in immunization of mice against *Coccidioides posadasii*. J. Immunol. **175:** 3900–3906.

54. AWASTHI, S. & D.M. MAGEE. 2004. Differences in expression of cell surface co-stimulatory molecules, Toll-like receptor genes and secretion of IL-12 by bone marrow-derived dendritic cells from susceptible and resistant mouse strains in response to *Coccidioides posadasii*. Cell. Immunol. **231:** 49–55.

55. MAGEE, D.M. *et al.* 2005. Role of B cells in vaccine-induced immunity against coccidioidomycosis. Infect. Immun. **73:** 7011–7013.

56. COLE, G.T. *et al.* 2004. A vaccine against coccidioidomycosis is justified and attainable. Med. Mycol. **42:** 189–216.

57. DIXON, D.M. 2001. *Coccidioides immitis* as a select agent of bioterrorism. J. Appl. Microbiol. **91:** 602–605.

58. DIXON, D.M. *et al.* 1998. Development of vaccines and their use in the prevention of fungal infections. Med. Mycol. **36:** 57–67.

59. LEVINE, H.B., D. PAPPAGIANIS & J.M. COBB. 1970. Development of vaccines for coccidioidomycosis. Mycopathol. Mycol. Appl. **41:** 177–185.

60. PAPPAGIANIS, D. & H.B. LEVINE. 1975. The present status of vaccination against coccidioidomycosis in man. Am. J. Epidemiol. **102:** 30–41.

61. CASTLEBERRY, M.W., J.L. CONVERSE & P.J. SOTO JR. 1964. Antibiotic control of tissue reactions in dogs vaccinated with viable cells of *Coccidioides immitis*. J. Bacteriol. **87:** 1216–1220.

62. PAPPAGIANIS, D. *et al.* 1960. Response of monkeys to respiratory challenge following subcutaneous inoculation with *Coccidioides immitis*. Am. Rev. Respir. Dis. **82:** 244–250.

63. LEVINE, H.B. & Y.C. KONG. 1966. Immunologic impairment in mice treated intravenously with killed *Coccidioides immitis* spherules: suppressed response to intramuscular doses. J. Immunol. **97:** 297–305.

64. LEVINE, H.B. & G.M. SCALARONE. 1971. Deficient resistance to *Coccidioides immitis* following intravenous vaccination. 3. Humoral and cellular responses to intravenous and intramuscular doses. Sabouraudia **9:** 97–108.

65. SCALARONE, G.M. & H.B. LEVINE. 1969. Attributes of deficient immunity in mice receiving *Coccidioides immitis* spherule vaccine by the intravenous route. Sabouraudia **7:** 169–177.

66. SCALARONE, G.M. & H.B. LEVINE. 1971. Deficient resistance to *Coccidioides immitis* following intravenous vaccination. I. Distribution of spherules after intravenous and intramuscular doses. Sabouraudia **9:** 81–89.

67. SCALARONE, G.M. & H.B. LEVINE. 1971. Deficient resistance to *Coccidioides immitis* following intravenous vaccination. II. Evidence against an immune tolerance mechanism. Sabouraudia **9:** 90–6.

68. PAPPAGIANIS, D. 1993. Evaluation of the protective efficacy of the killed *Coccidioides immitis* spherule vaccine in humans. The Valley Fever Vaccine Study Group. Am. Rev. Respir. Dis. **148:** 656–660.

69. ABUODEH, R.O. *et al.* 1999. Resistance to *Coccidioides immitis* in mice after immunization with recombinant protein or a DNA vaccine of a proline-rich antigen. Infect. Immun. **67:** 2935–2940.

70. JIANG, C., D.M. MAGEE & R.A. COX. 1999. Coadministration of interleukin 12 expression vector with antigen 2 cDNA enhances induction of protective immunity against *Coccidioides immitis*. Infect. Immun. **67:** 5848–5853.

71. JIANG, C. *et al.* 2002. Role of signal sequence in vaccine-induced protection against experimental coccidioidomycosis. Infect. Immun. **70:** 3539–3545.

72. JIANG, C. *et al.* 1999. Genetic vaccination against *Coccidioides immitis*: comparison of vaccine efficacy of recombinant antigen 2 and antigen 2 cDNA. Infect. Immun. **67:** 630–635.

73. KIRKLAND, T.N. *et al.* 1998. Evaluation of the proline-rich antigen of *Coccidioides immitis* as a vaccine candidate in mice. Infect. Immun. **66:** 3519–3522.

74. KIRKLAND, T.N. *et al.* 1998. Immunogenicity of a 48-kilodalton recombinant T-cell-reactive protein of *Coccidioides immitis*. Infect. Immun. **66:** 424–431.

75. LECARA, G., R.A. COX & R.B. SIMPSON. 1983. *Coccidioides immitis* vaccine: potential of an alkali-soluble, water-soluble cell wall antigen. Infect. Immun. **39:** 473–475.

76. PAPPAGIANIS, D. *et al.* 1979. Immunization of mice against coccidioidomycosis with a subcellular vaccine. Infect. Immun. **25:** 440–445.

77. PENG, T. *et al.* 1999. Proline-rich vaccine candidate antigen of *Coccidioides immitis*: conservation among isolates and differential expression with spherule maturation. J. Infect. Dis. **179:** 518–521.

78. PENG, T. *et al.* 2002. Localization within a proline-rich antigen (Ag2/PRA) of protective antigenicity against infection with *Coccidioides immitis* in mice. Infect. Immun. **70:** 3330–3335.

79. SHUBITZ, L. *et al.* 2002. Protection of mice against *Coccidioides immitis* intranasal infection by vaccination with recombinant antigen 2/PRA. Infect. Immun. **70:** 3287–3289.

80. TARCHA, E.J. *et al.* 2006. A recombinant aspartyl protease of *Coccidioides posadasii* induces protection against pulmonary coccidioidomycosis in mice. Infect. Immun. **74:** 516–527.

81. TARCHA, E.J. *et al.* 2006. Multivalent recombinant protein vaccine against coccidioidomycosis. Infect. Immun. **74:** 5802–5813.

82. XUE, J. *et al.* 2005. Immune response of vaccinated and non-vaccinated mice to *Coccidioides posadasii* infection. Vaccine **23:** 3535–3544.

83. ZIMMERMANN, C.R. *et al.* 1998. Protection against lethal murine coccidioidomycosis by a soluble vaccine from spherules. Infect. Immun. **66:** 2342–2345.

84. ALBERT, M.M. *et al.* 1994. Efficacy of AmBisome in murine coccidioidomycosis. J. Med. Vet. Mycol. **32:** 467–71.

85. CLEMONS, K.V. *et al.* 1990. Efficacy of SCH39304 and fluconazole in a murine model of disseminated coccidioidomycosis. Antimicrob. Agents Chemother. **34:** 928–30.

86. CLEMONS, K.V., M.E. HOMOLA & D.A. STEVENS. 1995. Activities of the triazole SCH 51048 against *Coccidioides immitis in vitro* and *in vivo*. Antimicrob. Agents Chemother. **39:** 1169–72.

87. CLEMONS, K.V., D.F. RANNEY & D.A. STEVENS. 2001. A novel heparin-coated hydrophilic preparation of amphotericin B hydrosomes. Curr. Opin. Investig. Drugs **2:** 480–487.

88. CLEMONS, K.V. & D.A. STEVENS. 1991. Comparative efficacy of amphotericin B colloidal dispersion and amphotericin B deoxycholate suspension in treatment of murine coccidioidomycosis. Antimicrob. Agents Chemother. **35:** 1829–1833.

89. CLEMONS, K.V. & D.A. STEVENS. 1992. Efficacies of amphotericin B lipid complex
    (ABLC) and conventional amphotericin B against murine coccidioidomycosis.
    J. Antimicrob. Chemother. **30:** 353–363.
90. CLEMONS, K.V. & D.A. STEVENS. 1994. Utility of the triazole D0870 in the treatment
    of experimental systemic coccidioidomycosis. J. Med. Vet. Mycol. **32:** 323–326.
91. CLEMONS, K.V. & D.A. STEVENS. 1997. Efficacies of two novel azole derivatives
    each containing a morpholine ring, UR-9746 and UR-9751, against systemic
    murine coccidioidomycosis. Antimicrob. Agents Chemother. **41:** 200–203.
92. CLEMONS, K.V. & D.A. STEVENS. 2000. Efficacies of sordarin derivatives
    GM193663, GM211676, and GM237354 in a murine model of systemic coccid-
    ioidomycosis. Antimicrob. Agents Chemother. **44:** 1874–1877.
93. FIERER, J., T. KIRKLAND & F. FINLEY. 1990. Comparison of fluconazole and SDZ89-
    485 for therapy of experimental murine coccidioidomycosis. Antimicrob. Agents
    Chemother. **34:** 13–16.
94. GALGIANI, J.N. et al. 1990. Activity of cilofungin against *Coccidioides immitis*:
    differential *in vitro* effects on mycelia and spherules correlated with *in vivo*
    studies. J. Infect. Dis. **162:** 944–948.
95. GONZALEZ, G.M. et al. 2001. Correlation between antifungal susceptibilities of
    Coccidioides immitis in vitro and antifungal treatment with caspofungin in a
    mouse model. Antimicrob. Agents Chemother. **45:** 1854–1859.
96. GONZALEZ, G.M. et al. 2002. *In vitro* and *in vivo* activities of posaconazole against
    Coccidioides immitis. Antimicrob. Agents Chemother. **46:** 1352–1356.
97. GONZALEZ, G.M. et al. 2004. Efficacies of amphotericin B (AMB) lipid complex,
    AMB colloidal dispersion, liposomal AMB, and conventional AMB in treatment
    of murine coccidioidomycosis. Antimicrob. Agents Chemother. **48:** 2140–2143.
98. GRAYBILL, J.R., L. GRIFFITH & S.H. SUN. 1990. Fluconazole therapy for coc-
    cidioidomycosis in Japanese macaques. Rev. Infect. Dis. **12**(Suppl 3): S286–
    S290.
99. HECTOR, R.F., B.L. ZIMMER & D. PAPPAGIANIS. 1990. Evaluation of nikkomycins
    X and Z in murine models of coccidioidomycosis, histoplasmosis, and blasto-
    mycosis. Antimicrob. Agents Chemother. **34:** 587–593.
100. HOEPRICH, P. 1982. Amphotericin B methyl ester and leukoencephalopathy: the
    other side of the coin. J. Infect. Dis. **146:** 173–176.
101. HOEPRICH, P.D. & J.M. MERRY. 1985. Activity of BAY n 7133 and BAY 1 9139
    in vitro and in experimental murine coccidioidomycosis. Eur. J. Clin. Microbiol.
    4: 400–403.
102. HOEPRICH, P.D. & J.M. MERRY. 1987. Comparative efficacy of forphenicinol,
    cyclosporine, and amphotericin B in experimental murine coccidioidomycosis.
    Diagn. Microbiol. Infect. Dis. **6:** 287–292.
103. HOEPRICH, P.D. & J.M. MERRY. 1988. Effect of recombinant human interleukin 2
    in experimental murine coccidioidomycosis. Diagn. Microbiol. Infect. Dis. **9:**
    115–118.
104. HUPPERT, M., S.H. SUN & A.J. GROSS. 1972. Evaluation of an experimental animal
    model for testing antifungal substances. Antimicrob. Agents Chemother. **1:**
    367–372.
105. HUPPERT, M., S.H. SUN & K.R. VUKOVICH. 1974. Combined amphotericin B-
    tetracycline therapy for experimental coccidioidomycosis. Antimicrob. Agents
    Chemother. **5:** 473–478.
106. KIRKLAND, T.N. & J. FIERER. 1983. Cyclosporin A inhibits *Coccidioides immitis*
    in vitro and *in vivo*. Antimicrob. Agents Chemother. **24:** 921–924.

107. Levine, H.B. 1976. R34000, a dioxolane imidazole in the therapy for experimental coccidioidomycosis. Comparison with miconazole and econazole. Chest **70:** 755–759.

108. Levine, H.B. 1984. A direct comparison of oral treatments with BAY-n-7133, BAY-1-9139 and ketoconazole in experimental murine coccidioidomycosis. Sabouraudia **22:** 37–46.

109. Levine, H.B. & J.M. Cobb. 1978. Oral therapy for experimental coccidioidomycosis with R41 400 (ketoconazole), a new imidazole. Am. Rev. Respir. Dis. **118:** 715–721.

110. Levine, H.B. & J.M. Cobb. 1980. Ketoconazole in early and late murine coccidioidomycosis. Rev. Infect. Dis. **2:** 546–550.

111. Levine, H.B., S.M. Ringel & J.M. Cobb. 1978. Therapeutic properties of oral ambruticin (W7783) in experimental pulmonary coccidioidomycosis of mice. Chest **73:** 202–206.

112. Levine, H.B. *et al.* 1975. Miconazole in coccidioidomycosis. I. Assays of activity in mice and *in vitro.* J. Infect. Dis. **132:** 407–414.

113. Lones, G.W. 1970. 5–fluorocytosine in experimental coccidioidomycosis. Am. Rev. Respir. Dis. **102:** 128.

114. Lutz, J.E. *et al.* 1997. Activity of the triazole SCH 56592 against disseminated murine coccidioidomycosis. Antimicrob. Agents Chemother. **41:** 1558–1561.

115. Pappagianis, D. *et al.* 1990. Therapeutic effect of the triazole Bay R 3783 in mouse models of coccidioidomycosis, blastomycosis, and histoplasmosis. Antimicrob. Agents Chemother. **34:** 1132–1138.

116. Shubitz, L.F. *et al.* 2006. Efficacy of ambruticin analogs in a murine model of coccidioidomycosis. Antimicrob. Agents Chemother. **50:** 3467–3469.

117. Williams, P.L. *et al.* 1998. A model of coccidioidal meningoencephalitis and cerebrospinal vasculitis in the rabbit. J. Infect. Dis. **178:** 1217–1221.

118. Williams, P.L. 2001. Vasculitic complications associated with coccidioidal meningitis. Semin. Respir. Infect. **16:** 270–279.

119. Williams, P.L. *et al.* 1992. Vasculitic and encephalitic complications associated with *Coccidioides immitis* infection of the central nervous system in humans: report of 10 cases and review. Clin. Infect. Dis. **14:** 673–682.

120. Clemons, K.V. *et al.* 2002. Efficacy of intravenous liposomal amphotericin B (AmBisome) against coccidioidal meningitis in rabbits. Antimicrob. Agents Chemother. **46:** 2420–2426.

121. Sorensen, K.N. *et al.* 2000. Comparative efficacies of terbinafine and fluconazole in treatment of experimental coccidioidal meningitis in a rabbit model. Antimicrob. Agents Chemother. **44:** 3087–3091.

122. Sorensen, K.N. *et al.* 2000. Comparison of fluconazole and itraconazole in a rabbit model of coccidioidal meningitis. Antimicrob. Agents Chemother. **44:** 1512–1517.

123. Williams, P.L. *et al.* 2002. Levels of matrix metalloproteinase-9 within cerebrospinal fluid in a rabbit model of coccidioidal meningitis and vasculitis. J. Infect. Dis. **186:** 1692–1695.

124. Zucker, K.E. *et al.* 2006. Temporal expression of inflammatory mediators in brain basilar artery vasculitis and cerebrospinal fluid of rabbits with coccidioidal meningitis. Clin. Exp. Immunol. **143:** 458–466.

125. Capilla, J. *et al.* 2006. Efficacy of Abelcet in a rabbit model of coccidioidal meningitis. 46th Interscience Conference on Antimicrobial Agents and

Chemotherapy, Am. Soc. Microbiol., San Francisco, Sept. 2006, Abst. M-1748, p. 431.

126. CLEMONS, K.V. *et al.* 2001. Comparative toxicities and pharmacokinetics of intrathecal lipid (amphotericin B colloidal dispersion) and conventional deoxycholate formulations of amphotericin B in rabbits. Antimicrob. Agents Chemother. **45:** 612–615.

127. JOHNSON, R.H. & H.E. EINSTEIN. 2006. Coccidioidal meningitis. Clin. Infect. Dise. **42:** 103–107.

128. LEPAGE, E. 1993. Using a ventricular reservoir to instill amphotericin B. J. Neurosci. Nurs. **25:** 212–217.

129. PEREZ, J.A. JR. *et al.* 1995. Fluconazole therapy in coccidioidal meningitis maintained with intrathecal amphotericin B. Arch. Intern. Med. **155:** 1665–1668.

130. STEVENS, D.A. & S.A. SHATSKY. 2001. Intrathecal amphotericin in the management of coccidioidal meningitis. Semin. Respir. Infect. **16:** 263–269.

Virulence Mechanisms of *Coccidioides*

CHIUNG-YU HUNG, JIANMIN XUE, AND GARRY T. COLE

Department of Biology and South Texas Center for Emerging Infectious Diseases, University of Texas at San Antonio, San Antonio, Texas, USA

ABSTRACT: *Coccidioides* is a fungal respiratory pathogen of humans that can cause disease in both immunosuppressed and immunocompetent individuals. We describe here three mechanisms by which the pathogen survives in the hostile host environment: production of a dominant spherule outer wall glycoprotein (SOWgp) that modulates host immune response and results in compromised cell-mediated immunity to coccidioidal infection, depletion of SOWgp presentation on the surface of endospores, which prevents host recognition of the pathogen when the fungal cells are most vulnerable to phagocytic defenses, and induction of elevated production of host arginase I and coccidioidal urease, which contribute to tissue damage at sites of infection. Arginase I competes with inducible nitric oxide synthase (iNOS) in macrophages for the common substrate, L-arginine, and thereby reduces nitric oxide (NO) production and increases the synthesis of host orinithine and urea. Host-derived L-ornithine may promote pathogen growth and proliferation by providing a pool of the monoamine, which could be taken up and used for synthesis of polyamines via metabolic pathways of the parasitic cells. We have shown that high concentrations of *Coccidioides*- and host-derived urea at infection sites in the presence of urease produced and released by the pathogen, results in secretion of ammonia and contributes to alkalinization of the microenvironment. We propose that ammonia and enzymatically active urease released from spherules during the parasitic cycle of *Coccidioides* exacerbate the severity of coccidioidal infection by contributing to a compromised immune response to infection and damage of host tissue at foci of infection.

KEYWORDS: coccidioidomycosis; virulence factors; pathogenesis; gene expression

INTRODUCTION

Coccidioides is a human respiratory pathogen characterized by a parasitic cycle that is unique among medically important fungi that cause systemic mycoses. Air-dispersed spores (arthroconidia) are released from saprobic mycelia,

Address for correspondence: Chiung-Yu Hung, Department of Biology, University of Texas at San Antonio, One UTSA Circle, San Antonio, TX 78249-0662. Voice: 210-458-7018; fax: 210-458-7015. chiungyu.hung@utsa.edu

Ann. N.Y. Acad. Sci. 1111: 225–235 (2007). © 2007 New York Academy of Sciences. doi: 10.1196/annals.1406.020

which reside in the soil of semiarid regions of the southwestern United States, as well as parts of northern Mexico and Central and South America.[35] The tiny spores (2×4 μm) convert into multinucleate round cells (spherules) within the lungs of the host and undergo isotropic growth to produce large parasitic cells (60 to > 100 μm in diameter). The latter undergo an elaborate process of wall growth and cytoplasmic compartmentalization to form and release a multitude of endospores (4–10 μm in diameter). Each endospore grows and differentiates into a second-generation spherule, which is again able to yield an average of 200 to 300 endospores. Mature spherules most likely escape phagocytosis simply because they are too large to be ingested by neutrophils, macrophages, and dendritic cells. Although endospores that are newly released from maternal spherules can be engulfed by host phagocytes, *in vitro* studies of host–pathogen interactions suggest that many of these fungal cells survive.[14] *Coccidioides* is a successful pathogen, which is able to establish disease both in immunocompromised and immunocompetent hosts. Coccidioidal infection is characterized by both an intracellular and extracellular relationship with the host. Both clinical and experimental evidence has demonstrated that T cell immunity is pivotal for defense against coccidioidomycosis. Two functionally distinct subsets of CD4[+] T cells have been identified and are distinguished by different patterns of secreted cytokines;[31] T helper 1 (Th1) lymphocytes are characterized by production of interleukin (IL)-2, IL-12, tumor necrosis factor (TNF)-α, and gamma interferon (IFN-γ), while T helper 2 (Th2) cells secrete IL-4, IL-5, IL-6, and IL-10. Cytokines and chemokines are host factors that guide the differentiation of Th1 and Th2 cells, and many of the key T cell receptors and signal transduction events associated with Th1/Th2 cell differentiation have been characterized.[18] These two cytokine profiles correlate with resistance and susceptibility to *Coccidioides* infection, respectively.[12,13,26,27] The ability of *Coccidioides* to establish disease in a host is a function of the virulence traits that are expressed by the pathogen.[7,10] This article is focused on three mechanisms used by *Coccidioides* to survive within the hostile environment of the host.

IMMUNOMODULATION OF HOST RESPONSE TO COCCIDIOIDAL INFECTION

Studies of murine models of coccidioidomycosis have revealed that Th1 immune response is pivotal for defense against coccidioidal infection.[13,25,45] Results of clinical studies have also shown that patients recovering from coccidioidal infection without antifungal treatment develop delayed-type hypersensitivity, but generate only low titers of antibody against *Coccidioides*.[37] On the other hand, most patients who do not recover spontaneously may develop both pulmonary and extrapulmonary infections, and usually have negative skin tests with high titers of pathogen-specific antibody.[1,37,42] These data suggest that the susceptible host, which ultimately contracts disseminated disease, mounts an early and sustained Th2 pathway of immune response to coccidioidal infection. We have reported an immunodominant, cell-surface-localized coccidioidal antigen, which is capable of driving a Th2 response.[10] As isotropic growth of the spherule progresses, a membranous layer composed of polysaccharides, lipids, and proteins is deposited at the cell surface. When *Coccidioides* is grown in liquid culture in a shaking incubator, this phase-specific spherule outer wall (SOW) material is released from the parasitic cells into the medium as membranous sheets.[9] It has also been shown that the SOW layer is engulfed by phagocytic cells at infection sites. The SOW fraction isolated from *Coccidioides* cultures is characterized by high levels of immunoreactivity in both cellular and humoral assays. SOW has been shown to react with patient anti-*Coccidioides* antibody based on immunofluorescence and immunodiffusion-tube precipitin assays.[8] The component of SOW responsible for this immunoreactivity is a single glycoprotein, designated SOWgp.[19,21] SOWgp has been identified as an immunodominant antigen, which elicits both antibody-mediated and cellular immune responses in patients with coccidioidal infection.[19] This glycoprotein consists of a signal peptide and propeptide, a proline- and aspartic acid-rich tandem repeat motif, and a glycosylphosphatidylinositol (GPI) anchor signal consensus sequence.[21,23] The repeat domain contains three to six copies of proline- and aspartic acid-rich sequences (strain-dependent), ranging from 41 to 47 amino acids in length, and is responsible for patient antibody reactivity with SOWgp. Enzyme-linked immunosorbent assays (ELISA) of sera from patients with confirmed coccidioidal infection have revealed that high titers of antibody are present, which react with a bacterial-expressed, recombinant repeat fragment of SOWgp.[10] We have also shown that high titers of patient anti-SOWgp antibody correlate with severity of disease. In addition, SOWgp stimulates a proliferative response of immune CD4[+] T cells *in vitro*, suggesting that this immunodominant antigen may direct Th2 cells to secrete cytokines that promote B cells to produce antibody. The role of B cells in stimulation of protective immunity against coccidioidal infection is undefined. Magee and co-workers (2005) proposed that B cells and/or antibody play an important role in acquired protective immunity against *Coccidioides* infection,[28] but there is no direct evidence to support this proposal. Protective immunity against coccidioidomycosis in mice has been induced by vaccination with formalin-killed spherules (FKS).[36] The concentrations of both Th1 and Th2-related cytokines, such as IFN-γ and IL-4, were shown to increase in the vaccinated and infected mice, while identically treated MuMT mice deficient in mature B cells were significantly less protected than the wild-type mice. These data suggest that both T and B cells, and perhaps balanced Th1- and Th-2 responses, are essential for activation of protective immunity against coccidioidal infection. We have conducted two sets of experiments to test whether host exposure to SOWgp modulates the immune response toward an unbalanced Th2 response and exacerbates the course of disease. First, we generated a Δsowgp mutant strain of *C. posadasii* by deleting the SOWgp gene using a homologous recombination strategy. A revertant strain was also produced, which expresses the wild-type *SOWgp* gene in the mutant background. Three groups of C57 BL/6 mice were infected intranasally with an equal number of viable arthroconidia derived from either the parental strain of *C. posadasii* the revertant strain, or the Δsowgp mutant. The Δsowgp mutant showed attenuated virulence in the murine model of coccidioidomycosis, while the revertant strain containing the restored wild-type *SOWgp* gene was as virulent as the parental strain.[21] Mice infected with the Δsowgp mutant showed a significantly lower level of IL-5 production in bronchoalveolar lavage fluids (BALFs) and produced higher amounts of IFN-γ at 7–8 days post infection compared to mice infected with the parental strain. These results suggest that host infection with the parental strain of *C. posadasii* producing intact SOWgp results in a Th2-biased immune response.[10] In the second experiment, B cell knockout mice (*Igh-6*$^{-/-}$) and normal C57 BL/6 mice (*Igh-6*$^{+/+}$) were challenged intranasally with a sublethal inoculum of *C. posadasii* arthroconidia. At 30 days post challenge, total T cells were separately isolated from the spleens of each strain of mice and stimulated with the bacterial-expressed recombinant protein containing the repeat sequence of SOWgp as described above. The T cells from both groups of mice proliferated in response to *in vitro* exposure of the recombinant peptide, but the supernatants of the responding T cells isolated from normal mice revealed significantly higher levels of production of inflammatory and Th2-type cytokines (IL-6 and IL-10) compared to the supernatants of T cells isolated from the B cell knockout mice. These data also suggest that host exposure to SOWgp modulates host immunity in a Th2-biased manner. Persistence of an immune response to coccidioidal infection in which the balance is shifted in the direction of the Th2 pathway offers a distinct advantage to the pathogen by compromising the host cellular immune defenses.

Persistently high levels of IL-6 have been observed both in BALFs of C57 BL/6 mice infected mice with the parental strain, and the supernatant of the responding T cells from these infected mice stimulated with recombinant SOWgp *in vitro* as described above. IL-6 plays a key role in acute inflammation.[24] IL-6 and TGF-β together induce the differentiation of naïve T lymphocytes into IL-17-producing T cells (Th17 pathway), which is an inflammatory immune response that is distinct from those involving Th1 and Th2 activation and production of regulatory T cells (Treg).[4] Th17 cells are associated with the pathogenesis of autoimmune disease and persistent activation of the Th17 pathway can result in significant degrees of host tissue damage. Treg cells, on the other hand, prevent tissue injury during the onset of autoimmune disease. The generation of Th17 and Treg cells represents distinct and opposing pathways of immune response. Whether Th17 cells are involved in the pathogenesis of *Coccidioides* is unknown. Experiments have been designed to explore this possibility.

Immune modulation is an effective mechanism for pathogens to combat host defenses. A 19-kDa lipoprotein of *Mycobacterium tuberculosis* has been shown to be capable of eliciting potent humoral and cell-mediated immune recall response in immunized mice.[6] The purified 19-kDa protein also induces macrophages to produce high levels of IL-12, thereby contributing to the activation of the Th-1 pathway of immunity. However, recent studies have shown this same antigen is involved in several immune modulation functions, which downregulate the immune-effector mechanisms of the host.[34,38,40] It has been demonstrated that the 19-kDa antigen modulates the host-signaling cascade to inhibit IFN-γ-induced expression of the major histocompatibility (MHC) class II transactivator, which is required for chromatin remodeling and expression of MHC class II molecules. These mechanisms may diminish class II MHC expression by infected macrophages, thus contributing to immune evasion by *Mycobacterium*.[38] It is tempting to speculate that SOWgp may be capable of a similar mechanism of immune suppression during the course of coccidioidomycosis. Secreted endosporulation antigens (EA) of *Coccidioides* prepared from the culture media have been reported to stimulate and suppress cell-mediated immunity, depending on the antigen preparation method and antigen concentration to which T cells are exposed *in vitro*.[5] High concentrations of the EA were shown to produce a suppressive or toxic effect on lymphocyte transformation *in vitro*, while heating and dialysis abolished this suppressive effect. These results suggest that there are both stimulatory and suppressive substances in EA, and the outcome of lymphocyte transformation is dependent on the ratio of the two biologically active components. It is not known whether SOWgp is a component of the EA. Identification and characterization of these stimulatory and suppressive substances will advance our understanding of the nature of host immunity to coccidioidomycosis.

EVASION OF HOST DETECTION

Since SOWgp is exposed on the cell surface, the glycoprotein would be expected to bind antibody and enhance opsonization, phagocytosis, and clearance of the pathogen from sites of infection. The opportunities for phagocytes to engulf *Coccidioides* are restricted during the parasitic cycle to stages of initial round cell formation from germinated arthroconidia and endospore differentiation and release from ruptured spherules. Mature spherules are too large to be phagocytosed (approximately 80–100 μm in diameter.). In addition, the pathogen has developed a mechanism that prevents host recognition of endospores. Examination of *SOWgp* gene expression and glycoprotein production during the first generation of parasitic phase *in vitro* has revealed that the peak of SOWgp production occurs during isotropic growth of the spherules, but then sharply decreases during endospore formation.[21] The crude SOW fraction isolated from the media of endosporulating cultures lacks detectable

SOWgp, but instead reveals a catalytically active, 34-kDa metalloproteinase (Mep1) that belongs to the metzincin superfamily.[20,41] *In vitro* studies have demonstrated that the recombinant Mep1 enzyme can efficiently digest purified SOWgp. Temporal expression of the *C. posadasii MEP1* gene peaks at the early stage of endospore formation. It appears that gene expression and protein production of SOWgp are highly regulated during the parasitic cycle as a result of both transcriptional control of *SOWgp* gene expression, and rapid degradation of the glycoprotein during the endosporulation phase. To evaluate whether the absence of Mep1 activity and retention of SOWgp at the surface of the endospores influence the outcome of coccidioidal infection in mice, we vaccinated C57 BL/6 mice with recombinant SOWgp and then challenged the animals intranasally with a mutant strain of *C. posadasii* in which the *MEP*1 gene was disrupted. We had previously shown that the Δmep1 strain had not lost virulence in mice compared to the wild-type strain. Although patients with coccidioidal infection produce high titers of anti-SOWgp antibody during the course of disease, mice typically become moribund within 2 weeks post challenge and die 1 to 2 weeks later, thus preventing production of comparable high anti-SOWgp titers. Therefore, vaccination with recombinant SOWgp prior to infection was necessary to generate levels of anti-SOWgp antibody that simulated the human condition. In this respect, the murine model of coccidioidomycosis is not an ideal simulation of the human disease and immune response to *Coccidioides* infection. Nevertheless, C57 BL/6 mice vaccinated with recombinant SOWgp and then challenged with the Δmep1 mutant strain of *C. posadasii* revealed a significant increase in the percentage of survival compared to SOWgp-immunized mice which were challenged with the parental strain. These results support our hypothesis that retention of SOWgp on the surface of endospores of the Δmep1 mutant strain in the presence of high titers of antibody to the immunodominant antigen contribute to opsonization, increased phagocytosis, and killing of the fungal cells.[20] We suggest that Mep1 plays a key role in the virulence of *Coccidioides*.

INDUCTION OF HOST ARGINASE I PRODUCTION COMPROMISES HOST DEFENSE AGAINST COCCIDIOIDAL INFECTION

We applied a microarray approach to explore the differences between the murine host responses to infection with the parental versus the Δsowgp mutant strain using the U74 A chip (Affymetrix; Santa Clara, CA, USA). Total RNA samples were isolated at 6 days after intranasal challenge from lung homogenates of two groups of C57 BL/6 mice, each infected with equal numbers of arthroconidia isolated from either the parental or Δsowgp mutant strain. Gene transcript levels were determined from data image files using algorithms in the Microarray Suite version 5.0 software (Affymetrix). The data indicated significantly higher degrees of upregulation of expression of arginase I and IL-6 in mice infected with the parental strain compared to animals infected with the Δsowgp mutant. As previously discussed, high amounts of IL-6 correlate with intense inflammatory response to *Coccidioides* infection, which may contribute to host tissue damage and exacerbation of disease. Macrophages express two arginase isoforms, arginase I and arginase II.[44] Arginase I is located in the cytosol of macrophages, while arginase II is a mitochondrial enzyme. The hydrolysis of L-arginine to L-ornithine by arginase I provides the substrate for ornithine decarboxylase (ODC;),[17] which is a key enzyme in the polyamine biosynthetic pathway and putative regulator of *Coccidioides* parasitic cell differentiation.[17,22] The availability of host-derived L-ornithine at sites of fungal infection may promote pathogen growth and proliferation by providing a pool of the monoamine, which could be used for synthesis of L-glutamine, L-proline, and polyamines via metabolic pathways of the parasitic cells.[22] Arginase I competes with inducible nitric oxide synthase (iNOS) in macrophages for the common substrate, L-arginine.[32] The activities of these two enzymes are regulated by Th1 and Th2 T cells via their secreted cytokines. Arginase I expression has been shown to be induced in macrophages upon exposure to Th2-type cytokines, such as IL-4 and IL-10 during infection with *Entamoeba histolytica, Trypanosoma cruzi,* or *Leishmania major,*[15,16,22] while the synthesis of NO is simultaneously reduced on account of decreased amounts of L-arginine substrate. Dendritic cells have also been shown to upregulate arginase I expression and arginase activity upon Th2 stimulation.[32] Although this regulatory pathway during host–pathogen interactions has been extensively investigated in parasitic diseases, no comparable studies of fungal–host interactions have been conducted. FIGURE 1 summarizes the sequence of events involving arginase I production by host inflammatory cells in murine lungs infected with *C. posadasii.*

We have shown that host *iNOS* gene expression in lung homogenates of mice infected with the parental strain of *C. posadasii* at 7 days post challenge is very low.[10] On the other hand, the level of *iNOS* expression in mice infected with the Δsowgp mutant is approximately ninefold higher, which is consistent with our argument that host exposure to SOWgp contributes to a Th2-biased response to infection and skews the balance between arginase and nitric oxide synthase activities in favor of the former. Given that host arginase activation may result in decreased levels of NO production in macrophages and permit intracellular survival of the fungal pathogen, it follows that a physiological inhibitor of arginase (e.g., N^{ω}-hydroxy-nor-L-arginine [nor-LOHA]) should contribute to pathogen clearance and host survival.[22,43] Intraperitoneal treatment of *Coccidioides*-infected BALB/c mice with nor-LOHA significantly increased host survival, suggesting that arginase activity plays an important role in disease development.[10]

Accumulation of urea results from the high level of expression of arginase I at sites of coccidioidal infection. Urea is hydrolyzed by coccidioidal urease

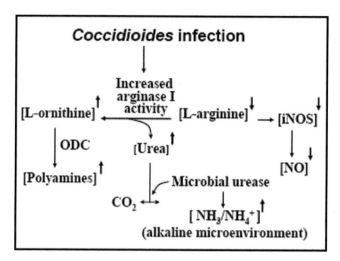

FIGURE 1. Summary of the arginase I pathway that occurs in murine macrophages in response to coccidioidal infection. Elevated production of arginase I results in increased polyamine synthesis, decreased NO production, and alkalinization of the microenvironment as a consequence of increased urea concentration and microbial urease activity at sites of infection. These metabolic events are suggested to contribute to the survival of *Coccidioides* in the hostile environment of the host. iNOS = inducible nitric oxide synthase; NO = nitric oxide; ODC = ornithine decarboxylase.

to yield carbonic acid and ammonia.[46] At physiological pH, the carbonic acid proton dissociates and ammonia molecules become protonated with a resultant increase in pH.[30] Urease has been shown to be an important virulence factor in bacterial and fungal pathogens.[11,33,39] Mice infected with a urease knockout strain of *Cryptococcus neoformans* showed significant reduction in inflammatory response at sites of blood-to-brain invasion.[33] The authors suggested that the enzyme contributes to central nervous system invasion by enhancing yeast sequestration within microcapillary beds in the brain during hematogenous dissemination. Deletion of the *URE* gene from *C. posadasii* also resulted in a significant reduction in pathogenicity of the mutant strain in mice.[29] It has been suggested that ammonia and enzymatically active urease released from spherules during the parasitic cycle of *C. posadasii* contribute to host tissue damage, which further exacerbates the severity of coccidioidal infection and enhances the virulence of this human respiratory pathogen. Increase in pH at infection sites due to urease activity may also contribute to the elevation of pH of the endospore-containing phagosomes. Early data showed that the endospore-containing phagosomes of peritoneal macrophages were not stained with acridine orange, a fluorescent cationic dye, which is an indicator of acidic organelles.[2] In contrast, macrophages obtained from mice vaccinated with formalin-killed spherules stained positively with acridine orange and contained endospores with reduced fungal viability *in vitro*.[3] The inability of macrophages from nonvaccinated mice to kill endospores has been suggested to be at least partly due to the inhibition of lysosomal fusion, which may explain the negative results of acridine orange staining of the phagocytes. The Δure mutant strain of *Coccidioides is* a valuable tool for studies of the effect of pH on host–pathogen interaction during coccidioidomycosis.

CONCLUSION

Little is known about the virulence mechanisms of *Coccidioides*. However, with the recent completion of genome sequencing projects of *Coccidioides* spp. and the application of genomics-based methods (e.g., microarray analyses) to studies of host–pathogen interactions, we will be able to rapidly expand the set of candidate genes which express virulence factors. These technological advances can now be combined with the availability of genetic manipulation methods to knockout and restore selected virulence genes of *Coccidioides* cells in order to determine their precise biological function during the course of infection.

ACKNOWLEDGMENTS

The authors are grateful for grant support from the National Institute of Allergy and Infectious Diseases, National Institutes of Health (AI19149 and AI37232), and the California HealthCare Foundation.

REFERENCES

1. AMPEL, N.M. & L. CHRISTIAN. 1997. In vitro modulation of proliferation and cytokine production by human peripheral blood mononuclear cells from subjects with various forms of coccidioidomycosis. Infect. Immun. **65:** 4483–4487.
2. BEAMAN, L. *et al.* 1981. Role of lymphocytes in macrophage-induced killing of *Coccidioides immitis in vitro*. Infect. Immun. **34:** 347–353.
3. BEAMAN, L. *et al.* 1983. Activation of macrophages by lymphokines: enhancement of phagosome-lysosome fusion and killing of *Coccidioides immmitis*. Infect. Immun. **39:** 1201–1207.
4. BETTELLI, E. *et al.* 2006. Reciprocal developmental pathways for the generation of pathogenic effector Th17 and regulatory T cells. Nature **441:** 235–238.
5. BRASS, C. *et al.* 1982. Stimulation and suppression of cell-mediated immunity by endosporulation antigens of *Coccidioides immitis*. Infect. Immun. **35:** 431–436.
6. BRIGHTBILL, H.D. *et al.* 1999. Host defense mechanisms triggered by microbial lipoproteins through toll-like receptors. Science **285:** 732–736.
7. CASADEVALL, A. 2006. Cards of virulence and the global virulome for human. Microbe **1:** 359–364.

8. COLE, G.T. *et al.* 1988. Immunoreactivity of a surface wall fraction produced by spherules of *Coccidioides immitis.* Infect. Immun. **56:** 2695–2701.

9. COLE, G.T. *et al.* 1988. Isolation and morphology of an immunoreactive outer wall fraction produced by spherules of *Coccidioides immitis.* Infect. Immun. **56:** 2686–2694.

10. COLE, G.T. *et al.* 2006. Virulence mechanisms of *Coccidioides. In* Molecular Principles of Fungal Pathogenesis. J. Heitman *et al.*, Eds.:363–391. ASM Press. Washington, DC

11. COX, G.M. *et al.* 2000. Urease as a virulence factor in experimental cryptococcosis. Infect. Immun. **68:** 443–448.

12. COX, R.A. & D.M. MAGEE. 1995. Production of tumor necrosis factor alpha, interleukin-1 alpha, and interleukin-6 during murine coccidioidomycosis. Infect. Immun. **63:** 4178–4180.

13. COX, R.A. & D.M. MAGEE. 1998. Protective immunity in coccidioidomycosis. Res. Immunol. **149:** 417–428; discussion 506–417.

14. DRUTZ, D.J. & M. HUPPERT. 1983. Coccidioidomycosis: factors affecting the host-parasite interaction. J. Infect. Dis. **147:** 372–390.

15. ELNEKAVE, K. *et al.* 2003. Consumption of L-arginine mediated by *Entamoeba histolytica* L-arginase (EhArg) inhibits amoebicidal activity and nitric oxide production by activated macrophages. Parasite Immunol. **25:** 597–608.

16. GIORDANENGO, L. *et al.* 2002. Cruzipain, a major *Trypanosoma cruzi* antigen, conditions the host immune response in favor of parasite. Eur. J. Immunol. **32:** 1003–1011.

17. GUEVARA-OLVERA, L. *et al.* 2000. Sequence, expression and functional analysis of the *Coccidioides immitis* ODC (ornithine decarboxylase) gene. Gene **242:** 437–448.

18. HATTON, R.D. & C.T. WEAVER. 2003. T-bet or not T-bet. Science **302:** 993–994.

19. HUNG, C.-Y. *et al.* 2000. A major cell surface antigen of *Coccidioides immitis* which elicits both humoral and cellular immune responses. Infect. Immun. **68:** 584–593.

20. HUNG, C.-Y. *et al.* 2005. A metalloproteinase of *Coccidioides posadasii* contributes to evasion of host detection. Infect. Immun. **73:** 6689–6703.

21. HUNG, C.-Y. *et al.* 2002. A parasitic phase-specific adhesin of *Coccidioides immitis* contributes to the virulence of this respiratory fungal pathogen. Infect. Immun. **70:** 3443–3456.

22. INIESTA, V. *et al.* 2002. Arginase I induction in macrophages, triggered by Th2-type cytokines, supports the growth of intracellular *Leishmania* parasites. Parasite Immunol. **24:** 113–118.

23. JOHANNESSON, H. *et al.* 2005. Concerted evolution in the repeats of an immunomodulating cell surface protein, SOWgp, of the human pathogenic fungi *Coccidioides immitis* and *C. posadasii.* Genetics **171:** 109–117.

24. KAPLANSKI, G. *et al.* 2003. IL-6: a regulator of the transition from neutrophil to monocyte recruitment during inflammation. Trends Immunol. **24:** 25–29.

25. LI, K. *et al.* 2001. Recombinant urease and urease DNA of *Coccidioides immitis* elicit an immunoprotective response against coccidioidomycosis in mice. Infect. Immun. **69:** 2878–2887.

26. MAGEE, D.M., & R.A. COX. 1995. Roles of gamma interferon and interleukin-4 in genetically determined resistance to *Coccidioides immitis.* Infect. Immun. **63:** 3514–3519.

27. MAGEE, D.M., & R.A. COX. 1996. Interleukin-12 regulation of host defenses against *Coccidioides immitis*. Infect. Immun. **64:** 3609–3613.
28. MAGEE, D.M. *et al.* 2005. Role of B cells in vaccine-induced immunity against coccidioidomycosis. Infect. Immun. **73:** 7011–7013.
29. MIRBOD-DONOVAN, F. *et al.* 2006. Urease produced by *Coccidioides posadasii* contributes to the virulence of this respiratory pathogen. Infect. Immun. **74:** 504–515.
30. MOBLEY, H.L. *et al.* 1995. Molecular biology of microbial ureases. Microbiol. Rev. **59:** 451–480.
31. MOSMANN, T.R., & R.L. COFFMAN. 1989. Heterogeneity of cytokine secretion patterns and functions of helper T cells. Advances Immunol. **46:** 111–147.
32. MUNDER, M. *et al.* 1999. Th1/Th2-regulated expression of arginase isoforms in murine macrophages and dendritic cells. J. Immunol. **163:** 3771–3777.
33. OLSZEWSKI, M.A. *et al.* 2004. Urease expression by *Cryptococcus neoformans* promotes microvascular sequestration, thereby enhancing central nervous system invasion. Am. J. Pathol. **164:** 1761–1771.
34. PAI, R.K. *et al.* 2003. Inhibition of IFN-gamma-induced class II transactivator expression by a 19-kDa lipoprotein from *Mycobacterium tuberculosis*: a potential mechanism for immune evasion. J. Immunol. **171:** 175–184.
35. PAPPAGIANIS, D. 1988. Epidemiology of coccidioidomycosis. Cur. Top. Med. Mycol. **2:** 199–238.
36. PAPPAGIANIS, D. 1993. Evaluation of the protective efficacy of the killed *Coccidioides immitis* spherule vaccine in humans. Am. Rev. Respir. Dis. **148:** 656–660.
37. PAPPAGIANIS, D., & B.L. ZIMMER. 1990. Serology of coccidioidomycosis. Clin. Microbiol. Rev. **3:** 247–268.
38. PENNINI, M.E. *et al.* 2006. *Mycobacterium tuberculosis* 19-kDa lipoprotein inhibits IFN-gamma-induced chromatin remodeling of MHC2TA by TLR2 and MAPK signaling. J. Immunol. **176:** 4323–4330.
39. RADOSZ-KOMONIEWSKA, H. *et al.* 2005. Pathogenicity of *Helicobacter pylori* infection.Clin. Microbiol. Infect. **11:** 602–610.
40. RAO, V. *et al.* 2005. Increased expression of *Mycobacterium tuberculosis* 19 kDa lipoprotein obliterates the protective efficacy of BCG by polarizing host immune responses to the Th2 subtype. Scand. J. Immunol. **61:** 410–417.
41. RAWLINGS, N.D. *et al.* 2004. MEROPS: the peptidase database. Nucleic Acids Res. **32** Database issue:D160-D164.
42. SMITH, C. *et al.* 1956. Patterns of 39,500 serologic tests in coccidioidomycosis. JAMA **160:** 546–552.
43. STEMPIN, C. *et al.* 2004. Arginase induction promotes *Trypanosoma cruzi* intracellular replication in *Cruzipain*-treated J774 cells through the activation of multiple signaling pathways. Eur. J. Immunol. **34:** 200–209.
44. VINCENDEAU, P. *et al.* 2003. Arginases in parasitic diseases. Trends Parasitol. **19:** 9–12.
45. XUE, J. *et al.* 2005. Immune response of vaccinated and non-vaccinated mice to *Coccidioides posadasii* infection. Vaccine **23:** 3535–3544.
46. YU, J.-J. *et al.* 1997. Isolation and characterization of the urease gene (URE) from the pathogenic fungus *Coccidioides immitis*. Gene **198:** 387–391.

The Role of IL-10 in Genetic Susceptibility to Coccidioidomycosis on Mice

JOSHUA FIERER

Division of Infectious Diseases, VA Healthcare San Diego, and UC San Diego School of Medicine, San Diego, California 92161, USA

ABSTRACT: Epidemiological and clinical studies have confirmed that coccidioidomycosis is more severe in African American and Filipino patients than in Caucasians, suggesting a genetic basis for susceptibility in humans. We discovered that inbred strains of mice also vary greatly in their susceptibility to *Coccidioides immitis* infections, and although resistance is the dominant phenotype, it is a multigenic trait in mice. We found a strong direct correlation between susceptibility in mice and the amount of IL-10 made in response to infection. We then showed that IL-10-deficient mice are much more resistant to infection than the parent C57BL/6 strain. Finally, we showed that genetically resistant mice that are transgenic for IL-10 and so overproduce that cytokine are more susceptible to *C. immitis*. This is in part due to suppression of NOS2 expression by IL-10.

KEYWORDS: interleukin; nitric oxide; fungus

INTRODUCTION

Most infections with *Coccidioides immitis* and *C. posadasii* produce mild or even asymptomatic self-limited illnesses.[1] Only a small percentage of those infections result in disseminated coccidioidomycosis, which is the most severe form of this disease.[1] African American and Filipino patients are 5–10 times more likely to have disseminated coccidioidomycosis than are similarly exposed Caucasians.[2,3] Because these patients were previously healthy and do not seem predisposed to other infections, this implies there is a genetic basis for susceptibility to coccidioidomycosis that may be unique to this fungus. Because the nature of the immune response to the fungus is the chief determinant of whether the infection will be progressive or self-limited, there is a connection between genetic susceptibility and the immune response to the fungus.[4] People

Address for correspondence: Joshua Fierer, Division of Infectious Diseases, VA Healthcare San Diego and UC San Diego School of Medicine, 3350 La Jolla Village Drive, San Diego, CA 92161. Voice: 858-552-7446; fax: 858-552-4398.
 jfierer@ucsd.edu

Ann. N.Y. Acad. Sci. 1111: 236–244 (2007). © 2007 New York Academy of Sciences.
doi: 10.1196/annals.1406.048

who have disseminated infections have high titers of antibodies against fungal antigens, but do not develop delayed-type hypersensitivity as manifested by a positive skin test reaction to fungal antigens. The opposite is true of patients who have self-limited infections.[4]

GENETICS OF SUSCEPTIBILITY IN INBRED STRAINS OF MICE

To try to understand the genetic basis for susceptibility to coccidioidomycosis we began to study the immunological basis for inherited susceptibly using inbred mice, and we discovered that different mouse strains vary over 10,000-fold in their susceptibility to i.p. infection with *C. immitis* (TABLE 1), and that resistance is the dominant phenotype.[5,6] We also found that the offspring of the susceptible BALB/c X C57BL/6 strains are also susceptible, implying that those two strains have the same genetic defect. DBA/2 mice are by far the most resistant strain among those that we studied. While this was fairly convincing evidence that there was a genetic basis for resistance to coccidioidomycosis in mice, we were not sure whether resistance was a Mendelian dominant trait so we infected 26 recombinant inbred (RI) BXD strains i.p. and we found that some were more resistant than the parental DBA/2, while others were more susceptible than the parental C57BL/6 (B6) (FIG. 1).[7] Because RI lines are inbred mice with reassortments of the genomes of the two parental strains, this result was not compatible with a single dominant resistance gene, and it suggested that we were studying a multigenic trait.

IMMUNE RESPONSES OF INBRED MICE TO *C. IMMITIS* INFECTION

In order to understand the immunological basis for the increased susceptibility of B6 mice we compared their immune responses to those of the resistant

TABLE 1. LD$_{50}$ values of *C. immitis* infections in inbred mice 28 days after i.p. infection

Mouse strain	LD$_{50}$ $\pm$ SEM (log$_{10}$)
BALB/cAnN	1.67 (0.60)
C57BL/6N	2.83 (0.23)
(BALB/c × C57BL/6)Fl	1.82 (0.45)
AKR /N	<2.0
A/J	2.30 (0.68)
CBA/J	3.32 (0.21)
DBA/lJ	3.71 (0.55)
C3H/HeN	3.85 (0.24)
DBA/2N	5.25 (0.36)
(C57BL/6 × DBA/2)Fl	4.20 (0.19)
(BALB/c × DBA/2)F1	4.95 (0.18)

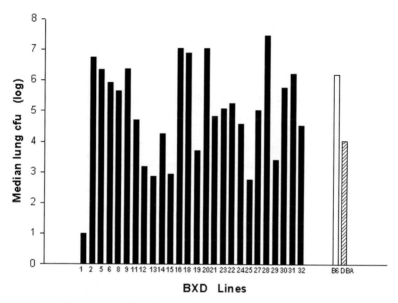

FIGURE 1. The susceptibility of recombinant inbred BXD lines to i.p. infection with *C. immitis*. Mice were infected i.p with ~500 RS arthroconidia and sacrificed 14 days later for quantitative mycology. There were 6–10 mice in each BXD line. We show here the median number of viable fungi recovered from the lungs, as nearly all mice died of pneumonia.

DBA/2 strain. We measured cytokine-specific mRNA in the lungs of the two strains at various times after infection and we found that the most striking difference was in the amount of IL-10 mRNA made by the two strains 14 days after infection, at the same time that B6 mice have increased fungal growth in their lungs (FIG. 2A). In addition, we showed that cells taken from the lungs and spleens of infected B6 mice and placed in culture for 6 h produced more IL-10 than did cells from DBA/2 mice (FIG. 2B).[8] We were able to confirm the discovery of Magee *et al.* that susceptible mice make more IL-4,[9] but that difference was not as great as the difference in IL-10. We repeated this experiment, infecting the mice intra-nasally (i.n.), and had similar results (FIG. 3). In this experiment we found that by day 16 after infection the difference in IL-4 mRNA levels between the two strains was no longer apparent, even though the difference in colony counts was even greater on day 16 (B6 mice had already begun to die). In contrast the difference in IL-10 mRNA was even greater on day 16 than on day 10 after infection.[10] We also compared the amount of IL-10 mRNA in the lungs of BXD strains with the colony counts in their lungs and there was a highly significant correlation, further evidence that the amount of IL-10 made is an important aspect of the genetic susceptibility to *C. immitis* in these strains.

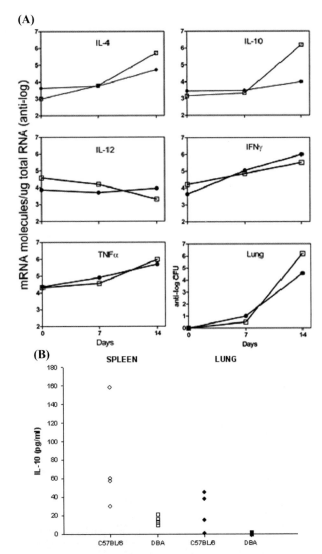

FIGURE 2. B6 lungs have higher levels of IL-10 mRNA and secrete more IL-10. (**A**) We extracted RNA from the lungs of mice before, 7, and 14 days after they were infected i.p with *C. immitis* RS arthroconidia. We measured cytokine-specific mRNA by competitive RT-PCR. Each point represents the geometric mean value from four mice. The *open squares* represent B6 mice and the *closed circles* show DBA/2 mice. Both IL-4 and IL-10 showed significant differences on day 14. (**B**) The spleen and one lung were removed from mice 14 days after infection and made into single-cell suspensions. Lung cells (1×10^7) and spleen cells (2×10^7) were placed in 12-and 24-well Costar plates, respectively, and incubated at $37°C$ for 6 h in RPMI containing 5% endotoxin-free fetal calf serum. Supernatants were frozen at $-70°C$ until assayed with an ELISA kit from R&D Systems (Minneapolis, MN, USA).

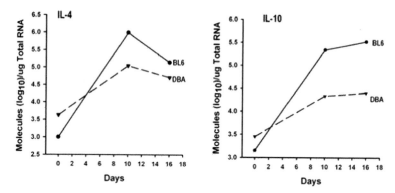

FIGURE 3. Time course of IL-10 and IL-4 mRNA in DBA/2 and B6 mice with *C. immitis* pneumonia. Mice were infected i.n. and their lungs were removed before and 10 and 16 days after i.n. infection. We measured IL-4 and IL-10 mRNA by competitive RT-PCR. There were three mice/group/day and the geometric means are shown for the mRNA. These measurements were repeated on the same samples with similar results.

These results did not answer the question of whether the high level of IL-10 was the cause or the effect of a more severe infection, resulting in a large antigenic load. To answer that question we infected IL-10-deficient B6 mice either i.n. or i.p. and compared them to control B6 mice. In both cases the

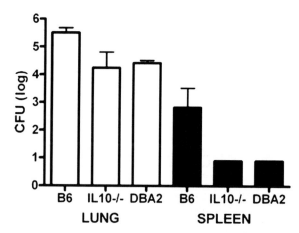

FIGURE 4. IL-10-deficient B6 mice are more resistant to *C. immitis* infection. Mice were infected i.n. with 100 arthroconidia and sacrificed 15 days later. The difference between the median CFU/lung in B6 and B6.IL-10$^{-/-}$ mice was not quite statistically significant ($P = 0.056$) because of the large standard deviation in the IL-10$^{-/-}$ group, but the difference in spleen CFUs was significant ($P = 0.034$), showing that IL-10 deficient mice are less likely to have a disseminated infection. DBA/2 mice were included only as a resistant control, and no statistical comparisons were made between IL-10$^{-/-}$ and DBA/2 mice in this experiment. There were 10 mice/group.

IL-10-deficient mice were significantly more resistant to the infection than control B6 mice; in the i.n. infection model this was most manifest as a decrease in extra-pulmonary dissemination (FIG. 4). In the i.p. infection experiment we also included IL-4 mutants on the same B6 background, and while both were more resistant, the deletion of IL-10 had a greater effect.[8] Because IL-10 mutants were more resistant to *C. immitis* than the control B6 mice, we concluded that too much IL-10 was a cause of their susceptibility, not a consequence.

We then asked whether increasing the amount of IL-10 would make resistant mice susceptible. Because *C. immitis* is a biohazard and we keep the infected

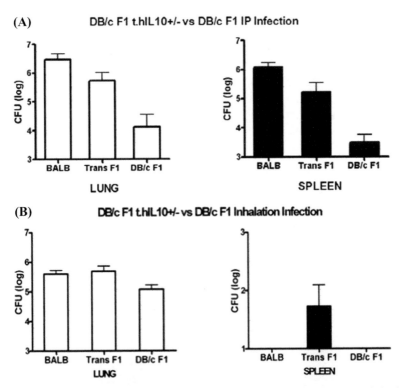

FIGURE 5. Human IL-10 transgenic mice are more susceptible to *C. immitis* infection. (**A**) (DBA/2 X BALB/c) F_1.h.IL-10$^{+/-}$ transgenic mice (Trans F1) and control (DBA/2 X BALB/c) F 1 (DB/c F 1) mice were infected i.p. with 310 arthroconidia and sacrificed 14 days later. Spleens and lungs were removed for quantitative culture. The F 1 mice expressing the transgene had more organisms in their lungs and spleens than the genetically resistant DB/c F 1 control mice ($P < 0.01$). BALB/c mice were included as susceptible controls, but no statistical comparisons were made between them and the other two groups of mice. (**B**) The same groups of mice were infected i.n. with 50 arthroconidia and were sacrificed 15 days after infection for quantitative mycology. The difference between median lung CFUs in DB/c h.IL-10$^{+/-}$ and DB/c F 1 mice was nearly 10-fold ($P = 0.02$), and only the IL-10 transgenic mice had *C. immitis* in their spleens. There were 10 mice/group.

mice inside a HEPA-filtered glove box, it is not possible to repeatedly inject them with IL-10, which has a short half-life *in vivo*. Therefore, we chose a genetic approach to answer this question. Amy Beebe at DNAX generously provided us with human IL-10 transgenic mice that expressed the transgene behind the strong MHC class II Ea (a class II histocompatability gene) promoter. Because the transgenic mice were on a BALB/c background, and those mice are already susceptible to *C. immitis*, we crossed them with DBA/2, making F1 transgenic mice that should be genetically resistant to *C. immitis* (TABLE 1). We found that the h.IL-10 transgenic mice infected i.p had more organisms in their lung and spleens than the resistant F1 controls, but fewer than the susceptible BALB/c. Transgenic mice that were infected i.n. had similar numbers of organisms in their lungs as did the control animals (BALB/CxDBA/2)F1, but dissemination to the spleen could not be prevented in the h.IL-10 transgenic mice.(FIG. 5) This is similar to what happens in humans; everyone will get pneumonia, but only susceptible individuals will have clinically significant extra-pulmonary dissemination.[11]

IL-10 has many effects on the immune system, nearly all of which lead to downregulation of inflammation.[12] We analyzed the mRNA responses of the transgenic mice and the controls and we found that the control F1 mice had 10 times more mRNA for NOS2 (iNOS) (not shown), the inducible form of nitric oxide synthetase that is present in macrophages.[13] Because NO has potent antimicrobial activity we considered the possibility that one way that IL-10 increased susceptibility to *C. immitis* was to suppress synthesis of NO. To test that hypothesis we inhibited NO synthesis in DBA/2 mice and then infected them with *C. immitis*.[10] We first showed that aminoguanidine (AMG)[14] in the

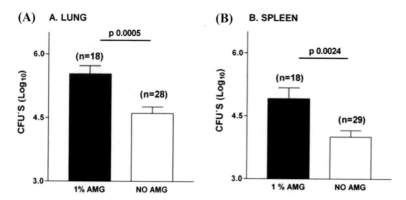

FIGURE 6. Aminoguanidine treatment increases the susceptibility of DBA/2 mice to *C. immitis* infection. We measured CFUs/lung and spleen 14 days after i.p. infection in mice treated with 1% AMG in their drinking water and untreated controls. The differences between the means of the groups are shown. These data are the composite of two separate experiments, and the total number of mice in each group is in parentheses.

drinking water reduced the urinary excretion nitrate (an excellent reflection of the amount of NO made by the mice[15]) by infected DBA/2 mice. This in turn increased the severity of infection as measured by colony counts in the lungs and spleens, 14 days after i.p. infection (FIG. 6).

We cannot say for certain how our findings in mice relate to the human response to this fungus, but there are some parallels between human and mouse responses to the fungus. Ampel and Kramer analyzed the responses of peripheral blood mononuclear cells to T27K, a soluble antigen from spherules.[16] They found that healthy immune donors respond by making IFNγ, and that exogenous IL-10 inhibits that response, which is what we found in the IL-10 transgenic mice. Interestingly, if NK cells were depleted from the mononuclear cells, a source of IFNγ was eliminated and the remaining stimulated cells made more IL-10, indicating there is a dynamic reciprocal relationship between those two cytokines. Similar studies have not been done comparing patients with disseminated infections and healthy immune controls, and of course in human studies the responses in infected tissues are not assayed directly.

In summary, we have shown that there is a strong correlation between the amount of IL-10 made in response to *C. immitis* infection, and that IL-10-deficient mice are more resistant to that infection. We also showed that increasing the amount of IL-10 produced can make genetically resistant mice more susceptible. One of the explanations for the deleterious affect of IL-10 in this infection is the downregulation of NOS2. We have not yet established why infected B6 and B10 mice produce more IL-10.

REFERENCES

1. CHILLER, T.M., J.N. GALGIANI & D.A. STEVENS. 2003. Coccidioidomycosis. Infect. Dis. Clin. North Am. **17:** 41–57.
2. JOHNSON, W.M. 1982. Racial factors in coccidioidomycosis: mortality experience in Arizona. A review of the literature. Ariz. Med. **39:** 18–24.
3. PAPPAGIANIS, D. 1988. Epidemiology of coccidioidomycosis. *In* Current Topics in Medical Mycology. M.R. McGinnis, Ed.: 199–238. Springer-Verlag. New York.
4. PAPPAGIANIS, D. & B.L. ZIMMER. 1990. Serology of coccidioidomycosis. Clin. Microbiol. Rev. **3:** 247–268.
5. KIRKLAND, T.N. & J. FIERER. 1983. Inbred mouse strains differ in resistance to lethal *Coccidioides immitis* infection. Infect. Immun. **40:** 912–916.
6. KIRKLAND, T.N. & J. FIERER. 1985. Genetic control of resistance to *Coccidioides immitis*: a single gene that is expressed in spleen cells determines resistance. J. Immunol. **135:** 548–552.
7. FIERER, J., L. WALLS, F. WRIGHT & T.N. KIRKLAND. 1999. Genes influencing resistance to *Coccidioides immitis* and the interleukin-10 response map to chromosomes 4 and 6 in mice. Infect. Immun. **67:** 2916–2919.
8. FIERER, J., L. WALLS, L. ECKMANN, *et al.* 1998. Importance of interleukin-10 in genetic susceptibility of mice to *Coccidioides immitis*. Infect. Immun. **66:** 4397–4402.

9. MAGEE, D.M. & R.A. COX. 1995. Roles of gamma interferon and interleukin-4 in genetically determined resistance to *Coccidioides immitis*. Infect. Immun. **63:** 3514–3519.
10. JIMENEZ, Md. P., L. WALLS & J. FIERER. 2006. High levels of interleukin-10 impair resistance to pulmonary coccidioidomycosis in mice in part through control of nitric oxide synthase 2 expression. Infect. Immun. **74:** 3387–3395.
11. DRUTZ, D.J. & A. CATANZARO. 1978. Coccidioidomycosis. Part II. Am. Rev. Respir. Dis. **117:** 727–771.
12. MOORE, K.W., R. DE WAAL MALEFYT, R.L. COFFMAN & A. O'GARRA. 2001. Interleukin-10 and the interleukin-10 receptor. Annu. Rev. Immunol. **19:** 683–765.
13. NATHAN, C. 1995. Natural resistance and nitric oxide. Cell **82:** 873–876.
14. CHAN, J. & J. FLYNN 1999. Nitric oxide in Mycobacterium tuberculosis infection. *In* Nitric Oxide and Infection. F.C. Fang, Ed.: 281–310. Kluwer Academic/Plenum Publishers. New York.
15. BOOCKVAR, K.S., M. MAYBODI, R.M. POSTON, et al. 1999. Nitric oxide in Listeriosis. *In* Nitric Oxide and Infection. F.C. Fang, Ed.: 447–471. Kluwer Academic/Plenum Publishers. New York.
16. AMPEL, N.M. & L.A. KRAMER. 2003. In vitro modulation of cytokine production by lymphocytes in human coccidioidomycosis. Cell Immunol. **221:** 115–121.

The Complex Immunology of Human Coccidioidomycosis

NEIL M. AMPEL

Medical Service, 1-111, SAVAHCS, Tucson, Arizona 85723

ABSTRACT: The human immune response during coccidioidomycosis is intimately involved with the development of delayed-type hypersensitivity and cellular immunity. Sixty percent of those infected have no symptoms and benign outcome is generally associated with a specific cellular immune response to coccidioidal antigens. We have recently teased out the human pulmonary granulomatous response during coccidioidomycosis and noted that there are perigranulomatous clusters of lymphocytes consisting predominantly of B lymphocytes and CD4$^+$ T lymphocytes. In other work, we have found that the mannose receptor as well as the toll-like receptors TLR2 and TLR4 may have a role in recognizing glycosylated coccidioidal antigens. In addition, the IL-12 receptor axis appears to be operative during antigen recognition and IL-12p40 may be the active moiety. Finally, peripheral blood mononuclear cells from persons with disseminated coccidioidomycosis are able to respond to coccidioidal antigen when it is presented by a mature monocyte-derived IL-4-generated dendritic cell (DC). These observations could be useful in the development of a human vaccine against coccidiodomycosis.

KEYWORDS: coccidioidomycosis; cellular immunity; human

INTRODUCTION

This article reviews material that has led me to pursue study of the human cellular immune response to coccidioidomycosis, and reviews of my own recent work in this area.

HISTORICAL BACKGROUND

Recognition of Clinical Spectrum of Disease

Coccidioidomycosis was first described in the late 19th century as a disfiguring and frequently fatal granulomatous process due to the mould *Coccidioides*.

Address for correspondence: Neil M. Ampel, M.D., Medical Service, 1-111, SAVAHCS, 3601 S. Sixth Avenue, Tucson, AZ 85723. Voice: 520-792-1450, ext: 6186; fax: 520-629-4793.
nampel@email.arizona.edu

Ann. N.Y. Acad. Sci. 1111: 245–258 (2007). © 2007 New York Academy of Sciences.
doi: 10.1196/annals.1406.032

Subsequently, clinicians recognized that infection could present as a benign pulmonary process frequently associated with cutaneous manifestations of delayed-type hypersensitivity.[1] Charles E. Smith and his colleagues noted that coccidioidomycosis had a broad clinical spectrum, including asymptomatic infection, which could be detected by the development of a cutaneous delayed-type hypersensitivity response after injection of coccidioidal antigen. They also posited that infection could serve as an immunization that would result in resistance to reinfection and reduce the risk for the development of a more severe disease.[2]

The Development of Coccidioidin

To clinically detect cutaneous delayed-type hypersensitivity, Smith and his colleagues developed coccidioidin, using tuberculin as their model. A filtrate was obtained from coccidioidal mycelia grown on an artificial liquid medium for 8 weeks. Thimerosal was then added to preserve and sterilize the filtrate. The resultant material, called coccidioidin, was found to be heat-stable, and consisted predominantly of polysaccharides by weight. However, when the protein content was destroyed, the ability of coccidioidin to induce delayed-type hypersensitivity was lost.

Using this material, Smith and his colleagues performed a study using coccidioidin at three Army air bases in the San Joaquin Valley during the 1940s. Airmen were tested upon arrival and those who were negative were periodically retested. Those who presented with positive skin tests or whose tests converted to positive over the course of the study were asked to complete a questionnaire. Among the 1,351 individuals who were found to react to coccidioidin, 60% did not recall any symptoms.[2]

Smith and colleagues made a number of additional observations.[2] For example, while some patients developed a wheal at the site of injection within 6 h, induration usually only occurred after 24 h. Persons with true "Valley Fever," a pulmonary infection with the cutaneous manifestations of delayed-type hypersensitivity erythema nodosum or erythema multiforme, invariably reacted to coccidioidin, while individuals with erythema nodosum and erythema multiforme due to other etiologies did not. Persons with coccidioidomycosis that had clinically disseminated beyond the thoracic cavity generally did not respond to coccidioidin, although those with meningitis were an exception. Finally, persons who persistently reacted to coccidioidin tended to have a benign clinical course.

CLINICAL EXPRESSION OF DISEASE AND ITS RELATIONSHIP TO IMMUNITY

As noted above, 60% of all those infected with *Coccidioides* are completely asymptomatic. The other 40% have some form of pulmonary disease.

Approximately 5% of these will have concomitant expression of dermal hypersensitivity manifesting either as erythema nodosum or erythema multiforme. A very small percentage of individuals, probably fewer than 1% of all infections, develop clinically apparent extrathoracic dissemination.[3]

Cellular immunity, as manifested by delayed-type hypersensitivity, appears to be very long-lived and protective in humans since clinical illness resulting from reinfection has not been described. Moreover, reactivation of clinical infection after initial resolution appears to only occur in those who subsequently develop profound immune suppression, such as those receiving corticosteroids or anticancer therapy,[4] antibody therapy directed against TNF-α,[5] or from HIV infection.[6]

This protective effect of cellular immunity in human coccidioidomycosis is similar to observations for other granulomatous diseases. Drutz and Catanzaro pointed this out graphically in their 1978 monograph on coccidioidomycosis[7] when they used a figure by Bullock portraying declining cellular immunity and increasing antibody response linked to disseminated leprosy, while robust cellular immunity and minimal antibody response were related to localized disease.

DISSECTING THE HUMAN GRANULOMATOUS RESPONSE

To more closely examine the response of humans to coccidioidal infection, we embarked on a study of the *in situ* pulmonary granulomatous reaction. Necrotizing granulomata have long been recognized as a hallmark of the immune response to coccidioidomycosis. However, only one study had specifically explored this[8] and all but one sample examined in that report were from cutaneous disseminated lesions.

We decided to examine tissue from pulmonary lesions containing coccidioidal necrotizing granulomata and to describe their histological appearance as well as determine the cytokine expression and enumerate the type of immune cells involved.[9] Using hematoxylin–eosin staining, perigranulomatous clusters of lymphocytes were observed. These clusters were found to consist predominantly of B lymphocytes, interspersed with CD4+ T lymphocytes and surrounded by a smaller number of CD8+ T lymphocytes.

In addition, the suppressive cytokine IL-10 was found both in the mantle region surrounding the necrotic center of the granuloma as well as in the perigranulomatous clusters. However, the presence of the stimulatory cytokine interferon-γ (IFN-γ) was significantly reduced in the clusters compared to the mantle. When confocal microscopy was employed, the IL-10 produced in the perigranulomatous lymphocyte clusters co-localized with B lymphocytes and CD4+ T lymphocytes. In the mantle, IL-10 appeared to be produced by macrophages.

These data suggest that both suppressive and stimulatory cytokines are produced within coccidioidal granuloma. While the role of the perigranulomatous lymphocytes could not be ascertained from these studies, a reasonable hypothesis is that they represent focal areas of immune suppression and perhaps antibody production.

IN VITRO STUDIES

Antigens and Human in Vitro Studies

A variety of antigens have been identified with *Coccidioides* that are capable of inducing cellular immune response *in vitro* using human peripheral blood mononuclear cells (PBMCs). In early work done in collaboration with Dr. John Galgiani, we found that a preparation called toluene spherule lysate (TSL) stimulated PBMCs from immune but not from nonimmune donors.[10] More recently, we have examined T27K, an antigen preparation made by mechanically disrupting thimerosal-killed spherules and obtaining the water-soluble supernatant after centrifugation at $27,000 \times g$.[11] We have found that T27K stimulates the production of both IFN-γ and IL-2 by PBMCs from coccidioidal immune but not nonimmune donors.[12] Results of IFN-γ release by PBMCs after incubation with T27K significantly correlate with the degree of the delayed-type hypersensitivity reaction observed after the intradermal injection of the skin-test reagent coccidioidin.[13] In addition, we have observed a correlation between *in vitro* response to T27K and the clinical expression of coccidioidomycosis. Intracellular expression of IFN-γ by CD3$^+$ lymphocytes in response to T27K was increased in samples from healthy immune donors compared to nonimmune donors and compared to donors with chronic active pulmonary coccidioidomycosis, disseminated coccidioidomycosis, as well as subjects with HIV infection and coccidioidomycosis.[14]

T27K is highly glycosylated. Monosaccharide analysis reveals it to contain 115.1 nM mannose, 38.0 nM glucose, and 16.4 nM galactose.[15] Moreover, T27K contains numerous discrete antigens. Two of these, Ag2/PRA and aspartyl protease, when expressed as recombinant proteins and used as vaccines have afforded protection to mice challenged with *Coccidioides*.[16,17]

Pattern Recognition Receptors and Immunity

In 1989, Janeway articulated a novel concept[18] suggesting that the evolutionarily old innate immune system plays an important role in adaptive host defense. He proposed that invariant molecular patterns found on a variety of pathogens, called pathogen-associated molecular patterns (PAMPs), could be distinguished from self through a set of ligands, called pattern recognition receptors (PRRs), located on antigen-presenting cells.[19]

C-Lectin Type Receptors (CLRs)

The C-lectin type receptors represent one of the major classes of PRRs. CLRs may be membrane-bound or secreted as soluble proteins. They have the ability to recognize a variety of carbohydrate structures through at least one carbohydrate-binding domain and are calcium-dependent.[20] Mannose receptor (MR, CD206), DEC205 (CD205), Dectin-1, Dectin-2, langerin (CD207), DC-SIGN (CD209), BDCA-2, DCIR, DLEC, and CLEC-1 have been identified among the membrane-bound CLRs.[21–25] The ligands for these receptors are not fully defined, but mannose and fucose are known to bind the MR[26] as well as DC-SIGN,[27] while 1,3 β-glucan is the ligand for Dectin-1.[28,29]

We have examined the role of MRs in human coccidioidomycosis using an *in vitro* model of cytokine release from PBMCs from coccidioidal-immune donors after incubation with T27K.[15] First, we measured the expression of MRs on the monocyte cell fraction of PBMCs. Mannan, a polysaccharide of mannose, caused a dose-dependent reduction of MR expression on monocytes such that 3.0 mg/mL of mannan reduced MR expression by more than 80% when compared to control samples. In a parallel manner, 3.0 mg/mL of mannan caused significant reductions in the release of both IFN-γ ($P < 0.05$) and IL-2 ($P < 0.02$) by PBMCs in response to T27K. In another model, Dionne and colleagues demonstrated that the phagocytosis of autoclave-killed spherules by human monocyte-derived dendritic cells (DCs) was significantly reduced in the presence of mannan.[30]

Mannose-binding lectin (MBL) is a soluble, circulating CLR. It binds D-mannose, L-fucose, and *N*-acetylglucosamine and then activates the complement system. In humans, MBL is encoded by one gene, *Mbl2*, which exists in multiple haplotypes that are associated with marked differences in circulating levels of MBL.[31] There is speculation that MBL may be an important host response factor and that low levels in the serum are associated with an increased risk of infection. For example, the risk of cryptosporidiosis in Haitian children was found to be increased among those with low circulating MBL levels.[32] While there are no published data, we have preliminary data that low levels are associated with active and disseminated coccidioidomycosis (Ampel, unpublished observations). Further examination of MBL in human coccidioidomycosis is warranted.

Toll-Like Receptors

A second major class of receptors are the toll-like receptors (TLRs). There are currently at least 13 identified mammalian homologues of the *Drosophila* toll family. TLRs are not only important in establishing an early innate immune response to control initial infection,[33] but are also critical in inducing adaptive immunity by priming T cells.[34] TLRs recognize a broad array of ligands. The first defined TLR ligand was LPS for TLR4. Other known ligands are dsRNA for TLR3, unmethylated CpG DNA for TLR9, and flagellin for TLR5. TLR2 has the largest number of known ligands, including peptidoglycan and zymosan.[19]

We have explored the effect of blocking TLR2 and TLR4 on adherent monocytes[35] derived from PBMCs using the method of Flo et al.[36] In brief: PBMCs from healthy, immune donors were allowed to adhere for 24 h. The monolayers were then washed and 20 μg/mL anti-TLR2 (αTLR2) or 1 μg/mL anti-TLR4 (αTLR4) was added for 30 min. Subsequently, 20 μg of T27K was added to wells. In addition to T27K, we used 0.1 μg/mL of peptidoglycan (PG) derived from *Staphylococcus aureus* (Sigma Chemical Co., St. Louis, MO, USA) for a TLR2 control ligand and LPS (*E. coli*-derived; Sigma) at a concentration of 0.02 μg/mL was used as a TLR4 control ligand. After 6 h, the supernatant was harvested and enzyme-linked immunosorbent assay (ELISA) was performed for TNF-α concentration.

Results are displayed in TABLE 1 and demonstrate a significant reduction of TNF-α concentrations in T27K-stimulated samples incubated with anti-TLR2 ($P < 0.010$). As expected, there was also reduced production of TNF-α for PG when incubated with anti-TLR2. There was a quantitatively less but still significant ($P < 0.050$) decrease in TNF-α when cells were incubated with T27K and anti-TLR2, while incubation of LPS with anti-TLR4 also resulted in significant reduction of TNF-α concentration. Incubation with isotype antibody for anti-TLR2 and anti-TLR4 did not result in TNF-α inhibition for any ligands (data not shown). On the basis of these results, both TLR2 and TLR4 appear to mediate the *in vitro* cellular immune response in human coccidioidomycosis.

TABLE 1. TNF-α concentrations (pg/mL) in cultured supernatants from adherent peripheral blood mononuclear cells (PBMCs) incubated for 24 h with 20 μg/mL T27K, 0.1 μg/mL peptidoglycan (PG), or with 0.02 μg/mL LPS with or without 30 min prior incubation with 20 μg/mL anti-TLR2 or 1 μg/mL anti-TLR4

	T27K (6)	PG (5)	LPS (5)
Neat	192 ± 37	739 ± 118	$10,153 \pm 2,902$
+anti-TLR2	62 ± 20^a	392 ± 88^b	$6,267 \pm 1,234$
+anti-TLR4	102 ± 22^b	743 ± 111	$2,892 \pm 769^b$

Results expressed as mean ± SEM. *P*-values determined based on paired *t*-test. Numbers in parentheses indicate number of subjects.
$^a P < 0.010$; $^b P < 0.050$.

THE IL-12 AXIS AND HUMAN COCCIDIOIDOMYCOSIS

IL-12 is a critical cytokine in the induction of the Th-1 cellular immune response. It consists of p35 and p40 subunits that in combination yield the IL-12p70 heterodimer.[37] Through the heterodimeric IL-12 receptor (IL-12R), IL-12 induces IFN-γ transcription through signal transduction and activator of transcription (STAT)-4. The role of the IL-12 receptor signaling system in human granulomatous diseases, such as coccidioidomycosis, is assumed, but has not been extensively explored.[38] Because inhibitors of this system have been proposed as therapies for psoriasis[39] and Crohn's disease[40] and because such signaling may play an important part in inducing protective immunity, we elected to examine this further in human coccidioidomycosis.[41]

In previous work, we have demonstrated that IL-12 is secreted by PBMCs obtained from immune donors after *in vitro* incubation with a coccidioidal antigen.[42] However, these assays detected both the p40 monomer and the p70 heterodimer portions of IL-12. In this study, we employed two different ELISAs that individually detected either IL-12p40 or IL-12p70. Because IL-23 is a heterodimer consisting of IL-12p40 and a p19 monomer,[43] we also measured the IL-23 concentration in PBMC supernatants. The median concentration of IL-12p40 from PBMCs obtained from seven coccidioidal immune donors after incubation with the coccidioidal antigen preparation T27K was 76.4 pg/mL (10.2–481), significantly above the median of 0.0 pg/mL (0.0–13.6) seen in samples from six nonimmune donors ($P = 0.003$). In addition, the levels of IL-12p40 released were highly and significantly correlated with the concentration of IFN-γ released by PBMCs incubated with T27K from immune donors

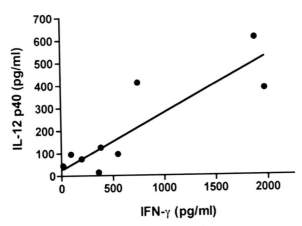

FIGURE 1. Relationship between release of IL-12p40 into culture medium and IFN-γ by peripheral blood mononuclear cells (PBMCs) from coccidioidal immune donors after incubation with T27K.

(FIG. 1, $r^2 = 0.857$, $P = 0.002$). The concentration of IL-12p40 from PBMCs incubated with cell culture medium alone was 0.0 pg/mL in all samples from both immune and nonimmune donors. No IL-12p70 or IL-23 was detected in PBMC supernatants obtained from immune donors incubated with T27K, but levels were detected after incubation with the mitogen phytohemagglutinin (data not shown).

We next wished to examine how IL-12 engages the immune cell. The IL-12 receptor consists of two subunits, β1 and β2.[37,44] Because we were unable to detect these subunits on the surface of immune response cells by labeled antibody, we examined their mRNA expression. RNA was isolated from PBMCs after 3 days of incubation with T27K and the gene products generated using polymerase chain reaction. The relative amounts of mRNA for IL-12Rβ1 and IL-12Rβ2 were subsequently determined as the ratio of the intensity of these products after gel electrophoresis to that of the intensity of 18S rRNA isolated from the same sample. Incubation of PBMCs with T27K resulted in significant increases in expression of IL-12Rβ1 mRNA among both the six immune and five nonimmune donors (for both, $P < 0.050$), but IL-12Rβ2 mRNA expression only increased significantly when PBMCs from the six immune donors were incubated with T27K ($P = 0.028$).

Similarly, the levels of nuclear-associated active STAT-4, measured as optical density, increased significantly only in PBMCs from immune donors after incubation with T27K. Specifically, among five nonimmune donors, levels of active STAT-4 for the control samples incubated with medium only was 0.030 (0.011–0.052) compared to 0.047 (0.019–0.095) for samples incubated with T27K ($P = 0.080$). Among seven immune donors, the median level was 0.015 (0.000–0.060) in samples incubated in medium alone compared to 0.144 (0.004–0.431) for those incubated with T27K (FIG. 2, $P = 0.028$).

These data indicate that the IL-12/IL-12 receptor/STAT-4 axis is functional during the human immune response in coccidioidomycosis. Moreover, they suggest that IL-12p40, heretofore considered a suppressive cytokine fraction, at least as a homodimer in mice,[45] may be the active element in upregulating this axis.

MODULATION OF THE IMMUNE RESPONSE WITH DCs

DCs are critical antigen-presenting cells in that they have the ability to process antigen and stimulate naive lymphocytes that have not previously responded to antigen.[46] Dionne et al. have shown that immature DCs can ingest killed coccidioidal spherules and eventually present antigen in association with HLA-DR on the cell surface.[30]

Individuals who develop coccidioidomycosis that has disseminated beyond the thoracic cavity have a poor prognosis. In addition, their condition is frequently associated with a lack of specific cellular immunity. We have made observations that, at least in some patients, a cellular immune response directed against coccidioidal antigens may occur with successful antifungal therapy (Ampel, unpublished observations). This suggests that the immunological defect observed in these patients is not fixed. However, the site of this defect has not been defined.

To examine this issue further, we explored whether incubation of lymphocytes from individuals with disseminated coccidioidomycosis could recognize antigen if it was presented *in vitro* by DCs. To achieve this, immature DCs were generated from peripheral blood mononuclear cells using granulocyte/monocyte colony-stimulating factor (GM-CSF) and IL-4. They were then incubated further with T27K, the coccidioidal antigen used in the studies described above, matured using TNF-α and prostaglandin E_2, and then incubated with lymphocytes from these same patients. These mature DCs were normal with regard to expression of accessory molecules and in their ability to drive a mixed-lymphocyte response. While PBMCs from these subjects responded minimally to T27K, there was significant lymphocyte transformation when incubated with the mature DCs that were loaded with T27K.[47]

These data demonstrate that neither lymphocytes nor DCs of patients with disseminated coccidioidomycosis are suppressed in their ability to react to or present coccidioidal antigen *in vitro*. The results indicate that the inability of individuals with disseminated coccidioidomycosis to mount an appropriate immune response must lie elsewhere. Possible etiologies are *in vivo* production

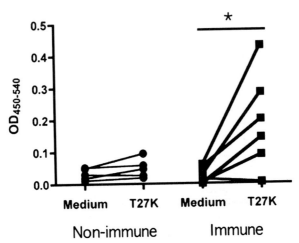

FIGURE 2. Relative amount of active STAT-4 in PBMCs measured as optical density ($OD_{450-540}$) for cells incubated with medium only and with 10 μg/mL T27K for 120 h. During the final 30 min of incubation, 10 ng/mL of rhuIL-12 was added to all samples. Data represent values for five nonimmune and seven immune donors. *$P = 0.028$.

of suppressive cytokines, such as IL-4 or IL-10, or by *in vivo* suppression by antigen. Another possibility is the *in vivo* development of peripheral tolerance by immature DCs.[46] Further assessment of these possibilities is warranted.

THE PURSUIT OF A HUMAN VACCINE FOR COCCIDIOIDOMYCOSIS

Given what we know about the human immune response to coccidioidomycosis, what can be opined with regard to a possible vaccine? Previously, a trial of a killed spherule vaccine was inconclusive.[48] Since then, there has been much interest in using recombinant antigens and these have demonstrated efficacy in murine models.[49,50] However, we have not been able to consistently demonstrate that recombinant protein antigens are recognized in our human *in vitro* system. For example, IL-2 release by PBMCs from healthy coccidioidal immune donors after incubation with T27K was significantly higher than that observed after incubation with recombinant Ag2/PRA comprising the first 106 amino acids ($rAg2/PRA_{1-106}$) or recombinant *Coccidioides*-specific antigen (rCS-Ag) (FIG. 3). The latter two antigens did not stimulate release of IL-2 significantly above control (Ampel, unpublished observations). On the other hand, after demonstrating that native glycoproteins derived from the spherule outer wall (SOW) were stimulatory to immune human PBMCs,[51] a recombinant peptide derived from this and provided by Chung-Yu Hung and Garry Cole was also found to be stimulatory *in vitro* (Ampel, unpublished observations). These results suggest that further screening of recombinant peptides

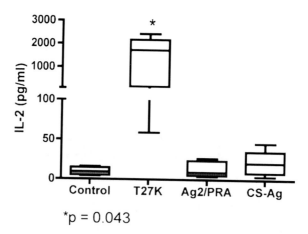

FIGURE 3. Results using 2×10^6 PBMCs from five healthy coccidioidal immune donors incubated with nothing (control), T27K at 20 μg/mL, recombinant $Ag2/PRA_{1-106}$ (Ag2/PRA), or recombinant *Coccidioides*-specific antigen (CS-Ag), both at 5 μg/mL, for 24 h. The supernatant was then assayed for IL-2.

may still be a useful endeavor in the search for a human vaccine against coccidioidomycosis.

An attenuated live strain of *Coccidioides* could also serve as a vaccine. However, concerns about reversion to pathogenicity as well as about genetic drift, as reported for bacille Calmette Guérin (BCG) in tuberculosis,[52] would have to be resolved prior to human use. Jiang and colleagues have demonstrated that a cDNA vaccine using the Ag2/PRA sequence protects mice from coccidioidal challenge.[53] However, many uncertainties, including the route of inoculation, choice of antigen, reproducibility of results and whether it would be used therapeutically or for disease prevention,[54] would have to be resolved before considering a human DNA vaccine against coccidioidomycosis.

Increasing evidence indicates that glycosylation plays a part in both recognition and presentation of antigen.[55,56] Given recent advances in glycology, it should be possible in the near future to develop a synthetic coccidioidal antigen that contains appropriate glycosylation to induce or enhance immunogenicity.

The recognition of the role of IL-12 in inducing a cellular immune response is growing. Our data demonstrating that IL-12p40 may be an active cytokine, at least *in vitro*, is intriguing and merits further study. Whether recombinant IL-12 could serve as an adjuvant to increase the cellular immune response to a vaccine has been explored in mice,[57] but not in humans.

Finally, DCs are another promising tool for inducing a cellular immune response, and significant work is under way in cancer research to use DCs loaded with tumor antigen as a therapeutic maneuver.[46] DCs loaded with a defined or a complex coccidioidal antigen to induce an appropriate cellular immune response in individuals with disseminated disease could be a strategy to shorten the time of antifungal therapy in such patients.

ACKNOWLEDGMENTS

This work presented above was supported by a Merit Review grant from the Department of Veterans Affairs and the NIH NIAID Grant 1PO1AI061310–01. We thank Demosthenes Pappagianis, M.D., Ph.D. and his laboratory for their collaboration and for providing the coccidioidal antigen preparation T27K. The author also specifically acknowledges the work and collaboration of the following persons: Lijin Li, Suzette Chavez, Daniel K. Nelson, Sara O. Dionne, Abigail Podany, Douglas F. Lake, John O. Richards, John N. Galgiani, and Suzanne M. Johnson. All work with human subjects was done with the approval of the Human Subjects Protection Program of the University of Arizona.

REFERENCES

1. DICKSON, E.C. 1938. Coccidioidomycosis. The preliminary acute infection with fungus *Coccidioides*. JAMA **111:** 1362–1365.

2. SMITH, C.E. & R. BEARD. 1946. Varieties of coccidioidal infection in relation to the epidemiology and control of the diseases. Am. J. Public Health **36:** 1394–1402.
3. KIRKLAND, T.N. & J. FIERER. 1996. Coccidioidomycosis: a reemerging infectious disease. Emerg. Infect. Dis. **2:** 192–199.
4. AMPEL, N.M., K.J. RYAN, P.J. CARRY, *et al.* 1986. Fungemia due to *Coccidioides immitis*. An analysis of 16 episodes in 15 patients and a review of the literature. Medicine (Balt.) **65:** 312–321.
5. BERGSTROM, L., D.E. YOCUM, N.M. AMPEL, *et al.* 2004. Increased risk of coccidioidomycosis in patients treated with tumor necrosis factor alpha antagonists. Arthritis Rheum. **50:** 1959–1966.
6. HERNANDEZ, J.L., S. ECHEVARRIA, A. GARCIA-VALTUILLE, *et al.* 1997. Atypical coccidioidomycosis in an AIDS patient successfully treated with fluconazole. Eur. J. Clin. Microbiol. Infect. Dis. **16:** 592–594.
7. DRUTZ, D.J. & A. CATANZARO. 1978. Coccidioidomycosis. Part I. Am. Rev. Respir. Dis. **117:** 559–585.
8. MODLIN, R.L., G.P. SEGAL, F.M. HOFMAN, *et al.* 1985. *In situ* localization of T lymphocytes in disseminated coccidioidomycosis. J. Infect. Dis. **151:** 314–319.
9. LI, L., S.M. DIAL, M. SCHMELZ, *et al.* 2005. Cellular immune suppressor activity resides in lymphocyte cell clusters adjacent to granulomata in human coccidioidomycosis. Infect. Immun. **73:** 3923–3928.
10. GALGIANI, J.N., K.O. DUGGER, N.M. AMPEL, *et al.* 1988. Extraction of serologic and delayed hypersensitivity antigens from spherules of *Coccidioides immitis*. Diagn. Microbiol. Infect. Dis. **11:** 65–80.
11. ZIMMERMANN, C.R., S.M. JOHNSON, G.W. MARTENS, *et al.* 1998. Protection against lethal murine coccidioidomycosis by a soluble vaccine from spherules. Infect. Immun. **66:** 2342–2345.
12. AMPEL, N.M., L.A. KRAMER, L. LI, *et al.* 2002. *In vitro* whole-blood analysis of cellular immunity in patients with active coccidioidomycosis by using the antigen preparation T27K. Clin. Diagn. Lab. Immunol. **9:** 1039–1043.
13. AMPEL, N.M., R.F. HECTOR, C.P. LINDAN & G.W. RUTHERFORD. 2006. An archived lot of coccidioidin induces specific coccidioidal delayed-type hypersensitivity and correlates with *in vitro* assays of coccidioidal cellular immune response. Mycopathologia **161:** 67–72.
14. AMPEL, N.M. & L. CHRISTIAN. 2000. Flow cytometric assessment of human peripheral blood mononuclear cells in response to a coccidioidal antigen. Med. Mycol. **38:** 127–132.
15. AMPEL, N.M., D.K. NELSON, L. LI, *et al.* 2005. The mannose receptor mediates the cellular immune response in human coccidioidomycosis. Infect. Immun. **73:** 2554–2555.
16. SHUBITZ, L., T. PENG, R. PERRILL, *et al.* 2002. Protection of mice against *Coccidioides immitis* intranasal infection by vaccination with recombinant antigen 2/PRA. Infect Immun. **70:** 3287–3289.
17. TARCHA, E.J., V. BASRUR, C.Y. HUNG, *et al.* 2006. A recombinant aspartyl protease of *Coccidioides posadasii* induces protection against pulmonary coccidioidomycosis in mice. Infect. Immun. **74:** 516–527.
18. JANEWAY, C.A. Jr. 1989. Approaching the asymptote? Evolution and revolution in immunology. Cold Spring Harb. Symp. Quant. Biol. **54**(Pt 1): 1–13.
19. JANEWAY, C.A. Jr. & R. MEDZHITOV. 2002. Innate immune recognition. Annu. Rev. Immunol. **20:**197–216.

20. GEIJTENBEEK, T.B., S.J. VAN VLIET, A. ENGERING, *et al.* 2004. Self- and nonself-recognition by C-type lectins on dendritic cells. Annu. Rev. Immunol. **22:** 33–54.

21. ARCE, I., P. RODA-NAVARRO, M.C. MONTOYA, *et al.* 2001. Molecular and genomic characterization of human DLEC, a novel member of the C-type lectin receptor gene family preferentially expressed on monocyte-derived dendritic cells. Eur. J. Immunol. **31:** 2733–2740.

22. BATES, E.E., N. FOURNIER, E. GARCIA, *et al.* 1999. APCs express DCIR, a novel C-type lectin surface receptor containing an immunoreceptor tyrosine-based inhibitory motif. J. Immunol. **163:** 1973–1983.

23. EBNER, S., Z. EHAMMER, S. HOLZMANN, *et al.* 2004. Expression of C-type lectin receptors by subsets of dendritic cells in human skin. Int. Immunol. **16:** 877–887.

24. MCGREAL, E.P., M. ROSAS, G.D. BROWN, *et al.* 2006. The carbohydrate-recognition domain of Dectin-2 is a C-type lectin with specificity for high mannose. Glycobiology **16:** 422–430.

25. TAKAHARA, K., Y. OMATSU, Y. YASHIMA, *et al.* 2002. Identification and expression of mouse Langerin (CD207) in dendritic cells. Int. Immunol. **14:** 433–444.

26. FIGDOR, C.G., Y. VAN KOOYK & G.J. ADEMA. 2002. C-type lectin receptors on dendritic cells and Langerhans cells. Nat. Rev. Immunol. **2:** 77–84.

27. CAMBI, A., K. GIJZEN, J.M. DE VRIES, *et al.* 2003. The C-type lectin DC-SIGN (CD209) is an antigen-uptake receptor for *Candida albicans* on dendritic cells. Eur. J. Immunol. **33:** 532–538.

28. BROWN, G.D., P.R. TAYLOR, D.M. REID, *et al.* 2002. Dectin-1 is a major beta-glucan receptor on macrophages. J. Exp. Med. **196:** 407–412.

29. TAYLOR, P.R., G.D. BROWN, D.M. REID, *et al.* 2002. The beta-glucan receptor, dectin-1, is predominantly expressed on the surface of cells of the monocyte/macrophage and neutrophil lineages. J. Immunol. **169:** 3876–3882.

30. DIONNE, S.O., A.B. PODANY, Y.W. RUIZ, *et al.* 2006. Spherules derived from *Coccidioides posadasii* promote human dendritic cell maturation and activation. Infect. Immun. **74:** 2415–2422.

31. TAKAHASHI, K., W.E. IP, I.C. MICHELOW & R.A. EZEKOWITZ. 2006. The mannose-binding lectin: a prototypic pattern recognition molecule. Curr. Opin. Immunol. **18:** 16–23.

32. KIRKPATRICK, B.D., C.D. HUSTON, D. WAGNER, *et al.* 2006. Serum mannose-binding lectin deficiency is associated with cryptosporidiosis in young Haitian children. Clin. Infect. Dis. **43:** 289–294.

33. DOHERTY, T.M. & M. ARDITI. 2004. TB, or not TB: that is the question—does TLR signaling hold the answer? J. Clin. Invest. **114:** 1699–1703.

34. PASARE, C. & R. MEDZHITOV. 2003. Toll-like receptors: balancing host resistance with immune tolerance. Curr. Opin. Immunol. **15:** 677–682.

35. AMPEL, N.M. 2005. TLR2 is required for recognition of the coccidioidal antigen T27K in human coccidioidomycosis. Annual Meeting of the Infectious Diseases Society of America. San Francisco.

36. FLO, T.H., O. HALAAS, E. LIEN, *et al.* 2000. Human toll-like receptor 2 mediates monocyte activation by *Listeria* monocytogenes, but not by group B streptococci or lipopolysaccharide. J. Immunol. **164:** 2064–2069.

37. TRINCHIERI, G. 2003. Interleukin-12 and the regulation of innate resistance and adaptive immunity. Nat. Rev. Immunol. **3:** 133–146.

38. ROMANO, C.C., M.J. DE MENS-GIANNINI, A.J. DUARTE & G. BENARD. 2005. The role of interleukin-10 in the differential expression of interleukin-12p70 and its

beta2 receptor on patients with active or treated paracoccidioidomycosis and healthy infected subjects. Clin. Immunol. **114:** 86–94.

39. ROSMARIN, D. & B.E. STROBER. 2005. The potential of interleukin 12 inhibition in the treatment of psoriasis. J. Drugs Dermatol. **4:** 318–325.

40. FUSS, I.J., C. BECKER, Z. YANG, *et al.* 2006. Both IL-12p70 and IL-23 are synthesized during active Crohn's disease and are down-regulated by treatment with anti-IL-12 p40 monoclonal antibody. Inflamm. Bowel Dis. **12:** 9–15.

41. LI, L. & N.M. AMPEL. 2004. Induction of the IL-12R/STAT-4 pathway in human coccidioidomycosis. Annual meeting of the Infectious Diseases Society of America. Boston, MA.

42. AMPEL, N.M. & L. CHRISTIAN. 1997. *In vitro* modulation of proliferation and cytokine production by human peripheral blood mononuclear cells from subjects with various forms of coccidioidomycosis. Infect. Immun. **65:** 4483–4487.

43. HUNTER, C.A. 2005. New IL-12-family members: IL-23 and IL-27, cytokines with divergent functions. Nat. Rev. Immunol. **5:** 521–531.

44. LANGRISH, C.L., B.S. MCKENZIE, N.J. WILSON, *et al.* 2004. IL-12 and IL-23: master regulators of innate and adaptive immunity. Immunol. Rev. **202:** 96–105.

45. SCHMITT, D.A. & S.E. ULLRICH. 2000. Exposure to ultraviolet radiation causes dendritic cells/macrophages to secrete immune-suppressive IL-12p40 homodimers. J. Immunol. **165:** 3162–3167.

46. BANCHEREAU, J. & A.K. PALUCKA. 2005. Dendritic cells as therapeutic vaccines against cancer. Nat. Rev. Immunol. **5:** 296–306.

47. RICHARDS, J.O., N.M. AMPEL & D.F. LAKE. 2002. Reversal of coccidioidal anergy *in vitro* by dendritic cells from patients with disseminated coccidioidomycosis. J Immunol. **169:** 2020–2025.

48. PAPPAGIANIS, D. 1993. Evaluation of the protective efficacy of the killed *Coccidioides immitis* spherule vaccine in humans. The Valley Fever Vaccine Study Group. Am. Rev. Respir. Dis. **148:** 656–660.

49. SHUBITZ, L.F., J.J. YU, C.Y. HUNG, *et al.* 2006. Improved protection of mice against lethal respiratory infection with *Coccidioides posadasii* using two recombinant antigens expressed as a single protein. Vaccine **24:** 5904–5911.

50. TARCHA, E.J., V. BASRUR, C.Y. HUNG, *et al.* 2006. Multivalent recombinant protein vaccine against coccidioidomycosis. Infect. Immun. **74:** 5802–5813.

51. HUNG, C.Y., N.M. AMPEL, L. CHRISTIAN, *et al.* 2000. A major cell surface antigen of *Coccidioides immitis* which elicits both humoral and cellular immune responses. Infect. Immun. **68:** 584–593.

52. BEHR, M.A. & P.M. SMALL. 1999. A historical and molecular phylogeny of BCG strains. Vaccine **17:** 915–922.

53. JIANG, C., D.M. MAGEE, T.N. QUITUGUA & R.A. COX. 1999. Genetic vaccination against *Coccidioides immitis*: comparison of vaccine efficacy of recombinant antigen 2 and antigen 2 cDNA. Infect. Immun. **67:** 630–635.

54. LI, J.M. & D.Y. ZHU. 2006. Therapeutic DNA vaccines against tuberculosis: a promising but arduous task. Chin. Med. J. (Engl) **119:** 1103–1107.

55. JENSEN, T., P. HANSEN, L. GALLI-STAMPINO, *et al.* 1997. Carbohydrate and peptide specificity of MHC class II-restricted T cell hybridomas raised against an O-glycosylated self peptide. J. Immunol. **158:** 3769–3778.

56. VLAD, A.M. & O.J. FINN. 2004. Glycoprotein tumor antigens for immunotherapy of breast cancer. Breast Dis. **20:** 73–79.

57. JIANG, C., D.M. MAGEE & R.A. COX. 1999. Coadministration of interleukin 12 expression vector with antigen 2 cDNA enhances induction of protective immunity against *Coccidioides immitis*. Infect. Immun. **67**(11): 5848–5853.

The Public Health Need and Present Status of a Vaccine for the Prevention of Coccidioidomycosis

RICHARD HECTOR[a,b] AND GEORGE W. RUTHERFORD[a,b]

[a] The Institute for Global Health, University of California, San Francisco, California, USA

[b] Department of Epidemiology and Biostatistics, University of California, San Francisco, California, USA

ABSTRACT: Although the epidemiology of coccidioidomycosis has been well described, there is a paucity of recent data on the public health burden associated with this disease. Accordingly, California's Inpatient Hospital Discharge Data Set from 1997 to 2002 was used to calculate the incidence of hospitalization for coccidioidomycosis by county, year, age, race, ethnicity, and gender. The overall finding that coccidioidomycosis has a significant impact in endemic areas supports the conclusion that the need for a preventive vaccine is great. Investigators of the Valley Fever Vaccine Project (VFVP) have successfully identified a number of recombinant coccidioidal protein antigens and two attenuated mutant strains that have been evaluated as vaccines, demonstrating protective responses in murine models. Efforts to select and develop a vaccine for human clinical trials are in progress.

KEYWORDS: coccidioidomycosis; Coccidioides; vaccine; epidemiology; incidence; hospitalization; recombinant; immunity

INTRODUCTION

Coccidioidomycosis is a disease caused by *Coccidioides immitis* and *Coccidioides posadasii*, fungi that are found only in areas of the Western Hemisphere known ecologically as the Lower Sonoran Life Zone.[1] In the southwest United States, regions endemic for *Coccidioides* spp. are home to approximately 20% of the U.S. population; an estimated 5 million persons live in the areas of highest endemicity.[2–4]

During the past decade, the incidence of documented coccidioidal disease in California and Arizona has increased, possibly as a result of variations in

Address for correspondence: Dr. Richard Hector, Institute for Global Health, University of California, San Francisco, 50 Beale St., Suite 1200, San Francisco, CA 94105. Voice: 415-597-9251; fax: 415-476-6106.

rhector@psg.ucsf.edu

Ann. N.Y. Acad. Sci. 1111: 259–268 (2007). © 2007 New York Academy of Sciences.
doi: 10.1196/annals.1406.035

climate and/or changes in the demographics of the populace of the endemic area.[5,6] The population in endemic areas increased significantly, and much of the increase was due to immigration of adults from nonendemic areas who had no prior exposure.[7] Changes in the ethnic distribution of the immigrant population may also be contributory to the increased numbers of reported cases as previous studies found that blacks and Filipinos, in particular, are at increased risk of hospitalization for coccidioidomycosis.[8–10] Finally, immuno-compromised individuals are at greatly increased risk of coccidioidomycosis, and this population expanded dramatically in the region in the 1980s and 1990s primarily because of infection by human immunodeficiency virus (HIV).[11,12] It is presently estimated that there are over 150,000 primary coccidioidal infections each year, resulting in several thousand serious or disseminated cases.[13] Thus, despite the inherent difficulties in developing a vaccine for a disease mediated by cellular immunity, the public health need for a preventive vaccine compels efforts to this end.

BURDEN OF DISEASE IN CALIFORNIA

Much of our fundamental knowledge of the epidemiology of coccidioidomycosis dates to the 1940s and 1950s, when extensive studies of civilian and military populations were conducted.[14–17] As the manifestations of the disease vary widely, ranging from transient asymptomatic infection to severe, chronic disseminated disease, estimating the current incidence and severity of coccidioidomycosis has been challenging.[18] Studies have shown that among study subjects with skin test conversion, 60% were entirely asymptomatic.[16] A study in the civilian population of Kern County, California, showed that among those who developed symptomatic disease, 85% had a benign clinical course, resembling that of ordinary influenza, while 8% developed severe acute pulmonary disease requiring hospitalization, and an additional 7% developed disseminated coccidioidomycosis, including meningitis and lesions of the skin and bone.[8] Furthermore, although coccidioidomycosis is a reportable disease in California, Arizona, and nationally, the wide spectrum of clinical presentation increases the likelihood that many asymptomatic and mildly symptomatic cases will be neither diagnosed nor reported.[18]

Among the most important gaps in our knowledge of the epidemiology of coccidioidomycosis is an estimate of the current burden of disease. In order to estimate this in California, information from California's Inpatient Hospital Discharge Data Set from 1997 to 2002 was used to calculate the incidence of hospitalization for coccidioidomycosis by county, year, age, race, ethnicity, and gender (Valerie J. Flaherman, personal communication). The 7,457 inpatient hospitalizations for coccidioidomycosis in nonfederal institutions in California resulted in an average annual incidence of hospitalization for coccidioidomycosis of 3.7 per 100,000, with Kern, Los Angeles, and San Diego counties accounting for 47% of the total hospitalizations. The relative risk ratio for hospitalization of subjects in Kern County was 43.5, compared to 9.6 in Merced County. Eleven of 58 California counties had hospitalization rates above the California state annual average of 3.7 per 100,000. Coccidioidal meningitis accounted for 19% of all hospitalizations of patients with a diagnosis of coccidioidomycosis. Further, it was determined that 8.9% of all individuals hospitalized with a diagnosis of coccidioidomycosis had a fatal outcome, with a statewide mortality rate of 2.1 deaths per 1,000,000 residents per year. Of note, the risk of hospitalization from coccidioidomycosis far exceeds the risk of hospitalization from tuberculosis among a very large segment of the California population, including both white and the black populations.[19]

Confirming previous studies concerning genetic predisposition to dissemination,[8–10,20,21] data from the present study provide that while black race and male gender groups had increased rates of hospitalization, Native American and Asian-Pacific Islanders had rates lower than those of whites (see TABLE 1). Age appears to be a risk factor, with incidence of hospitalization <1 per 100,000 for those in the age group of 0–14 years, while the rate steadily increased to 7.2 per 100,000 for those 50 years and older, also confirming earlier studies.[8]

Regarding immunocompromised status as a risk factor for hospitalization, persons with AIDS had a 35-fold higher risk of hospitalization compared to all Californians. Persons with HIV had a 14-fold higher rate of hospitalization compared to all Californians. Examining only records with a repeat

TABLE 1. Incidence rates of hospitalization for California with diagnosis of coccidioidomycosis, 1997–2002

Population	Incidence of hospitalization[a]
Race	
White	3.6
Hispanic	3.4
Black	8
Native American	1.4
Asian-Pacific Islander	2
Gender	
Female	2.3
Male	5
Age (in years)	
0–14	0.5
15–49	3.5
50–69	7.1
70 or older	7.3
Special conditions	
Pregnancy	3.8
AIDS	127
All HIV	51

[a]crude incidence per 100,000 residents.

hospitalization, 24% of those admitted with coccidioidomycosis with HIV co-infection died during hospitalization, compared to 8.2% of individuals admitted with coccidioidomycosis without HIV co-infection ($P < 0.001$ by chi-square analysis). With respect to pregnancy and coccidioidomycosis, during the 6-year period the statewide hospitalization rate was 3.8 per 100,000 for pregnant women, while the rate was 1.5 hospitalizations per 100,000 nonpregnant women of childbearing age.

The incidence of coccidioidal disease in the military has taken on renewed interest, given the increased role of desert warfare training in the endemic area.[22–24] In a recent skin test study using mycelial-derived coccidioidin administered to soldiers engaged in field training exercises at Fort Irwin, CA, results provided an estimate of an annualized incidence rate of 4.9%, confirming that such soldiers are at risk for exposure and would be a target population that may benefit from an effective vaccine.[25]

RESEARCH EFFORTS TO IDENTIFY AND DEVELOP A VACCINE FOR COCCIDIOIDOMYCOSIS

Because the level of morbidity associated with the primary and disseminated forms of this disease can be severe, a vaccine for the prevention of coccidioidomycosis has long been sought.[26,27] Although a study of the cost-effectiveness of a coccidioidal vaccine projected that the savings for increased quality-adjusted life expectancy declines with the age of the vaccinee, the authors concluded that the aggregate health benefits and economic savings at the county and state level would nevertheless be substantial.[28]

Although experimental efforts to create a vaccine date back several decades, the ability to propagate the parasitic phase *in vitro* led to the creation of a formalin-killed spherule (FKS) vaccine that proved highly effective in a mouse model, and was able to reduce the severity of disease in a primate model.[26,27] A phase III trial was conducted during the early 1980s in humans using the FKS vaccine, but it proved too reactogenic for human use.[29] Subsequent efforts focused on the use and evaluation of purified fractions obtained from the cell wall as potential immunizing antigens.[30,31] Now, modern molecular biology, genomics, and an ever-increasing understanding of the immunologic response to coccidioidomycosis has raised the possibility of efficient antigen discovery, cloning, and production, leading coccidioidomycologists to renew a cooperative effort to identify and develop a safe and effective vaccine for this disease.[32]

The Valley Fever Vaccine Project (VFVP) is an academic consortium that originated, in part, as a response to a major epidemic of coccidioidomycosis that plagued parts of California in the early 1990s.[33] The lack of effective therapeutics, coupled with the great public health burden of disease, resulted in efforts to create the VFVP, with an overall goal of identifying commercially viable vaccine candidates for the prevention of coccidioidomycosis. Because coccidioidomycosis is a disease that has largely been ignored by the pharmaceutical industry, the project committed to move the most promising candidate into clinical trials. The staff and laboratories of Garry Cole (Medical University of Ohio), Rebecca Cox (University of Texas/San Antonio), John Galgiani (University of Arizona), Theo Kirkland (University of California/San Diego), and Demosthenes Pappagianis (University of California/Davis) conducted the research discovery component of the project. George Rutherford and Richard Hector (UCSF) have provided project oversight, while California State University, Bakersfield, Foundation has served as the primary contractor.

The research approach taken by this group has been (1) antigen discovery, (2) efficacy evaluation in murine models, (3) secondary evaluation in large animal models, and (4) antigen production and characterization. In what proved to be a strength of the program, the individual laboratories conducted the antigen discovery process using a diversity of approaches, leading to important discoveries in both the vaccinology and immunology of coccidioidomycosis.

Through the efforts of the investigators, a large number of recombinant and cellular-derived antigens, as well as two live attenuated mutant vaccines have been isolated and evaluated in a variety of murine models, leading to 10 patent applications. In particular, the recombinant antigens $Ag2/PRA_{1-106}$,[34] Csa, ELI1,[35] and Gel1[36] showed early promise. Subsequent experiments with immunizations by mixtures of recombinant antigens showed that the combination of $rAg2/PRA_{1-106}$ and rCsa used with monophosphoryl lipid A (MPL) adjuvant provided superior protection in C56BL/6 mice against an otherwise lethal challenge compared to mice immunized with the single antigens (TABLE 2).[37] Based on these encouraging results, a single chimeric fusion protein comprising $Ag2/PRA_{1-106}$ plus Csa was created and introduced into a *Saccharomyces cerevisiae* expression system, and the resulting chimeric fusion protein was evaluated in the same mouse model. The results in the murine survival model showed enhanced activity compared to the single proteins when used in conjunction with MPL adjuvant.[37] Accordingly, the fusion protein was selected by the VFVP as the lead recombinant vaccine candidate for further evaluation.

The safety and efficacy of the chimeric fusion protein was evaluated in a nonhuman primate model of coccidioidomycosis using female cynomolgus (*Macaca fascicularis*) monkeys.[38] Overall, the experimental results confirmed that vaccination resulted in sensitization to the antigen and a reduction in the burden of disease in the high-dose vaccinated group, compared to adjuvant controls, but did not prevent pulmonary disease.

Initial efforts to develop pilot-scale methods for the production and recovery of the chimeric fusion protein were based on fermentations using a transformed *Saccharomyces cerevisiae* host. Poor control over expression during fermentation and low production yields led to the creation of a new host construct, based on *Pichia pastoris*. While expression and yields were improved, other

TABLE 2. Survival of C57BL/6 mice immunized with single or a mixture of recombinant antigens or adjuvant alone, followed by lethal intranasal challenge

Group	% Survival[a]	Significance
Adjuvant	0	—
rCsa	24[b]	
rAg2/PRA1-106	83[b]	
rAg2/PRA1-106 + rCsa	97[b,c]	

[a]Results are a combination of three experiments.
[b]$P < 0.05$ compared to adjuvant control (Kaplan–Meier with Mantel–Haenszel log-rank test).
[c]$P < 0.05$ compared to rAg2/PRA1-106 and rCsa single antigen groups.

manufacturing issues were identified. Given the complexity of the manufacturing and the unavailability of a suitable adjuvant led the VFVP to consider other approaches. Current efforts are focused on linking the recombinant protein antigens to proprietary antigen delivery systems, which would obviate the need for an adjuvant in the vaccine formulation and may also lead to resolution of the more vexing manufacturing issues. If such an approach proves effective, it would increase the likelihood of moving a vaccine into pharmaceutical development and clinical trials in the near term.

In addition to the recombinant protein vaccines, the VFVP has investigated the use of live attenuated mutant strains of *C. posadasii* as vaccines. Live vaccines offer the theoretical advantage of containing a fuller range of immunizing antigens presented in their native conformation and with any associated glycosylation. To this end, two live attenuated mutant strains have been created: $\Delta chs5$ and $\Delta cts2/\Delta ard1/\Delta cts3$.[39–41] Both were created by knocking out one or more chitin-related genes, that is, chitin synthase or chitinase, involved in cellwall biosynthesis by double cross-over mutations of wild-type *C. posadasii* with specifically constructed plasmids. While the $\Delta chs5$ strain is able to grow in the mycelial phase, it is unable to undergo transformation into the parasitic spherule-endospore phase. In contrast, the $\Delta cts2/\Delta ard1/\Delta cts3$ strain can transform into first-generation spherules, but is incapable of endosporulation. The net result is that neither of the strains is capable of reproducing in the parasitic, spherule form of the fungus. Since the strains cannot reproduce, they are theoretically incapable of causing disease.

The lack of virulence in the mutant strains was confirmed by infecting groups of naïve BALB/c mice intranasally with high doses of the respective strains or the parent wild-type and monitoring the mice for evidence of disease. For example, in the case of the $\Delta cts2/\Delta ard1/\Delta cts3$ strain, mice infected with doses as high as 5,000 arthroconidia had 100% survival and no evidence of residual disease, while the corresponding mice infected with the wild-type parent strain all succumbed to infection. Importantly, the experimental confirmation of the lack of virulence enabled both strains to be categorized as excluded from select agent status by the Centers for Disease Control.[42]

With proof of attenuation in hand, the strains were then evaluated for immunization potential in vaccination experiments. In one experiment, using BALB/c mice and prime/boost immunizations with a total of 7.5×10^3 arthroconidia of the $\Delta cts2/\Delta ard1/\Delta cts3$ strain, vaccinated mice challenged intranasally with 80 arthroconidia had 100% survival 75 days post infection, while all negative controls succumbed to the same challenge within 16 days. Quantitative cultures of organs from the vaccinated mice resulted in recovery of approximately 3 $\log_{10}$ colony-forming units in the lungs, and no evidence of dissemination to the spleens.

While the experiments demonstrate the feasibility of using attenuated strains as vaccines, there remain additional regulatory safety and manufacturing challenges that would have to be overcome before a live vaccine could be evaluated in humans. Accordingly, the VFVP is presently evaluating whether additional mutations can be introduced into the strains to increase the safety margin, thereby meeting stringent FDA requirements for human clinical trials. Because of the containment issues associated with the production and collection of arthroconidia using dry cultures, efforts are under way to evaluate the use of broth cultures followed by fragmentation of hyphae as an alternate method of formulating the live vaccine.

Overall, the investigators of the VFVP conducted or collaborated in efforts that resulted in the sequencing of the *C. posadasii* genome, the identification and expression of dozens of coccidioidal recombinant antigens, the evaluation of multiple antigens and live mutant strains in efficacy models leading to the filing of 10 patent applications (see TABLE 3) and the selection of lead vaccine candidates; they conducted experimentation that refined animal models and increased the understanding of the immune response to coccidioidomycosis, and conducted human clinical trials with the coccidioidin skin test antigen.

TABLE 3. List of U.S. patents and applications based on VFVP-sponsored research

Antigen/Strain	USPTO Number
rCsa	10/794,287
Chs5 attenuated strain	11/102,217
Cts attenuated strain	11/292,431
rELI-1	10/985,853
rGel1	10/417,997
Overexpression mutant	60/568,609
RAg2/Pra106+rCsa	7,078,037
Ag2 Signal peptide	6,923,973
rSOWgp	09/850,677
rUre	10/418,962

MOVING A VACCINE FOR COCCIDIOIDOMYCOSIS TOWARD CLINICAL TRIALS

Because of the rapidly escalating costs and stringent regulatory requirements for drug and vaccine development, the pharmaceutical industry has become increasingly selective with respect to the preclinical candidates that are selected for commercialization, with a focus on high-value market products. Accordingly, the VFVP began with the knowledge that human safety and efficacy data would likely have to be obtained by the project in order to generate interest in a coccidioidomycosis vaccine in the increasingly risk-averse pharmaceutical industry. The VFVP believes that the increasing burden of coccidioidal disease underscores the public health need for a preventive vaccine and justifies the costs and efforts required to obtain such data. With the recent advances made by the project investigators, the VFVP anticipates that a vaccine candidate will be moved into the initial phases of pharmaceutical development, including manufacturing and toxicology safety testing in the near term, as a prelude to human clinical trials.

ACKNOWLEDGMENTS

This project has been supported by grants from the California HealthCare Foundation (99-1017), California Department of Health Services (00-90236 and 01-16362), and the California State University, Bakersfield, Foundation.

REFERENCES

1. HECTOR, R.F. & R. LANIADO-LABORIN. 2005. Coccidioidomycosis—a fungal disease of the Americas. PLoS Med. **2**: e2.
2. MADDY, K.T. 1958. The geographic distribution of *Coccidioides immitis* and possible ecologic implications. Ariz. Med. **15**: 178.
3. PAPPAGIANIS, D. 1980. Epidemiology of coccidioidomycosis. *In* Coccidioidomycosis: A Text. D.A. Stevens, Ed. Plenum Publishing Corporation. New York, NY.
4. U.S BUREAU OF THE CENSUS. Population estimates. Available at: http://quickfacts.census.gov/qfd/. Accessed 4 January 2007.
5. PARK, B.J., K. SIGEL, V. VAZ, *et al.* 2005. An epidemic of coccidioidomycosis in Arizona associated with climatic changes, 1998-2001. J. Infect. Dis. **191**: 1981–1987.
6. COMRIE, A.C. 2005. Climate factors influencing coccidioidomycosis seasonality and outbreaks. Environ. Health Perspect. **113**: 688–692.
7. FORSTALL, R.L. California, Population of Counties by Decennial Census. U.S. Bureau of the Census. Available at: www.census.gov/population/cencounts/ca190090.txt. Accessed September 2, 2005.

8. ROSENSTEIN, N.E., K.W. EMERY, S.B. WERNER, *et al.* 2001. Risk factors for severe pulmonary and disseminated coccidioidomycosis: Kern County, California, 1995-1996. Clin. Infect. Dis. **32:** 708–715.

9. FLYNN, N.M., P.D. HOEPRICH, M.M. KAWACHI, *et al.* 1979. An unusual outbreak of windborne coccidioidomycosis. N. Engl. J. Med. **301:** 358–361.

10. PAPPAGIANIS, D. & H. EINSTEIN. 1978. Tempest from Tehachapi takes toll or *Coccidioides* conveyed aloft and afar. West. J. Med. **129:** 527–530.

11. AMPEL, N.M. 2005. Coccidioidomycosis in persons infected with HIV type 1. Clin. Infect. Dis. **41:** 1174–1178.

12. WOODS, C.W., MCRILL C., PLIKAYTIS B.D., *et al.* 2000. Coccidioidomycosis in human immunodeficiency virus-infected persons in Arizona, 1994–1997: incidence, risk factors, and prevention. J. Infect. Dis. **181:** 1428–1434.

13. GALGIANI, J.N., AMPEL N.M., BLAIR J.E., *et al.* 2005. Coccidioidomycosis. Clin. Infect. Dis. **41:** 1217–1223.

14. SMITH, C.E. 1958. Coccidioidomycosis. *In* Preventive Medicine in World War II, Volume IV: Communicable Diseases Chiefly Transmitted through Respiratory and Alimentary Tracts. J. Lada, Ed.: 285–316.U.S. Government Printing Office. Washington, DC.

15. EGEBERG, R.O. 1968. Coccidioidomycosis. *In* Internal Medicine in World War II, Volume III: Infectious Diseases and General Medicine. R.S. Anderson, W.P. Havens Jr, L.D. Heaton & S. Bayne-Jones, Eds. Office of the Surgeon General, Department of the Army Washington, DC. Available at: http://history. amedd.army.mil/booksdocs/wwii/internalmedicinevolIII/frameindex.html. Accessed 26 December 2006.

16. SMITH, C.E., R.R. BEARD, E.G. WHITING & H.G. ROSENBERGER. 1946. Varieties of coccidioidal infection in relation to the epidemiology and control of the diseases. Am J. Public Health **36:** 1394–1402.

17. SMITH, C.E., M.T. SAITO & S.A. SIMONS. 1956. Pattern of 39,500 serologic tests in coccidioidomycosis. JAMA **160:** 546–552.

18. VALDIVIA, L., NIX D., WRIGHT M., *et al.* 2006. Coccidioidomycosis as a common cause of community-acquired pneumonia. Emerg. Infect. Dis. **12:** 958–962.

19. CALIFORNIA DEPARTMENT OF HEALTH SERVICES, DIVISION OF COMMUNICABLE DISEASE CONTROL, TUBERCULOSIS CONTROL BRANCH. Report on Tuberculosis in California, 2003. California Department of Health Services. Richmond, CA, January 2005.

20. SMITH, C.E. & R.R. BEARD. 1946. Varieties of coccidiodal infection in relation to the epidemiology and control of diseases. Am. J. Public Health **36:** 1394–1402.

21. PAPPAGIANIS, D., S. LINDSAY, S. BEALL & P. WILLIAMS. 1979. Ethnic background and the clinical course of coccidioidomycosis [letter]. Am. Rev. Resp. Dis. **120:** 959–961.

22. CRUM, N.F., M. POTTER & D. PAPPAGIANIS. 2004. Seroincidence of coccidioidomycosis during military desert training exercises. J. Clin. Microbiol. **42:** 4552–4555.

23. CRUM, N.F., E.R. LEDERMAN, B.R. HALE, *et al.* 2003. A cluster of disseminated coccidioidomycosis cases at a U.S. military hospital. Mil. Med. **168:** 460–464.

24. CRUM, N., C. LAMB, G. UTZ, *et al.* 2002. Coccidioidomycosis outbreak among United States Navy SEALs training in a *Coccidioides immitis*-endemic area-Coalinga, California. J. Infect. Dis. **186:** 865–868.

25. WHEELER, G., A. FARAVARDEH, K. DAILY, *et al.* 2006. A phase II study to evaluate the reactivity and tolerability of intradermally administered doses of coccidioidin

in human subjects in a target population. Presented at the American College of Physicians, Army Chapter, National Meeting. Arlington, VA, Nov. 15-18, 2006.

26. LEVINE, H.B., Y.-C KING & C.E. SMITH. 1965. Immunization of mice to *Coccidioides immitis*: dose, regimen and spherulation stage of killed spherule vaccines. J. Immunol. **94:** 132–142.

27. LEVINE, H.B., R.L. MILLER & C.E. SMITH. 1962. Influence of vaccination on respiratory coccidioidal disease in cynomologous monkeys. J. Immunol. **89:** 242–251.

28. BARNATO, A.E., G.D. SANDERS & D.K. OWENS. 2001. Cost-effectiveness of a potential vaccine for *Coccidioides immitis*. Emerg. Infect. Dis. **7:** 797–806.

29. PAPPAGIANIS, D. VALLEY FEVER VACCINE STUDY GROUP. 1993. Evaluation of the protective efficacy of the killed *Coccidioides immitis* spherule vaccine in humans. Am. Rev. Resp. Dis. **148:** 656–660.

30. DUGGER, K.O., J.N. GALGIANI & N.M. AMPEL. 1991. An immunoreactive apoglycoprotein purified from *Coccidioides immitis*. Infect. Immun. **59:** 2245–2251.

31. LECARA, G., R.A. COX & R.B. SIMPSON. 1983. *Coccidioides immitis* vaccine: potential of an alkali-soluble, water-soluble cell wall antigen. Infect. Immun. **39:** 473–475.

32. COX, R.A. & D.M. MAGEE. 2004. Coccidioidomycosis: host response and vaccine development. Clin. Microbiol. Rev. **17:** 804–839.

33. PAPPAGIANIS, D. 1994. Marked increase in cases of coccidioidomycosis in California: 1991, 1992, and 1993. Clin. Infect. Dis. **19:** S14–S18.

34. PENG, T., L. SHUBITZ, J. SIMONS, *et al.* 2002. Localization within a proline-rich antigen (Ag2/PRA) of protective antigenicity against infection with *Coccidioides immitis* in mice. Infect. Immun. **70:** 3330–3335.

35. IVEY, F.D., D.M. MAGEE, M.D. WOITASKE, *et al.* 2003. Identification of a protective antigen of *Coccidioides immitis* by expression library immunization. Vaccine **21:** 4359–4367.

36. DELGADO, N., XUE J., YU J.J., *et al.* 2003. A recombinant beta-1,3-glucanosyltransferase homolog of *Coccidioides posadasii* protects mice against coccidioidomycosis. Infect. Immun. **71:** 3010–3019.

37. SHUBITZ, L.F., J.J. YU, C.Y. HUNG, *et al.* 2006. Improved protection of mice against lethal respiratory infection with *Coccidioides posadasii* using two recombinant antigens expressed as a single protein. Vaccine **24:** 5904–5911.

38. LERCHE, N.W., D. PAPPAGIANIS, S.M. JOHNSON, *et al.* 2006. Safety, antigenicity, and efficacy of a recombinant coccidioidomycosis vaccine in cynomolgus macaques (*Macaca fascicularis*). Abstracts of the Sixth International Symposium on Coccidoidomycosis, Stanford, CA, August 23-26.

39. OKEKE, C.N. & G.T. COLE. 2003. Morphological defects and loss of virulence of *Coccidioides posadasii* caused by disruption of CHS5, encoding a chitin synthase with a myosin motor-like domain [abstract]. Annual Meeting of the American Society of Microbiology, Washington, DC, May 18-22.

40. CHEN, X., C.Y. HUNG & G.T. COLE. 2003. Isolation and characterization of the chitinase gene family of *Coccidioides posadasii* [abstract]. Annual Meeting of the American Society of Microbiology, Washington, DC, May 18-22.

41. COLE, G.T., J. XUE, C.N. OKEKE, *et al.* 2004. A vaccine against coccidioidomycosis is justified and attainable. Med. Mycol. **42:** 189–216.

42. Notification of Exclusion. Select Agent Program. Centers for Disease Control and Prevention. http://www.cdc.gov/od/sap/sap/exclusion.htm

Dendritic Cell-Based Vaccine against *Coccidioides* Infection

SHANJANA AWASTHI

Department of Pharmaceutical Sciences, University of
Oklahoma Health Science Center, Oklahoma City, Oklahoma, USA

ABSTRACT: *Coccidioides* causes coccidioidomycosis in the southwestern United States. Its clinical manifestations range from the primary asymptomatic to progressive pulmonary and extrapulmonary disease. Because of endemicity, frequent relapse, and virulent nature of *Coccidioides*, there is an urgent need for the development of effective therapy or vaccine. It has been recognized from studies in human patients and in murine models that the divergence in their susceptibility to *Coccidioides* infection is related to differences in T cell response. Dendritic cells (DCs) are most potent antigen-presenting cells that play a critical role in activating naïve T cells. On account of their unique immunostimulatory capacity, DCs have been used for the development of immunotherapy and vaccines against cancer and infectious diseases. We recently investigated the immunostimulatory potential of a DC-based vaccine in a murine model against *Coccidioides posadasii* (*C. posadasii*). Our results suggest that DCs act as a potent adjuvant and activate protective responses in mice against *C. posadasii*.

KEYWORDS: dendritic cell; vaccine; immunotherapy; *Coccidioides*

INTRODUCTION

Coccidioidomycosis or Valley Fever is an infectious disease caused by the highly virulent, dimorphic, soil fungus *Coccidioides immitis* (*C. immitis*) or *Coccidioides posadasii* (*C. posadasii*). The fungus is endemic in the southwestern regions of the USA, Mexico, and Central America. Nearly 25,000–100,000 new cases are reported every year in the endemic areas of the United States.[1] However, isolated cases of coccidioidomycosis have been reported worldwide and in nonendemic areas of the U.S., mainly related to travel to endemic areas.[2,3] The dissemination of endospores to pulmonary and extrapulmonary organs leads to severe pulmonary infection and/or the progressive disseminated

Address for correspondence: Shanjana Awasthi, Ph.D., Department of Pharmaceutical Sciences, 1110 North Stonewall Avenue, Oklahoma City, OK 73117. Voice: 1-405-271-6593; ext: 47332; fax: 1-405-271-7505.
Shanjana-Awasthi@ouhsc.edu

Ann. N.Y. Acad. Sci. 1111: 269–274 (2007). © 2007 New York Academy of Sciences.
doi: 10.1196/annals.1406.013

form of the disease involving skin, bones, central nervous system, or other organ systems, respectively. Most of the patients recover after the primary exposure to *Coccidioides* arthroconidia, but 5–10% of the patients develop active disease.[4]

A VACCINE IS REQUIRED FOR THE PREVENTION OF COCCIDIOIDOMYCOSIS

Nearly all cases of progressive coccidioidomycosis require an aggressive treatment with antifungal agents for a prolonged period of time.[5] The effectiveness of treatment with antifungal agents depends on dose, duration, stage of infection, and immune status of the patients.[6–9] In most of the cases, treatment with antifungal agents has not been effective against *Coccidioides* because of relapse.[5] Since studies in both animal models and patients suggest that the deficient Th1 immune response is related to their increased susceptibility and clinical remission, it is recognized that a vaccine or immunotherapy will be very useful for the prevention and/or treatment of coccidioidomycosis.

THE SUSCEPTIBILITY TO *COCCIDIOIDES* INFECTION IS ASSOCIATED WITH DEFICIENT TH1 RESPONSE

Among different mouse models, BALB/c and C57BL6 mouse strains are susceptible and DBA/2 mouse strain is resistant to coccidioidomycosis.[10,11] The difference in their susceptibility is related to differences in Th1 response against *Coccidioides*.[11] Similar to mouse strains, it has been established in humans that the susceptibility to coccidioidomycosis is determined by racial or genetic predispositions, age, pregnancy, and immunological status of people in the endemic areas.[6] For example, patients with AIDS,[12] pregnant women,[13] and patients with deficient T cell reactivity[14] are more at risk of developing severe disseminated form of coccidioidomycosis.

TH1 RESPONSE IS PROTECTIVE AGAINST COCCIDIOIDES INFECTION

Coccidioides is a highly immunogenic fungus and induces a variety of cellular and humoral immune responses. A relationship has been found to exist between severity of the disease and depressed cellular immune response to coccidioidal antigens.[1] Primary infection is associated with acquisition of cellular response and low antibody response, while the chronic and progressive disease is associated with depressed cellular response and high antibody levels. Low titers ($<1:16$) of antibodies by complement fixation test have been linked to primary pulmonary infection in patients, high titers ($> 1{:}32$) with the severe disseminated form of the disease.[15] Also, treatment with recombinant IFN-γ and Th1-stimulating cytokine IL-12 has been found protective.[14,16]

DENDRITIC CELLS ARE A POTENT ADJUVANT AND PLAY AN IMPORTANT ROLE IN ACTIVATING TH1 RESPONSE

Dendritic cells are the most potent antigen-presenting cells. The DCs are able to capture antigens and then move to local lymph nodes, where they express the antigenic epitope to the naïve T cells and activate long-term specific Th1 responses.[17,18] Because of this unique capability, DCs have been used to develop vaccine and immunotherapy. Autologous DCs provide a good choice of an immunostimulatory adjuvant and have been used against cancer and infectious diseases.[19–23] Dendritic cell–based preparations are already in clinical trials against cancer.[24,25] The preclinical results of DC-based vaccines in animal models of fungal infection have been reviewed elsewhere.[26,27] Earlier, it was demonstrated that DCs are activated after interacting with fungal pathogens, such as *Aspergillus conidia*,[28] *Candida* yeasts,[29] and coccidioidal ligands, such as T27K antigen.[30] Dendritic cells have also been shown to interact with *Coccidioides* arthroconidia[31] and spherules[32] under *in vitro* conditions. In another study, Richards *et al.*[33] showed that the anergic T cells isolated from patients with disseminated coccidioidomycosis are activated when co-incubated with coccidioidal antigen-pulsed DCs.

DEVELOPMENT OF DC-BASED VACCINE AGAINST COCCIDIOIDES

Recently, we developed a DC-based vaccine against *Coccidioides* in mouse model. The vaccine was prepared by transfecting murine bone marrow–derived DCs (JAWS II cells derived from C57BL6 mice) with a plasmid DNA coding for *Coccidioides*-Ag2/proline-rich antigen (PRA).[34] The syngeneic mice (C57BL6) were immunized intranasally with *Coccidioides*-Ag2/PRA-cDNA-transfected DCs. The migration and biodistribution of DCs were studied in mice using advanced nuclear imaging techniques.[35] Post immunization, mice were challenged with a lethal dose of live *C. posadasii* arthroconidia through the intraperitoneal or intranasal route. After 10 days of infection, the mice were sacrificed and fungal burden, IFN-γ release, and histopathology of the lung were assessed. After intranasal injection, most of the DCs migrated to the gastrointestinal tract, but a significant amount remained in the lung. The fungal burden in lung and spleen of immunized mice was significantly reduced compared to that of the control animals. The immunized animals had higher levels of IFN-γ in lung tissue homogenates. These animals also showed improved lung condition compared to control animals. Our results suggest that a DC-based vaccine is effective in mounting a protective Th1 response and in reducing the fungal burden.[35]

This is the first study where DC-based vaccination has been used against *Coccidioides* infection.[35] Although our preliminary results look promising, we need to evaluate several aspects including the dosage, route of administration of the vaccine, efficacy of other coccidioidal epitopes, and mechanism of activation of *Coccidioides*-specific immune responses in the murine model after challenge with different lethal doses of *Coccidioides*. A successful vaccine needs to be very efficacious in mounting protective responses specific to *Coccidiodes* and in reducing the fungal burden and clinical symptoms. Therefore, a comprehensive study is under way to assess the efficacy of a DC-based vaccine in a murine model before it can be further evaluated for use against *Coccidioides* in humans.

REFERENCES

1. Cox, R.A. & D.M. Magee. 2004. Coccidioidomycosis: host response and vaccine development. Clin. Microbiol. Rev. **17:** 804–839.
2. Hughes, C.W. & P.A. Kvale. 1989. Pleural effusion in Michigan caused by *Coccidioides immitis* after travel to an endemic area. Henry Ford Hosp. Med. J. **37:** 47–49.
3. Pappagianis, D. 1988. Epidemiology of coccidioidomycosis. Curr. Top. Med. Mycol. **2:** 199–238.
4. Cole, G.T. *et al.* 2004. A vaccine against coccidioidomycosis is justified and attainable. Med. Mycol. **42:** 189–216.
5. Galgiani, J.N. *et al.* 2000. Practice guideline for the treatment of coccidioidomycosis. Infectious Diseases Society of America. Clin. Infect. Dis. **30:** 658–661.
6. Rosenstein, N.E. *et al.* 2001. Risk factors for severe pulmonary and disseminated coccidioidomycosis: Kern County, California, 1995–1996. Clin. Infect. Dis. **32:** 708–715.
7. Ampel, N.M. 2003. Measurement of cellular immunity in human coccidioidomycosis. Mycopathologia **156:** 247–262.
8. Arguinchona, H.L. *et al.* 1995. Persistent coccidioidal seropositivity without clinical evidence of active coccidioidomycosis in patients infected with human immunodeficiency virus. Clin. Infect. Dis. **20:** 1281–1285.
9. Barbee, R.A. *et al.* 1991. The maternal immune response in coccidioidomycosis. Is pregnancy a risk factor for serious infection? Chest **100:** 709–715.
10. Kirkland, T.N. & J. Fierer. 1983. Inbred mouse strains differ in resistance to lethal *Coccidioides immitis* infection. Infect. Immun. **40:** 912–916.
11. Magee, D.M. & R.A. Cox. 1995. Roles of gamma interferon and interleukin-4 in genetically determined resistance to *Coccidioides immitis*. Infect. Immun. **63:** 3514–3519.
12. Fish, D.G. *et al.* 1990. Coccidioidomycosis during human immunodeficiency virus infection. A review of 77 patients. Medicine (Baltimore) **69:** 384–391.
13. Powell, B.L. *et al.* 1983. Relationship of progesterone- and estradiol-binding proteins in *Coccidioides immitis* to coccidioidal dissemination in pregnancy. Infect. Immun. **40:** 478–485.

14. KUBERSKI, T.T., R.J. SERVI & P.J. RUBIN. 2004. Successful treatment of a critically ill patient with disseminated coccidioidomycosis, using adjunctive interferon-gamma. Clin. Infect. Dis. **38:** 910–912.

15. COX, R.A. *et al.* 1976. *In vivo* and *in vitro* cell-mediated responses in coccidioidomycosis. I. Immunologic responses of persons with primary, asymptomatic infections. Am. Rev. Respir. Dis. **114:** 937–943.

16. MAGEE, D.M. & R.A. COX. 1996. Interleukin-12 regulation of host defenses against *Coccidioides immitis*. Infect. Immun. **64:** 3609–3613.

17. PULENDRAN, B. & R. AHMED. 2006. Translating innate immunity into immunological memory: implications for vaccine development. Cell **124:** 849–863.

18. STEINMAN, R.M. & H. HEMMI. 2006. Dendritic cells: translating innate to adaptive immunity. Curr. Top. Microbiol. Immunol. **311:** 17–58.

19. RIDGWAY, D. 2003. The first 1000 dendritic cell vaccinees. Cancer Invest. **21:** 873–886.

20. RIDOLFI, R. *et al.* 2004. Evaluation of *in vivo* labelled dendritic cell migration in cancer patients. J. Transl. Med. **2:** 27.

21. SCHEDING, S. *et al.* 1998. *Ex vivo* expansion of hematopoietic progenitor cells for clinical use. Semin. Hematol. **35:** 232–240.

22. SHAO, C. *et al.* 2005. Dendritic cells transduced with an adenovirus vector encoding interleukin-12 are a potent vaccine for invasive pulmonary aspergillosis. Genes Immun. **6:** 103–114.

23. STEINMAN, R.M. 2003. Some interfaces of dendritic cell biology. Apmis **111:** 675–697.

24. DURRANT, L.G. & I. SPENDLOVE. 2003. Cancer vaccines entering phase III clinical trials. Expert Opin. Emerg. Drugs **8:** 489–500.

25. VOSS, C.Y., M.R. ALBERTINI & J.S. MALTER. 2004. Dendritic cell-based immunotherapy for cancer and relevant challenges for transfusion medicine. Transfus. Med. Rev. **18:** 189–202.

26. BOZZA, S. *et al.* 2004. Dendritic cell-based vaccination against opportunistic fungi. Vaccine **22:** 857–864.

27. DAN, J.M. & S.M. LEVITZ. 2006. Prospects for development of vaccines against fungal diseases. Drug Resist. Updat. **9:** 105–110.

28. BOZZA, S. *et al.* 2002. Dendritic cells transport conidia and hyphae of *Aspergillus fumigatus* from the airways to the draining lymph nodes and initiate disparate Th responses to the fungus. J. Immunol. **168:** 1362–1371.

29. ROMAGNOLI, G. *et al.* 2004. The interaction of human dendritic cells with yeast and germ-tube forms of *Candida albicans* leads to efficient fungal processing, dendritic cell maturation, and acquisition of a Th1 response-promoting function. J. Leukoc. Biol. **75:** 117–126.

30. RICHARDS, J.O. *et al.* 2001. Dendritic cells pulsed with *Coccidioides immitis* lysate induce antigen-specific naive T cell activation. J. Infect. Dis. **184:** 1220–1224.

31. AWASTHI, S. & D.M. MAGEE. 2004. Differences in expression of cell surface co-stimulatory molecules, toll-like receptor genes and secretion of IL-12 by bone marrow-derived dendritic cells from susceptible and resistant mouse strains in response to *Coccidioides posadasii*. Cell Immunol. **231:** 49–55.

32. DIONNE, S.O. *et al.* 2006. Spherules derived from *Coccidioides posadasii* promote human dendritic cell maturation and activation. Infect. Immun. **74:** 2415–2422.

33. RICHARDS, J.O., N.M. AMPEL & D.F. LAKE. 2002. Reversal of coccidioidal anergy *in vitro* by dendritic cells from patients with disseminated coccidioidomycosis. J. Immunol. **169:** 2020–2025.

34. AWASTHI, S. & R.A. COX. 2003. Transfection of murine dendritic cell line (JAWS II) by a nonviral transfection reagent. Biotechniques **35:** 600– 602; 604.

35. AWASTHI, S. *et al.* 2005. Efficacy of antigen 2/proline-rich antigen cDNA-transfected dendritic cells in immunization of mice against *Coccidioides posadasii.* J. Immunol. **175:** 3900–3906.

Characteristics of the Protective Subcellular Coccidioidal T27K Vaccine

SUZANNE M. JOHNSON, KATHLEEN M. KEREKES,
JENNINE M. LUNETTA, AND DEMOSTHENES PAPPAGIANIS

*Department of Medical Microbiology and Immunology, School of Medicine,
University of California, Davis, California 95616, USA*

ABSTRACT: While the whole killed spherule vaccine, protective in mice
and monkeys, did not prevent coccidioidal disease in humans, the 27K
vaccine, a soluble derivative, retains protective activity in mice with little
irritant action. Gel filtration and anion exchange fractions of thimerosal-
inactivated spherules (T27K), when administered with alum adjuvant,
also protect mice against lethal respiratory coccidioidal challenge. However, the superb protection afforded by T27K antigens is maintained
for some 3 months, but may then diminish. This appears unrelated to
the aging of the mice. Prolongation of the protective action may require
addition of a different adjuvant or administration of booster doses of
vaccine.

KEYWORDS: vaccine; coccidioidomycosis; immunization route; FPLC;
gel filtration; anion exchange; adjuvant

INTRODUCTION

It is well established that recovery from prior coccidioidal infection confers
lifelong protection against reinfection[1-3] and therefore a vaccine is feasible
for the prevention or amelioration of coccidioidomycosis. The whole killed
spherule vaccine of Levine *et al.*,[4] although protective against respiratory or
intravenous challenge when given to experimental animals,[5] was, however,
nonefficacious when evaluated in humans.[6] This failure was attributed to the
vaccines' irritant properties, which limited the tolerable human dose to approximately 1/1000 (body weight basis) of that used in prior experimental
studies in mice.[6] Following safety trials on the whole killed spherule vaccine
in humans, it was concluded that a solubilized vaccine may be needed to provide adequate antigens for immunization.[7] A subcellular vaccine administered

Address for correspondence: Suzanne M. Johnson, Ph.D., Department of Medical Microbiology and
Immunology, School of Medicine, University of California, Davis, CA 95616. Voice: 530-752-7214;
fax: 530-752-8692.
smjohnson@ucdavis.edu

Ann. N.Y. Acad. Sci. 1111: 275–289 (2007). © 2007 New York Academy of Sciences.
doi: 10.1196/annals.1406.016

with alum adjuvant was shown to be protective in mice challenged by respiratory (intranasal, IN) route.[8] Our efforts have, therefore, focused on a subcellular vaccine that retains its immunoprotective capacity while devoid of irritants.

Zimmer et al.[9] reported the preparation of a soluble vaccine, 27K, from mechanically disrupted formalin-killed spherules (FKS) that, when administered with alum, would protect mice against a lethal systemic (IV) coccidioidal challenge. Later, the 27K vaccine provided protection against respiratory (IN) challenge equal to that provided by the FKS vaccine.[10] However, the formalin-derived 27K preparation did not fractionate well by traditional methods, such as polyacrylamide gel electrophoresis (PAGE), and this limited the ability to isolate and identify the protective components. Zimmermann et al.[11] then showed that a similar vaccine prepared from thimerosal-inactivated spherules (T27K) also offered protection, but was more readily fractionated, perhaps because of less chemical cross-linking. We have continued to fractionate the T27K vaccine to assist with isolation of protective components. We here report fractionation of the T27K vaccine using gel filtration and anion exchange chromatography and testing of the resulting pools as protective antigens. Additionally, studies to compare immunization routes, subcutaneous and intradermal, as well as the effect of mouse aging on survival are included.

METHODS

Abbreviations Used and Experiments Performed

A listing of the abbreviations used as well as a description of methods and vaccines is presented in TABLE 1. An outline of the experiments performed and their outcomes is presented in TABLE 2.

Mouse Vaccine Studies

Female Swiss Webster mice (Charles River Laboratories, Wilmington, MA, USA), 16–18 g (28–35 days old), received three immunizations (0.3 mL), subcutaneously (SC) at weekly intervals unless otherwise noted. All vaccines except killed spherules included alum adjuvant. Mice were challenged with C. posadasii strain Silveira arthroconidia by the respiratory route (IN) 4 weeks after the last immunization. Animals were followed for survival for a minimum of 80 days. Each group initially contained 7 mice. Mice that died within 2 days following challenge were excluded from the experiment.

TABLE 1. List of abbreviations used

Abbreviation	Description
Route of immunization or infection	
ID	Intradermal
IN	Intranasal
IV	Intravenous
SC	Subcutaneous
Methods	
AE	Anion exchange
FPLC	Fast protein liquid chromatography
GF	Gel filtration
MWCO	Molecular weight cutoff
PBS	Phosphate-buffered saline
PAGE	Polyacrylamide gel electrophoresis
PAS	Periodic acid Schiff
SDS–PAGE	Sodium dodecyl sulfate–polyacrylamide electrophoresis
Vaccines	
27K	Soluble vaccine prepared from mechanically disrupted formalin-killed spherules
AE1–AE6	Vaccines prepared following separation of GFI using anion exchange chromatography
AER	Vaccine obtained from regeneration of anion exchange column
CW	Vaccine obtained from washing anion exchange column
FKS	Formalin-killed endosporulating spherules
GFI–GFVII	Vaccines prepared following separation of T27K using gel filtration
T27K	Soluble vaccine prepared from mechanically disrupted thimerosal-killed spherules
TKS	Thimerosal-killed endosporulating spherules

Statistical Evaluation

GraphPad Prism 4.0 (GraphPad software, San Diego, CA, USA) was used to prepare Kaplan–Meier survival curves. Curves were compared by means of the log-rank test.

T27K Antigen Preparation

The soluble, aqueous antigen, T27K, was prepared from thimerosal-inactivated mature endosporulating spherules (TKS) of *C. posadasii* strain Silveira as previously described.[9,10] In brief, mature endosporulating spherules, were inactivated with thimerosal (1:10,000 final concentration), washed with endotoxin-free Cellgro PBS, pH 7.4 (Mediatech, Herndon, VA, USA), and stored at 4°C in phosphate-buffered saline (PBS) + thimerosal (1:10,000 concentration) until use. Spherules were mechanically disrupted with glass beads in a bead beater (Biospec Products, Bartlesville, OK, USA) and then centrifuged for 30 min at 2 K × *g*. The supernatant was collected and

TABLE 2. List of experiments and studies performed

	Fractionation	
Parent material	Method	Result
T27K	Gel filtration	7 pools: GFI–GFVII
GFI	Anion exchange	8 pools: AE1–AE6, AER, CW

Vaccine survival studies	
Purpose	Vaccines evaluated
Vaccine comparison	GFI–GFVII, TKS
Vaccine comparison	AE3, AE6, AER, FKS
Immunization route comparison	GFI
Young and old mice comparison	GFI

subjected to additional centrifugation for 30 min at 27 K $\times$ g. This supernatant was then collected, passed through 0.2-μm pore-size filter, concentrated using a stirred pressure cell (Millipore, Billerica, MA, USA) fitted with a YM10 (10,000 MWCO) membrane, dialyzed against water, lyophilized, and stored at –20°C.

Gel Filtration Fractionation

The T27K was fractionated by passage through a Superdex 200 HiLoad 16/60 gel filtration column using an AKTA FPLC System (GE Healthcare, Piscataway, NJ, USA). Fifteen milligrams (dry weight) of T27K, suspended in endotoxin-free Cellgro PBS at 3 mg (dry weight)/mL, was applied to the column at 1 mL/min and 5-mL fractions were collected. The absorbency at 280 nm was monitored and fractions pooled, corresponding to chromatographic peaks and valleys. These seven pools, labeled GFI through GFVII, were dialyzed against water, lyophilized, and stored at –20°C.

Anion Exchange Fractionation

GFI was further fractionated using the AKTA FPLC system and a Resource Q column. Resource Q is a quaternary ammonium strong anion exchanger and binds negatively charged proteins. Fifteen milligrams of GFI (dry weight), suspended in 30 mL of 50 mM Tris, pH 7.0, 50 mM NaCl, was applied to the 6-mL column at a flow rate of 1 mL/min. Bound proteins were eluted by applying a linear gradient of 50 mM to 1 M NaCl in 50 mM Tris, pH 7.0. Fractions were pooled on the basis of chromatographic peaks and were labeled AE1 through AE6. The column was regenerated by passage of low-salt followed by high-salt buffer. Fractions were pooled and labeled AER. The column was then washed with base (1 M NaOH) followed by acid (1 M HCl), each alternating with high salt (1 M NaCl) to removed tightly bound proteins. This material was pooled and labeled CW (column wash). Pools were dialyzed against water, lyophilized, and stored at 4°C until used.

Electrophoresis

Pooled fractions were evaluated for protein composition following SDS-PAGE. Fractions were denatured and reduced by boiling in the presence of β-mercaptoethanol prior to electrophoresis. Gels were stained with Coomassie blue and/or silver.

Protection of Mice by T27K Gel Filtration Fractions

Vaccines were formulated to provide 400 μg alum adjuvant, predetermined to be an effective dose,[12] and either 250 μg (dry weight) T27K, 100 μg (dry weight) of the pooled gel filtration fractions (GFI–GFVII), or saline per 0.3 mL immunization. The thimerosal-killed whole spherule (TKS) vaccine contained 1 mg dry weight/0.3 mL and no adjuvant. Mice were challenged with either 500 or 1,500 arthroconidia and followed for survival for 173 days.

Protection of Mice by T27K-GFI Anion Exchange Fractions

Vaccines were formulated to provide 50 μg dry weight of the anion exchange fraction pools and 400 μg alum adjuvant in 0.3 mL. The formalin-killed whole spherule vaccine (FKS) provided 1 mg dry weight/0.3 mL and no adjuvant. Mice were challenged with 5,000 arthroconidia and followed for survival for 220 days.

Comparison of Immunization Routes

Mice were vaccinated with GFI (100 μg dry weight) plus alum or saline by either SC or intradermal (ID) routes. Mice were challenged with either 500 or 1,500 arthroconidia and followed for survival for 167 days. On day 100, 2 mice from each vaccine group with survivors were euthanitized and organs were harvested for future experimental comparisons.

Comparison of Survival with Young and Old Mice

Young mice (6 weeks old) and old mice (23 weeks old) were vaccinated with 100 μg GFI or saline. Mice were challenged with 500 or 1,500 arthroconidia and followed for survival for 154 days.

RESULTS

T27K Gel Filtration Fractionation

Passage of T27K through the gel filtration column using the AKTA fast protein liquid chromatography (FPLC) system yielded a reproducible chromatograph containing characteristic peaks and valleys. This chromatograph was then used to distribute collected fractions into seven pools designated GFI–GFVII (FIG. 1). Since gel filtration separates on the basis of size, those components in GFI were the largest, while those in GFVII were the smallest. SDS-PAGE was used to visualize the complexity of each of the pools (FIG. 1 inset). The banding pattern of each of the fractions was generally unique, although some bands were shared with neighboring pools.

Protection of Mice with T27K Gel Filtration Fractions

Mice immunized with T27K gel filtration fractions (GFI–GFVII) were observed for 173 days after challenge. TABLE 3 shows the percentage of animals surviving in each group at 30, 80, and 173 days after challenge. Survival curves were prepared for these time points and the survival of each vaccine

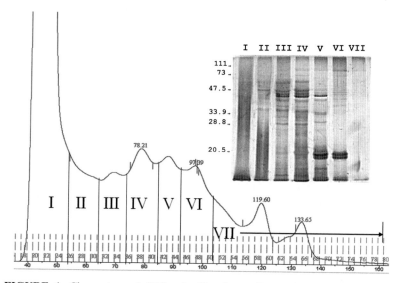

FIGURE 1. Chromatograph (280 nm) of fractions collected following T27K gel filtration. Fractions constituting the pools GFI–GFVII are indicated on the graph. *Inset*: SDS-PAGE comparison of GF pools. Lanes contained 25 μg (dry weight) of the respective pools which were reduced and denatured prior to electrophoresis. The gel was stained with Coomassie blue and silver. Molecular-size markers (kDa) are shown on the left of the gel.

TABLE 3. Survival of mice immunized with T27K gel filtration fractions following coccidioidal challenge

Vaccine dose formulation[a]		500 Arthroconidia				1,500 Arthroconidia			
			Percentage of mice surviving[b]				Percentage of mice surviving[b]		
Designation	Vaccine[c]	mice[d]	30 days	80 days	173 days	mice[d]	30 days	80 days	173 days
TKS	1 mg	5	80* (0.0215)	80* (0.0215)	60* (0.0048)	7	57	43	14
T27K	250 µg	5	80* (0.0386)	80* (0.0386)	60* (0.0094)	7	86* (0.0090)	57* (0.0298)	28
GFI	100 µg	6	100* (0.0032)	100* (0.0032)	83* (0.0004)	7	100* (0.0013)	100* (0.0013)	14
GFII	100 µg	4	100* (0.0139)	100* (0.0139)	25* (0.0219)	7	42	14	0
GFIII	100 µg	7	28	14	0	7	14	14	0
GFIV	100 µg	7	28	28	0	7	14	14	14
GFV	100 µg	7	14	14	0	6	14	14	14
GFVI	100 µg	7	43	28	0	7	0	0	0
GFVII	100 µg	7	28	14	0	7	14	14	0
Saline	none	7	14	14	0	7	14	14	14

[a] All vaccines except TKS were formulated with alum adjuvant, 400 µg per dose.
[b] Days following challenge.
[c] Dry weight per dose.
[d] Mice that died within 2 days following challenge were not included.
* Statistically significant difference compared to saline; *P*-value shown in parentheses.

group compared to saline. Vaccines GFIII, GFIV, GFV, GFVI, and GFVII did not show protection, with the majority of the mice succumbing by day 30. In contrast, GFI and GFII showed 100% protection through 80 days, and TKS and T27K showed 80% survival at the 500 arthroconidia challenge level (FIG. 2). Lower protection was observed at the 1,500 arthroconidia challenge level in all but GFI (100% survival). However, it was observed that seemingly protected mice (in all groups) died beginning 80 to 100 days following challenge.

T27K-GFI Anion Exchange Fractionation

Separation of the T27K-GFI fraction using anion exchange chromatography also yielded a reproducible and characteristic chromatogram (FIG. 3). Pool AE1 was composed of neutral and positively charged proteins that passed unbound through the column, whereas AE2 were proteins that were nonspecifically bound and then washed from the column. Pools AE3–AE6 contained bound

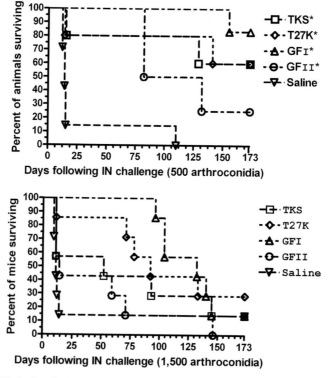

FIGURE 2. Survival of mice immunized with gel filtration fractions or the parent material TKS or T27K. Mice were challenged IN with 500 or 1,500 arthroconidia. *Survival statistically different compared with saline at experiment end.

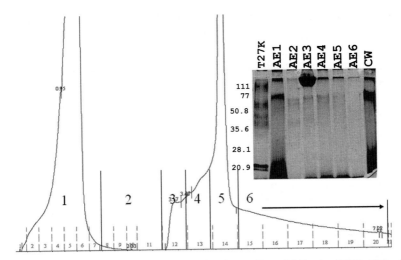

FIGURE 3. Chromatograph (280 nm) of fractions collected following T27K-GFI anion exchange separation. Fractions making up the pools AE1–AE6 are indicated on the graph. *Inset*: SDS-PAGE comparison of T27K and AE pools. Lanes contained 25 μg (dry weight) of each respective pool or T27K and were reduced and denatured prior to electrophoresis. The gel was stained with Coomassie blue. Molecular-size markers are shown on the left.

proteins that eluted following the application of the NaCl gradient. CW and AER were collected following the column wash and regeneration steps when preparing for column reuse. The composition and complexity of the pooled fractions were evaluated by SDS-PAGE followed by staining with Coomassie blue. While staining of T27K showed discrete bands, staining of the many AE fractions showed smearing (FIG. 3, inset). This smearing was also observed to stain with PAS, indicating the presence of glycoproteins (not shown). Additionally, high molecular weight material was observed to remain in the stacking gel.

Protection of Mice Immunized with T27K-GFI Anion Exchange Fractions

Previous experiments (data not shown) indicated a lack of immunoprotection with vaccines containing AE1, AE2, and AE4. Therefore only AE3, AE6, and AER are included in this study. Survival after challenge with 5,000 arthroconidia was followed for 220 days. All but one of the saline adjuvant mice died prior to day 30 following challenge, while between 67% (FKS) and 86% (AER) of the immune mice survived through the same period (TABLE 4 and FIG. 4). Protection held though 80 days for those immunized with AE6 and AER; however, some of this protection waned and late deaths were observed in these groups.

TABLE 4. Survival of mice immunized with T27K-GFI anion exchange fractions following coccidioidal challenge with 5,000 arthroconidia

Vaccine dose formulation				Percentage of mice surviving[a]		
Designation	Vaccine[b]	Alum	mice[c]	30 days	80 days	220 days
FKS	1 mg	none	6	67* (0.0138)	33	0
AE3	50 μg	400 μg	7	43* (0.0272)	43	0
AE6	50 μg	400 μg	5	80* (0.161)	80* (0.0161)	40
AER	50 μg	400 μg	7	86* (0.0071)	71* (0.0139)	57* (0.0265)
Saline	none	400 μg	7	14	14	14

[a]Days following challenge.
[b]Dry weight.
[c]Mice that died within 2 days following challenge were excluded. Only 5 mice were vaccinated with AE6.
*Statistically significant difference compared with saline; *P*-value shown in parentheses.

Comparison of Immunization Routes

Mice immunized with GFI or saline with alum by either the subcutaneous (SC) or intradermal (ID) routes were followed for 167 days after challenge with either 500 or 1,500 arthroconidia. Protection of GFI-vaccinated mice was comparable between routes with no significant difference in survival (FIG. 5).

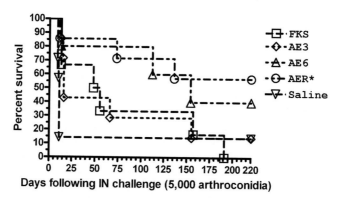

FIGURE 4. Survival of mice immunized with anion exchange fractions (AE3, AE6, or AER) or FKS. Mice were challenged IN with 5,000 arthroconidia. *Survival statistically different compared with saline at experiment end.

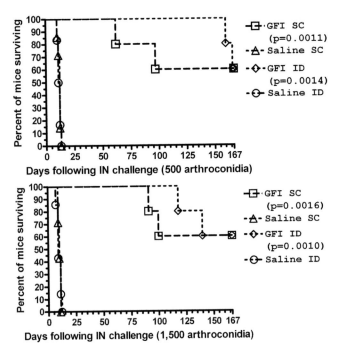

FIGURE 5. Survival of mice immunized with GFI by either the subcutaneous (SC) or intradermal (ID) routes. Mice were challenged IN with either 500 or 1,500 arthroconidia. Survival of GFI-immunized mice were compared to saline-immunized mice given by the same route and *P*-values shown in parentheses.

Comparison of Survival with Young and Old Immunized Mice

In previous studies, including those presented above, deaths of seemingly protected mice have occurred beginning 80 to 100 days following challenge. To ascertain whether this phenomenon was related to mouse aging, two age groups of mice were evaluated. The young mice were the age usually used and the old ones approximately 119 days older. Mice were immunized with GFI or saline and followed for 154 days after challenge with 1,500 or 5,000 arthroconidia (FIG. 6). While there was no statistical difference in the survival between the old and young mice, the survival of the GFI old mice challenged with 5,000 arthroconidia was lower than that of the other groups.

DISCUSSION

We previously showed that the subcellular fraction, 27K, prepared from formadehyde-killed spherules, protected mice against lethal intravenous (IV) and intranasal (IN) challenge with *Coccidioides* when given with alum

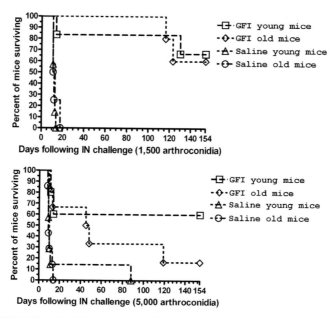

FIGURE 6. Survival of young and old mice immunized with GFI and challenged IN with either 1,500 or 5,000 arthroconidia.

adjuvant,[10] and that a similar subcellular fraction,T27K, prepared from TKS, when given with alum, also protected mice against IN challenge.[11] While each of these subcellular preparations stimulated immunoprotection in mice and was less irritating than the whole spherule vaccine, our goal is to identify individual protein or glycoprotein immunoprotective components.

Fractionation of the T27K by gel filtration under native conditions yielded a fraction pool, GFI, that, when given with alum adjuvant, afforded protection comparable to that of the parent vaccine, TKS. This pool, prepared with those fractions of the first and largest absorbance peak, contained large proteins or glycoproteins. However, these proteins may represent multipeptide aggregates since electrophoresis of reduced-denatured material results in staining almost the entire length of the lane (FIG. 1). While some bands are apparent, most of the stained region represents a smear. This material is observed to stain with PAS (not shown), indicating the presence of glycoslyated proteins or other glycosylated material.

The T-27K-GFI was further fractionated using anion exchange chromatography. Both AE6 and AER showed protection through 80 days (80% and 71% survival, respectively), although the protection waned somewhat after 100 days. It is interesting to note that AE6 shows very little material that stains with Coomassie blue in contrast to AE1 and AE3, perhaps also suggesting the presence of glycosylated material.

We compared the efficacy of the GFI vaccine administered through either the subcutaneous route (SC) or the intradermal (ID) route and found no difference. The dermis is reported to contain large numbers of dendritic cells, which are thought to induce cell-mediated immune responses and act as natural adjuvants.[13,14] Numerous studies have reported equivalent immune responses with lower doses administered ID compared to higher doses given SC, and in some cases ID resulted in fewer systemic reactions.[14–16] In our study, both routes received the same quantity of antigen and adjuvant although the administration volume was less for ID. It will be important to compare varying doses to determine whether one route emerges superior.

Previous vaccine studies from our group and others have reported survival studies that were terminated prior to 100 days following challenge.[10,17–19] We noticed frequently that vaccine-protected mice often survived at least 80 days following challenge, but might succumb later. These "late deaths" were apparent in all of the studies presented and prompted a study using mice of different ages. The study comparing the long-term vaccine protection using "young" and "old" mice demonstrated no difference in survival between the groups. And, in fact, mice in both groups immunized with GFI died after 100 days. This may perhaps be a result of waning of the immune response, resulting in an uncontrolled fungal growth in the host or perhaps is a result of a slow accumulation of fungal material, reaching a critical mass such that the immune response is overwhelmed.

We suggest that a booster dose may be required to increase the long-term vaccine protection. Indeed, Kong *et al.*[20] showed with FKS that immunity was enhanced by the injection of a booster dose of spherules. The timing of the boost was important, with the most pronounced enhancement observed when given greater than 2 months post vaccination and 6 to 7 days prior to challenge. Additionally, it was observed that the booster not only limited fungal growth following challenge, but also prevented the establishment of a continuing fungal infection in greater than 60% of the mice.

The majority of vaccine studies completed in our laboratory have used alum as the adjuvant. While we recognize that alum may enhance a Th2 humoral rather than the desired Th1 cell-mediated response, we have previously shown that in the absence of alum, T27K lacks the ability to induce protection.[12] Thus, the protective action may include humoral responses toward polysaccharide-containing antigens (e.g., opsonization of endospores), and cellular Th1 action toward spherules. However, it may be necessary to include an additional or different adjuvant to prolong the protection.

The GFI and AE6 and AER represent protective semicrude subcomponents of T27K. Each likely contains numerous, albeit reduced, components, some of which engender protection. Studies to identify single components in these preparations using proteomic, chromatographic, and electrophoretic methods are currently ongoing. Additionally, combinations of adjuvants, routes, and number and timing of immunizations will be evaluated to yield a vaccine capable of generating long-lived immunity.

ACKNOWLEDGMENTS

This project was supported by a grant from the California HealthCare Foundation and the California State University Bakersfield Foundation.

REFERENCES

1. SMITH, C.E. & D. PAPPAGIANIS. 1961. Human coccidioidomycosis. Bact. Rev. **25:** 310–320.
2. PAPPAGIANIS, D. 1999. *Coccidioides immitis* antigen. J. Infect. Dis. **180:** 243–244.
3. PAPPAGIANIS, D. 2001. Seeking a vaccine against *Coccidioides immitis* and serologic studies: expectations and realities. Fungal Genet. **32:** 1–9.
4. LEVINE, H.B., Y.-C.M. KONG & C.E. SMITH. 1965. Immunization of mice to *Coccidioides immitis*: dose, regimen and spherulation stage of the killed spherule vaccine. J. Immunol. **94:** 132–142.
5. LEVINE, H.B., R.L. MILLER, & C.E. SMITH. 1962. Influence of vaccination on respiratory coccidioidal disease in cynomolgous monkeys. J. Immunol. **89:** 242–251.
6. PAPPAGIANIS, D. & THE VALLEY FEVER VACCINE GROUP. 1991. Evaluation of the protective efficacy of the killed *Coccidioides immitis* spherule vaccine in humans. Am. Rev. Resp. Dis. **148:** 656–660.
7. PAPPAGIANIS, D. & H.B. LEVINE. 1975. The present status of vaccination against coccidioidomycosis in man. Am. J. Epidemiol. **102:** 30–41.
8. PAPPAGIANIS, D., R. HECTOR, H.B. LEVINE, *et al.* 1979. Immunization of mice against coccidioidomycosis with a subcellular vaccine. Infect. Immun. **25:** 440–445.
9. ZIMMER, B.L., S.M. JOHNSON, K. VAN HOOSER, *et al.* 1990. Immunization of mice with extracts of *Coccidioides immitis*. *In* Abstracts of the 90th Annual Meeting of the American Society for Microbiology. Abstract F-26. American Society for Microbiology. Washington, DC
10. ZIMMERMANN, C.R., S.M. JOHNSON, G.W. MARTENS, *et al.* 1998. Protection against lethal murine coccidioidomycosis by a soluble vaccine from spherules. Infect. Immun. **66:** 2342–2345.
11. ZIMMERMANN, C.R., S.M. JOHNSON, K. M. KEREKES, *et al.* 1999. Development and characterization of a new soluble, subcellular vaccine against infection in mice by *Coccidioides immitis*. *In* Proceedings of the Annual Coccidioidomycosis Study Group Meeting,Tijuana, B.C., Mexico. Abstract 21. The Valley Fever Center for Excellence, Tucson, AZ.
12. JOHNSON, S.M., K.M. KEREKES, R.H. WILLIAMS, *et al.* 2000. Immunoprotective efficacy of the 27K vaccine fractions in experimental coccidioidomycosis. *In* Abstracts of the 38th Annual Meeting of the IDSA, New Orleans, LA. IDSA. Alexandria, VA.
13. STEINMAN, R.M. & M. POPE. 2002. Exploiting dendritic cells to improve vaccine efficacy. J. Clin. Invest. **109:** 1519–1526.

14. LaMontagne, J.R. & A. Fauci. 2004. Intradermal influenza vaccination: Can less be more? N. Engl. J. Med. **351:** 2330–2332.

15. Halperin, W., W.I. Weiss, R. Altman, *et al.* 1979. A comparison of the intradermal and subcutaneous routes of influenza vaccination with A/New Jersey/76 (swine flu) and A/Victoria/75: Report of a study and review of the literature. Am. J. Public Health **69:** 1247–1251.

16. McBean, A.M., A.N. Agle, R. Compaore, *et al.* 1972. Comparison of intradermal and subcutaneous routes of cholera vaccine administration. Lancet March **4:** 527–529.

17. Tarcha, E.J., V. Basrur, C-Y Hung, *et al.* 2006. A recombinant aspartyl protease of *Coccidioides posadasii* induces protection against pulmonary coccidioidomycosis in mice. Infect. Immun. **74:** 516–527.

18. Shubitz, L.F., J-J Yu, C-Y Hung, *et al.* 2006. Improved protection of mice against lethal respiratory infection with *Coccidioides posadasii* using two recombinant antigens expressed as a single protein. Vaccine **24:** 5904–5911.

19. Jiang, C., D.M. Magee, F.D. Ivey, *et al.* 2002. Role of signal sequence in vaccine-induced protection against experimental coccidioidomycosis. Infect. Immun. **70:** 3539–3545.

20. Kong, Y-C.M., D.C. Savage & H.B. Levine. 1966. Enhancement of immune responses in mice by a booster injection of *Coccidioides* spherules. J. Immun. **95:** 1048–1056.

Safety, Antigenicity, and Efficacy of a Recombinant Coccidioidomycosis Vaccine in Cynomolgus Macaques (*Macaca fascicularis*)

SUZANNE M. JOHNSON,[a] NICHOLAS W. LERCHE,[b]
DEMOSTHENES PAPPAGIANIS,[a] JOANN L. YEE,[b] JOHN N. GALGIANI,[c]
AND RICHARD F. HECTOR[d]

[a]*Department of Medical Microbiology and Immunology, School of Medicine, University of California, Davis, California 95616, USA*

[b]*California National Primate Research Center, University of California, Davis, California 95616, USA*

[c]*Valley Fever Center for Excellence, University of Arizona and Southern Arizona, VA Health Care System, Tucson, Arizona 85723, USA*

[d]*Institute for Global Health, University of California, San Francisco 94105, California, USA*

ABSTRACT: The safety, immunogenicity and efficacy of recombinant Ag2/PRA106 + CSA chimeric fusion protein (CFP) vaccine in ISS/Montanide adjuvant–administered intramuscular (IM) was assessed in adult female cynomolgus macaques challenged with *Coccidioides posadasii*. Animals received three immunizations with either 5 µg CFP, 50-µg CFP, or adjuvant alone and were challenged 4 weeks following the final immunization. Although significant antibody response was produced in response to vaccination, there were no discernable adverse effects, suggesting that the vaccine was well tolerated. Upon intratracheal challenge, all animals showed evidence of disease. Two animals that received 5-µg doses of CFP were euthanatized prior to the study's end because of severe symptoms. Animals vaccinated with 50-µg doses of CFP showed evidence of enhanced sensitization compared to adjuvant controls and animals vaccinated with 5-µg doses of CFP. This was based on higher serum anti-CFP titers, enhanced secretion of interferon-gamma (IFN-γ) from stimulated bronchoalveolar lavage mononuclear cells (BALMC), reduced pulmonary radiologic findings following intratracheal challenge, reduced terminal complement fixation titers, and reduced necropsy findings. Overall the vaccine was well tolerated, induced sensitization, and resulted in a protective response when given at the higher 50-µg dose.

Address for correspondence: Nicholas W. Lerche, D.V.M. M.P.V.M., California National Primate Research Center, University of California, Davis, CA 95616. Voice: 530-752-6490; fax: 530-752-4816.

nwlerche@ucdavis.edu

Ann. N.Y. Acad. Sci. 1111: 290–300 (2007). © 2007 New York Academy of Sciences.
doi: 10.1196/annals.1406.042

Additional experiments may be needed to optimize the vaccination and to confer greater protection against lethal challenge.

KEYWORDS: *Coccidioides*; vaccine; primate; coccidioidomycosis; immunity; infection; experimental model

INTRODUCTION

Coccidioidomycosis results in a variety of clinical conditions ranging from mild transient illness to potentially life-threatening chronic pulmonary and systemic disease in both humans and animals. Fortunately, in most instances, the illness resolves concomitantly with the appearance of markers of a cell-mediated immune response.[1] This complete immunity following recovery from infection has provided the motivation for studies to identify coccidioidal-specific antigens capable of producing similar immunity when administered as a vaccine. Ag2/PRA, a coccidioidal cell wall protein independently described by separate research groups,[2–5] has been cloned and expressed in *Escherichia coli* and *Saccharomyces* systems. The recombinant protein was shown to elicit a protective response when used as a vaccine in the mouse model.[6–10] A truncated Ag2/PRA construct containing only amino acids 1–106 (Ag2/PRA106) was found to retain the efficacy of the full-length protein. A second recombinant antigen, the *Coccidioides*-specific antigen (CSA) was protective against intraperitoneal (IP) infection but less effective against intranasal (IN) challenge.[11] However, improved protection was demonstrated by combining the two recombinant antigens, Ag2/PRA106 and CSA, or expressing the two as a single chimeric fusion protein (CFP).[10] This recombinant antigen was therefore selected for evaluation of safety, immunogenicity and efficacy in a nonhuman primate model of coccidioidomycosis.

METHODS

Vaccine

The vaccine employed in the study consisted of the recombinant Ag2/PRA106 and CSA CFP expressed in *Saccharomyces cerevesiae* as previously described.[10] Briefly, the chimeric construct transformant was grown for 48 to 72 h at 37°C in 30 mL of synthetic complete medium (SCM) containing 1.0 g/L yeast synthetic drop-out media supplement without tryptophan (Sigma, St. Louis, MO), 6.7 g/L yeast nitrogen base without amino acids (Difco, Detroit, MI), and 2% glucose. The CFP was purified using anti-FLAG-M1 affinity gel (Sigma) with 0.1 M glycine according to the manufacturer's instructions and then dialyzed against phosphate-buffered saline at 4°C. Purity and molecular weight were assessed following Coomassie blue R250 staining of gels subjected to reduced, denatured sodium dodecyl sulfate-polyacrylamide electrophoresis (SDS-PAGE). Protein concentration was estimated using BCA (Pierce, Rockford, IL). Preparations were stored at $-70°$C until use. The CFP was combined with ISS (Dynavax, Berkeley, CA), and Montanide ISA51 (Seppic, S.A., Paris, France). Each vaccine dose contained 0.1 mL ISS (1 mg), 0.3 mL ISA 51, and either 5 or 50-μg CFP.

Animals, Study Design, and Challenge

Fourteen adult female cynomolgus macaques (*Macaca fascicularis*), 2.0 – 4.5 kg, and negative for preexisting coccidioidal-specific antibodies were selected from the California National Primate Research Center (CNPRC, Davis, CA) breeding colony for use in the project. Animals were randomly assigned to treatment groups that included: ISS/Montanide adjuvant control ($n = 4$); 5 μg CFP dose + adjuvant ($n = 5$); and 50-μg CFP dose + adjuvant ($n = 5$). Animals were immunized intramuscularly (i.m.) with the respective treatments on days 0, 28, and 112. Adjuvant controls and CFP-vaccinated animals were challenged with 2,500 arthroconidia (*C. posadasii* strain Silveira) 28 days after the final immunization (day 140). The dosage was determined by a preliminary study (unpublished). The inoculum was delivered intratracheally in 1 mL of saline followed by a 1-mL saline flush at the level of the major bronchial bifurcation while the animal was under Telazol anesthesia. Animals were observed daily and examined at 2-week intervals by physical examination, which included body weight, complete blood count (CBC), serum chemistry, and thoracic radiographs. Animals were followed for 13 weeks postchallenge and were then euthanatized and necropsy was performed. Animals moribund prior to this time were immediately euthanatized and necropsy was performed.

Immunization and Immune Response

At 24, 48, and 72 h pos–timmunization, the injection site body temperature was measured and the site evaluated for vaccine adverse effects. Blood samples were collected pre- and post–immunization for CBC and serum chemistry evaluation as well as to assess humoral response. The presence of serum anti-CFP antibodies was determined by enzyme immunoassay (EIA). Microtiter plates were coated with 30 ng/well of CFP. After blocking, the plates were incubated with serum diluted between $1:10^2$ and $1:10^6$. Following washing and subsequent incubation with horseradish peroxidase (HRP)– conjugated goat anti-monkey IgG, antibody was visualized with 2,2'-azino-bis(3-ethylbenzthiazoline-6-sulphonic acid) substrate. A positive signal was defined as 0.5 OD units above the animal's pre–immunization sample diluted $1:10^2$. The antibody titer was the highest dilution yielding a positive signal. The titers were compared using Kruskal–Wallace analysis of variance (ANOVA) with Dunn's multiple comparison test using GraphPad Prism 4.0 (GraphPad Software, San Diego, CA). Subtypes of IgG were not determined.

The presence of coccidioidal complement fixation (CF) antibodies (IgG) in serum from blood samples collected at necropsy was determined by double immunodiffusion (ID) at the UCD Coccidioidal Serology Laboratory (Davis, CA). Serum, neat or diluted between 1:2 and 1:512, was added to wells without prediffusion. An unheated coccidioidal mycelial antigen was used as the reference antigen.

Release of interferon-gamma (IFN-γ) following CFP stimulation of bronchial alveolar lavage mononuclear cells (BALMCs) collected pre–challenge and 3, 7, and 10 days following challenge was measured. Bronchial alveolar lavage was performed on animals under ketamine anesthesia by instilling sterile phosphate-buffered saline (PBS), 2 mL/kg body weight, into the lower part of the lung at each of three separate areas (three flushes) followed by aspiration of the fluid. For each animal, the fluid from the three lavages was pooled into a single sample for analysis. The isolated mononuclear cells were suspended in RPMI medium at 2×10^6 cells/mL and then stimulated with either 10 μg/mL CFP or PBS (negative control). Supernatants were harvested after incubation for 24 h at 37°C and secreted IFN-γ was quantified by EIA (U-Cy-Tech, Utrecht, the Netherlands). In brief: wells coated with anti-monkey IFN-γ antibody captured IFN-γ that was then bound with a detector-biotinylated anti-monkey IFN-γ. The complex was then visualized by binding with strepavidin-peroxidase and tetramethylbenzidine (TMB) substrate. The amount of IFN-γ was estimated using standards containing 5–320 pg/mL.

Radiographic Assessment

A board-certified veterinary radiologist evaluated pre- and post–challenge radiographs in a blinded fashion. Radiographs, taken at 2-week intervals following challenge, were assessed, and numerical scores assigned for hilar lymph node enlargement, distribution pattern (normal, bronchial, unstructured interstitial, nodular interstitial, alveolar), and severity (normal, mild, moderate, severe) of pulmonary disease. This scoring system was used to determine a cumulative radiographic score (right lung + left lung + hilar node) over three post–challenge time points for each animal. The mean cumulative radiographic score was then calculated for each treatment group and compared using the nonparametric Wilcoxon Ranked Sum test (STATA software, College Station, TX).

RESULTS

Immunization and Immune Response

There was no evidence of reactogenicity or systemic effects encountered as a consequence of immunization, suggesting that the vaccine was well tolerated.

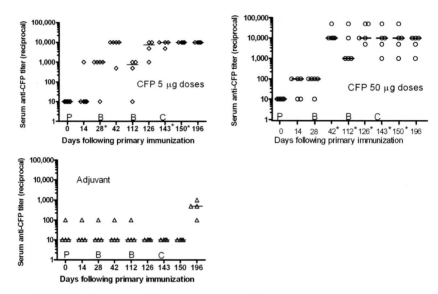

FIGURE 1. Serum anti-CFP titer measured by EIA. Samples whose reactivity indicated titers less than 100 (lowest dilution measured) were assigned a titer of 10. Line indicates median titer. Animals were primed (P) on day 0, received boosters (B) on day 28 and day 112, and were challenged (C) on day 140. *Titer on indicated day statistically different from adjuvant immunized group, P value <0.05.

One animal (CFP 5-μg immunization group) was euthanatized for an unrelated condition following the first immunization boost.

Anti-CFP antibody (IgG) was detected as early as day 14, although high titers were not apparent until day 42 (FIG. 1). After the final boost, animals immunized with 50-μg CFP doses had higher titers than those immunized with 5-μg CFP doses. While animals immunized with adjuvant alone had median titers of $<1:100$ following immunization, the anti-CFP titer began to increase on day 196 (8 weeks following challenge) in response to infection.

Measurement of IFN-γ Following Challenge

Secreted IFN-γ from PBS or CFP-stimulated BALMCs collected 0, 3, 7, and 10 days post–challenge (days 140, 143, 147, and 150, respectively) is shown in FIGURE 2. Although in all groups, the secreted IFN-γ following CFP stimulation of BALMCs was observed to increase following challenge, IFN-γ was highest in the group immunized with 50-μg CFP doses. Non–specific stimulation was indicated by secretion of IFN-γ following PBS stimulation. In a few instances the cells stimulated with PBS secreted greater amounts of IFN-γ than those stimulated with CFP. The reasons for these anomalous data are unclear.

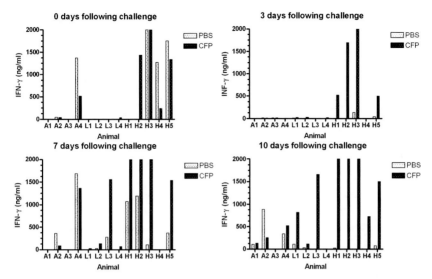

FIGURE 2. Secreted IFN-γ from PBS- or CFP-stimulated BALMCs collected prior to and following challenge. Animals A1–A4 were immunized with adjuvant alone, animals L1–L4 were immunized with 5-μg doses of CFP, and animals H1–H5 were immunized with 50-μg doses of CFP.

Measurement of Coccidioidal IDCF Antibodies

At necropsy, coccidioidal IDCF antibodies were present in serum from all animals with the exception of one immunized with 50-μg doses of CFP (FIG. 3). Although not statistically significant, IDCF titer of the animals that received 50-μg CFP doses was lower than that of the other groups.

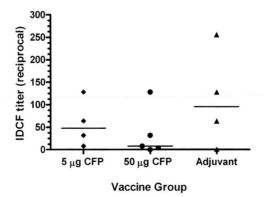

FIGURE 3. Complement fixation titer (IDCF) in serum collected prior to necropsy measured by quantitative ID. Line indicates median titer.

TABLE 1. Radiographic scores of vaccinated and challenged animals

Vaccine	Mean radiographic score	Range	P value[a]
Adjuvant	57.25	52–62	–
5 μg CFP	56.25	22–103	0.243
50-μg CFP	33.60	22–52	0.019[b]

[a]The mean cumulative radiographic score for each vaccine group was compared to the adjuvant group and tested for significance using the Wilcoxon ranked sum test.
[b]Statistically significant.

Radiographic Evaluation

Radiographic changes in the lung were assessed using radiographs collected pre- and post–challenge. The mean cumulative radiographic score of animals receiving 50-μg CFP doses was 33.6, a 1.7-fold reduction compared to the mean score of animals receiving adjuvant alone. This difference was statistically significant (TABLE 1). In contrast, there was little difference between the mean score of animals receiving 5-μg CFP doses and those that received adjuvant. A representative radiograph from an animal immunized with adjuvant alone is shown in FIGURES 4A and 4B.

Necropsy

Two animals vaccinated with 5-μg CFP doses were euthanized prior the study termination because of weight loss and their moribund state. On gross examination, all animals showed signs of pneumonia and/or had granulomatous disease in the lungs. In addition, most animals showed tracheobronchial lymph node involvement. A single animal that received 50-μg CFP doses had evidence of hepatic lesions on gross necropsy, possibly consistent with coccidioidomycosis; however, this was not confirmed by histopathology. No animals had documented dissemination. FIGURES 4C and 4D show the gross lung pathology of two animals at necropsy. The lungs of the animal that received adjuvant alone (FIG. 4C) shows consolidation and extensive lung involvement. In contrast, the lungs from an animal immunized with 50-μg doses of CFP (FIG. 4D) have only small lesions visible.

DISCUSSION

While numerous coccidioidal vaccines have been evaluated in mice, only the formalin-killed spherule (FKS) vaccine has been previously studied using the cynomolgus macaque model.[12] In that study, immunized animals challenged by the respiratory route were followed for 265 days. During that period, 5 of the 10 placebo-immunized animals died while only 1 of the 9 FKS-vaccinated died (a second FKS-vaccinated animal was inadvertently killed during a cardiac

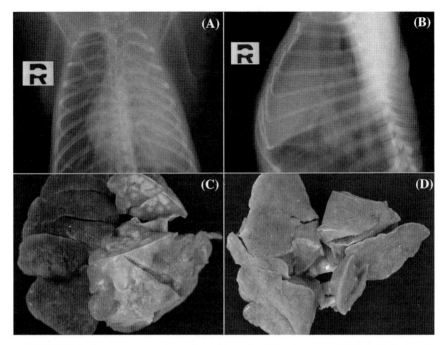

FIGURE 4. Representative thoracic radiographs following coccidioidal challenge (**A** and **B**) and gross lung pathology at necropsy (**C** and **D**). **A** (postero-anterior view) and **B** (lateral view): adjuvant-immunized animal shows severe pneumonia and consolidation. **C**: gross lung pathology of animal immunized with adjuvant alone, showing consolidation and extensive coccidioidal involvement. **D**: animal immunized with CFP, 50-μg doses, with relatively few lesions evident.

puncture procedure). At necropsy, both FKS and placebo-vaccinated animals showed evidence of pulmonary and disseminated disease, although the severity was less pronounced in the FKS-vaccinated animals.

In the present study, generation of anti-CFP (IgG) antibody following vaccination was used as an indicator of sensitization. Results demonstrated that animals immunized with CFP produced measurable titers that generally increased following each boost until challenge and then remaining relatively constant. Animals immunized with the higher dose of CFP had a higher anti-CFP antibody titer when compared with adjuvant-immunized animals, although the differences did not achieve statistical significance between the high and low doses of CFP. The anti-CFP antibody titer was observed to rise in adjuvant-immunized animals approximately 9 weeks following the coccidioidal challenge (day 196), but it remained lower than the immunized groups. The benefit of antibody with regard to protection against coccidioidal disease remains unclear. In patients with coccidioidal disease, Smith *et al.*[13,14] demonstrated a correlation between complement fixing (CF) antibody titer and disease severity, with higher titers reflecting more serious disease. However, antibody may provide assistance by virtue of opsonization, resulting in enhanced phagocytosis.[15,16]

In the present study, although all animals became ill following the coccidioidal challenge, only two animals succumbed to infection. Because of the low mortality rate, indicators of disease severity such as radiographic lesions and CF titer were used to assess the vaccine's protective capacity.

Animals immunized with 50-μg doses of CFP appeared to have less severe disease compared to other groups, as indicated by reduced radiographic changes and lower CF titers measured by quantitative ID. Accordingly, the data suggest that administration of the high dose of CFP results in sensitization and a reduced burden of disease.

Evidence indicates that cellular immunity is critical for resistance to coccidioidal infection.[11] Early studies showed that congenitally athymic, and therefore T-cell deficient, nude mice were more susceptible to coccidioidal disease.[17] Further studies in mice have shown that increased levels of Th1-associated cytokines (IFN-γ) correlate with greater resistance while Th2-associated cytokines (IL-4) correlate with susceptibility.[11,18] Therefore an adjuvant was selected that would promote a Th1 immune response. Previous studies have demonstrated that CpG adjuvants, including ISS, result in the secretion of Th1 cytokines IFN-γ and IL-12 and increased vaccine efficacy when used in combination with Montanide ISA-51, a water-in-oil emulsion.[19,20] Results from the present study demonstrated that vaccination with 50-μg doses of CFP with ISS and Montanide ISA 51 resulted in pronounced release of IFN-γ from mononuclear cells following stimulation. This suggests that vaccination in this group resulted in an enhanced Th1 response compared to that of adjuvant controls and animals receiving low-dose CFP, and contributed to the greater degree of protection.

It is important to note that this is the first study to evaluate a purified recombinant coccidioidal antigen as a vaccine in primates. Hence, conditions of dose response and dose interval may not have been optimal. Overall, the experimental results confirm that the vaccination was well tolerated and resulted in sensitization to the antigen. Additionally, there was a reduction in the burden of disease in those animals immunized with 50-μg doses of CFP compared with adjuvant controls, supporting the conclusion that non–human primates can be protected by immunization with acellular protein vaccines. Accordingly, the model may prove useful for the optimization of recombinant vaccines prior to evaluation in humans.

ACKNOWLEDGMENTS

This project was supported by a grant from the California HealthCare Foundation, the California State University Bakersfield Foundation, and CNPRC. The studies were supported in part by the U.S. Office of Veterans Affairs.

We thank Dr. Lisa Shubitz for her suggestions during the design of the study.

REFERENCES

1. VALDIVIA, L. *et al.* 2006. Coccidioidomycosis as a common cause of community-acquired pneumonia. Emerg. Infect. Dis. **12:** 958–962.
2. ZHU, Y. *et al.* 1996. *Coccidioides immitis* Antigen 2: analysis of gene and protein. Gene **181:** 121–125.
3. ZHU, Y. *et al.* 1996. Molecular cloning and characterization of *Coccidioides immitis* Antigen 2 cDNA. Infect. Immun. **64:** 2695–2696.
4. DUGGER, K.O. *et al.* 1996. Cloning and sequence analysis of the cDNA for a protein from *Coccidioides immitis* with immunogenic potential. Biochem. Biophys. Res. Commun. **218:** 485–489.
5. DUGGER, K.O. *et al.* 1991. An immunoreactive apoglycoprotein purified from *Coccidioides immitis*. Infect. Immun. **59:** 2245–2241.
6. KIRKLAND, T.N. *et al.* 1998. Evaluation of the proline-rich antigen of *Coccidioides immitis* as a vaccine candidate in mice. Infect. Immun. **66:** 3519–3922.
7. SHUBITZ, L.F. *et al.* 2002. Protection of mice against *Coccidioides immitis* intranasal infection by vaccination with recombinant Antigen 2/PRA. Infect. Immun. **70:** 3287–3289.
8. JIANG, C. *et al.* 1999. Genetic vaccination against *Coccidioides immitis*: comparison of vaccine efficacy of recombinant Antigen 2 and Antigen 2 cDNA. Infect. Immun. **67:** 630–635.
9. ABUODEH, R.O. *et al.* 1999. Resistance to *Coccidioides immitis* in mice after immunization with recombinant protein or DNA vaccine of a proline-rich antigen. Infect. Immun. **67:** 2935–2940.
10. SHUBITZ, L.F. *et al.* 2006. Improved protection of mice against lethal respiratory infection with *Coccidioides posadasii* using two recombinant antigens expressed as a single protein. Vaccine **24:** 5904–5911.
11. COX, R.A. & D.M. MAGEE. 2004. Coccidioidomycosis: host response and vaccine development. Clin. Microbiol. Rev. **17:** 804–839.
12. LEVINE, H.B., R.L. MILLER & C.E. SMITH. 1962. Influence of vaccination on respiratory coccidioidal disease in cynomologous monkey. J. Immun. **89:** 242–251.
13. SMITH, C.E., M.T. SAITO & S.A. SIMONS. 1956. Pattern of 39,000 serologic tests in coccidioidomycosis. J. Am. Med. Assoc. **160:** 546–552.
14. SMITH, C.E. *et al.* 1950. Serological tests in the diagnosis of coccidioidomycosis. Am. J. Hyg. **52:** 1–21.
15. WEGNER, T.N. *et al.* 1972. Some evidence for the development of a phagocytic response by polymorphonuclear leukocytes recovered from the venous blood of dogs inoculated with *Coccidioides immitis* or vaccinated with irradiated spherule vaccine. Am. Rev. Resp. Dis. **105:** 845–849.
16. PAPPAGIANIS, D. & B.L. ZIMMER. 1990. Serology of coccidioidomycosis. Clin. Microbiol. Rev. **3:** 247–268.
17. CLEMONS, K.V., C.R. LEATHERS & K.W. LEE. 1985. Systemic *Coccidioides immitis* infection in nude and beige mice. Infect. Immun. **47:** 814–821.

18. MAGEE, D.M. & R.A. COX. 1985. Roles of gamma interferon and interleukin-4 in genetically determined resistance to *Coccidioides immitis*. Infect. Immun. **63:** 3519–3519.
19. KLINMAN, D.M. *et al.* 1996. CpG motifs present in bacterial DNA rapidly induce lymphocytes to secrete interleukin 6, interleukin 12, and interferon γ. Proc. Natl. Acad. Sci. **93:** 2879–2883.
20. KUMAR, S. *et al.* 2004. CpG oligodeoxynucleotide and montanide ISA 51 adjuvant combination enhance the protective efficacy of a subunit malaria vaccine. Infect. Immun. **72:** 949–957.

Laboratory Aspects in the Diagnosis of Coccidioidomycosis

MICHAEL A. SAUBOLLE

Division of Infectious Diseases, Laboratory Sciences of Arizona/Sonora Quest Laboratories, Banner Health, Phoenix/Tucson, Arizona, USA

Department of Medicine, University of Arizona College of Medicine, Phoenix/Tucson, Arizona, USA

ABSTRACT: *Coccidioides immitis* and *Coccidioides posadasii*, the two recognized causes of coccidioidomycosis, may be detected by direct microscopy, culture, and serologic documentation. Two useful stains include the Grocott methenamine silver (GMS) and the calcofluor white (CFW). Other useful stains used in histopathologic studies include hematoxylineosin (H&E) and periodic acid Schiff (PAS). Nucleic acid amplification tests (NAATs) have been introduced for detection of *Coccidioides* spp. in specimens, but are not yet commercially available. Isolation of *Coccidioides* spp. by culture is not difficult as many fungal as well as routine bacteriologic media are available. For the safe isolation of *Coccidioides* spp., the laboratory should maintain a biological safety level 2 or 3. Identification of *Coccidioides* spp. uses the organisms' phenotypic or genotypic characteristics. Phenotypic identification to genus level may be achieved by visualization of spherules in specimens and/or by the presence of arthroconidia in culture. Isolates may be confirmed as *Coccidioides* spp. by molecular probes. Separation of species into *C. immitis and C. posadasii* is best achieved by specialized molecular techniques which are not normally available in routine clinical laboratories. Humoral antibodies can be used for the diagnosis and prognosis of coccidioidomycosis. Although positive serologic results may be helpful in the diagnosis of coccidioidomycosis, negative serologic results cannot be used to rule out the disease. Enzyme immunoassays (EIA) and immunodiffusion methods are commonly used for detection of both IgM and IgG antibody groups. Sequential complement fixation (CF) studies for IgG class of antibody are useful for the prognosis of coccidioidomycosis.

KEYWORDS: coccidioidomycosis; Valley Fever; *Coccidioides immitis*; *Coccidioides posadasii*; laboratory diagnosis of Coccidioidomycosis

Address for correspondence: Dr. Michael A. Saubolle, Department of Clinical Pathology, Banner Good Samaritan Medical Center, 1111 E. McDowell Road, Phoenix, AZ 85006. Voice: 602-239-3485; fax: 602-239-5605.

mike.saubolle@bannerhealth.com

Ann. N.Y. Acad. Sci. 1111: 301–314 (2007). © 2007 New York Academy of Sciences.
doi: 10.1196/annals.1406.049

INTRODUCTION

Coccidioidomycosis is now known to be caused by two species within the fungal genus *Coccidioides*.[1] Phylogenetic studies have shown definite differences between *Coccidioides* spp. from California (CA), and those from Arizona (AZ), Texas (TX), Mexico, and South America (SA). Isolates from the various geographic areas fall into two phylogenetic clades representing separate evolutionary species; those from CA make up one group (Grp II), while those from outside of CA make up a second group (Grp I).[1] The CA group (Grp II) remained as *C. immitis*, while the non-CA group (Grp I) was named *C. posadasii*.[1,2]

Coccidioides spp. are found in the hot, dry areas of the southwestern United States, often being associated with the Lower Sonora Life zone that extends into northern Mexico.[3–5] Presence of fine sand and silt in the soil seems to be a characteristic common to areas harboring the organism. *Coccidioides* spp. are most highly concentrated in the San Joaquin Valley of California (Kern County) and in south-central Arizona, with more than 95% of cases being reported from the two states. The organism is also found in smaller numbers in southern New Mexico, Texas, northern Mexico, areas of South America, and sporadically in areas of Utah and Nevada.[3] The incidence of coccidioidomycosis is increasing, and may be dependent in part on continued regional influx of susceptible hosts and on disruption and aerosolization of the desert surface by construction, wildfires, and earthquakes.[3–5] The chance of patients being diagnosed with coccidioidomycosis outside of endemic areas is increasing as travel to those areas continues to grow. Travel history should always be sought in patients.[6]

Coccidioides spp. are dimorphic, producing septate hyphae and thick-walled arthroconidia while growing in the environment, and spherical forms known as spherules when invading an animal host.[3,5] Arthroconidia are separated by empty, thin-walled cells (disjunctors), and are easily dislodged into the air. Upon inhalation or (rarely) upon percutaneous implantation into tissue, each arthroconidium transforms into a spherule. The spherule increases in size, forms a thick outer wall, and begins to divide internally, culminating in the production of numerous uninucleated endospores.[5,7]

Coccidioides spp. produce a very broad spectrum of illness, with about 60% of cases being asymptomatic or unrecognizable. Symptomatic patients present with acute or subacute illness, ranging from viral-like to acute pneumonia. The most common presentations are usually self-limited and commonly misdiagnosed. Disseminated coccidioidomycosis occurs in fewer than 5% of symptomatic patients. Patients of ethnic backgrounds (e.g., Black and Asian), women in their third trimester, and immunocompromised patients are at significant risk for disseminated disease. Dissemination may occur to any organ system; skin, lymph nodes, the skeletal system, and extremities are more commonly involved, but central nervous system (CNS) disease is less commonly seen. Skin and soft-tissue infection after percutaneous inoculation may occur on rare occasion.[8]

Therapy may not be necessary in most cases, because most of these patients are asymptomatic or present with only mild symptoms. Full discussion of therapeutic approaches in different clinical scenarios may be found elsewhere in these reviews.

LABORATORY DIAGNOSIS

Safety

Coccidioides spp. pose a high risk in the clinical laboratory. Arthroconidia are so easily aerosolized that special precautions and considerations should be given to any manipulation of colonies. Negative air pressure must be maintained in the work area. Minimally, Standard Precautions should be exercised in handling cultures. Biosafety Level 2 Practices and a Class II Biological Safety Cabinet are mandatory. Biosafety Level 3 practices in the mycology laboratory have been suggested by some. Because *Coccidioides* spp. grow well on routine bacterial media, care must be given to all mould isolated in the routine laboratory, and work should be performed under the laminar flow hood. Culture plates with suspected *Coccidioides* spp. should not be saved for more than 3–5 days, as the chance of production of arthroconidia increases; the lids of all such plates should be taped.

Select Agent Status

Coccidioides immitis and *C. posadasii* are considered by the United States to be select agents of bioterrorism. Laboratories isolating *Coccidioides* spp. have to follow strict rules mandated by the federal government.[9] They must either register with the government to work with and save select agents, or, if not registered, must follow strict criteria in reporting and destroying such isolates. Registered laboratories must follow defined work practices, document and provide inventory controls of select isolates, adhere to background checks of all involved employees, strictly restrict access to work areas, include monthly reporting to the Centers for Disease Control and Prevention (CDC), and pay registration fees. Unregistered laboratories must destroy all isolates and report to the CDC each isolate's disposition in writing within 7 days of identification. Once identified, isolates cannot be transferred for any reason without prior written permission from the CDC. Patient specimens and isolates that have not yet been identified may be sent to a reference laboratory for identification, using appropriate criteria for postal transport.

Hematology

Elevated erythrocyte sedimentation rate and eosinophilia may be seen in coccidioidomycosis and the latter should especially heighten suspicion. In coccidioidal meningitis, the CSF has a variable overall increase in cell count, a predominance of lymphocytes over polymorphonuclear cells, a low to moderate elevation in protein, and a moderate decline in glucose.

Histopathology

Inflammatory Reaction

In immunocompetent patients, inflammatory changes normally occur in two phases, early and late. The early phase includes a mixed acute reaction consisting of an influx of polymorphonuclear neutrophils (PMNs) and, to a far lesser degree, granulomatous cells.[10] Tissue eosinophilia may be present. As endospores mature into spherules, a granulomatous reaction predominates, with an influx of lymphocytes, plasma cells, and multinucleated giant cells. Mixed inflammatory reactions may occur as spherules release endospores, thereby precipitating the reoccurrence of a PMN response.

Microscopy

Microscopic visualization of endospore-containing spherules in infected material is diagnostic of coccidioidomycosis. The organism may be detected in potassium hydroxide (KOH) wet mounts, the calcofluor white (CFW) fluorescent stain, and histopathologic preparations using the Grocott methenamine silver (GMS), the periodic acid Schiff (PAS) and the hematoxylin-eosin (H&E) stains (TABLE 1). The KOH wet mount is easy to prepare and may be performed rapidly, but is rather difficult to read and interpret, thus lacking the sensitivity and specificity of the CFW and GMS stains. The CFW fluorescent stain is one of the most sensitive, detecting fungi by binding the chitin and cellulose in their cell wall.[10] It may be used on tissue, body fluids, as well as lower respiratory secretions and can be set up and read more rapidly than the histopathologic stains. However, it also may nonspecifically bind plant as well as fatty acid material and requires a fluorescent microscope. Overall, its sensitivity was 22% in 374 specimens from which *Coccidioides* spp. were isolated.[8] The GMS stain is considered the most sensitive stain in detecting fungi in histopathologic preparations, but may obfuscate internal structures such as endospores within spherules by overstaining some fungal and tissue material. Stains such as the PAS and H&E are usually workhorses in histopathology and often detect coccidioidal structures, but, because they are less sensitive, are not considered as good as GMS in the diagnosis of coccidioidomycosis. Other stains such as the Giemsa, Papanicolau (PAP) and mucicarmine, may identify *Coccidioides* spp., but are not considered as efficient as the first stains described.[10] The Gram-stain does not stain fungal material well and should never be used as a primary stain.

TABLE 1.   Direct microscopic detection of *Coccidioides* spp. in specimens

Characteristic Structures
  Round spherules (10–100μm) in various stages of development, with or without
    internal endospores (2–5μm); endospores may be released and thus external to
    spherules; mycelia may be found at perimeter of cavitary lesions or in skin lesions.
Microbiologic Stains
  Calcofluor white fluorescent (CFW-primary): binds chitin and cellulose in fungal
    cell wall; requires fluorescent microscope; sensitive but may stain plant material.
  KOH wet mount (infrequent): not as sensitive as CFW; requires well-trained
    personnel.
Histopathologic Stains
  Grocott methenamine silver (GMS-primary): most sensitive of histological stains;
    black on green counterstain; may stain other tissue elements.
  Periodic acid Schiff (PAS-common): fungi stained red; delineates fungi and their
    morphology; more sensitive than H&E.
  Hematoxylin-eosin (H&E-common): versatile but may be more difficult to
    differentiate fungi from tissue.

Spherules (10–100 μm) in various stages of development and often containing endospores (2–5 μm) are the most common forms seen in clinical material. Spherules may be seen in giant cells, microabscesses, and in acute presentations. They are less likely to be found in caseous, calcified, or liquefaceous foci. They may also be seen in rare cases of smear positive CSF from patients with meningitis. Recognition of endospores within spherules is considered diagnostic, whereas presence of a few spherules without internal structures is often considered presumptive evidence for coccidioidomycosis. Immature spherules touching each other may be confused with *Blastomyces*. Furthermore, the presence of endospores without spherules (especially in CSF) may be confused with *Histoplasma*, *Cryptococcus*, or *Candida*.[5, 10] Mycelia may also be present in clinical specimens from boundaries of old cavitary lesions in the lung, in skin lesions (FIG. 1), and in ventricular fluid during CNS infections (especially if drainage tubes are present in the latter). Although characteristic arthroconidia are not normally associated with these mycelia, arthroconidia-like structures may be noted in some. Presence of mycelia without concomitant spherules is not diagnostic.

Molecular

Molecular probes such as those used for identification of *Coccidioides* spp. in culture are not commercially available for direct detection in specimens. Identification of *Coccidioides* to species level (*C. immitis* and *C. posadasii*) using molecular methods have been reported.[2] There is a paucity of data on use of nucleic acid amplification test (NAAT) methods to detect *Coccidioides* spp. directly in specimens.[11] A recent study by Binnicker and colleagues, however, used a real-time polymerase chain reaction (PCR) assay to detect the organisms

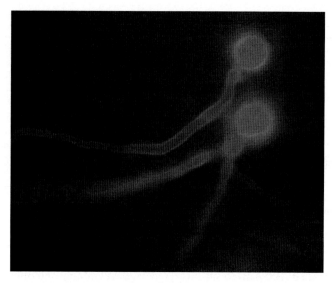

FIGURE 1. Evaginating spherules detected by CFW fluorescent stain in tissue collected from a skin lesion in a case of disseminated coccidioidomycosis (original magnification ×1,000).

directly in clinical specimens, including respiratory secretions, fresh tissue, and formalin- fixed, paraffin-embedded tissue.[12] In comparison to culture, results of the rapid PCR method showed 100% sensitivity and 98.4% specificity for *Coccidioides* spp. when testing respiratory specimens ($n = 266$), and 92.9% sensitivity, 98.1% specificity testing fresh tissue ($n = 66$). In testing paraffin-embedded tissue ($n = 148$), sensitivity and specificity of the assay was 73.4% and 100%, respectively. Use of real-time PCR makes it probable that rapid detection of *Coccidioides* spp. in fresh specimens or those fixed in formalin and paraffin-embedded may soon be feasible in the clinical laboratory.

Culture and Isolation

Clinical material from which *Coccidioides* spp. can be recovered includes a broad array of specimens whose efficacy of testing depends on the clinical presentation at the time of sampling. Not surprisingly, the respiratory tract yielded the highest recovery rate of 8.3% in one study (TABLE 2), while cultures of blood yielded the lowest (0.4%).[8] Recovery is enhanced by a high level of suspicion and availability of appropriate specimens. Although exact data are not available, the concentration of organisms in CSF is usually very low, and increased recovery from that source requires a high volume of specimen.[8] Cultures of CSF have been reported to yield the organism in only approximately

TABLE 2. Recovery rate of *Coccidioides* spp. from specimens submitted for fungal culture to a Phoenix Metropolitan Area laboratory over a 6-year period[a,b]

Specimen source	Total no. of specimens submitted	Recovery rate of *Coccidioides* spp.
Respiratory tract	10,372	8.3%
Bone marrow	267	2.6%
Nonsterile body site (other than respiratory and urine)	25,628	2.5%
Other normally sterile body site	11,566	2.1%
CNS	2,280	0.9%
Urinary tract	649	0.6%
Blood	5,026	0.4%

[a] Adapted from Ref. 8.
[b] As a whole, 55,788 specimens were submitted to the Mycology Section of Laboratory Sciences of Arizona in Phoenix over a 6-year period (1998–2003); overall recovery rate was 3.2%.

one-third of the patients; even with high volumes of CSF (of approximately 10 mL), recovery rate of organisms in cultures is only about 50%.[7,13] Coccidioidal meningitis is therefore more amenable to diagnosis by serologic documentation in the CSF.[7,8,13]

Coccidioides spp. can be grown on most selective and nonselective fungal media (brain–heart infusion agar, potato dextrose or potato flakes agar, and Sabouraud dextrose agar, with and without cycloheximide), as well as bacterial media (sheep blood and chocolate agars).[5] The species will also grow well on media formulated for isolation of *Legionella* (buffered charcoal-yeast extract) and *Bordetella* (Bordet-Gengou and Regan-Lowe) and are occasionally and unexpectedly recovered on such media (TABLE 3). Because *Coccidioides* spp. are poor competitors with other fungal and bacterial species, specimens with potentially mixed flora should be always inoculated onto selective (containing cycloheximide and other antimicrobial agents) as well as nonselective media.

TABLE 3. Media for isolation of *Coccidioides* spp.

Mycology Media
   Brain–heart infusion agar (BHI)
   Potato dextrose agar (PDA) or potato flakes agar (PFA)
   Sabouraud dextrose agar (SDA)
     (selective and nonselective)
Bacterial Media
   Blood agar
   Chocolate agar
[a] Buffered charcoal-yeast extract (selective and nonselective)
[b] Bordet-Gengou and Regan-Lowe (selective and nonselective)

[a] *Legionella* formulation
[b] *Bortetella* formulation

Specimens from sterile sites may be inoculated to nonselective fungal media alone.[8]

Colonies of *Coccidioides* spp. may be detected on the average between 4 and 5 days after plate inoculation.[8] The time to detection may range between 2 and 16 days. Young colonies (within 2–3 days of detection) are commonly gray, membranous, and difficult to recognize. Arthroconidia are not produced until the colonies lose their membranous characteristic (4–5 day or later); still, all colonies should be manipulated only under a laminar-flow biological safety cabinet. As they age, colonies start to take on a white to buff color, but may present with many different colors. Older colonies may be of variable texture and are usually floccose.

Identification

Identification of *Coccidioides* spp. uses the organisms' phenotypic and genotypic characteristics. The phenotypic characteristic of the production of arthroconidia is used to presumptively identify an isolate as *Coccidioides* spp. Presumptive identification is adequate to make the diagnosis of coccidioidomycosis if spherules with endospores have already been detected in the clinical specimen or if serologic conversion is documented. In these circumstances, confirmation of species by another means is unnecessary. However, if microscopic or serologic documentation is lacking, the isolate's identity should be further confirmed.

Several phenotypic methods of identity-confirmation of *Coccidioides* spp. have been used in the past. These include studies by immunodiffusion of exoantigen for lines of identity (serologic), propagation of the spherule phase in Converse medium (conversion from mycelial to spherule phase), or mouse inoculation studies (conversion from mycelial to spherule phase).[5] These older studies are tedious, time-consuming, increase risk of acquisition of infection by laboratory personnel, and are not routinely used in the clinical laboratory. A molecular-based, genus-specific genetic probe (ACCUPROBE®, GenProbe, San Diego, CA), which recognizes but does not differentiate both *Coccidioides* species, is now the method of choice for identification of the fungal isolates to genus level. Nonviable controls of previously confirmed isolates may be prepared by heat inactivation and stored at $-200°C$ or lower. Autoclaving and formalin should not be used to kill an isolate as these might cause false-negative reactions to occur with the probes.[14] Alternatively, a noninvasive *Coccidioides* spp. strain has been introduced and, if available, may be used for control functions.[15] This strain is not considered a select agent and may be maintained in the laboratory. The introduction of the molecular-based probe method of identification has made the older methods of identification obsolete in most laboratories.

Phenotypically, growth differences have been noted between *C. immitis* and *C. posadasii*. *C. posadasii* has been reported to grow significantly faster at

37°C *in vitro* than does *C. immitis*.[16] Otherwise, it is very difficult at this time to phenotypically differentiate between the two species. Identification of isolates into the species *C. immitis* and *C. posadasii* is practically only possible by specialized molecular techniques such as PCR and molecular sequencing and is available only in research-oriented laboratories. However, at present it is clinically unnecessary to routinely differentiate between the two species, as the two seem to have almost identical clinical presentations and antifungal susceptibility profiles. Further differentiation to species level may become more necessary as more is learned about the two species and their clinical presentations.

In vitro *Susceptibility Studies*

In vitro susceptibility testing of the fungi has progressed substantially within the last two decades, moving from testing of yeasts to that of moulds. In 2002, the Clinical Laboratory Standards Institute published Standard M38-A for *in vitro* susceptibility testing of the filamentous, conidia-forming fungi.[17] The standard facilitated interlaboratory agreement on defining of minimum inhibitory concentrations (MICs) and reproducibility of results has also made it possible to compare efficacy of other testing methods such as the quantitative gradient strip (E-test, AB Biodisk, Sölna, Sweden), commercially available microdilution trays, disk diffusion, or more recently, flow cytometry and fluorimetry using fluorescent probes.[18]

The goal of *in vitro* susceptibility testing of organisms is to predict therapeutic outcomes for organisms that have variably susceptible/resistant strains. Interpretive criteria have been provided and validated for some antifungal agents available for certain clinical syndromes caused by *Candida* and other yeast species.[19] Validations of the criteria were made possible by evaluation of known resistance mechanisms, appropriate *in vitro* susceptibility data, pharmacokinetic (PK) and pharmacodynamic (PD) parameters, as well as by clinical outcome studies in animal models and preferably in human cases.[19] Although these data are available for many yeast and mould infections, they are noticeably scarce in regard to coccidioidomycosis. Thus, although it is evident that routine *in vitro* susceptibility testing should be recommended for some yeast-associated syndromes, efficacy of testing is far less evident and more controversial when therapeutic decisions have to be made during coccidioidomycosis. Efficacy of *in vitro* susceptibility testing of moulds is influenced by and dependent on variability of host factors, paucity of clinical outcome data as compared to *in vitro* susceptibility data, as well as on laboratory technical issues.[20] In many of these fungal genera, including *Coccidioides*, resistance to specific antifungal agents seems to be intrinsic and not associated with resistance arising during therapy. Because of these issues, few regular laboratories are capable of proficiently providing such testing.

In vitro susceptibility testing of *Coccidioides* spp. need not be performed routinely in clinical laboratories. Such testing should be performed in proficient, specialized reference laboratories that have validated their test performances. *In vitro* susceptibility testing during coccidioidomycosis may be limited to surveillance studies, research protocols evaluating clinical applicability of *in vitro* susceptibilities and patient outcomes, and perhaps in cases in which patients are not responding to seemingly appropriate therapy.

Serologic Testing

Cell-Mediated Response

Cell-mediated immunity is protective in the host and can be measured by skin testing for cell-mediated cellular response using spherulin or coccidioidin (fungus-specific antigen preparations). Historically, skin testing was an important component of surveillance studies in the epidemiology of coccidioidomycosis. Unfortunately, skin testing is no longer practiced in the United States because the antigenic preparations are not commercially available.[8]

Humoral Antibody Response

Anticoccidioidal humoral antibody (IgM and IgG) in an infected host does not provide protection against the fungus, but indicates the organism's level of pathogenic activity. Measure of humoral antibodies can therefore be used for the diagnosis and prognosis of coccidioidomycosis. The serologic response may include the production of early (IgM) and late (IgG) antibodies.[21–24] Serologic responses are often, however, compromised or missing in the immunosuppressed patient. Early IgM (often called tube precipitin or TP) becomes measurable earlier in the acute phase of infection, usually between the first [50%] and third [90%] weeks of onset. The IgG (often called complement fixation or CF) antibody class responds later in the infectious process, becoming measurable sometime between the second and 28th week post onset. This class may remain for several months, but is usually related to the level of activity of infection.

Serologic studies have now been recognized as being less sensitive then previously thought (especially in self-limited clinical cases or in the immunocompromised patient). Thus, although positive serologic results may be helpful in the diagnosis of coccidioidomycosis, negative serologic results cannot be used to rule out the disease (especially early in the acute phase).

Three major techniques are now available for measure of the serologic response to coccidioidomycosis in the clinical laboratory.[21–23] These include enzyme immunoassays (EIAs), immunodiffusion (IMDF) and CF studies. EIAs are now available for detection of both the IgM and IgG classes and are probably the most sensitive of the methods. As such, negative results by EIA do not have to be confirmed by any of the other methods. However, positive EIA results, especially those for the IgM class of antibody, may be less specific and may require confirmation by IMDF or CF tests if clinically incompatible with coccidioidomycosis. Immunodiffusion methods are available for detection of both antibody groups as well, but require longer incubation periods (up to 4 days) to rule out negatives. Neither the EIA nor the IMDF tests are considered quantitative measurements of antibody, but the latter can be modified to quantify titers.[25] Such application of IMDF is useful in testing sera that have anticomplementary activity, thus precluding use of the CF test to obtain titers.

The CF test (together with the skin test) has historically played an enormous role in the understanding of the epidemiology, pathogenesis, and clinical diversity of coccidioidomycosis. As described by Smith and his colleagues in 1950, it has remained one of the primary diagnostic and prognostic tools since the late 1930s.[24] CF studies for IgG class of antibody are useful, primarily for diagnosis of more severe forms of coccidioidomycosis, serial titration in the evaluation of fungal activity, and the prognosis of the disease in an individual, and in evaluation of efficacy of antifungal therapy. Increasing CF titers in sequentially collected serum specimens tested in parallel or those above 1:32 may suggest dissemination; decreasing titers may indicate the amelioration of disease activity.[21,22,24] It must be recognized that presently there are several testing protocols and a variety of reagents available for the tests being performed in different laboratories and that very different results may be obtained by these different laboratories; it is imperative that for CF testing, consecutive specimens from the same patient be evaluated in parallel by the same laboratory. The CF test is not as sensitive as the EIA or IMDF tests (especially because the latter two can also be directed against the earlier appearing IgM).[26,27] This is especially true for more benign presentations of coccidioidomycosis in which symptoms are mild or transitory. Thus, it probably should not be used alone for the diagnosis of more mild forms of the disease or in its early acute stages. The CF test is, however, a primary serologic test for the diagnosis of meningeal disease and remains the standard in that capacity. Pleural and synovial fluids may also be tested by the CF test, but their diagnostic efficacy is still controversial; in this setting, CF results must be interpreted cautiously and only in the context of the whole clinical picture. There may also be significant discrepancies between EIA, immunodiffusion and CF study results; this is especially true between IgM results using either EIA or immunodiffusion methods.[8,26]

In one study of patients documented nonserologically to have coccidioidomycosis, the overall sensitivity of anticoccidioidal serologic studies was only 82%, confirming that serologic tests alone cannot rule out the disease. Individually, the sensitivities for EIA, IMDF, and CF were 83%, 71%, and 56%, respectively.[27] This study did not evaluate the temporal characteristics of the serologic conversions; neither did it correlate the various serologic conversions to severity of disease, except to a positive culture for the etiologic agent; there were no follow-up determinations to evaluate serologic conversion of various methods at later points in the disease process. There is a paucity of comparative data as to how quickly the various classes of antibody convert to positive by the differing methods of testing now available, and how long they remain detectable. It has, however, been documented that both CF and IMDF titers decrease as clinical disease abates.[21,24]

Recently it was found that a second-generation histoplasma urine antigen test, through a possible cross-reaction between the histoplasma and coccidioides antigens, was reactive in approximately 58% of patients with systemic coccidioidomycosis.[28] The sensitivity of the test was reportedly greatest (79%) in acute cases. These findings are at present being evaluated for possible application to the diagnosis of coccidioidomycosis.

SUMMARY

As with most infectious diseases, laboratory studies on patients with coccidioidomycosis provide additional pieces of information which in conjunction with the other clinical, radiographic, and historic pieces of information can provide or confirm the diagnosis. The number of coccidioidomycosis cases is rising rapidly in the endemic areas of the United States and, with the advent of increasing travel, may be seen more and more often in other parts of the country. A high level of suspicion should be placed on clinically compatible cases without documented etiologies.[29] It is critical to understand the appropriate laboratory studies to order and to be able to interpret their results.

REFERENCES

1. FISHER, M.C. et al. 2002. Molecular and phenotypic description of Coccidoides posadasii sp. nov., previously recognized as the non-California population of Coccidioides immitis. Mycologia 94: 73–84.
2. BIALEK, R.J. et al. 2004. PCR assays for identification of Coccidioides posadasii based on the nucleotide sequence of the antigen 2/ proline-rich antigen. J. Clin. Microbiol. 42: 778–783.
3. KIRKLAND, T.N. & J. FIERER. 1996. Coccidioidomycosis: a reemerging infectious disease. Emerg. Infect. Dis. 3: 192–199.
4. CRUM, N.F. et al. 2004. Coccidioidomycosis: a descriptive survey of a reemerging disease. Clinical characteristics and current controversies. Medicine 83: 149–175.
5. SAUBOLLE, M.A. 2000. Mycology and the clinical laboratory in the diagnosis of respiratory mycoses. In Fungal Diseases of the Lung. G.A. Sarosi & S.F. Davies, Eds.: 1–16. Lipincott Williams & Wilkins. Philadelphia, PA.

6. CHATURVEDI, V. *et al.* 2000. Coccidioidomycosis in New York State. Emerg. Infect. Dis. **6:** 25–29.

7. AMPEL, N. 2000. Coccidioidomycosis. *In* Fungal Diseases of the Lung. G.A. Sarosi & S.F. Davies, Eds.: 59–77. Lipincott Williams & Wilkins. Philadelphia, PA.

8. SAUBOLLE, M.A. *et al.* 2007. Epidemiologic, clinical and diagnostic aspects of coccidioidomycosis. J. Clin. Microbiol. **45:** 26–30.

9. MILLER, J.M. 2006. The select agent rule and its impact on clinical laboratories. Clin. Microbiol. Newsl. **28:** 57–63.

10. WIEDEN, M.A. & M.A. SAUBOLLE. 1996. The histopathology of coccidioidomycosis. *In* Coccidioidomycosis: Proceedings of the 5th International Conference. H.E. Einstein & A. Catanzaro, Eds.: 12–17. National Foundation for Infectious Diseases. Washington, DC.

11. JOHNSON, S.M. *et al.* 2004. Amplification of coccidioidal DNA in clinical specimens by PCR. J. Clin. Microbiol. **42:** 1982–1985.

12. BINNICKER, M.J. *et al.* 2007. Detection of *Coccidioides* species in clinical specimens by real-time PCR. J. Clin. Microbiol. **45:** 173–178.

13. BARENFANGER, J. *et al.* 2004. Nonvalue of culturing cerebrospinal fluid for fungi. J. Clin. Microbiol. **42:** 236–238.

14. GROMADZKI, S.G. & V. CHATURVEDI. 2000. Limitation of the AccuProbe *Coccidioides immitis* culture identification test: false-negative results with formaldehyde-killed cultures. J. Clin. Microbiol. **38:** 2427–2428.

15. MCGINNIS, M.R. *et al.* 2006. Use of *Coccidioides posadasii* Δchs5 strain for quality control in the ACCUPROBE culture identification test for *Coccidioides immitis*. J. Clin. Microbiol. **44:** 4250–4251.

16. BARKER, B. *et al.* S. 2006. Some like it hot: differences in the thermotolerance of *Coccidioides* spp.

17. CLINICAL AND LABORATORY STANDARDS INSTITUTE. 2002. Reference method for broth-dilution antifungal susceptibility testing of conidia-forming filamentous fungi: Approved Standard M38-A. CLSI, Wayne, PA

18. NYE, M.B. *et al.* 2006. Diagnostic mycology: controversies and consensus—what should laboratories do? Part II. Clin. Microbiol. Newsl. **28:** 129–134.

19. PFALLER, M.A. *et al.* 2006. Interpretative breakpoints for fluconazole and *Candida* revisited: a blueprint for the future of antifungal susceptibility testing. Clin. Microbiol. Rev. **19:** 435–447.

20. FORREST, G. 2006. Role of antifungal susceptibility testing in patient management. Curr. Opin. Infect. Dis. **19:** 538–543.

21. PAPPAGIANIS, D. 1996. Serology of coccidioidomycosis. *In* Coccidioidomycosis: Proceedings of the 5th International Conference. H.E. Einstein & A. Catanzaro, Eds.: 33–35. National Foundation for Infectious Diseases. Washington, DC.

22. PAPPAGIANIS, D. & B.L. ZIMMER. 1990. Serology of coccidioidomycosis. Clin. Microbiol. Rev. 3: 247–268.

23. YEO, S.F. & B. WONG. 2002. Current status of nonculture methods for diagnosis of invasive fungal infections. Clin. Microbiol. Rev. **15:** 465–484.

24. SMITH, C.E. *et al.* 1956. Pattern of 39,500 serologic tests in coccidioidomycosis. JAMA **160:** 546–552.

25. WIEDEN, M.J. *et al.* 1983. Comparison of immunodiffusion techniques with standard complement fixation assay for quantification of coccidioidal antibodies. J. Clin. Microbiol. **18:** 529–534.

26. KAUFMAN, L. *et al.* 1995. Comparative evaluation of commercial Premier EIA and microimmunodiffusion and complement fixation tests for *Coccidioides immitis* antibodies. J. Clin. Micobiol. **33:** 618–619.

27. POLLAGE, C.R. *et al.* 2006. Revisiting the sensitivity of serologic testing in culture positive coccidioidomycosis. Presented at the 106th Annual Meeting of the American Society of Microbiology. Orlando, FL.

28. KUBERSKI, T. *et al.* 2007. Diagnosis of coccidioidomycosis by antigen detection using cross-reaction with *Histoplasma* antigen. Clin. Infect. Dis. **44:** 50–54.

29. GALGIANI, J.N. *et al.* 2005. Coccidioidomycosis. Clin. Infect. Dis. **41:** 1217–1223.

Diagnosis of Coccidioidomycosis by Culture

Safety Considerations, Traditional Methods, and Susceptibility Testing

DEANNA A. SUTTON

Fungus Testing Laboratory, Department of Pathology, University of Texas Health Science Center at San Antonio, San Antonio, Texas 78229, USA

ABSTRACT: The recovery of *Coccidioides* spp. by culture and confirmation utilizing the AccuProbe nucleic acid hybridization method by Gen-Probe remain the definitive diagnostic method. Biosafety considerations from specimen collection through culture confirmation in the mycology laboratory are critical, as acquisition of coccidioidomycosis by laboratory workers is well documented. The designation of *Coccidioides* spp. as select agents of potential bioterrorism has mandated strict regulation of their transport and inventory. The genus appears generally susceptible, *in vitro*, although no defined breakpoints exist. Susceptibility testing may assist in documenting treatment failures.

KEYWORDS: *Coccidioides*; culture; susceptibility; select agent

INTRODUCTION

The three mainstays of laboratory diagnosis of coccidioidomycosis include culture, histopathology, and serology.[1] Because a thorough review of histopathological findings and immunologic methods may be found elsewhere in these proceedings, the objective of this brief article is to review traditional culture methods available to routine mycology laboratories. Although recent PCR procedures that amplify coccidioidal DNA from clinical specimens show diagnostic promise,[2,3] the recovery of *Coccidioides* spp. by culture continues to be the most definitive diagnostic method.[4] Molecular characterization of the genus has now identified two species, *Coccidioides immitis* and *Coccidioides posadasii*; however, the morphology of the two species, the clinical manifestations of the disease, and the therapeutic options are essentially identical.[5]

Address for correspondence: Deanna A. Sutton, Ph.D., M.T., S.M. (A.S.C.P.), R.M., S.M. (N.R.M.), Department of Pathology—MSC 7750, 7703 Floyd Curl Drive, San Antonio, TX 78229-3900. Voice: 210-567-4032; fax: 210-567-4076.

suttond@uthscsa.edu

Ann. N.Y. Acad. Sci. 1111: 315–325 (2007). © 2007 New York Academy of Sciences.
doi: 10.1196/annals.1406.005

Coccidioides immitis is indigenous to California and *C. posadasii* to other areas, although animal isolates appear to delineate this distinction more accurately than do patient cases.[6] With the advent of *Coccidioides* spp. being included in the select agent list of potential bioterrorism agents, the additional safeguards and requirements regarding biosafety, transport, storage, and reporting are in effect. While the clinical utility of antifungal susceptibility testing is not universally embraced, numerous clinical isolates have been tested and continue to be submitted for *in vitro* evaluation.

SPECIMEN COLLECTION

The collection, transport, and processing of appropriate clinical specimens is an important consideration in the recovery of *Coccidioides* spp., and several specific guidelines apply to specimens collected for the recovery of dimorphic pathogens. Common respiratory sites include sputum, bronchoalveolar lavage, transtracheal aspirates, and lung biopsies. Sputum is typically contaminated with bacteria, yeasts from the oral cavity, and inhaled conidia from a variety of saprobic moulds, and should be transported and processed as quickly as possible to curtail overgrowth by these more rapidly growing bacteria and endogenous flora. Recovery from blood, bone marrow, cerebrospinal fluid, and other body fluids, as well as cutaneous and/or soft tissue biopsy specimens is common when there is dissemination to extrapulmonary sites. Aspirates or pus should be submitted in syringes, when possible. Syringes must have the needle removed and be recapped prior to transport.[7,8] Several sources provide a more detailed guide to specimen collection.[7,9] All specimens should be considered potentially hazardous, and universal precautions should be observed during collection and processing.[7] Specimens should be accompanied by a properly completed requisition, and alerting the laboratory regarding potential cases of coccidioidomycosis is always advised.

Although it is recommended that all direct specimens be plated within 24 h, laboratories that must ship these samples to other sites for culture may do so by adhering to the Category B guidelines for biological substances as defined by the U.S. hazardous materials regulations and packaged according to the Department of Transportation guidelines for Transportation of Hazardous Materials. Shipment of filamentous cultures of suspected or known *Coccidioides* spp. is addressed later.

THE MYCOLOGY LABORATORY

Hyperendemic areas for coccidioidomycosis in the U.S. include Kern County in the San Joaquin Valley of California, and Maricopa (Phoenix) and Pima (Tucson) counties in Arizona. The risk of acquisition in these areas increases with events such as dust storms, earthquakes, and droughts that disrupt the soil and thereby disperse high concentrations of arthroconidia into the atmosphere, facilitating airborne transmission of the fungus.[10–15] Handling of clinical specimens with the subsequent growth and recovery of *Coccidioides* spp. in laboratory cultures provides a scenario in which arthroconidia could potentially be present several orders of magnitude higher than in the air of endemic areas, and thus the concern for laboratory-acquired infections among healthcare workers. This method of acquisition is well documented.[16–28] Laboratory workers are also at risk of accidental percutaneous inoculation of the spherule form of the fungus while they are processing direct specimens and handling unfixed histology specimens.[29,30] Some reports stating that 90% of laboratory-acquired infections result in clinical disease[31] further support the need for clinical laboratories to adhere to the safety standards promulgated by the Centers for Disease Control Office of Health and Safety in their 5th edition of *Biosafety in Microbiological and Biomedical Laboratories* (BMBL).[32] This document is available online at http://www.cdc.gov/od/ohs/biosfty/bmbl5/bmbl5toc.htm and summarizes the four different recommended biosafety levels for infectious agents, and addresses which types of biological safety cabinets are appropriate for each biosafety level. Biosafety Level 2 practices and facilities are recommended for handling and processing clinical specimens and identifying isolates of *Coccidioides* spp. Biosafety Level 2 practices differ from Biosafety Level 1 practices in that (1) laboratory personnel have specific training in handling pathogenic agents and are directed by qualified scientists; (2) access to the laboratory is limited when work is being conducted; (3) extreme precautions are taken with contaminated sharp items; and (4) procedures in which infectious aerosols or splashes may be created are conducted in Class II biological safety cabinets. Class II biological safety cabinets have vertical laminar airflow with HEPA (high-efficiency particulate air) filtered supply and exhaust air. They protect the worker, the product, and the environment.[32]

DIRECT MICROSCOPY

Clinical specimens such as fresh wet sputa, centrifuged bronchoalveolar lavage fluids, cerebrospinal fluid, aspirates, exudates, tissue biopsies, skin scrapings, and other similar specimens should also be examined microscopically. Direct microscopy often provides a rapid, presumptive diagnosis of coccidioidomycosis when intact or partially intact spherules are observed, and this procedure carries much less risk to laboratory workers than do cultural methods. Mature spherules are thick-walled, may measure up to 80 μm in diameter, and are filled with endospores. These spherules rupture, releasing small endospores approximately 2–4 μm in diameter that in turn undergo progressive maturation to variously sized spherules (FIG. 1). A preliminary diagnosis is more problematic when only small spherules (10–20 μm) lacking

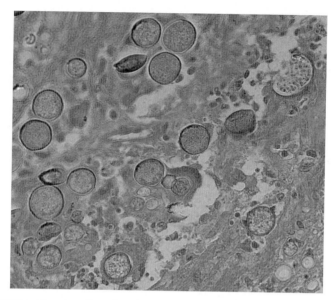

FIGURE 1. Hematoxylin and eosin (H&E) stain of spherules of *Coccidioides immitis* seen in various stages of maturation. Original magnification approximately 960×.

endospores or a few endospores of varying sizes are seen. These may often be mistaken for phagocytic cells, artifacts, or other fungi, particularly atypical forms of *Blastomyces dermatitidis* and yeasts such as *Histoplasma capsulatum, Candida glabrata*, and *Cryptococcus* species. A simple procedure to rule out artifacts such as pollen and other spherical fungi is to place a small amount of the primary specimen suspected of containing endospores or small spherules in a drop of sterile saline on a glass slide and then seal the specimen with a petroleum-edged coverslip. This provides a microculture in which spherules put out tubular structures and become hyphae within 2 to 3 days at room temperature.[33] Hyphal forms have also been observed in chronic cavitary and granulomatous lesions and in cerebrospinal fluid.[34–36] Several direct microscopy methods may be employed such as the Gram stain, a KOH preparation, and the calcofluor white stain.[37] The cytopathological Papanicolaou stain is also particularly useful for the detection of spherules.[38]

CULTURE SETUP AND MEDIA CONSIDERATIONS

All direct inoculation procedures for specimens suspected of containing *Coccidioides* spp. should be performed only within a biological safety cabinet and be cultured onto tubes rather than plates. Select bloody, purulent, or caseous portions from respiratory specimens, and mince tissue biopsies with a scalpel; do *not* grind. *Coccidioides* spp. are not particularly fastidious, and a variety of primary isolation media may be used. Common mycologic media include brain–heart infusion agar, Sabouraud dextrose agar, potato dextrose agar, and Sabouraud dextrose agar with cycloheximide. For non-sterile sites, prevention of overgrowth by bacteria and more rapidly growing saprobic fungi is necessary. Antibiotics commonly added to media to suppress bacterial growth include chloramphenicol, gentamicin, streptomycin, and penicillin. The addition of cycloheximide inhibits some saprobic fungi and is useful as a selective medium as it supports the growth of all dimorphic fungi. Media without cycloheximide should also be included in the battery so as not to inhibit other opportunistic pathogens.[8] Cultures are typically incubated at 30°C; however, room temperature is also suitable. *Coccidioides* spp. are also commonly recovered on a variety of bacterial media. Unopened tubes should be examined daily for the first 10 days and weekly thereafter for 3 weeks. It is seldom necessary to hold cultures for longer than 3 weeks to recover spp. of *Coccidioides* unless patients are receiving antifungal agents at the time of specimen collection.

CULTURAL FEATURES

Colonies develop readily, usually within 3–5 days, and are initially white to cream glistening, glabrous and tenacious, adhering to the medium. As they mature, discrete concentric rings develop and filamentous areas containing arthroconidia become visible. While species of *Coccidoidies* are generally white to off-white, various other colors have been observed, ranging from tan to yellow, rose to lavender, and pale brown to gray-brown. Tease-mounts for observation of microscopic morphology should be made in lactophenol cotton blue and the coverslips should be sealed with a mounting medium and all manipulations of cultures must be performed within a Class II biological safety cabinet. Hyphae are thin, hyaline, and septate. Thicker side branches give rise to unicellular, barrel-shaped arthroconidia (3–4 × 3–6 μm) that alternate with thin-walled, empty disjunctor cells (FIG. 2). At maturity the arthroconidia are released, and remnants of the disjunctor cells are seen as "annular frills" on the ends of individual arthroconidia.[8, 39] The arthroconidia of *Coccidoides* species must be differentiated from some "look-alike" genera with alternating arthroconidia such as *Malbranchea*, *Chrysosporium*, and *Geomyces*, and from *Sporotrichum pruinosum*. Isolates morphologically consistent with *Coccidioides* spp. are commonly confirmed in clinical laboratories using the AccuProbe nucleic acid hybridization method by GenProbe, San Diego, California.[40–42] This method is based on the ability of complementary nucleic acid strands to specifically align and associate to form stable double-stranded complexes. The method uses a chemiluminescent labeled, single-stranded DNA probe complementary to rRNA of the target organisms to form a DNA:RNA hybrid that is measured in a luminometer. A selection reagent differentiates

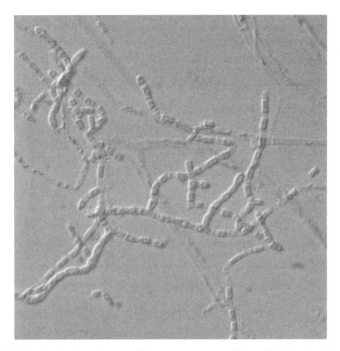

FIGURE 2. Lactophenol cotton blue tease-mount of *Coccidioides immitis* showing chains of arthroconidia separated by disjunctor cells. Original magnification approximately 960×.

the non-hybridized from the hybridized probe. The cutoff for a positive test is 50,000 RLU (relative light units). Repeat testing should be performed for isolates with RLU between 40,000 and 49,999. False negative results may occur with formaldehyde-killed cultures.[43] Exoantigen testing may also be used for confirmation. The method detects the presence of cell-free antigens, known as exoantigens. Mature slants are overlaid with merthiolate 1:5,000 overnight and extracts sterilized through a membrane filter.[44] Exoantigens are demonstrated by immunodiffusion of antigens forming precipitin lines of identity with reference antisera. This method has mostly been replaced by the AccuProbe method. Conversion of arthroconidia to spherules in Converse medium is not performed in routine clinical laboratories.[45,46]

BIOTERRORISM AND REQUIREMENTS FOR REPORTING

Coccidioides immitis and *C. posadasii* are now classified as select agents of potential bioterrorism and their recovery from clinical samples must be documented and reported. Clinical or diagnostic laboratories that have identified *Coccidioides* species must report their findings to the Centers for Disease Control within seven calendar days on CDC Form 4 and either destroy cultures by a recognized method, or forward these isolates to a select agent laboratory registered with CDC select agent programs. Shipment of these isolates previously identified as *Coccidioides immitis* or *C. posadasii* must be through approved hazardous materials channels for infectious agents and may be shipped as Category A biological substances. Select agent laboratories may keep culture collections, but must follow stringent documentation and inventory control of cultures and adhere to the guidelines for handling select agents. Clinical laboratories sending isolates suspected of being *Coccidioides* spp. to select agent laboratories for purposes of identification and/or susceptibility testing may ship these isolates as Category B biological substances. Complete handling and transfer regulations for select agents are found in the *Federal Register 42 CFR Part 1003, Possession, Use, and Transfer of Select Agents and Toxins; Final Rule, March 18, 2005* at http://www.cdc.gov/od/sap/pdfs/42_cfr_73_final_rule.pdf

SUSCEPTIBILITY TESTING

In vitro antifungal susceptibility testing for *Coccidioides* spp. remains unstandardized outside specialized laboratories experienced in this testing. However, modifications of the Clinical Laboratory Standards Institute M38-A method for filamentous fungi have been used for testing the arthroconidial form of the fungus.[47] All testing must be performed within a biological safety cabinet. The Fungus Testing Laboratory utilizes an enclosed macrobroth modification of the M38-A method with spectrophotometric standardization of the inoculum to 95% T at 530 nm (1×10^4 to 5×10^4 CFU/mL), incubation at 35°C, and visual endpoint readings. Because no defined breakpoints exist, "susceptibility" may be viewed as minimum inhibitory concentrations, minimum lethal concentrations, or minimal effective concentrations of the agent being tested that are within safely achievable levels in the patient. *In vitro* MICs/MECs beyond levels attainable would suggest potential resistance. *Coccidioides* spp., *in vitro,* are generally "susceptible" to amphotericin B, fluconazole, itraconazole, voriconazole, posazonazole[48] as well as the echinocandins.[49] Fluconazole treatment failures have been reported, however,[50,51] and may, in some cases, be partially attributed to resistant organisms displaying MICs $\geq$ 64 μg/mL. It should be noted, however, that there may be a lack of correlation between *in vitro* activity and clinical efficacy for coccidioidomycosis,[52] and that factors such as the pharmacokinetics of the drug, general host factors, and site of infection all influence clinical outcomes.[53]

SUMMARY

The recovery of *Coccidioides immitis* or *Coccidioides posadasii* by culture remains the "gold standard" for a definitive diagnosis. Appropriate specimens must be submitted, and those from non-sterile sites should be plated on media that will support growth as well as inhibit bacteria and rapidly growing saprobic fungi. Direct microscopy should always be included in the culture process, as a preliminary diagnosis may be made with the observance of spherules and/or endospores. Although colonies of *Coccidioides* spp. are generally white to cream to buff, many other colors have been reported and should be expected occasionally. Confirmation of suspected *Coccidioides* isolates is commonly made using the AccuProbe DNA probe. While *in vitro* susceptibility testing remains non-standardized, it may be useful in detecting potential resistance and documenting treatment failures.

REFERENCES

1. AMPEL, N.M. 2003. Coccidioidomycosis. *In* Clinical Mycology. W.E. Dismukes, P.G. Pappas & J.D. Sobel, Eds.: 311–327. Oxford University Press. New York, NY.
2. BIALIK, R. *et al.* 2004. PCR assays for identification of *Coccidioides posadasii* based on the nucleotide sequence of the antigen 2/proline-rich antigen. J. Clin. Microbiol. **42:** 778–783.
3. BIALEK, R., S.M. JOHNSON & K.A. PAPPAGIANIS. 2005. Amplification of coccidioidal DNA in clinical specimens by PCR. J. Clin. Microbiol. **43:** 1492–1493.
4. PAPPAGIANIS, D. 2005. Coccidioidomycosis. *In* Topley and Wilson's Medical Mycology, 10th edition. W. Merz & R.J. Hay, Eds.: 502–518. ASM Press. Washington, DC.
5. FISHER, M.C. *et al.* 2002. Molecular and phenotypic description of *Coccidioides posadasii* sp. nov., previously recognized as the non-California population of *Coccidioides immitis*. Mycologia **94:** 73–84.
6. BARKER, B., S. KROKEN & M. ORBACH. 2006. Microsatellite analysis of environmental isolates of the human pathogenic fungus *Coccidioides* validates the genetic and geographical isolation among populations and species. Presented at the Sixth International Symposium on Coccidioidomycosis. Stanford, CA, August 25.
7. SUTTON, D.A. 2003. Specimen collection, transport, and processing: mycology. *In* Manual of Clinical Microbiology, 8th edition. P.R. Murray, E.J. Baron, J.H. Jorgensen, M.A. Pfaller & R.H. Yolken, Eds.: 1659–1667. ASM Press. Washington, DC.
8. WALSH, T.J. *et al.* 2003. *Histoplasma, Blastomyces, Coccidioides*, and other dimorphic fungi causing systemic mycoses. *In* Manual of Clinical Microbioloty, 8th edition. P.R. Murray, E.J. Baron, J.H. Jorgensen, M.A. Pfatter, & R.H. Yolken, Eds.: 1781–1797. ASM Press. Washington, DC.
9. PFALLER, M.A. & M.R. MCGINNIS. 2003. The laboratory and clinical mycology. *In* Clinical Mycology. E.J. Anaissie, M.R. McGinnis, & M.A. Pfaller, Eds.: 67–79. Churchill Livingstone. Philadelphia, PA.
10. RIPPON, J.W. 1988. Coccidioidomycosis. *In* Medical Mycology. The Pathogenic Fungi and the Pathogenic Actinomycetes, 3rd edition.:433–473. W. B. Saunders Company. Philadelphia, PA.

11. ANSTEAD, G.M. & J.R. GRAYBILL. 2006. Coccidioidomycosis. *In* Fungal Infections. Infectious Disease Clinics of North America. T.F. Patterson & R.C. Moellering, Jr. Eds.: 621–643. W. B. Saunders Company. Philadelphia, PA.

12. CRUM, N.G. *et al.* 2004. Coccidioidomycosis: a descriptive survey of a reemerging disease. Clinical characteristics and emerging controversies. Medicine **83:** 149–175.

13. FLYNN, N. *et al.* 1979. An unusual outbreak of windborne coccidioidomycosis. N. Engl. J. Med. **301:** 358–361.

14. SCHNEIDER, E. *et al.* 1997. A coccidioidomycosis outbreak following Northridge, Calif. earthquake. JAMA **277:** 904–908.

15. WILLIAMS, P.L. *et al.* 1979. Symptomatic coccidioidomycosis following a severe natural dust storm. An outbreak at the Naval Air station, Lemoore, Calif. Chest **76:** 566–570.

16. BUSH, J.D. 1943. Coccidioidomycosis. J. Med. Assoc. Alabama **13:** 159–166.

17. COLLINS, C.H. 1993. Laboratory-Acquired Infections: History, Incidence, Causes, and Prevention. Butterworth-Heinemann Ltd. Oxford, United Kindgom.

18. CONANT, N.F. 1955. Development of a method for immunizing man against coccidioidomycosis. Third Quarterly Progress Report. Contract DA-18-064-CML-2563. Duke University, Durham, NC. Available from Defense Documents Center, AD 121–600.

19. DICKSON, E.C. 1937. *Coccidioides* infection: part I. Arch. Intern. Med. **59:** 1029–1044.

20. DICKSON, E.C. 1937. "Valley fever" of the San Joaquin Valley and fungus *Coccidioides*. Calif. Western Med. **47:** 151–155.

21. DICKSON, E.C. & M.A. GIFFORD. 1938. *Coccidioides* infection (coccidioidomycosis): II. The primary type of infection. Arch. Intern. Med. **62:** 853–871.

22. JOHNSON, J.E. III *et al.* 1964. Laboratory-acquired coccidioidomycosis. Ann. Intern. Med. **60:** 941–956.

23. LOONEY, J.M. & T. STEIN. 1950. Coccidioidmycosis. The hazard involved in diagnostic procedures, with report of a case. N. Engl. J. Med. **242:** 77–82.

24. NABARRO, J.D.N. 1948. Primary pulmonary coccidioidomycosis: case of laboratory infection in England. Lancet **1:** 982–987.

25. PIKE, R.M. 1976. Laboratory-associated infections. Summary and analysis of 3,921 cases. Health Lab. Sci. **13:** 105–114.

26. SEWELL, D.L. 1995. Laboratory-associated infections and biosafety. Clin. Microbiol. Rev. **8:** 389–405.

27. SMITH, C.E. 1950. The hazard of acquiring mycotic infections in the laboratory. Presented at the 78th Annual Meeting of the American Public Health Association. St. Louis, MO.

28. SMITH, D.T. & E.R. HARRELL, Jr. 1948. Fatal coccidioidomycosis: a case of laboratory infection. Am. Rev. Tuberc. **57:** 368–374.

29. KOHN, G.J. *et al.* 1992. Acquisition of coccidiodomycosis at necropsy by inhalation of coccidioidal endospores. Diagn. Microbiol. Infect. Dis. **15:** 527–530.

30. TOMILSON, C.C. & P. BANCROFT. 1928. Granuloma coccidioides: report of a case responding favorably to antimony and potassium tartrate. JAMA **91:** 947–951.

31. WILSON, J.W., C.E. SMITH & O.A. PLUNKETT. 1953. Primary cutaneous coccidioidomycosis: the criteria for diagnosis and a report of a case. Calif. Med. **79:** 233–239.

32. U. S. DEPARTMENT OF HEALTH AND HUMAN SERVICES. 2007. Biosafety in Micro-biological and Biomedical Laboratories, 5th edition. U. S. Government Printing Office. Washington, DC.
33. WILSON, J.W. & O.A. PLUNKETT. 1974. Coccidioidomycosis. *In* The Fungous Dis-eases of Man: 25–48. University of California Press. Berkeley, CA.
34. DOLAN, M.J. *et al.* 1992. *Coccidioides immitis* presenting as a mycelial pathogen with empyema and hydropneumothorax. J. Med. Vet. Mycol. **30:** 249–255.
35. HAGMAN, H.M. *et al.* 2000. Hyphal forms in the central nervous system of patients with coccidioidomycosis. Infect. Dis. **30:** 349–353.
36. WAGES, D.D., L. HELFEND & H. FINKLE. 1995. *Coccidioides immitis* presenting as a hyphal form in a ventriculoperitoneal shunt. Arch. Pathol. Lab. Med. **119:** 91–93.
37. HAGEAGE, G.J. & B.J. HARRINGTON. 1984. Use of calcofluor white in clinical mycology. Lab. Med. **15:** 109–112.
38. SANDERS, J.S. *et al.* 1977. Exfoliative cytology in the rapid diagnosis of pulmonary blastomycosis. Chest **72:** 193–196.
39. PEREA, S. & T.F. PATTERSON. 2003. The laboratory and clinical mycology. *In* Clinical Mycology. E.J. Anaissie, M.R. McGinnis & M.A. Pfaller, Eds.: 352–369. Churchill Livingstone. Philadelphia, PA.
40. PADHYE, A.A. *et al.* 1994. Comparative evaluation of chemiluminescent DNA probe assays and exoantigen tests for rapid identification of *Blastomyces der-matitidis* and *Coccidioides immitis*. J. Clin. Microbiol. **32:** 867–870.
41. STOCKMAN, L. *et al.* 1993. Evaluation of commercially available acridinium ester-labeled chemiluminescent DNA probes for culture identification of *Blastomyces dermatitidis*, *Coccidioides immitis*, *Cryptococcus neoformans*, and *Histoplasma capsulatum*. J. Clin. Micribiol. **31:** 845–850.
42. VALESCO, M. & K. JOHNSTON. 1997. Stability of hydridization activity of *Coccidioides immitis* in live and heat-killed frozen cultures tested by Ac-cuProbe *Coccidioides immitis* culture identification test. J. Clin. Microbiol. **35:** 736–737.
43. GRAMADZKI, S.G. & V. CHATURVEDI. 2000. Limitation of the AccuProbe *Coccidioides immitis* culture identification test: false-negative results with formaldehyde-killed cultures. J. Clin. Microbiol. **38:** 2427–2428.
44. STANDARD, P.G. & L. KAUFMAN. 1982. Safety considerations in handling exoanti-gen extracts from pathogenic fungi. J. Clin. Microbiol. **15:** 663–667.
45. CONVERSE, J.L. 1956. Effect of physico-chemical environment on spherulation of *Coccidioides immitis* in a chemically defined medium. J. Bacteriol. **72:** 784–792.
46. SUN, S.H., M. HUPPERT & K.R. VUKOVICH. 1976. Rapid *in vitro* conversion and identification of *Coccidioides immitis*. J. Clin. Microbiol. **3:** 186–190.
47. Clinical Laboratory Standards Institute (formerly National Committee for Clinical Laboratory Standards). 2002. Reference method for broth dilution antifungal susceptibility testing of filamentous fungi; approved standard. CLSI document M38-A. Clinical Laboratory Standards Institute, Wayne, PA.
48. GONZALEZ, G.M. *et al.* 2002. *In vitro* and *in vivo* activities of posaconazole against *Coccidioides immitis*. Antimicrob. Agents Chemother. **46:** 1350–1356.
49. GONZALEZ, G.M. *et al.* 2001. Correlation between antifungal susceptibilities of *Coccidioides immitis in vitro* and antifungal treatment with caspofungin in a mouse model. Antimicrob. Agents Chemother. **45:** 1854–1859.
50. CORTEZ, K.J., T.J. WALSH, & J.E. BENNETT. 2003. Successful treatment of coccid-ioidal meningitis with voriconazole. Clin. Infect. Dis. **36:** 1619–1622.

51. PROIA, L.A. & A.R. TENORIO. 2004. Successful use of voriconazole for treatment of *Coccidioides* meningitis. Antimicrob. Agents Chemother. **48:** 2341.
52. ANSTEAD, G. *et al.* 2005. Refractory coccidioidomycosis treated with posaconazole. Clin. Infect. Dis. **40:** 1770–1776.
53. REX, J.H. *et al.* 1997. Development of interpretive breakpoints for antifungal susceptibility testing: conceptual framework and analysis of *in vitro-in vivo* correlation data for fluconazole, itraconazole, and *Candida* infections. Clin. Infect. Dis. **24:** 235–247.

Molecular Identification of *Coccidioides* Isolates from Mexican Patients

LAURA ROCÍO CASTAÑÓN-OLIVARES,[a] DIEGO GÜEREÑA-ELIZALDE,[b] MARISELA ROCÍO GONZÁLEZ-MARTÍNEZ,[c] ALEXEI FEDOROVICH LICEA-NAVARRO,[d] GLORIA MARÍA GONZÁLEZ-GONZÁLEZ,[e] AND ARTURO AROCH-CALDERÓN[a]

[a]*Facultad de Medicina, Universidad Nacional Autónoma de México, Ciudad Universitaria, Distrito Federal, México*

[b]*Laboratorio de Soluciones Genéticas, Tijuana, Baja California, México*

[c]*Unidad Médica de Atención Especializada No. 71, Instituto Mexicano del Seguro Social, Torreón, Coahuila, México*

[d]*Biotecnología Marina, Centro de Investigación Científica y de Educación Superior de Ensenada, Ensenada, Baja California, México*

[e]*Facultad de Medicina, Universidad Autónoma de Nuevo León, Monterrey, Nueva Léon, México*

ABSTRACT: Molecular studies of the genome of the fungus *Coccidioides* have demonstrated two nearly identical, but well-identified species, *Coccidioides immitis* and *C. posadasii*, known as "California" and "non-California" species, respectively. The objective of this study was to determine, through molecular methods, whether both species of *Coccidioides* are present in Mexican patients with coccidioidomycosis and to estimate, their geographical distribution in Mexico. We analyzed 56 clinical isolates of *Coccidioides* spp. from Mexican patients. Molecular identification of each strain was done by means of real time PCR using TaqMan® probes to amplify single nucleotide polymorphisms (SNPs) in four target sequences, *loci*, named proline 157, proline 174, hexokinase 149 and glucose-synthase 192. SNP analysis identified two of the 56 isolates as *Coccidioides immitis* and the remaining 54 as *C. posadasii*. The dual probe assay that included proline 157, proline 174 and glucose-synthase 192 gave consistent results on SNP differentiation between the two species. In contrast, the template matching hexokinase 149 gave negative results for any species in 34 samples. Our results did not show geographical overlap of the species, and they also confirmed that *C. posadasii* is the

Address for correspondence: Laura Rocío Castañón-Olivares, Facultad de Medicina Edificio A 2⁰ piso, Laboratorio de Micología Médica, Ciudad Universitaria, Delegación Coyoacán CP 04510, México D. F. Voice: +52-55-5623-2458 or +52-55-5623-2459; fax: +52-55-5623-2458.
lrcastao@servidor.unam.mx

Ann. N.Y. Acad. Sci. 1111: 326–335 (2007). © 2007 New York Academy of Sciences.
doi: 10.1196/annals.1406.047

most frequent species in Mexico. A vast majority of *C. posadasii* strains were localized in the north-central region of the country.

KEYWORDS: *Coccidioides posadasii*; *Coccidioides immitis*; molecular identification of *Coccidioides*

INTRODUCTION

In 1994 Zimmerman *et al.*, using a restriction fragment length polymorphism (RFLP) analysis, demonstrated that *Coccidioides immitis* could be differentiated in two well-defined groups: Group I and Group II.[1] Three years later, Koufopanou *et al.*,[2] using sequences of 350–650-bp fragments from five nuclear genes (*CHS1, pyrG, tcrP, serine-proteinase antigen* and *CTS2*), confirmed that *C. immitis* could be divided in two defined groups: these groups could be geographically differentiated and they were designated as the non-Californian species (corresponding to Zimmermann's Group I) and the Californian species (Zimmermann's Group II).[2] Later, using microsatellite *loci*, Fisher *et al.*, demonstrated that the genetic diversity of *Coccidioides* spp. had spread throughout Americas in a time span ranging from 8,940 to 134,000 years ago,[3,4] and that the isolates from South America had been derived from an original strain found in the center of the state of Texas in the United States of America. Finally, Fisher *et al.*, through phylogenetic analysis from polymerase chain reaction (PCR) of nine microsatellite-containing *loci* (*GAC, 621, GA37, GA1, ACJ, KO3, KO7, KO1* and *KO9*), demonstrated that *Coccidioides immitis* included two different species: *Coccidioides posadasii* sp. nov. (non-Californian) and the already well-known *C. immitis* (Californian), both being causal agents of clinical coccidioidomycosis.[5] Even though the geographical distribution of both species has been well defined, populations of *C. immits* and *C. posadasii* overlap, which might be due to patients having traveled between endemic regions, or due to the wind dispersion of infectious arthroconidia.[6]

The northern states in Estados Unidos Mexicanos (Mexico) have been considered traditionally as endemic regions for coccidioidomycosis. However, epidemiologic studies on coccidioidomycosis in Mexico are scarce.[7,8] Fisher *et al.*[5] analyzed 36 strains of *Coccidioides* spp. from Mexico and identified 28 as *C. posadasii* and 8 as *C. immitis*. In 2004, Bialek *et al.* analyzed 120 isolates from Mexican patients (using PCR based on the nucleotide sequence of antigen2/proline-rich antigen), all of which were identified as *C. posadasii*.[9] In order to expand the knowledge of this mycosis in Mexico we studied 56 isolates of *Coccidioides* spp. obtained from Mexican patients in an attempt to identify both species of *Coccidioides* employing the molecular method of single nucleotide polymorphisms (SNPs), and to estimate their geographical distribution.

MATERIALS AND METHODS

Patients

Patients were diagnosed in different hospitals located in the states of Baja California, Coahuila, Nuevo León and Distrito Federal. Clinical samples from which *Coccidioides* spp. was isolated included sputum, cerebrospinal fluid, bronchoalveolar lavage, pus or tissues, and the tip of a peritoneal derivation catheter from one patient. A total of 56 isolates were recovered in culture. The clinical variables extracted from patient files included gender, age, occupation, place of residence, and migratory history. When information about place of residence and travel history was available, it was then correlated with the isolates obtained.

Strains

Strains 5273 and 5252 were used as reference type-strains of *Coccidioides immitis*,[5] and strains 1 and 2 as reference type-strains for *C. posadasii*.[9]

Biotype Identification

In accordance to Centers for Disease Control guidelines,[10] all procedures involved in the isolation and conservation of the fungal isolates were carried out in a level 3 biosafety cabinet.

For the clinical isolates, standard mycological techniques (direct examination and culture on BBL™ Mycosel™ agar [Becton Dickinson, Detroit, MI, USA]) were used in every patient's specimen. To corroborate the identification as *Coccidioides* spp., all 56 isolates were tested using an exoantigens technique. Twenty-three isolates (those from Nuevo León) were also tested with AccuProbe® (Gen-Probe, Inc., San Diego, CA, USA). The morphology and serological identification were corroborated in the four type-strains, and only those *C. posadasii* were also tested with AccuProbe®.

DNA Isolation

The four reference strains and the 56 clinical isolates from Mexican patients were cultured in Sabouraud dextrose broth with streptomycin, penicillin and chloramphenicol in 50-mL plastic tubes incubated at 30°C until growth was observed (4 days on average). Once grown, cultures were heat-killed at 100°C for 15 min. To ensure sterility, a portion was inoculated on BBL™ Mycosel™ agar with cycloheximide and incubated at 22–25°C; the remainder of the sample was frozen at $-70°C$. If, after a week, the sterility control did not show any growth, then the respective frozen sample was considered sterile and consequently safe for proceeding with the isolation of DNA.

Each sample of *Coccidioides* spp. was transferred to a 1.5-mL Eppendorf tube containing 100 μL of 5 mM magnesium chloride and with 5 μL of proteinase K (10 mg/mL) added. The tube was incubated for 1 hour at 55°C, then heated to 95°C for 15 min, removed and immediately frozen to $-70°C$ for 10 min. These cycles of heat and freezing were repeated three times and 200 μL of TE buffer (10 mM Tris HCl 1 mM EDTA; pH 8.0–8.5) was added. DNA extraction was done by adding 500 μL of TRI-REAGENT® (Sigma-Aldrich, St. Louis, MO, USA) following the protocol recommended by the supplier. In brief: the sample was homogenized using vortex, and stored for 20 min at room temperature. The aqueous phase was removed and DNA precipitated from the interphase with 100% ethanol at room temperature. The sample was centrifuged at $12,000 \times g$ for 5 min at 4°C, and supernatant was removed to obtain the pellet containing the DNA. The DNA pellet was washed by centrifugation in 0.1 M trisodium citrate in 10% ethanol. The ethanol was removed and the DNA pellet was briefly air-dried for 5 min at room temperature. Finally, the DNA was dissolved in 8 mM NaOH.

Probes for SNPs

Taking into account the available published information on the genetic sequences of *Coccidioides* (GenBank), test probes were designed for multiple sequence targets. After a careful scrutiny of potential targets, four sequences were decided upon, two of them *loci* corresponding to genes that code for proline rich-antigen (proline 157 and proline 174), another for hexokinase (149), and the last for glucose-synthase (192) (TABLE 1), representing the best potential for a successful identification of the corresponding SNPs. The four designed and synthesized probes (SNP TaqMan® dual probe assays) were named: proline 157, proline 174, hexokinase 149, and glucose-synthase 192.

DNA Amplification

In all samples, the DNA was marked with VIC™ fluorophore to identify the amplified SNPs corresponding to *C. immitis* and FAM™ (6-carboxyfluorescein) fluorescent for *C. posadasii* (TABLE 1). The amplification reactions were carried out using an ABI 7500 Real Time PCR and ABI TaqMan PCR Master mix (Applied Biosystems® de México) with ROX (passive reference dye for normalization of reporter signal). Real-time PCR assay was performed as follows: 10 min at 94°C and subsequently 50 cycles of denaturation at 92°C for 15 sec and annealing and extension at 60°C for 1 min. The

TABLE 1. Primer and probe sequences for each of the four *loci* tested.

Gene	Forward primer	Reverse primer	Probe		Description	Amplicon size (bp)
Glucose synthase (192)	5'-CCGACGGGCTG GCCATAT-3'	5'-CCAACGCCCAGTA GTATGGT-3'	VIC	5'-CCTGGAAA ACCAC-3'	*C. immitis*	72
			FAM	5'-CCTGGGAA ACCAC-3'	*C. posadasii*	
Proline-rich antigen (157)	5'-GTCGTTGACCAG TGCTCCAA-3'	5'-CGGCGGTGGT GTCAACT-3'	VIC	5'-CCCAATTGA GATCCCA-3'	*C. immitis*	63
			FAM	5'-CCCAATTGAC ATCCCA-3'	*C. posadasii*	
Proline-rich antigen (174)	5'-CCGCTGAGCC GACTCA-3'	5'-CGGTTGGGAC GGCAGT-3'	VIC	5'-TCCTCCGTA GGCTCA-3'	*C. immitis*	50
			FAM	5'-CTCCGTGG GCTCA-3'	*C. posadasii*	
Hexokinase (149)	5'-GCCACCAAGCC TGAGCTT-3'	5'-ACCGCAAGCG GAGAGG-3'	VIC	5'-CGCCTGGC AGAAC-3'	*C. immitis*	80
			FAM	5'-CCGCCTAGC AGAAC-3'	*C. posadasii*	

PCR mixtures, in a final volume of 20 μL, contained 2.5 mM MgCl$_2$, 200 mM of each deoxynucleoside triphosphate, a 1-mM concentration of each primer, a 0.2-mM concentration of each probe,1 U Taq DNA polymerase (Promega, Madison, WI, USA), and as a template, and 2 μL extracted DNA were added. All samples were run in triplicate.

RESULTS

Laboratory Identification of Coccidioides spp.

We examined 56 clinical isolates of *Coccidioides* spp. In the initial direct examinations of clinical specimens spherules or endospores were observed in 55 of the 56 cases. Only a single case, the catheter-tip case, was included in the study without showing the direct presence of spherules or endospores. After culture, all isolates showed growth of white and cottony colonies, which were seen to be composed of thin hyaline hyphae on microscopic evaluation. However, some cultures had no arthroconidia present.

Isolates were identified using an exoantigen test, which was positive for *Coccidioides* spp. for all 60 isolates (56 clinical isolates and four reference strains). In addition, the Accuprobe® test for *Coccidioides* was positive for 23 isolates (those from Nuevo León), as well as the two reference strains of *C. posadasii*.

SNP Analysis for Determination of Species

Our SNP technique using four carefully chosen loci was applied to all 60 isolates. SNP analysis identified 54 of the 56 clinical isolates as *C. posadasii* and 2 of the 56 as *C. immitis*; the four reference strains also were correctly identified to species (TABLE 1). The assays with the proline 157 and 174 targets and those with the glucose-synthase 192 target had consistent results when differentiating between the two species of *Coccidioides*. In contrast, the template for hexokinase 149 did not result in a product of amplifcation for either species with 34 DNA samples (TABLE 2).

TABLE 2. Molecular identification with SNPs of 56 isolates of *Coccidioides* spp

Amplification with	Probe test			
	Proline 157	Proline 174	Glucose-synthase 192	Hexokinase 149
FAM marker	54[a] = *C. posadasii*	54 = *C. posadasii*	54 = *C. posadasii*	22 = *C. posadasii*
VIC marker	2 = *C. immitis*	2 = *C. immitis*	2 = *C. immitis*	0
Without amplification	0	0	0	34 = Unidentified

[a]Number of isolates.

TABLE 3. Geographic locale of 19 out of 56 isolates of *Coccidioides* spp. examined in this study

City where the diagnosis was made	Institution providing isolates	Patient's residence state	No. of isolates	Region in the map (FIG. 1)
Distrito Federal (Mexico City)	Instituto Nacional de Neurología y Neurocirugía	Baja California	1	A
Tijuana	Hospital General	Baja California	1	A
Torreón	Unidad Médica de Atención Especializada	Coahuila de Zaragoza	11	B
		Chihuahua	2	B
		Durango	4	B

Patient's place of residence and travel data could be obtained from only 19 of the 56 patients (TABLE 3). Using that information a map of physical localization of each isolate was elaborated (FIG. 1). Two of those 19 isolates were identified as *C. immitis*, both of them ocurring in patients whose place of residence was the city of Tijuana, Baja California (Mexican northwest region, A in the map). These two patients declared they had traveled to the United States, but their activity was never outside California (southwest U.S. region, C in the map). They also declared they had never traveled to any other region inside México. Consequently both patients' trips were exclusively limited to regions A and C (FIG. 1). The remaining 17 patients came from the cities of Delicias, Chihuahua, Gómez-Palacios, Durango, and Torreón, Coahuila (north-central Mexican region, B in the map), the isolates corresponding to those 17 patients were identified as *C. posadasii*. In this group some patients had traveled near their place of residence and/or to some region in Texas or New Mexico (central-south U.S. region, D in the map). According to the information obtained from these 17 patients, travels in eastern Mexico or the United States were not mentioned, so their migration was limited to areas B and D (FIG. 1).

In the remaining 37 patients, with isolates identified as *C. posadasii*, we only knew the place where they were diagnosed: seven in Distrito Federal (Mexico City), seven in Coahuila, and 23 in Nuevo León.

DISCUSSION

The template that included proline 157, proline 174 and glucose-synthase allowed the molecular identification of both species in 56 isolates from Mexican patients. The inconsistent results observed with the hexokinase 149 probe could be attributed to the presence of more than one polymorphism in the design assay; in this regard, we suspect a low efficiency in the amplification on account of a poor correlation with the design sequence/oligonucleotide and the specificity of the TaqMan probes in that SNP, implying a second mutation

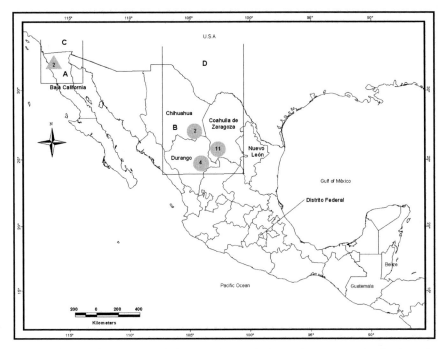

FIGURE 1. Distribution of *Coccidioides* species in Mexico. *C. posadasii* is shown in grayed circle and C. immitis is shown in grayed triangle. The number inside figure means the number of isolates of each species. Regions: A = Mexican northwest; B = Mexican north-central; C = USA southwest and D = USA south-central.

or a different nucleotide not reported previously. In consequence, we expect to further continue this study via DNA sequencing for this particular gene region.

Mexican patients were infected by both species of *Coccidioides*, predominantly *C. posadasii* in the north-central Mexican region (Chihuahua, Coahuila and Durango, region B in the map); also, some patients were infected by *C. immitis* in Baja California (northwest region, A in the map). This is similar to the distribution of the two species reported in the United States. This study appears to corroborate a previous literature report[5] stating that the predominant species of *Coccidioides* in Mexico is *C. posadasii*.

The hypothesis explaining the overlap of species due to the migration of human beings could not be confirmed in our study since migration of the 19 patients with travel history was only between homologue zones (A–C only; or B–D only).

The designed map shows only an approach of the *Coccidioides* species distribution in Mexico. Insufficient residence and migratory data in 37 patients diagnosed for *C. posadasii* did not allow us to establish a precise place of residence.

An hypothesis explains that the species overlap is due to the eolic dispersion of *Coccidioides* arthroconidia from one species-delimitated region to another. This hypothesis is unlikely in Mexico since the northwest (A in the map) and north-central regions (B in the map) are separated by mountains (Sierra Madre Occidental): this orographic situation blocks the dispersion of the winds coming from the west and northwest heading east. Also, according to meteorological studies, those regions near both sides of the Sierra Madre Occidental have low velocity winds (4–6 m/sec according to Beaufort scale).[12] Then the radius of propagation of *C. immitis* from northwest (A) to north-central (B) Mexican territory is very reduced. Winds blowing from east to west are insignificant, so the spread of *C. posadasii* conidia from north-central (B) to northwest (A) regions does not have solid meteorological support. We do not know whether meteorological conditions in the United States are similar to those in Mexico. Given these conditions, our explanation about the overlap of species of *Coccidioides* in human migration is stronger than that of wind dispersion, at least in our country.

The results obtained in this study are not enough to determine whether both species of *Coccidioides* are endemic in Mexico. According to Fiesse[11] and Baker *et al.*,[6] it will be necessary to get isolates of this fungus from soil and animals in those Mexican regions in which coccidioidomycosis is considered traditionally "endemic"; in this way we could accurately track the distribution of both *Coccidioides* species in Mexico.

ACKNOWLEDGMENTS

We thank Cudberto Contreras (Instituto Nacional de Referencia Epidemiológica, SSA) for the kind donation of *Coccidioides immitis* reference type-strains; Verónica Angeles (Instituto Nacional de Neurología y Neurocirugía, SSA) and Rafael Laniado-Laborín (Hospital General de Tijuana, ISESALUD) for contributing the isolates of *Coccidioides* spp.; Roberto Jiménez (Instituto de Geografía, UNAM) for his help in the map design; and Héctor Castañón for the English revision. We acknowledge DGAPA, UNAM for financial support of this investigation (Grant PAPIIT IN-209105-2).

REFERENCES

1. ZIMMERMANN, C.R., C.J. SNEDKER & D. PAPPAGIANIS. 1994. Characterization of *Coccidioides immitis* isolates by restriction fragment length polymorphisms. J. Clin. Microbiol. **32:** 3040–3042.
2. KOUFOPANOU, V., A. BURT & J.W. TAYLOR. 1997. Concordance of gene genealogies reveals reproductive isolation in the pathogenic fungus *Coccidioides immitis*. Proc. Natl. Acad. Sci. USA **94:** 5478–5482.

3. FISHER, M.C., T.J. WHITE & J.W. TAYLOR. 1999. Primers for genotyping single nucleotide polymorphism and microsatellites in the pathogenic fungus *Coccidioides immitis*. Mol. Ecol. **8**: 1082–1084.
4. FISHER, M.C., G.L. KOENIG, T.J. WHITE, *et al.* 2001. Biogeographic range expansion into South America by *Coccidioides immitis* mirrors New World patterns of human migration. Proc. Natl. Acad. Sci. USA **98**: 4558–4562.
5. FISHER, M.C., G.L. KOENIG, T.J. WHITE, *et al.* 2002. Molecular and phenotypic description of *Coccidioides posadasii* sp. nov., previously recognized as the non-California population of *Coccidioides immitis*. Mycologia **94**: 73–84.
6. BARKER, B., S. KROKEN & M. ORBACH. 2006. Microsatellite analysis of environmental isolates of the human pathogenic fungus *Coccidioides* validates the genetic and geographical isolation among populations and species. Presented at the Sixth International Symposium on Coccidioidomycosis. Stanford, CA, August 23–26.
7. CASTAÑÓN OLIVARES, L.R., A. AROCH CALDERÓN, E. BAZÁN MORA, *et al.* 2004. Coccidioidomicosis y su escaso conocimiento en nuestro país. Rev. Fac. Med. U. N. A. M. **47**: 145–148.
8. MUÑOZ, B., L.R. CASTAÑÓN, I. CALDERÓN, *et al.* 2004. Parasitic mycelial forms of *Coccidioides* species in Mexican patients. J. Clin. Microbiol. **42**: 1247–1249.
9. BIALEK, R., J. KERN, T. HERRMANN, *et al.* 2004. PCR assays for identification of *Coccidioides posadasii* based on the nucleotide sequence of the antigen 2/proline-rich antigen. J. Clin. Microbiol. **42**: 778–783.
10. Biosafety in Microbiological and Biomedical Laboratories. 1999. U.S. Department of Health and Human Services Centers for Disease Control and Prevention and National Institutes of Health, 4th edition. U.S. Government Printing Office. Washington, DC.
11. FIESE, M.J. 1958. Coccidioidomycosis. Charles C Thomas. Springfield, IL.
12. Atlas Nacional de México. 1989. Tomo II. Sección IV.4.2 Vientos dominantes durante el año. Instituto de Geografía. Universidad Nacional Autónoma de México. México, D. F.

Coccidioidomycosis in Persons Infected with HIV-1

NEIL M. AMPEL

Medical Service (1–111), SAVAHCS, Tucson, Arizona 85723

ABSTRACT: Coccidioidomycosis is a recognized opportunistic infection in those with HIV-1 infection. The major risk factor is immunodeficiency, particularly when the peripheral blood CD4 T lymphocyte count is below 250/μL. There are many manifestations of coccidioidomycosis during HIV-1 infection, including diffuse, reticulonodular pneumonia, focal primary pneumonia, and disease disseminated beyond the thoracic cavity. Diagnosis is based on serology, culture and histopathologic identification. Two therapeutic modalities are currently available, the polyene antifungal amphotericin B and the triazole antifungals. Of the latter, the most experience is with the triazoles fluconazole and itraconazole. There are increasing data regarding drug interactions between triazoles and antiretroviral agents. The duration of treatment of coccidioidomycosis in those with HIV-1 infection is not established and in many patients it is either prolonged or life-long. Adherence to antiretroviral therapy is important in preventing recurrence.

KEYWORDS: coccidioidomycosis; HIV-1; human

INTRODUCTION

While I have reviewed coccidioidomycosis during HIV-1 infection in the past,[1,2] the field continues to change. This paper represents a distillation of my experience managing patients with HIV-1 infection and coccidioidomycosis. As part of this review, a MEDLINE search of "coccidioidomycosis" with either "HIV" or "AIDS" in the title was performed; only 17 unique titles were displayed. Where pertinent, these have been cited in this review, along with others that I found to be clinically relevant.

In discussing the problem of HIV-1 infection and coccidioidomycosis, it is important to realize that there has been only one prospective study of any kind[3] and no studies comparing the clinical course of coccidioidomycosis in those with HIV-1 infection to other groups of patients. Moreover, because the presentation of coccidioidomycosis during HIV-1 infection varies depending

Address for correspondence: Neil M. Ampel, M.D., Medical Service (1–111), SAVAHCS, 3601 S. Sixth Avenue, Tucson, Arizona 85723, USA. Voice: 520-792-1450, ext. 6186; fax 520-629-4793.
nampel@email.arizona.edu

Ann. N.Y. Acad. Sci. 1111: 336–342 (2007). © 2007 New York Academy of Sciences.
doi: 10.1196/annals.1406.033

on the state of immunodeficiency, it is nearly impossible to compare groups. Therapeutic studies of coccidioidomycosis that have included patients with HIV-1, such as the study comparing fluconazole to itraconazole,[4] have not detected a difference in response.

EPIDEMIOLOGY

While symptomatic coccidioidomycosis may occur in the immunocompetent host, patients with deficiencies of cellular immunity are at increased risk for severe and disseminated disease.[5] Therefore, it is no surprise that coccidioidomycosis would become associated with the epidemic of HIV-1 infection that began in the early 1980s in the United States. The first cases of coccidioidomycosis associated with HIV infection were sporadic and occurred outside the coccidioidal endemic region.[6] These initial cases and some reported later[7] clearly represented reactivation of previously acquired infection associated with the severe immunodeficiency that occurs with advanced HIV-1 infection.

Later in the 1980s, most cases of symptomatic coccidioidomycosis were reported from the coccidioidal endemic regions, particularly Arizona,[8] leading to two large case series.[9,10] Since then, that trend has held, with most cases of coccidioidomycosis being reported from either Arizona or California.[11] During the 1980s, we were able to execute a prospective study of HIV-1 infected patients attending a single clinic in southern Arizona.[3] The results of this study demonstrated that the risk of symptomatic coccidioidomycosis was high. Nearly 25% of the cohort had developed some form of symptomatic coccidioidomycosis after about three and one-half years of follow-up. Moreover, symptomatic disease was associated with immunodeficiency, particularly with a peripheral blood CD4 T lymphocyte count below 250/μL. Surprisingly, measures of prior coccidioidal infection, such as a history of coccidioidomycosis, living for a longer period of time in the endemic area, or a history of a positive coccidioidal skin test, were not associated with the development of symptomatic disease. These data suggest that, within the endemic region, most cases of coccidioidomycosis in association with HIV-1 infection are due to acute infection, while those manifesting outside the endemic area are due to reactivation.

Beginning around 1996, a major change in the management of HIV-1 infection occurred. Initially termed highly active antiretroviral therapy or HAART, the use of at least three active antiretroviral medications resulted in marked decreases in plasma HIV RNA for many patients. Suppression of HIV-1 replication was associated with increases in peripheral blood CD4 T cell counts, a drop in the number of opportunistic infections[12] and concomitant decrease in mortality.[13] While my clinical experience suggests that there has been a marked decrease in the number of cases of symptomatic coccidioidomycosis in those with HIV-1 since the advent of potent antiretroviral therapy, few data exist. Woods and colleagues were able to demonstrate a decrease in the number of cases of coccidioidomycosis in those with HIV-1 infection in Arizona from 1995 to 1997.[14]

Recently, Carmichael has reported[15] that his large HIV-1 clinic in Tucson, AZ is experiencing a resurgence of coccidioidomycosis with resultant increases in mortality. Review of these cases suggests two factors. First, many of the patients with recrudescent coccidioidomycosis were not adherent to their potent antiretroviral therapy and again became immunodeficient. Second, the most severe cases were in those with coccidioidal meningitis who had also stopped antifungal therapy. These observations suggest that we should not be complacent in the era of potent antiretroviral therapy about the risk of coccidioidomycosis in the HIV-1 infected patient and we should be particularly vigilant with regard to patients who are medication-nonadherent.

CLINICAL PRESENTATION

Coccidioidomycosis in those with concomitant HIV-1 infection seen during the early years of the epidemic was frequently fatal and was associated with overwhelming pulmonary disease. Typically, chest radiographs were described as showing diffuse and "reticulonodular" disease, with a combination of nodules and interstitial infiltrates. In some cases, the pattern of pulmonary infiltration could be very fine and resemble pneumocystosis and, in some instances, both coccidioidomycosis and pneumocystosis simultaneously occurred.

Over time, a wider spectrum of illness was appreciated.[9] Patients with relatively higher peripheral blood CD4 T cell counts often presented with focal pulmonary infiltrates similar to those seen in immunocompetent patients with primary pulmonary coccidioidomycosis. Extrathoracic lymphadenopathy and liver disease were also reported. Meningitis was frequently observed but, for reasons that are unclear, bone and joint disease was less common. One unusual manifestation was the finding of a positive coccidioidal serologic test without a clinical site of disease. Follow-up of such patients in the pre-HAART era indicated that many such patients developed overt symptomatic coccidioidomycosis over time without intervention.[16]

DIAGNOSIS

The diagnosis of coccidioidomycosis depends on either obtaining a positive coccidioidal serologic test, directly observing the fungus in clinical specimens, or growing the fungus from a clinical site. Of these methods, serologic testing is the most widely used. There are reports suggesting that serologic tests are less likely to be positive in patients with immunodeficiency,[17] including those due to HIV-1 infection.[18] On the other hand, Pappagianis has not observed this loss of sensitivity, possibly because his laboratory concentrates serum specimens prior to testing. He notes a 5–7% rate of seronegativity among patients with AIDS when tests are performed in his laboratory compared to the 15–20% he has noted in the literature.[19] Once an IgG anticoccidioidal response is measurable in a patient with HIV-1 infection, following the titer of that response is useful for judging clinical response to therapy.

THERAPY

In general, the treatment of coccidioidomycosis during HIV infection is no different from that of other patients.[20] However, patients with immunodeficiency due to HIV-1 infection tend to have more severe disease than noncompromised hosts. Moreover, there are significant potential interactions between many antifungal agents and drugs used to treat HIV-1.

Two classes of drugs are currently used for therapy of coccidioidomycosis. The polyene antibiotics comprise formulations of amphotericin B. While their use has declined since the development of other antifungals, amphotericin B formulations are still recommended for the treatment of overwhelming coccidioidal infection. In the patient with HIV-1 infection, their use should be strongly considered as initial therapy in patients with diffuse pulmonary disease and in those with clinically severe, disseminated disease. The other class of agents is the azole antifungal drugs. These drugs are currently the mainstays of treatment of coccidioidomycosis because of their relative effectiveness and because of their oral bioavailability. Most experience has been with two triazole antifungals, fluconazole and itraconazole, but recently two newer agents, voriconazole and posaconazole, have become available in the United States. For these latter drugs, evidence for effectiveness in coccidioidomycosis is either based on case reports or on relatively small controlled studies.

One controversial issue with regard to therapy of coccidioidomycosis in patients with HIV-1 infection is whether therapy should be continued indefinitely or whether it can be safely stopped if the patient develops immune reconstitution in association with potent antiretroviral therapy. There are no clinical studies in this area. Certainly, any patient with coccidioidal meningitis should be treated life-long regardless of HIV-1 infection status.[21] Similarly, patients with disseminated coccidioidomycosis should be treated for prolonged periods. Given the data that patients with peripheral blood CD4 cell counts $<250/\mu L$ are at increased risk for symptomatic disease[3] and lose their ability to react to coccidioidal antigens,[22] it seems prudent to recommend continued therapy for such patients regardless of type of coccidioidomycosis. The most pressing concern is in regard to patients with primary pulmonary coccidioidomycosis and peripheral blood CD4 cell counts $\geq250/\mu L$. In my opinion, if a patient demonstrates adherence to antiretroviral therapy and has had resolution of clinically active coccidioidomycosis, stopping antifungal therapy is a reasonable option.

DRUG INTERACTIONS

Use of azole antifungals in the patient with HIV-1 infection may be complicated by interactions with antiretroviral therapy. Almost all HIV-1 protease inhibitors are both metabolized by and inhibit the cytochrome P-450 enzyme system, particularly 3A4.[23] The same is true of triazole antifungals. Unfortunately, there are relatively few published studies that have examined the interaction of these two drug classes. No clinically significant effect was found when fluconazole was combined either with indinavir[24] or with ritonavir.[25] Similarly, no interaction was noted between fluconazole and either ritonavir or saquinavir.[26] Commentuyn and colleagues explored the interaction of the HIV-1 protease inhibitor combination lopinavir/ritonavir with itraconazole in a patient with HIV-1 infection and disseminated histoplasmosis.[27] They found that lopinavir/ritonavir increased the plasma concentration of itraconazole while reducing that of its major active metabolite, hydroxyitraconazole. On the other hand, plasma concentration of lopinavir or ritonavir appeared to be comparable to those seen in patients not receiving itraconazole.

The use of ritonavir with voriconazole is considered contraindicated because of the lowering of plasma levels of voriconazole (VFEND product information). However, Mikus et al. noted that ritonavir prolongs the clearance of voriconazole.[28] Concomitant use of voriconazole and efavirenz, a nonnucleoside reverse transcriptase inhibitor that increases CYP 3A4 activity,[29] has also been recommended as contraindicated (VFEND® product information) because of the lowering of plasma levels of voriconazole and an increase in efavirenz levels.[30] Few data are available with regard to posaconazole, but no major interactions have been reported.

How should these data be interpreted? First, fluconazole and itraconazole are probably safe to use with combination antiretroviral therapy. Currently, voriconazole should be avoided in patients receiving either HIV-1 protease inhibitors or nonnucleoside reverse transcriptase inhibitors until more data are available. While posaconazole is probably safe, more information is needed before it can be recommended for use in HIV-1 infected patients with active coccidioidomycosis who are also receiving antiretroviral therapy.

REFERENCES

1. AMPEL, N.M. 2001. Coccidioidomycosis among persons with human immunodeficiency virus infection in the era of highly active antiretroviral therapy (HAART). Semin. Respir. Infect. **16:** 257–262.

2. AMPEL, N.M. 2005. Coccidioidomycosis in persons infected with HIV type 1. Clin. Infect. Dis. **41:** 1174–1178.

3. AMPEL, N.M., C.L. DOLS & J.N. GALGIANI. 1993. Coccidioidomycosis during human immunodeficiency virus infection: results of a prospective study in a coccidioidal endemic area. Am. J. Med. **94:** 235–240.

4. GALGIANI, J.N., A. CATANZARO, G.A. CLOUD, *et al.* 2000. Comparison of oral fluconazole and itraconazole for progressive, nonmeningeal coccidioidomycosis. A randomized, double-blind trial. Mycoses Study Group. Ann. Intern. Med. **133:** 676–686.

5. DRUTZ, D.J. & CATANZARO A. 1978. Coccidioidomycosis. Part I. Am. Rev. Respir. Dis. **117:** 559–585.

6. ABRAMS, D.I., M. ROBIA, W. BLUMENFELD, *et al.* 1984. Disseminated coccidioidomycosis in AIDS. N. Engl. J. Med. **310:** 986–987.

7. HERNANDEZ, J.L., S. ECHEVARRIA, A. GARCIA-VALTUILLE, *et al.* 1997. Atypical coccidioidomycosis in an AIDS patient successfully treated with fluconazole. Eur. J. Clin. Microbiol. Infect. Dis. **16:** 592–594.

8. BRONNIMANN, D.A., R.D. ADAM, J.N. GALGIANI, *et al.* 1987. Coccidioidomycosis in the acquired immunodeficiency syndrome. Ann. Intern. Med. **106:** 372–379.

9. FISH, D.G., N.M. AMPEL, J.N. GALGIANI, *et al.* 1990. Coccidioidomycosis during human immunodeficiency virus infection. A review of 77 patients. Medicine (Baltimore) **69:** 384–391.

10. SINGH, V.R., D.K. SMITH, J. LAWERENCE, *et al.* 1996. Coccidioidomycosis in patients infected with human immunodeficiency virus: review of 91 cases at a single institution. Clin. Infect. Dis. **23:** 563–568.

11. JONES, J.L., P.L. FLEMING, C.A. CIESIELSKI, *et al.* 1995. Coccidioidomycosis among persons with AIDS in the United States. J. Infect. Dis. **171:** 961–966.

12. KAPLAN, J.E., D. HANSON, DWORKIN M.S. *et al.* 2000. Epidemiology of human immunodeficiency virus-associated opportunistic infections in the United States in the era of highly active antiretroviral therapy. Clin. Infect. Dis. **30** (Suppl. 1): S5–S14.

13. PALELLA, F.J., JR., K.M. DELANEY, A.C. MOORMAN, *et al.* 1998. Declining morbidity and mortality among patients with advanced human immunodeficiency virus infection. HIV Outpatient Study Investigators. N. Engl. J. Med. **338:** 853–860.

14. WOODS, C.W., C. MCRILL, B.D. PLIKAYTIS, *et al.* 2000. Coccidioidomycosis in human immunodeficiency virus-infected persons in Arizona, 1994–1997: incidence, risk factors, and prevention. J. Infect. Dis. **181:** 1428–1434.

15. CARMICHAEL, J.K. 2006. Coccidioidomycosis in HIV-infected persons. Clin. Infect. Dis. **42:** 1059; author reply -60.

16. ARGUINCHONA, H.L., N.M. AMPEL, C.L. DOLS, *et al.* 1995. Persistent coccidioidal seropositivity without clinical evidence of active coccidioidomycosis in patients infected with human immunodeficiency virus. Clin. Infect. Dis. **20:** 1281–1285.

17. BLAIR, J.E., B. COAKLEY, A.C. SANTELLI, *et al.* 2006. Serologic testing for symptomatic coccidioidomycosis in immunocompetent and immunosuppressed hosts. Mycopathologia. **162:** 317–324.

18. ANTONISKIS, D., R.A. LARSEN, B. AKIL, *et al.* 1990. Seronegative disseminated coccidioidomycosis in patients with HIV infection. AIDS **4:** 691–693.

19. PAPPAGIANIS, D. 2001. Serologic studies in coccidioidomycosis. Semin. Respir. Infect. **16:** 242–250.

20. GALGIANI, J.N., N.M. AMPEL, J.E. BLAIR, *et al.* 2005. Coccidioidomycosis. Clin. Infect. Dis. **41:** 1217–1223.

21. DEWSNUP, D.H., J.N. GALGIANI, J.R. GRAYBILL, *et al.* 1996. Is it ever safe to stop azole therapy for *Coccidioides immitis* meningitis? Ann. Intern. Med. **124:** 305–310.

22. AMPEL, N.M. 1999. Delayed-type hypersensitivity, *in vitro* T-cell responsiveness and risk of active coccidioidomycosis among HIV-infected patients living in the coccidioidal endemic area. Med. Mycol. **37:** 245–250.

23. WINSTON, A. & M. BOFFITO. 2005. The management of HIV-1 protease inhibitor pharmacokinetic interactions. J Antimicrob Chemother. **56:** 1–5.

24. DE WIT, S., M. DEBIER, M. DE SMET, *et al.* 1998. Effect of fluconazole on indinavir pharmacokinetics in human immunodeficiency virus-infected patients. Antimicrob. Agents Chemother. **42:** 223–227.

25. CATO, A., 3RD, G. CAO, A. HSU, *et al.* 1997. Evaluation of the effect of fluconazole on the pharmacokinetics of ritonavir. Drug Metab. Dispos. **25:** 1104–1106.

26. KOKS, C.H., K.M. CROMMENTUYN, R.M. HOETELMANS, *et al.* 2001. The effect of fluconazole on ritonavir and saquinavir pharmacokinetics in HIV-1-infected individuals. Br. J. Clin. Pharmacol. **51:** 631–635.

27. CROMMENTUYN, K.M., J.W. MULDER, R.W. SPARIDANS, *et al.* 2004. Drug–drug interaction between itraconazole and the antiretroviral drug lopinavir/ritonavir in an HIV-1-infected patient with disseminated histoplasmosis. Clin. Infect. Dis. **38:** e73–e75.

28. MIKUS, G., V. SCHOWEL, M. DRZEWINSKA, *et al.* 2006. Potent cytochrome P450 2C19 genotype-related interaction between voriconazole and the cytochrome P450 3A4 inhibitor ritonavir. Clin. Pharmacol. Ther. **80:** 126–135.

29. MA, Q., O.O. OKUSANYA, P.F. SMITH, *et al.* 2005. Pharmacokinetic drug interactions with non-nucleoside reverse transcriptase inhibitors. Expert Opin. Drug Metab. Toxicol. **1:** 473–485.

30. THEURETZBACHER, U., F. IHLE, & H. DERENDORF. 2006. Pharmacokinetic/pharmacodynamic profile of voriconazole. Clin. Pharmacokinet. **45:** 649–663.

Coccidioidomycosis in Rheumatology Patients

Incidence and Potential Risk Factors

LESTER E. MERTZ[a] AND JANIS E. BLAIR[b]

[a]Division of Rheumatology, Mayo Clinic, Scottsdale, Arizona, USA

[b]Division of Infectious Diseases, Mayo Clinic, Scottsdale, Arizona, USA

ABSTRACT: Coccidioidomycosis is a potentially serious fungal infection contracted in endemic areas of the desert southwestern United States. Limited information exists about its incidence and clinical course in patients with rheumatic diseases, who may be at higher risk of symptomatic or disseminated coccidioidomycosis because of either the rheumatic disease itself or its treatment. We analyzed the incidence and risk factors for symptomatic and complicated coccidioidomycosis in our academic rheumatology practice in central Arizona. Between January 1, 2000, and June 30, 2006, coccidioidomycosis was diagnosed in 1.9% of the overall practice and in 3.1–3.6% of patients with rheumatoid arthritis (RA). The annual incidence was 1% in patients recently diagnosed with RA and 2% among patients with recently initiated infliximab treatment. Coccidioidomycosis was identified only in patients with inflammatory rheumatic diseases and extrathoracic dissemination occurred only to joints in two patients. Corticosteroids, immunosuppressive medications, and tumor necrosis factor inhibitors (TNFIs) appeared to be risk factors for symptomatic, but not disseminated coccidioidomycosis.

KEYWORDS: coccidioidomycosis; corticosteroids; disease-modifying antirheumatic drug; immunosuppression; incidence; rheumatoid arthritis; tumor necrosis factor inhibitors

INTRODUCTION

An increased risk of infection is associated with the use of disease-modifying antirheumatic drugs (DMARDs), immunosuppressive drugs (ISDs), and corticosteroids in the treatment of a number of different inflammatory rheumatic diseases, including rheumatoid arthritis (RA). Medications such as hydroxychloroquine and sulfasalazine are associated with little increased risk of infection, whereas corticosteroids, methotrexate, azathioprine, leflunomide,

Address for correspondence: Janis E. Blair, M.D., Division of Infectious Diseases, Mayo Clinic, 13400 East Shea Boulevard, Scottsdale, AZ 85259.

Ann. N.Y. Acad. Sci. 1111: 343–357 (2007). © 2007 New York Academy of Sciences.
doi: 10.1196/annals.1406.027

and the more recently introduced tumor necrosis factor inhibitor (TNFI) medications infliximab, etanercept, and adalimumab are associated with a higher risk. In particular, the substantial risk of reactivation of latent tuberculosis associated with the initiation of TNFI therapy[1] prompted the U.S. Food and Drug Administration (FDA) to issue pretreatment tuberculosis screening and treatment guidelines.[2]

Confounding the analysis of infectious disease risk attributable to a medication is the increased infectious disease risk related to the rheumatic disease itself. In a study of patients with rheumatic disease treated before the common use of TNFIs, the adjusted hazard ratios of 1.70, 1.83, and 1.45 were calculated for objectively confirmed infections, infections requiring hospitalization, and any documented infections, respectively, in RA patients compared with those of controls.[3] Predictors of infection included increasing age, the presence of extraarticular manifestations of RA, leukopenia, the use of corticosteroids, and the comorbid conditions of chronic lung disease, alcoholism, organic brain disease, and diabetes mellitus.[4]

The common practice of treating rheumatic diseases with several DMARDs or ISDs in close sequence or in combination simultaneously further complicates attempts to attribute the increased risk of an infectious disease to a particular medication. In particular, FDA guidelines require the concomitant administration of methotrexate with infliximab to reduce infusion reactions and improve efficacy.[2,5]

Coccidioidomycosis is a fungal infection caused by either of the two closely related dimorphic fungi *Coccidioides immitis* and *Coccidioides posadasii;* it is contracted by inhalation of airborne arthrospores in endemic areas.[6] An initial pneumonitis most often resolves spontaneously with rare dissemination to additional anatomical sites (e.g., the central nervous system, bone, joints, and skin). Case identification is difficult because as many as two-thirds of patients are minimally symptomatic and not brought to medical attention. Individual patients residing within an endemic area may be at risk for new infection, or they may have immunity from an undiagnosed prior infection or be at risk for reactivation when immunosuppressive therapy is administered.[7]

Infectious complications were evident in the early clinical experience with TNFI therapy, including cases of coccidioidomycosis in patients recruited from endemic areas. Two cases of coccidioidomycosis, one fatal, were identified in infliximab-treated RA patients from Arizona.[2] Symptomatic coccidioidomycosis occurring among residents of Arizona, California, and Nevada with inflammatory arthritis were recently reported.[8,9] A cumulative incidence of 1% during the 3-year study period was calculated. The use of infliximab increased the risk of symptomatic but not disseminated coccidioidomycosis.

Because the number of reported cases of coccidioidomycosis in patients with rheumatic disease is small and the effects of rheumatic disease and its treatment on the clinical course and treatment of coccidioidomycosis in these patients have yet to be defined, we performed a retrospective analysis of coccidioidomycosis in rheumatic patients treated in our academic medical practice. This article is a synopsis of this report.

PATIENTS AND METHODS

We retrospectively reviewed the medical records of patients with both coccidioidomycosis and rheumatic disease treated between January 1, 2000, and June 30, 2006. Patients were identified by using a combination of two databases.

A coccidioidomycosis database included all patients from our institution who were reported to the Arizona Department of Health Services as required by law for serologic, microbiologic, or histopathologic evidence of coccidioidomycosis. This database has been maintained since October 1998 and includes asymptomatic patients unexpectedly discovered to have positive coccidioidal serologic findings on screening studies performed before initiation of TNFI therapy or before organ transplantation and postoperative immunosuppressive therapy.

The second database is a rheumatic disease survey database comprising two different groups of patients treated in our rheumatology division. The first group, treated with infliximab infusion, is composed of all current and former patients who received such infusions at Mayo Clinic, Scottsdale, Arizona, during the study interval. The second group, a current patient group, is composed of all patients with rheumatic diseases who were actively receiving treatment between January 1, 2006, and June 30, 2006, and who had been treated in the practice for variable lengths of time. As noted in TABLE 1, all 854 patients were in the current rheumatology practice except for the 56 former patients with RA, all of whom had previously received infliximab. Two persons were included as current patients who were last seen on December 8, 2005, and December 21, 2005, respectively, rather than during the study interval because of temporary treatment discontinuation after development of active coccidioidomycosis.

These two groups were combined into a single rheumatic disease survey database for data collection purposes and analyzed similarly. Each patient is represented only once in the rheumatic disease database. The major clinical rheumatic diagnosis dictated in the most recent note became the patient's rheumatic disease diagnosis for study purposes. Relevant data about each patient that were retrospectively collected from the electronic medical record and tabulated included demographic information, rheumatic disease diagnosis and treatment history, outpatient and inpatient coccidioidomycosis diagnosis and treatment details, and pertinent laboratory and radiographic information.

During the study interval, a rheumatic disease patient in whom coccidioidomycosis developed was treated by discontinuing the DMARD or ISD,

TABLE 1. Occurrence of coccidioidomycosis by diagnosis in rheumatic disease survey database[a]

Rheumatic disease	No. of patients $(N = 854)$	Patients with CM	
		No. $(N = 16)^b$	%
Rheumatoid arthritis			
Current patients	287	9	3.1
Former patients	56	2	3.6
Unspecified inflammatory arthritis	91	1	1.1
Polymyalgia rheumatica or temporal arteritis	81	0	0
Scleroderma, mixed connective tissue disease	54	0	0
Psoriatic arthritis	52	1	1.9
Systemic lupus erythematosus	47	1	2.1
Vasculitis	39	0	0
Osteoarthritis, tendonitis, bursitis, fibromyalgia	36	0	0
Polymyositis, dermatomyositis	22	0	0
Gout, pseudogout	20	0	0
Undifferentiated collagen vascular disorders	15	0	0
Ankylosing spondylitis	14	2	14.3
Reactive arthritis	4	0	0
Miscellaneous inflammatory disorders	36	0	0

NOTE: CM, coccidioidomycosis.
[a] Between January 1, 2000, and June 30, 2006.
[b] 1.9% of all patients.

reducing prednisone to the lowest tolerable dose, and prescribing a prolonged course of an antifungal agent, generally an oral azole. If the rheumatic disease activity increased substantially, treatment was carefully reintroduced at the lowest possible dose while continuing the azole.

A study case was defined as a patient who had symptomatic or asymptomatic coccidioidomycosis diagnosed on the basis of serologic, microbiologic, or histopathologic evidence between January 1, 2000, and June 30, 2006, and who was also a current rheumatology division patient or the recipient of an infliximab infusion between January 1, 2000, and June 30, 2006.

This study was approved by the Mayo Clinic Institutional Review Board.

Statistical Methods

Analysis of the effect of prednisone or TNFI therapy on laboratory tests and disease parameters was done with the Fisher exact test. Incidence values for coccidioidomycosis as a function of years after RA diagnosis and incidence values for coccidioidomycosis as a function of years after initiation of infliximab therapy were calculated by the Kaplan–Meier method with point confidence intervals calculated by using the log-minus-log transformation.

RESULTS

Rheumatic Disease Survey Database

A total of 854 rheumatology patients was included in the rheumatic disease survey database. Of these, 798 were patients currently in the practice and 56 were former RA patients who previously received infliximab infusions. These 854 patients represented a number of rheumatic diseases (TABLE 1), with follow-up ranging from 2 months to more than 10 years.

Coccidioidomycosis was identified in 16 (1.9%) of these 854 patients (TABLE 1). Coccidioidomycosis occurred in 1.1–3.6% of the specific rheumatic diseases, except for ankylosing spondylitis at 14.3%. Current and former RA patients had occurrence rates of 3.1% and 3.6%, respectively.

The demographic characteristics, comorbid conditions, and infectious disease characteristics of these 16 patients are summarized in TABLE 2. Most of the 16 patients were older, white, full-time Arizona residents with current follow-up. Of 6 patients with chronic lung disease, 2 had interstitial pulmonary fibrosis caused by the rheumatic disease itself. Twelve patients had pulmonary coccidioidomycosis, 2 had articular coccidioidomycosis (1 elbow, 1 wrist), but no patient had clinical evidence of both simultaneously. No other anatomical sites in any patient were affected by coccidioidomycosis. No *Coccidioides*-infected arthroplasties were identified. Two patients with no symptoms had positive *Coccidioides* serologic findings on screening studies before initiation of TNFI therapy.

TABLE 2. **Characteristics of 16 patients with rheumatic disease and coccidioidomycosis: demographic details, comorbid conditions, and details of coccidioidal infections**

Characteristic	No. of patients
Demographic detail	
Age ≥60 y at diagnosis	11
Male sex	8
White race	16
Current patient	14
Full-time Arizona resident	13
Comorbid condition	
Diabetes mellitus	2
Chronic lung disease[a]	6
Cigarette use at diagnosis	2
Coccidioidal infection detail	
Pulmonary only	12
Asymptomatic seroconversion	2
Dissemination to joints	2
Dissemination to other sites	0

NOTE: y = years.
[a]1 asthma, 2 interstitial pulmonary fibrosis, 1 bronchiectasis, 1 chronic obstructive pulmonary disease, and 1 emphysema.

Coccidioidomycosis was diagnosed during most months of the year (TABLE 3). Eleven of the 16 cases occurred during 2005 and the first 6 months of 2006, and 8 of 16 cases occurred during the 4 months from November through February.

The DMARDs and ISDs administered at coccidioidomycosis diagnosis are detailed in TABLE 4. Prednisone, methotrexate, and infliximab were the most commonly used immunosuppressive medications. The prednisone dose was generally 10 mg/day or less. Among patients treated with TNFIs, 6 patients were treated with infliximab, 1 with etanercept, and none with adalimumab.

Combinations of DMARDs and ISDs that were administered to each of the 16 patients who had coccidioidomycosis are listed in TABLE 5. Most patients were receiving more than one immunosuppressive agent simultaneously; in general, RA patients were taking more potent immunosuppressive agents than patients with other diagnoses. The two cases of coccidioidomycosis that disseminated to joints occurred in the wrist in one patient with systemic lupus erythematosus who was receiving prednisone and hydroxychloroquine and in the elbow in one RA patient receiving prednisone and methotrexate. No disseminated cases were identified in patients taking TNFIs.

TABLE 3. Annual and seasonal incidence of coccidioidomycosis in 16 patients with rheumatic disease

Time frame	Incidence
Annual	
2000	1
2001	0
2002	1
2003	2
2004	1
2005	6
2006[a]	5
Monthly	
January	3
February	2
March	1
April	2
May	1
June	1
July	0
August	2
September	1
October	0
November	1
December	2

[a]First 6 months only.

TABLE 4. Use of disease-modifying antirheumatic medications and immunosuppressive medications at diagnosis of coccidioidomycosis in 16 patients with rheumatic disease

Medication[a]	No. of patients[b]
Prednisone	11
Methotrexate	8
Infliximab	6
Azathioprine	3
Leflunomide	2
Hydroxychloroquine	2
Etanercept	1
Adalimumab	0

[a]Most of the 16 patients were receiving more than one disease-modifying antirheumatic medication or immunosuppressive medication simultaneously.
[b]Receiving medication at diagnosis of coccidioidomycosis.

The characteristics of the 16 coccidioidal infections, and their *Coccidioides*-specific diagnostic tests are summarized in TABLE 6; these are tabulated according to whether prednisone or TNFI therapy was administered at diagnosis. No statistically significant differences were found. However, prednisone-treated patients were less likely than patients not treated with prednisone to have positive serologic results and more likely to require hospitalization for an invasive procedure leading to a positive coccidioidomycosis culture. The two cases of dissemination both occurred in patients treated with prednisone compared with

TABLE 5. Individual combinations of disease-modifying antirheumatic medications and immunosuppressive medications at diagnosis of coccidioidomycosis in 16 patients with rheumatic disease

Medication combination	No. of patients (N = 16)	Disseminated disease (N = 2)[a]	Specific rheumatic diagnosis (N = 16)
None	1	0	RA
Prednisone alone	1	0	AS
Etanercept alone	1	0	PsA
Prednisone, hydroxychloroquine	2	1	UIA, SLE
Prednisone, methotrexate	3	1	RA, RA, RA
Methotrexate, azathioprine	1	0	RA
Methotrexate, infliximab	2	0	AS, RA
Prednisone, azathioprine, infliximab	2	0	RA, RA
Prednisone, leflunomide, infliximab	1	0	RA
Prednisone, methotrexate, infliximab	1	0	RA
Prednisone, leflunomide, methotrexate	1	0	RA

NOTE: AS = ankylosing spondylitis; IS = inflammatory spondyloarthropathy; PsA = psoriatic arthritis; RA = rheumatoid arthritis; SLE = systemic lupus erythematosus; UIA = unspecified inflammatory arthritis.
[a]One elbow and one wrist.

TABLE 6. Effects of prednisone and TNFI therapy at time of coccidioidomycosis diagnosis on diagnostic test results and infectious disease characteristics in 16 patients[a]

Variable	Prednisone		TNFI	
	Yes	No	Yes	No
Diagnostic test				
EIA IgM positive	3 of 11 (27)	3 of 5 (60)	4 of 7 (57)	2 of 9 (22)
EIA IgG positive	6 of 11 (55)	5 of 5 (100)	6 of 7 (86)	5 of 9 (56)
ID IgM positive	4 of 10 (40)	2 of 4 (50)	3 of 6 (50)	3 of 8 (38)
ID IgG positive	3 of 10 (30)	2 of 5 (40)	3 of 7 (43)	2 of 8 (25)
CF positive	3 of 9 (33)	3 of 5 (60)	4 of 7 (57)	2 of 7 (29)
Infectious disease characteristic				
Hospitalization required	4 of 11 (36)	0 of 5 (0)	2 of 7 (29)	2 of 9 (22)
Invasive diagnostic procedure performed	7 of 11 (64)	0 of 5 (0)	3 of 7 (43)	4 of 9 (44)
Positive CM culture	8 of 11 (73)	0 of 5 (0)	2 of 7 (29)	6 of 9 (67)
Disseminated CM	2 of 11 (18)	0 of 5 (0)	0 of 7 (0)	2 of 9 (22)
Antifungal treatment	7 of 10 (70)	3 of 4 (75)	5 of 6 (83)	5 of 8 (62)

NOTES: CF = complement fixation; CM = coccidioidomycosis; EIA = enzyme-linked immunosorbent assay; ID = immunodiffusion; TNFI = tumor necrosis factor inhibitor.
[a]Values are number of total (percentage).

no dissemination in patients treated with a TNFI. Patients taking a TNFI when diagnosed with coccidioidomycosis appeared somewhat more likely to have a positive coccidioidal serologic test than patients not taking a TNFI. There were no discernible differences in the manifestations of the infection among patients in these two groups.

Current RA Patients

Coccidioidomycosis occurring in the current RA patient subgroup was analyzed separately, because RA represented the largest, most homogeneous and immunosuppressed group of patients containing the largest number of cases of coccidioidomycosis. TABLE 7 summarizes their demographic and comorbid variables, as well as RA disease severity factors and medication therapies. These data suggest that coccidioidomycosis is more likely to occur in men and in patients with lung disease or diabetes. A history of arthroplasty, the presence of extraarticular features, and the need for TNFI therapy were more common in patients with coccidioidomycosis. Approximately three-fourths of this cohort, regardless of the diagnosis of coccidioidomycosis, had taken prednisone at some time during the study interval. The use at any time of azathioprine and a TNFI was somewhat more common in patients with coccidioidomycosis. None of these results was statistically significant, however.

TABLE 8 summarizes combinations of DMARDs and ISDs taken by current RA patients at the time of diagnosis of coccidioidomycosis. Seven patients

TABLE 7. Demographic details, comorbid conditions, rheumatoid arthritis severity factors, and select medications at diagnosis of coccidioidomycosis in 287 current rheumatoid arthritis patients[a]

Characteristic	Coccidioidomycosis	
	Yes	No
Demographic details		
Age ≥60 y in 2006	8 of 9 (88.9)	215 of 278 (77.3)
Female sex	4 of 9 (44.4)	200 of 278 (71.9)
White race[b]	9 of 9 (100.0)	227 of 236 (96.2)
Full-time Arizona resident	6 of 9 (66.7)	214 of 278 (77.0)
Comorbid condition		
Diabetes mellitus[b]	2 of 9 (22.2)	27 of 278 (9.7)
Chronic lung disease[c]	5 of 9 (55.6)	29 of 278 (10.4)
Cigarette use[c]	0 of 9 (0)	31 of 273 (11.4)
RA severity factor		
Rheumatoid factor positive	4 of 5 (80.0)	176 of 220 (80.0)
History of arthroplasty[c]	4 of 9 (44.4)	74 of 278 (26.6)
Extraarticular features[c]	4 of 9 (44.4)	87 of 278 (31.3)
Select medication		
Methotrexate[d]	7 of 9 (77.8)	217 of 278 (78.0)
Azathioprine[d]	3 of 9 (33.3)	22 of 278 (7.9)
Leflunomide[d]	4 of 9 (44.4)	91 of 278 (32.7)
Prednisone[e]	7 of 9 (77.8)	199 of 278 (71.6)
TNFI[e]	6 of 9 (66.7)	111 of 278 (39.9)

NOTES: RA = rheumatoid arthritis; TNFI = tumor necrosis factor inhibitor; y = year.
[a] Values are number of total (percentage).
[b] Race unknown for some patients.
[c] Present any time between January 1, 2000, and June 30, 2006.
[d] Medication used any time in past.
[e] Medication used any time between January 1, 2000, and June 30, 2006.

were receiving prednisone, and three were receiving the TNFI infliximab. The only case of disseminated coccidioidal disease in this group occurred in a patient taking prednisone and methotrexate; no disseminated coccidioidal disease occurred in patients taking infliximab.

Of the 79 current RA patients whose initial RA diagnosis was made during the study period at our institution, 4 contracted coccidioidomycosis during follow-up. The incidence of coccidioidomycosis was 1% within 1 year of RA diagnosis and 9% within 5 years (TABLE 9). These patients were receiving various medications when diagnosed with coccidioidomycosis. Of the 121 current and former RA patients with infliximab therapy initiated at our institution during the study period, 6 had contracted coccidioidomycosis at follow-up. The incidence was 2% within 1 year of initiation of inflixmab and 12% within 5 years. The confidence intervals for the incidence calculations of initially diagnosed RA patients and for the initiation of infliximab therapy, as delineated above, are too wide to allow meaningful comparisons of these two groups.

TABLE 8. Individual combinations of disease-modifying antirheumatic and immunosuppressive medications at diagnosis of coccidioidomycosis in 9 patients with current rheumatoid arthritis

DMARD or ISD combination	No. of patients	Disseminated disease
None	1	0
Prednisone, methotrexate	3	1
Methotrexate, azathioprine	1	0
Prednisone, azathioprine, infliximab	2	0
Prednisone, methotrexate, infliximab	1	0
Prednisone, leflunomide, methotrexate	1	0

NOTES: DMARD = disease-modifying antirheumatic drug; ISD = immunosuppressive drug.

TABLE 10 lists the median and mean number of office visits per year for current RA patients, whether they were diagnosed with coccidioidomycosis or not; it demonstrates that patients diagnosed with coccidioidomycosis had approximately two more office visits per year after January 1, 2000, until the time of their diagnosis with coccidioidomycosis. It is unclear whether these extra visits were related to an increased level of surveillance because of worsening RA manifestations, increased monitoring of new or stronger DMARDs and ISDs, or other factors not analyzed.

TABLE 9. Incidence rates of coccidioidomycosis at 1 year and 5 years after initial rheumatoid arthritis diagnosis or initiation of infliximab therapy

	Incidence[a]	
Time frame	1 y (95% CI)	5 y (95% CI)
After initial RA diagnosis	1 (0–9)	9 (3–24)
After initiation of infliximab	2 (0–7)	12 (5–29)

NOTES: CI = confidence interval; RA = rheumatoid arthritis; y = year.
[a]Values are percentage (95% confidence interval).

TABLE 10. Frequency of office visits for 287 current rheumatoid arthritis patients with or without coccidioidomycosis

No. of visits	CM diagnosis[a]	No CM diagnosis[b]
Median	6.75 visits/year	4.50 visits/year
Mean	7.25 visits/year	5.30 visits/year

NOTES: CM = coccidioidomycosis.
[a]Frequency of office visits from January 1, 2000, to coccidioidomycosis diagnosis.
[b]Frequency of office visits between January 1, 2000, and June 30, 2006.

DISCUSSION

We retrospectively analyzed symptomatic and complicated cases of coccidioidomycosis that occurred between January 1, 2000, and June 30, 2006, among all patients currently in the practice and 56 former RA patients who previously received infliximab infusions. Coccidioidomycosis was diagnosed exclusively in patients with inflammatory arthritis, especially in the large group of RA patients, but not in patients with a wide variety of other inflammatory rheumatic diseases. During the study interval, about 2% of all patients and about 3–4% of RA patients were diagnosed with coccidioidomycosis. Actual annual incidence rates could not be calculated from these data, because not all patients were followed continuously in our practice during the study interval and because case ascertainment was incomplete. Previous infections in our patients due to *Coccidioides* spp. could not be determined, because we do not document this historical detail and because coccidioidal skin testing is no longer available. Coccidioidomycosis was diagnosed in a greater percentage of patients with ankylosing spondylitis, but the small number of patients and cases makes it impossible to draw firm conclusions about any increased risk in this patient group.

The demographic characteristics of the 16 patients with coccidioidomycosis are typical of those of patients seen in our rheumatology practice in general: largely Medicare-eligible, white residents of Arizona. The frequent presence of chronic lung disease may have contributed to the contraction of coccidioidomycosis in these patients. Disseminated coccidioidomycosis was unusual and occurred only in native joints. Beginning TNFI therapy during active coccidioidomycosis was avoided in two patients by pretreatment serologic screening.

Coccidioidomycosis was diagnosed much more commonly during the 18-month interval between January 2005 and June 2006 (11 of the 16 cases), compared with the 60-month interval between January 2000 and December 2004. This increased frequency is consistent with an increase in cases reported to the Arizona Department of Health Services, with a 5-year median for June of 199 cases compared with 235 cases for June 2005 and 464 for June 2006.[10] An increasing incidence in the general population might explain the increasing incidence in our patients with rheumatic disease, which would account for higher incidence values in our study compared with those in previous studies. In addition, awareness in the medical community of the increase in coccidioidomycosis cases during the past 2 years may have led to more frequent serologic testing and, as a result, to more frequent diagnosis. The increased incidence of coccidioidomycosis in Arizona during the 4 months between November and February[11] is also evident in our study, with 8 of the 16 cases occurring during these months. This conformity to the annual and seasonal patterns in the healthy population suggests that many observed cases represent newly acquired rather than reactivated infection.

The effect of rheumatic disease, its medication therapy, and its comorbid conditions on the development of symptomatic and disseminated coccidioidomycosis can be assessed from our data. Coccidioidomycosis occurred exclusively in patients with inflammatory rheumatic disease who demonstrated predominantly arthritic manifestations, which suggests that inflammatory arthritis or its specific treatment is a risk factor. Specifically, none of the 81 patients with polymyalgia rheumatica or temporal arteritis syndrome and only 1 of the 162 total patients with scleroderma, mixed connective tissue disease, systemic lupus erythematosus, vasculitis, polymyositis, or dermatomyositis were diagnosed with coccidioidomycosis, although these patients are often systemically ill and commonly treated with prednisone or an ISD.

Prednisone is frequently used in treating inflammatory rheumatic diseases and may be only a common treatment rather than a common risk factor in the 11 total patients who were receiving low-dose prednisone at diagnosis. This conclusion is supported by analysis of the current RA patient subgroup, in which the proportion of patients who ever used prednisone during the study interval was nearly the same for those with or without a diagnosis of coccidioidomycosis. Similar comments can be made about methotrexate but too few patients continued to take leflunomide or azathioprine to allow definitive comparisons.

Seven of the 16 patients with coccidioidomycosis were being treated with a TNFI at diagnosis; in the current RA patient subgroup, a slightly higher proportion of patients with than without coccidioidomycosis had used a TNFI during the study. This higher proportion may suggest either an increased risk of symptomatic coccidioidomycosis due to the TNFI or to more severe rheumatic disease treated with a TNFI. Separating the effect of treatment from the severity of disease was not possible. Also notable is the fact that TNFI therapy is given exclusively for inflammatory arthritis, which may explain the increased incidence of symptomatic coccidioidomycosis observed in these patients compared with those with other inflammatory rheumatic diseases.

Disseminated disease occurred in only two patients, both of whom were receiving low-dose prednisone at diagnosis of coccidioidomycosis. No coccidioidal dissemination was found in any patient receiving a TNFI, which suggests that more severe disease is not necessarily the outcome of TNFI therapy, although rheumatic disease patients are commonly taking these agents when diagnosed. The data suggest that male sex, diabetes, and severity of disease, measured by the presence of extraarticular features or need for arthroplasty, may be risk factors for symptomatic coccidioidomycosis, but these results were not statistically significant.

The incidence rates of coccidioidomycosis 1 year after diagnosis of RA or initiation of infliximab therapy were 1% and 2%, respectively. Although the confidence intervals are wide, these values exceed the annual incidence rate for spontaneously reported cases of coccidioidomycosis in Arizona, which was 7 per 100,000 population in 1990 (12), 14.9 per 100,000 in 1995, 33 per

100,000 in 1998, and 43 per 100,000 (0.043%) in 2001,[11] compared with 40 per 100,000 in 1996 for persons older than 65 years of age.[13] However, the annual incidence figures derived from the current study compare favorably with the 2–4% annual incidence of coccidioidomycosis calculated in an endemic area of California that was based on skin test results in that population during 1995.[14]

The only previous study of coccidioidomycosis in patients with rheumatic disease[9] reported an overall 1% cumulative incidence (between January 2000 and February 2003) for inflammatory arthritis patients with values of 0.5% or 2.8% for those taking or not taking infliximab. As a whole, these incidence values of coccidioidomycosis in immunosuppressed patients are much closer to those based on the more comprehensive skin test survey than they are to the incidence values based on spontaneously reported cases. A reasonable explanation for this difference would be that surveillance bias favors the detection of coccidioidomycosis in immunosuppressed patients who are more highly monitored than the general population and may have more severe disease manifestations on account of immunosuppression. The greater frequency of office visits for the subgroup of current RA patients diagnosed with coccidioidomycosis in this study tends to support such a conclusion as well.

Limitations of this study include the lack of detailed information and long-term follow-up data on all patients in the rheumatology practice, which might have revealed additional cases of coccidioidomycosis and risk factors not identified by focusing attention on inflammatory rheumatic diseases, especially RA. Our patients are predominantly older and white, so we cannot comment on the occurrence of coccidioidomycosis in younger persons with diverse racial backgrounds. The small number of cases limited our ability to perform a valid statistical analysis of numerous variables, and thus we can only tentatively infer certain conclusions. We collected no information about a previous diagnosis of coccidioidomycosis, which makes it impossible for us to comment on the risk of infections caused by reactivation of a previous infection. The data were too limited to clearly separate the effects of the rheumatic disease itself from the effects of medication on the risk for symptomatic or complicated coccidioidomycosis. Our data and conclusions are not strictly comparable with those reported previously,[9] perhaps because the incidence of coccidioidomycosis has increased in the general population and the usual dose of infliximab prescribed by rheumatologists has decreased since then. Finally, we do not yet have long-term follow-up for coccidioidomycosis patients that would allow an accurate determination of the ultimate outcome of their infection and rheumatic disease.

CONCLUSION

In spite of the common use of DMARDs, ISDs, and corticosteroids in this group of patients, coccidioidomycosis was relatively uncommon and not much more prevalent than in the general population, with surveillance bias possibly explaining incidence discrepancies among different populations. No indisputable risk factors emerged from this study, although male sex, chronic lung disease, diabetes, and disease severity may be important factors in the manifestation of coccidioidomycosis in patients diagnosed with inflammatory arthritis. TNFI therapy was not shown to increase the risk of disseminated coccidioidomycosis. Given the low rate of serious coccidioidal complications observed in our rheumatology practice, we will continue our current practice of treating coccidioidomycosis, namely, to temporarily discontinue the DMARD or ISD, reduce the dose of prednisone, and prolong the course of treatment with an antifungal antibiotic. Longer follow-up will clarify the clinical course of coccidioidomycosis and whether life-long antifungal therapy will be needed in rheumatic disease patients who need to restart a DMARD or an ISD to treat worsening rheumatic disease.

ACKNOWLEDGMENTS

Editing, proofreading, and reference verification were provided by the Section of Scientific Publications, Mayo Clinic.

REFERENCES

1. KEANE, J., S. GERSHON, R.P. WISE, *et al.* 2001. Tuberculosis associated with infliximab, a tumor necrosis factor alpha-neutralizing agent. N. Engl. J. Med. **345:** 1098–1104.
2. PHYSICIANS DESK REFERENCE ONLINE. Remicade for IV injection (Centocor). Available from http://www.pdr.net/login/Login.aspx.
3. DORAN, M.F., C.S. CROWSON, G.R. POND, *et al.* 2002. Frequency of infection in patients with rheumatoid arthritis compared with controls: a population-based study. Arthritis Rheum. **46:** 2287–2293.
4. DORAN, M.F., C.S. CROWSON, G.R. POND, *et al.* 2002. Predictors of infection in rheumatoid arthritis. Arthritis Rheum. **46:** 2294–2300.
5. ST. CLAIR, E.W., D.M. VAN DER HEIJDE, J.S. SMOLEN, *et al.* 2004. Combination of infliximab and methotrexate therapy for early rheumatoid arthritis: a randomized, controlled trial. Arthritis Rheum. **50:** 3432–3443.
6. GALGIANI, J. 2005. Coccidioides species. *In*: G.L. Mandell, J.E. BENNETT, R. DOLIN, Eds. Mandell, Douglas, and Bennett's Principles and Practice of Infectious Diseases, 6th edition. :3040–3051. Elsevier/Churchill Livingstone. Philadelphia, PA.
7. LOGAN, J.L., J.E. BLAIR & J.N. GALGIANI. 2001. Coccidioidomycosis complicating solid organ transplantation. Semin. Respir. Infect. **16:** 251–256.
8. BERGSTROM, L., D.E. YOCUM, J. TESSER, *et al.* 2002. Coccidiomycosis (Valley Fever) occuring during infliximab therapy [abstract]. Arthritis Rheum. **46** Suppl. 9: S169.

9. BERGSTROM, L., D.E. YOCUM, N.M. AMPEL, *et al.* 2004. Increased risk of coccidioidomycosis in patients treated with tumor necrosis factor alpha antagonists. Arthritis Rheum. **50:** 1959–1966.

10. ARIZONA, DEPARTMENT OF HEALTH SERVICES, DIVISION OF PUBLIC HEALTH SERVICES [homepage on the Internet]. Monthly summary of selected reportable diseases, September 2006 [cited 2006 Aug]. Available from http://www.azdhs.gov/phs/oids/stats/pdf/monthly.pdf.

11. KOMATSU, K., V. VAZ, C. MCRILL, *et al.* 2003. Increase in coccidioidomycosis: Arizona, 1998–2001. MMWR Morb. Mortal Wkly. Rep. **52:** 109–112.

12. MOSLEY, D., K. KOMATSU, V. VAZ, *et al.* 1996. Coccidioidomycosis: Arizona, 1990–1995. MMWR Morb. Mortal Wkly. Rep. **45:** 1069–1073.

13. LEAKE, J.A., D.G. MOSLEY, B. ENGLAND, *et al.* 2000. Risk factors for acute symptomatic coccidioidomycosis among elderly persons in Arizona, 1996–1997. J. Infect. Dis. **181:** 1435–1440.

14. LARWOOD, T.R. 2000. Coccidioidin skin testing in Kern County, California: decrease in infection rate over 58 years. Clin. Infect. Dis. **30:** 612–613.

Coccidioidomycosis and Pregnancy

A Review

IRENE M. SPINELLO, ROYCE H. JOHNSON, AND SHEHLA BAQI

*David Geffen School of Medicine at UCLA, Kern Medical Center,
Bakersfield, California 93305, USA*

ABSTRACT: Coccidioidomycosis (CM) is a fungal infection endemic to
the southwestern United States, northwestern Mexico, and parts of
Central and South America. CM has been recognized as a complicating
factor in pregnancy since at least the 1940s, and seems to be a relatively
uncommon infection during pregnancy. The disease presentation during pregnancy includes a wide clinical spectrum that ranges from mild
influenza-like illness and pneumonia, especially in the first two trimesters
of pregnancy. The third trimester of pregnancy is a time of high risk for
dissemination. Immunologic and hormonal changes during pregnancy
and the postpartum period may account for any increased frequency
and severity of disease observed during pregnancy. Early diagnosis and
appropriate aggressive therapeutic intervention with careful monitoring
usually result in good outcome.

KEYWORDS: coccidioidomycosis; pregnancy; dissemination; risk factors;
treatment; azoles

BACKGROUND

Coccidioidomycosis (CM) is a fungal infection endemic to the southwestern
United States, northwestern Mexico, and parts of Central and South America.
The inhalation of the infectious arthroconidia results in symptomatic illness in
approximately one-third of infected persons. Disease presentations include a
wide clinical spectrum that ranges from mild influenza-like illness and pneumonia to dissemination. The latter develops in approximately 1% of infected
persons.[1] CM has been recognized as a complicating factor in pregnancy since
at least the 1940s. It was noted to be the leading cause of maternal morbidity
and mortality in the Kern community for the next several decades.[2] Data from
the pre-antifungal era show that symptomatic pregnant women had a 10% risk
of dissemination.

Address for correspondence: Irene M. Spinello, M.D., F.C.C.P., David Geffen School of Medicine
at UCLA Chief, Critical Care and Pulmonary Services, Kern Medical Center, 1830 Flower Street,
Bakersfield, CA 93305. Voice: 661-326-2022; fax: 661-326-2950.
    spinelloi@kernmedctr.com

Ann. N.Y. Acad. Sci. 1111: 358–364 (2007). © 2007 New York Academy of Sciences.
doi: 10.1196/annals.1406.008

RISK FACTORS

At least two risk factors have been associated with dissemination in pregnant women: the third trimester and race. It is reported that 96% of cases diagnosed in women in the third trimester had dissemination and that there is a 13-fold increased risk of dissemination in non-whites, compared with whites.[3] Ethnicity is a risk factor for disseminated disease, but is not a risk factor for acquisition of disease; thus the number of individuals presenting with pulmonary CM is largely reflective of the population served at any given center. Investigators in the Tucson area have noted that there were approximately 10 cases of CM for 47,000 births for a frequency of 0.02%.[3] At our institution the frequency of CM during pregnancy is approximately 10 times higher (101 case). These data are not yet published. It is thought that the incidence of disease in the non-pregnant population in the two metropolitan areas is roughly similar. Whether this represents referral patterns is conjectural. Whether other risk factors such as ethnicity have any significance in the increased disseminated disease seen in pregnant women is unclear. At Kern Medical Center, the presentation of CM is evenly distributed through all three trimesters (unpublished data).

PATHOGENESIS

The increased severity and risk for dissemination in pregnant women is attributed to the depressed cell-mediated immunity of pregnancy and to the direct stimulation of growth of *Coccidioides* resulting from the high serum levels of estradiol and progesterone in pregnant women.[4] Immunosuppression in pregnancy is documented by a decrease in CD4+ T cells and increase in CD8+ T cells, causing a significant fall in the CD4+/CD8+ ratio in late pregnancy and into the postpartum period.[5] In addition, the rate of growth and maturation of *Coccidioides immitis* is stimulated by levels of 17-beta-estradiol, with the most striking effects seen at levels encountered in advanced pregnancy, possibly mediated by the sex hormone binding protein produced by the fungus.[4,6] In support of this hypothesis, risk of dissemination has been reported to be highest in the second and especially third trimesters and increases directly as a function of the duration of the pregnancy.[7,8]

INTRAUTERINE TRANSMISSION

It was thought that *Coccidioides* does not cross the placenta because of its large size (40–70 μm). Moreover, thrombotic and chronic granulomatous reactions in the placenta appear to wall off the infection from the villous circulation.[9] Fetal intrauterine infection rarely occurs, despite a large number of documented cases of *Coccidioides* placentitis. In rare neonatal cases of CM, the infection is acquired at birth by contact with or aspiration of infectious material originating in the mother's vagina.[10] Nevertheless, evidence of intrauterine acquisition has been reported.[4] Maternal CM may predispose to premature delivery and intrauterine growth retardation. Neonatal CM has been reported, though rarely.[10–12]

CLINICAL PRESENTATION

The majority of patients seen with coccidioidomycosis, both pregnant and non-pregnant, have pulmonary disease. The percentage of patients with pregnancy and disseminated disease as a presenting symptom is higher than in most other populations. Pulmonary presentations represent approximately 90% of disease and disseminated disease approximately 10% of disease.[13]

The presenting symptoms in those individuals with primary CM are cough and fever followed by erythema nodosum, fatigue, headache, chest pain, shortness of breath, rash, and weight loss. Pregnant patients with disseminated disease most frequently present with central nervous system disease followed by cutaneous soft tissue and osseous disease.[3] Skin involvement in CM is not just limited to being a site of dissemination. Erythema nodosum, appearing as multiple painful nodules limited to the extremities, is a common dermatological finding in primary CM infection and can be used as a strong prognostication tool. It is highly associated with the relatively benign nature of the CM. This association is truer for the pregnant patient population than for the general population. Patients without erythema nodosum have a higher dissemination rate and recover less frequently than patients with erythema nodosum.[14] It must be understood that CM can present in any organ system. Co-morbid conditions do not appear to be a common circumstance in usually young and otherwise healthy pregnant women who present with CM.[3]

Most pregnant patients who have primary CM present with pulmonary infiltrate. A significant fraction of individuals may have a negative chest X-ray film. This has been observed in the non-pregnant patient as well. Occasionally pleural effusion is the dominant clinical and radiographic presentation—isolated cavitary disease and isolated hilar and paratracheal lymphadenopathy also occur. Miliary presentations are uncommon, accounting only for 10% of all cases.[3]

Routine laboratory studies are of modest benefit in the evaluation of individuals with CM who are pregnant. Peripheral blood eosinophilia, accounting for two-thirds of the circulating leukocytes, may be present in pregnancy, although it is not an independent factor positively associated with infection.[15] Whether this is higher than in the non-pregnant population is unclear.

MORTALITY

Maternal mortality correlates with late recognition of the disease and delay in initiation of effective antifungal therapy. Historically, 65% of mothers survived, but this number decreased to 45% when patients are diagnosed in their third trimester. The current survival rates have considerably improved on account of effective antifungal therapy.[3]

TREATMENT

Treatment options for CM depend on the severity of the disease, dissemination and the site of dissemination, and trimester. Pregnant women with a history of prior CM or an inactive lesion are often treated by close observation with no anti-fungal therapy. For active disease, both pulmonary and extrapulmonary, the treatment should be prompt and extended not only through pregnancy, but also for some months thereafter.[16] The treatment generally consists of amphotericin B deoxycholate (or less toxic lipid-based amphotericin) and azoles or a combination of the two. Although a combination of polyenes and azoles is not formally recommended by the guidelines, it has occasionally been used in our institution for severe disseminated disease and meningitis, requiring initial treatment with amphotericin, concomitantly with fluconazole. For more than 20 years azoles have been part of the therapeutic armamentarium for CM. Ketoconazole was used first, but was later replaced by fluconazole and itraconazole. The latter has been used to a lesser extent largely because of problems with its bioavailability and teratogenicity.[17]

Pregnant women with CM require substantially different treatment standards than non-pregnant patients. The predominant reason for this is that azoles are contraindicated in pregnancy. Fluconazole exposure is teratogenic in the first trimester, and could possibly predispose to prematurity in the second trimester.[17] Whether azoles can be used safely in the latter parts of pregnancy is unclear. In a prospective assessment of pregnancy outcomes after first-trimester exposure to fluconazole, the most frequent adverse outcome was therapeutic abortion.[18] The data did not find any increased risk for miscarriages, congenital anomalies or low birth weight associated with low-dose fluconazole regimens (150–700 mg per week). But the data could not exclude an increased risk with prolonged high-dosage regimens, such as those employed in the three malformed infants reported by Pursley *et al.*[19] Both fluconazole and itraconazole have a risk of teratogenicity and these drugs should therefore be avoided in pregnancy unless absolutely necessary, as would be in the case of saving the life of the mother. Amphotericin B is the drug of choice especially in the early stages of pregnancy. Presumably, a switch to azoles can be safely made later in pregnancy—the later the switch, the lower the risk of teratogenicity.[20] A user-independent method of contraception should be advocated for maximum safety. Breast-feeding should also be avoided during azole therapy, while postpartum non-breast feeding mothers may be safely switched from amphotericin B to an azole preparation. Even though azoles are given safely to infants, the nursing infant can only safely ingest 5% of the usual pediatric dosage.[21] The concentration of fluconazole in maternal serum and in breast milk of breast-feeding mothers treated with usual doses of fluconazole for CM will exceed this safe dose.

Amphotericin B deoxycholate has been used in pregnant patients for approximately five decades without apparent significant adverse effect on the pregnancy or the fetus. In recent years, lipid preparations of amphotericin B are increasingly preferred over amphotericin B deoxycholate. The rationale for this is based on the demonstrable reduction in toxicity of lipid-based amphotericin. In the absence of comparative trials for *Coccidioides* infections in pregnant patients, the cumulative clinical experience with the lipid-based amphotericin preparations is now adequate to consider them as suitable replacements for amphotericin B deoxycholate.[22] There are new agents that act against *Coccidioides* like posaconazole, voriconazole, and caspofungin, but the evidence of their efficacy and safety is limited to case reports only.[23] Posaconazole's safety is similar to that of fluconazole. Caspofungin (category C) should be used during pregnancy only if the potential benefit justifies the potential risk to the fetus. Voriconazole (category D) certainly has the most unfavorable profile on account of documented fetal harm and teratogenicity and should be avoided in pregnancy.

Before the advent of antifungal therapy, it was estimated that pregnant women with symptomatic infections had at least a 10% risk of dissemination and a subsequent 90% mortality.[13] More recent evidence disputes these figures, stating that though diagnosis and maternal death from CM are rare, the women who develop CM late in pregnancy are at risk of developing severe disseminated infection.[24] During the CM epidemic in California's Central Valley at the beginning of the last decade, when there were eight times as many cases as usually reported, the incidence of dissemination was 9% and there were no cases of maternal death.[25] The explanation of this dramatically improved outcome most likely lies in modern diagnostic techniques and effective antifungal therapy. As with any other potentially lethal condition, public education and awareness for the endemic areas and early recognition and intervention may be life saving.

CONCLUSION

Coccidioidal infection before, during, and after pregnancy is unusual but not rare. The potential for devastating illness is well documented. Immunologic and hormonal changes during pregnancy and the postpartum period may account for the increased frequency and severity of disease during pregnancy. Maternal outcome has dramatically improved in the antimicrobial era. Therefore, early diagnosis and appropriate aggressive therapeutic intervention with careful monitoring will usually result in a good outcome.

[Competing interests: Royce H. Johnson receives research support and is a consultant/speaker for Merck & Co, Inc. and Pfizer, Inc.]

REFERENCES

1. EINSTEIN, H.E. & R.H. JOHNSON. 1993. Coccidioidomycosis: new aspects of epidemiology and therapy. Clin. Infect. Dis. **16:** 349–356.
2. CANTANZARO, A. 1984. Pulmonary mycosis in pregnant women. Chest **86** (Suppl): 14S–18S.
3. CRUM, N.F. & G. BALLON-LANDA. 2006. Coccidioidomycosis in pregnancy: case report and review of the literature. Am. J. Med. 119, 993.e11–993.e17.
4. POWELL, B.L., D.J. DRUTZ, M. HUPPERT, *et al.* 1983. Relationship of progesterone and estradiol-binding proteins in *Coccidioides immitis* to coccidioidal dissemination in pregnancy. Infect. Immunol. **40:** 478.
5. STAGNARO-GREEN, A., S.H. ROMAN, R.H. COBIN, *et al.* 1992. A prospective study of lymphocyte-initiated immunosuppression in normal pregnancy: evidence of a T-cell etiology for postpartum thyroid dysfunction. J. Clin. Endocrinol. Metab. **74:** 645–653.
6. DRUTZ, D.J., M. HUPPERT, S.H. SUN, *et al.* 1981. Human sex hormones stimulate the growth and maturation of *Coccidioides immitis*. Infect. Immun. **40:** 478–485.
7. PETERSON, C.M., K. SCHUPPERT, P.C. KELLY, *et al.* 1993. Coccidioidomycosis and pregnancy. Obstet. Gynecol. Surv. **48:** 149–156.
8. HARVEY, R.P. 1980. Coccidioidomycosis in Pregnancy. *In* D.A. Stevens, Ed. Coccididoidomycosis. A Text. Plenum. New York.
9. COHEN, R. 1951. Placental *coccidioides*. Proof that congenital *coccidioides* is nonexistent. Arch. Pediatr. **68:** 59–66.
10. LINSANGAN, L.C. & L.A. ROSS. 1999. *Coccidioides immitis* infection of the neonate: two route of infection. Pediatr. Infect. Dis. J. **18:** 171–173.
11. TOWNSEND, T. & R. McKEY.1953. Coccidioidomycosis in infants. Am. J. Dis. Child. **86:** 51–53.
12. CHARLTON, V., K. RAMSDELL & S. SEHRING. 1999. Intrauterine transmission of coccidioidomycosis. Pediatr. Infect. Dis. **18:** 561–563.
13. PAPPAGIANIS, D. 1980. Epidemiology of coccidioidomycosis. *In* Coccidioidomycosis. D. Stevens, Ed.: 63–85. Plenum. New York.
14. ARSURA, E.L., W.B. KILGORE & S.N. RATNAYAKE. 1998. Erythema nodosum in pregnant patients with coccidioidomycosis. Clin. Infect. Dis. **27:** 1201–1203.
15. YOZWIAK, M.L., L.L. LUNDERGAN, S.S. KERRICK, & J.N. GALGIANI. 1988. Symptoms and routine laboratory abnormalities associated with coccidioidomycosis. West. J. Med. **149:** 419–421.
16. ANSTEAD, G.M. & J.R. GRAYBILL. 2006. Coccidioidomycosis. Infect. Dis. Clin. N. Am. **20:** 621–643.
17. SPINELLO, I.M. & R.H. JOHNSON. 2006. A 19-year-old pregnant woman with a skin lesion and respiratory failure. Chest **130:** 611–615.
18. MASTROIACOVO, Pierpaolo, T. MAZZONE, L.D. BOTTO, *et al.* 1996. Prospective assessment of pregnancy outcomes after first-trimester exposure to fluconazole. Am. J. Obstet. Gynecol. **175:** 1645–1650.
19. PURSLEY, T.J., I.K. BLOMQUIST, J. ABRAHAM, *et al.* 1996. Fluconazole-induced congenital anomalies in three infants. Clin. Infect. Dis. **22:** 336–340.
20. BAR-OZ, B., M.E. MORETTI, R. BASHAI, *et al.* 2000. Pregnancy outcome after in utero exposure to itraconazole: a prospective cohort study. Am. J. Obstet. Gynecol. **183:** 617–620.
21. HALE, T.W. 1999. Medications and Mothers' Milk: 1999–2000, 8th ed. Pharmasoft Medical. Amarillo, Texas.

22. OSTROSKY-ZEICHNER, L., K.A. MARR, J.H. REX, *et al.* 2003. Amphotericin B: time for a new "gold standard." CID **37:** 415–425.
23. DERESINSKI, S.C. 2001. Coccidioidomycosis: efficacy of new agents and future prospects. Curr. Opin. Infect. Dis. **14:** 693–696.
24. WACK, E.E., N.M. AMPEL, J.N. GALGIANI, *et al.* 1988. Coccidioidomycosis during pregnancy. An analysis of ten cases among 47,120 pregnancies. Chest **94:** 376–379.
25. CALDWELL, J.W., E.L. ARSURA, W.B. KILGORE, *et al.* 2000. Coccidioidomycosis in pregnancy during an epidemic in California. Obstet. Gynecol. **95:** 236–239.

Coccidioidomycosis in Patients Who Have Undergone Transplantation

JANIS E. BLAIR

Division of Infectious Diseases, Mayo Clinic, Scottsdale, Arizona, USA

ABSTRACT: In the early years of transplantation in Arizona, coccidioidomycosis occurred in 7% to 9% of recipients, with a mortality rate as high as 72% in some cases. In current transplant programs, however, evolution of immunosuppression and institution of targeted prophylaxis have resulted in coccidioidal infection rates ranging from 1% to 2%. The clinical characteristics of this infection among transplant recipients range from asymptomatic to fulminant and fatal. Dissemination is common, and mortality is high (28%). Because serologic response is often absent or slow, diagnosis can be challenging and often requires invasive diagnostic procedures. Pharmacologic treatment follows the guidelines of the Infectious Diseases Society of America, but control of infection may also dictate a decrease in immunosuppressant treatment. After infection is controlled, secondary azole prophylaxis is recommended to prevent relapse. Patients with a history of coccidioidomycosis may undergo successful transplantation when disease is inactive and azole prophylaxis is instituted. The incidence of donor-derived coccidioidomycosis is not known. The risk of coccidioidal infection among transplant recipients visiting in or relocating to an endemic area is low, and routine prophylaxis for this group is not recommended.

KEYWORDS: coccidioidomycosis; fungal infections; organ transplantation

INTRODUCTION

Coccidioidomycosis is a fungal infection caused by *Coccidioides* species endemic to desert areas of the Sonoran life zone found in the southwestern United States, Mexico, and smaller areas of Central and South America.[1] In most cases, infection is initiated by inhalation of airborne arthrospores. Although 60% of infected persons experience no symptoms, most of the rest sustain a self-limited respiratory infection manifested by the protean symptoms of fever, headache, dry cough, and malaise. Rarely, the infection is complicated by extrapulmonary manifestations.[1]

Address for correspondence: Janis E. Blair, M.D., Division of Infectious Diseases, Mayo Clinic, 13400 East Shea Boulevard, Scottsdale, AZ 85259.

Ann. N.Y. Acad. Sci. 1111: 365–376 (2007). © 2007 New York Academy of Sciences.
doi: 10.1196/annals.1406.009

An intact immune system is essential to the control of coccidioidomycosis; persons who have impaired immunity, especially cellular immunity, such as that caused by the human immunodeficiency virus or organ transplantation, may contract an infection with severe or life-threatening manifestations.[1] Extrapulmonary infection is not uncommon in these patients. Because of these factors, coccidioidomycosis has been a concern for transplant programs in southern Arizona, which is an area where coccidioidomycosis is highly endemic.[2] This article is a synopsis of the current information about coccidioidomycosis in transplantation as presented at the Sixth International Symposium on Coccidioidomycosis, Stanford, California, August 23–26, 2006.

EARLY TRANSPLANTATION EXPERIENCE IN ARIZONA

Organ transplant programs have been active in southern Arizona since 1970. Over the years, these programs have expanded in size and scope. Current transplant programs differ greatly from the earliest programs. In earlier years, corticosteroids were the primary immunosuppressive medication, whereas steroid doses are smaller and often completely withdrawn in current programs. Corticosteroids are a known risk factor for disseminated and severe coccidioidal infection.[1] In addition, the only effective antifungal agent for treatment of coccidioidomycosis was amphotericin B deoxycholate. Because of the high risk of infusion-related and cumulative toxicity (chiefly nephrotoxicity) with amphotericin, it had no role in coccidioidal prophylaxis.

Much information was gained as a result of these early transplant programs. Coccidioidomycosis was not uncommon, with 18 cases identified among 260 (6.9%) kidney transplant recipients from 1970 to 1979[3]; dissemination was common (12 of 18 cases; 67%).[3] After coccidioidomycosis was treated and controlled, transplant recipients often sustained relapsed infection. Mortality was high (as much as 63% overall and 72% among transplant recipients with disseminated infection[3]). When azoles were introduced, azole prophylaxis was found to prevent infection.[4–6]

Observational studies conducted before the initiation of the Mayo Clinic transplant program in Arizona reported that risk factors for sustaining coccidioidomycosis after organ transplantation included high-dose corticosteroid[2, 7] treatment of rejection (either single or multiple episodes[5]), history of coccidioidal infection before transplantation, or seropositivity just before or at transplantation.[4, 6, 8, 9]

COCCIDIOIDOMYCOSIS IN ONE TRANSPLANT PROGRAM IN THE ENDEMIC AREA

In 1999, the solid-organ transplant program was established at Mayo Clinic, Scottsdale, Arizona, which is situated in the heart of the area where coccidioidomycosis is endemic. Since then, 1,188 transplant procedures have been performed through July 12, 2006. In contrast to transplantations in earlier coccidioidomycosis studies, the primary medication now used for initial immunosuppression is tacrolimus, which primarily impairs cellular, rather than global, immune processes. Corticosteroids are generally tapered over a few months and, in many cases, are discontinued within a few months. In kidney and heart recipients, immunosuppression may be initiated with rabbit antithymocyte globulin. Whether the evolution of antirejection therapy over time has affected the incidence or clinical manifestations of coccidioidomycosis is not known.

Because our program is located in an area highly endemic for coccidioidomycosis, we routinely perform coccidioidal serologic testing at the time of pretransplantation evaluation, at the time of transplantation surgery, 4 months after transplantation, and annually thereafter. Serologic testing is also repeated if clinically indicated. We use enzyme immunoassay, complement fixation, and immunodiffusion tests at each interval.

Prophylaxis

Within the endemic area, there is no consensus among different institutions about the best prophylaxis for coccidioidomycosis in transplant patients. My observation is that prophylaxis for this fungal infection ranges from universal and lifelong fluconazole for all transplant recipients to targeted prophylaxis, that is, the application of prophylaxis to persons with specific risk factors (e.g., prior history of coccidioidomycosis or positive serologic findings). Programs that use universal and lifelong fluconazole for all transplant recipients have not yet published the results of this prophylaxis.

All patients in the Mayo Clinic transplant program in Arizona, regardless of the presence or absence of risk factors for coccidioidomycosis, receive prophylaxis (generally fluconazole) to prevent fungal infections for a variable period of time (generally 1–3 months) after the transplant procedure. In addition to standard prophylaxis, patients receive prolonged and augmented prophylaxis if risk factors for posttransplantation coccidioidomycosis are present (TABLE 1). Such targeted prophylaxis is initiated when transplant recipients have a documented history of coccidioidomycosis, positive serologic findings, or recent coccidioidal infection, or when the donor is seropositive or was known or suspected to have transmitted to another recipient. In addition, prophylaxis for cytomegalovirus (CMV) and *Pneumocystis jiroveci* is routinely provided.

This prophylaxis schedule has resulted in an overall rate of coccidioidomycosis of 1.2% (hematologic) to 2.4% (liver) (TABLE 2). When asymptomatic seroconversion is included, 18 patients among 1,188 (1.5%) recipients of solid and hematopoietic transplants had complications from coccidioidomycosis.

TABLE 1. Targeted prophylaxis for coccidioidomycosis in solid-organ transplant recipients at Mayo Clinic, Scottsdale, Arizona

(1) Recipients with a history of coccidioidomycosis
  (A) Diagnosis by a physician is required for a patient to qualify for this prophylaxis schedule. The patient is usually able to describe a compatible clinical illness. Corroborating medical records are helpful but not required.
  (B) Patients do not receive this prophylaxis if they think they may have had coccidioidomycosis because of time spent in the endemic area, self-diagnosis, or granuloma on chest radiograph.
  (C) Chest radiograph and serologic findings must be negative for this prophylaxis schedule.
  (D) After transplantation, the recipient receives oral fluconazole 200 mg daily for 6 months.
(2) Recipients with positive serologic findings at transplantation evaluation or surgery
  (A) Any positive serologic finding by enzyme immunoassay, complement fixation, or immunodiffusion.
  (B) Fluconazole 400 mg daily by the oral route for the first year, then 200 to 400 mg daily thereafter for the duration of immunosuppression.
(3) Recipients with active coccidioidomycosis within 1 to 2 years before transplantation, or with active coccidioidomycosis at the time of transplantation evaluation.
  (A) Patients with a compatible clinical illness and positive serologic findings, chest radiograph, or asymptomatic seroconversion (documented negative serologic findings followed by positive serologic findings).
  (B) Infection must have resolved clinically, serologically, and radiographically. Patient must be cleared for transplantation by the infectious diseases expert in consultation with the transplant team.
  (C) Fluconazole 400 mg[a] daily for the first year after transplantation, then 200 to 400 mg daily thereafter.
(4) Active infection at transplantation surgery
  (A) If possible, defer transplantation until recipient candidate meets criteria 3B above
  (B) If infection is discovered after transplantation, treat initial infection, then give prophylaxis with fluconazole 400 mg[a] daily for first year, then lifelong prophylaxis with 200 to 400 mg daily thereafter, in conjunction with an infectious diseases consultation.
(5) Active coccidioidomycosis or positive serologic findings in donor.
  (A) Lifelong prophylaxis for the recipient with fluconazole 400 mg daily for the first year, then 200 mg daily thereafter.

[a]Unless higher doses were required to control the coccidioidal infection.

Thus far, we have not observed that coccidioidomycosis is more frequent among transplant recipients who continue to reside in the endemic area. The rate of coccidioidal infections in the transplant population remains similar to the estimated rate of coccidioidomycosis among all persons residing in the endemic area.

Clinical Manifestations

The clinical manifestations of coccidioidomycosis in transplant recipients vary widely in severity and extent. Many infections are asymptomatic, as

TABLE 2.   Coccidioidomycosis in transplant recipients

Type of transplant	Dates	Transplants (no.)	Coccidioidomyocosis (no.) (%)		
			Asymptomatic seroconversion[a]	Active infection	All coccidioidal infections[b]
Liver	6/1999-7/2006	324	4 (1.2)	4 (1.2)	8 (2.5)
Kidney	6/1999-7/2006	538	0	7 (1.3)	7 (1.3)
Pancreas	6/1999-7/2006	55	0	0	0
Heart	10/2005-7/2006	12	0	0	0
Hematologic	1/1993-7/2006				
Allogeneic		48	0	1 (2.1)	1 (2.1)
Autologous		211	0	2 (0.9)	2 (0.9)
Total		1,188	4 (0.3)	14 (1.2)	18 (1.5)

[a]Defined when a patient has previously had negative serologic findings by all methods, and a routine serologic test subsequently is positive by any or all methods, except enzyme immunoassay IgM alone.

[b]Active coccidioidomycosis and asymptomatic seroconversion with positive serologic findings.

illustrated by the seroconversion in patients identified on routine evaluation in 4 of 1,188 transplant recipients. On evaluation, such patients have no history of respiratory symptoms or febrile illness, they have no radiographic abnormalities, and they are otherwise well. Similar to many healthy patients with coccidioidomycosis, most patients with symptomatic infections do have a variable presentation, but common symptoms include fever, headache, myalgia, cough, dyspnea, and chest pain, and significant fatigue.

Extrapulmonary dissemination of coccidioidomycosis is common in transplant recipients, reported in as many as 75% of transplant recipients with coccidioidomycosis.[2] Multiple sites of dissemination are often found. In addition to the commonly identified sites of dissemination, such as skin, lymph nodes, skeleton, and meninges, dissemination in transplant recipients also commonly occurs in soft tissues, such as the spleen, genitourinary sites, the thyroid, brain, pancreas, and adrenal glands.[2] Dissemination to the transplanted organ is not uncommon.[2] My experience has been that dissemination is likely underdiagnosed; patients may manifest apparent radiographic abscesses within soft tissues or within the transplanted organ (such as the liver) that disappear with fluconazole treatment before an invasive procedure, and these are presumed, but not documented, extrapulmonary coccidioidal abscesses.

Mortality is high among transplant recipients whose course is complicated by coccidioidomycosis. Of 14 patients who had active infection, four died (29%) within 6 weeks of coccidioidomycosis onset; three of these deaths were reported earlier.[10] Only one death was directly attributed to the acute manifestations of coccidioidomycosis.

Diagnosis

Coccidioidomycosis presents with protean manifestations in both healthy and transplant recipient populations. Because the latter are prone to numerous infections that have similar symptoms, diagnostic testing is critical to establish the diagnosis and guide proper therapy. Multiple methods for serologic testing are available, including enzyme immunoassay, immunodiffusion, and complement fixation. Transplant recipients may have a decreased rate of seropositivity.[11] Because of delayed or decreased serologic results, diagnostic tests, such as cultures, biopsies, and bronchoalveolar lavage may be required. In immunosuppressed patient populations, such as this, a cerebrospinal fluid examination should be considered to exclude dissemination to the meninges.

Possible Role of Cytomegalovirus in Coccidioidomycosis

An interesting question yet to be answered is the interaction of CMV with coccidioidomycosis. CMV is commonly problematic in transplant recipients, either as primary infection resulting from the receipt by a seronegative recipient of an organ from a seropositive donor or as secondary infection resulting from reactivation of a previously latent virus.[12] The effects of CMV infection in the transplant recipient range from a mild febrile illness to severe infection involving the liver, kidney, colon, esophagus, bone marrow, lungs, or retina.[12] CMV may also contribute to allograft rejection and has been associated with secondary fungal infections in liver transplant recipients.[13] Thus far, no reports have been published to document an association between CMV infection and coccidioidomycosis. In the Mayo Clinic experience in Arizona, of the 14 patients with active coccidioidomycosis (not including those with asymptomatic seroconversion) after transplantation, 4 (29%) were CMV donor-seronegative and recipient-seropositive ("CMV mismatch"), a proportion similar to that of the population with transplants in this practice. Interestingly, among the 4 CMV-mismatched recipients with coccidioidomycosis, 3 (75%) had coccidioidal infections that occurred soon after an episode of active CMV infection. Although this fact is intriguing, the number of cases in these cohorts is too small to draw any definitive conclusions.

Treatment

The treatment of coccidioidomycosis in transplant recipients follows the recently updated treatment guidelines.[14] An important aspect of treatment is to define the extent of the infection, especially since transplant recipients are predisposed to extrapulmonary infection. Azoles such as fluconazole or itraconazole are usually adequate for treatment of coccidioidomycosis, but the treating physician should be aware that azoles interfere with the metabolism of tacrolimus and other calcineurin inhibitors, resulting in elevated levels of these medications, with resulting renal insufficiency. Therefore, when azoles are initiated or titrated, serum levels of the calcineurin inhibitors should be monitored. Amphotericin B, either deoxycholate or a lipid formulation, is usually reserved for severe clinical illness, and its use can be complicated by renal insufficiency and electrolyte disturbances. It is often necessary to decrease the level of immunosuppression to control the infection. Because this maneuver may jeopardize graft viability, treatment of coccidioidomycosis in transplant patients requires cooperative teamwork between the infectious diseases physician and the transplantation team.

After the coccidioidal infection has been successfully treated, a transplant recipient should receive continued treatment (secondary prophylaxis), generally with an azole, such as fluconazole 200 to 400 mg daily,[15] for the duration of the immunosuppression to prevent recrudescence. This recommendation is based on the observation that reactivated infection may occur after discontinuation of antifungal therapy.[5, 16, 17]

Outcomes of Recipients with Coccidioidomycosis before Transplantation

Since 1999, 47 persons in this transplant program have had a coccidioidal infection preceding successful transplantation of a liver, kidney, or pancreas allograft, and these results were recently reviewed.[18] At transplantation, 44 of the 47 had no evidence of any active coccidioidal infection. Two of the 47 had a history of extrapulmonary infection; thus far, no patients with meningeal coccidioidomycosis have proceeded to transplantation.

Among the 44 patients without active coccidioidomycosis at transplantation, 37 received and were compliant with prophylaxis according to the guidelines delineated in TABLE 1 and had no further problems with posttransplantation coccidioidal infection. Another three who never received prophylaxis for unknown reasons did not have any recrudescent infection. Four patients were noncompliant with the recommended prophylaxis, and two of these four had problems with relapsing or recurrent coccidioidomycosis.[18]

At transplantation, 3 of the 47 patients had evidence of some coccidioidal activity. One was asymptomatic without radiographic abnormalities but had a complement fixation titer of 1:8; another had a coccidioidal lung mass with a titer of 1:128. The latter patient normally would have been deferred for transplantation until the infection resolved, but because of underlying liver cancer had a prognosis of only weeks to live without transplantation; he proceeded to transplantation after written consent acknowledging the unknown behavior of the fungal infection in this particular circumstance. A third patient was identified with active pulmonary coccidioidomycosis with positive serologic findings after the patient had already begun the transplantation surgery. Her course was complicated by extrapulmonary dissemination, which was treated successfully.[19] All patients had their coccidioidal infection controlled and now receive ongoing fluconazole prophylaxis.[18]

On the basis of this experience in the transplant program at Mayo Clinic, Scottsdale, Arizona, I strongly recommend that any patient with pretransplantation coccidioidomycosis have complete resolution of infection before undergoing transplant surgery, as evidenced by a lack of symptoms, resolution, or stabilization of radiographic abnormalities, and low and stable serologic titer. Exceptions should be accompanied by written consent of the patient that acknowledges the unknown behavior of the fungus in this setting. Patients should receive anticoccidioidal prophylaxis, but the criteria for optimal selection of patients for such prophylaxis is not yet clearly defined and requires further investigation.

DONOR-DERIVED COCCIDIOIDOMYCOSIS

Donor-derived coccidioidomycosis was first recognized in organ recipients outside the endemic area, when organ recipients who had never lived in or traveled to endemic areas contracted coccidioidomycosis soon after receipt of a transplanted allograft.[20–22]

Two of the first three reports entailed the transfer of coccidioidomycosis from a quiescent lung infection in the donor (granuloma and lymph nodes),[20,21] causing a clinically significant respiratory infection in the recipient. Both recipients sustained widespread lung infection, and only one survived after diagnosis and treatment.

A subsequent report demonstrated that coccidioidomycosis could be transferred to the recipient by transplantation of organs other than lungs.[22] In this report, the donor had been remotely treated for disseminated coccidioidomycosis involving skin, vertebra, and sternum, which required multiple debridements and fluconazole, but he discontinued treatment a few years later and then died of coccidioidal meningitis, clinically occult at donor evaluation; at a subsequent autopsy of the donor, active brain and meningeal infection were present and the complement fixation titer was 1:32. Two organ recipients (one kidney and one liver) died of disseminated and fulminant coccidioidomycosis within 3 weeks of transplantation. When the donor-derived coccidioidal infection was recognized, the recipient of the second kidney was placed on itraconazole for 3 months and did not experience subsequent clinical infection.

How common is donor-derived coccidioidomycosis within the endemic area? The answer to this question is not known. Organ transplant recipients residing within the endemic area are always at some risk for coccidioidomycosis acquired from inhaling the ambient air, and most patients who present with acute coccidioidomycosis within the endemic area would be assumed to have acquired the infection by the respiratory route.

In an attempt to identify any donors infected with *Coccidioides* spp., we prospectively evaluate candidates for living kidney or liver transplantation using enzyme immunoassay, immunodiffusion, and complement fixation. Of 568 healthy, asymptomatic donor candidates screened prospectively, 12 (2%) had positive serologic findings indicative of recent or remote infection. Chest radiography was performed at the same time. Four of the 12 prospective donors proceeded to donor surgery. None of the four recipients of these donors had specific coccidioidomycosis prophylaxis, although two of the four received prolonged courses of fluconazole for other reasons. No recipient sustained any coccidioidal infection.[23] This information indicates that donor-derived coccidioidomycosis is not common in the liver and kidney recipients who are alive and that not all seropositive donors will transfer infection to recipients. However, because of the serious illness that may result from donor-derived coccidioidomycosis, donors from the endemic area should be serologically screened and recipients should be given prophylaxis when receiving an organ from a seropositive donor.

EXPOSURE OF TRANSPLANT RECIPIENTS TO ENDEMIC AREAS

Several case reports have detailed the visits of solid-organ transplant recipients to areas endemic for coccidioidomycosis, which later resulted in fulminant, disseminated, and in some instances, fatal cases of coccidioidomycosis.[24-26] These reports have prompted one author to discourage all organ recipients from such travel.[3]

A 2004 report summarized the Mayo Clinic experience in Arizona with 37 liver transplant recipients who had received an organ transplant in programs outside the endemic area and later relocated, either permanently or on a temporary but recurring basis (such as winter visitors) to Arizona, where they established care and follow-up in the transplant program.[27] All 37 had at least 1 year of follow-up with the program. No patients received any prophylaxis for coccidioidomycosis, nor was any formal education given to encourage them to avoid dust and dust-generating activities. At follow-up of 12 to 101 months (mean, 51 months), only a single patient had contracted coccidioidal infection, for an overall incidence of 1 in 37 (2.7%). The patient in whom coccidioidomycosis developed was a 35-year-old African-born woman who had undergone liver transplantation 1 year before relocating to Arizona. Soon after she established care with the Mayo Clinic liver transplant program, she was treated with three doses of methylprednisolone followed by a prednisone taper in addition to her previous cyclosporine monotherapy for biopsy-proved rejection. Although her liver tests subsequently normalized, she began experiencing an insidious onset of malaise, fatigue, and dyspnea with exertion, all due to coccidioidomycosis. A chest radiograph was normal, but a computed tomogram showed multiple pulmonary nodules, infiltrates, and adenopathy. Coccidioidal serologic findings were positive. She was treated with oral fluconazole and improved clinically, had no recurrent coccidioidal infection, and continued to take lifelong fluconazole.

Three years after the publication of the report about the 37 relocated transplant recipients,[27] the original cohort has not had any further episodes of coccidioidal infection. This cohort demonstrates that not all transplant recipients need prophylaxis for coccidioidomycosis if visiting or relocating to an endemic area. However, there may be circumstances in which it may be advisable for a particular patient (e.g., a lung recipient traveling to an area that is experiencing hyperendemic rates of coccidioidomycosis) to take such prophylaxis, but the decision should be made on a case-by-case basis. At this time, no further data exist to provide further guidance on this issue.

CONCLUSION

Organ transplant programs are flourishing in southern Arizona, an area highly endemic for coccidioidomycosis. With a program of targeted prophylaxis, the rate of infection is 1% to 2% among transplant recipients. Numerous questions remain about the prevention of this infection in transplant recipients. Because mortality is still high among patients whose course is complicated by this infection, further studies are required to develop tools for early recognition and for improving the treatment outcome.

ACKNOWLEDGMENT

Editing, proofreading, and reference verification were provided by the Section of Scientific Publications, Mayo Clinic.

REFERENCES

1. CHILLER, T.M., J.N. GALGIANI & D.A. STEVENS. 2003. Coccidioidomycosis. Infect. Dis. Clin. North Am. **17:** 41–57.
2. BLAIR, J.E. & J.L. LOGAN. 2001. Coccidioidomycosis in solid organ transplantation. Clin. Infect. Dis. **33:** 1536–1544.
3. COHEN, I.M., J.N. GALGIANI, D. POTTER & D.A. OGDEN. 1982. Coccidioidomycosis in renal replacement therapy. Arch. Intern. Med. **142:** 489–494.
4. HALL, K.A., J.G. COPELAND, C.F. ZUKOSKI, et al. 1993. Markers of coccidioidomycosis before cardiac or renal transplantation and the risk of recurrent infection. Transplantation **55:** 1422–1424.
5. SEROTA, A. 1996. The efficacy of fluconazole in prevention of coccidioidomycosis following renal transplantation. *In* Coccidioidomycosis. Proceedings of the 5th

International Conference on Coccidioidomycosis; 1994 Aug 24–27; Stanford University. H.E. Einstein & A. Catanzaro, Eds.: 248–264. National Foundation for Infectious Diseases. Washington, DC.

6. GALGIANI, J.N. 1997. Coccidioidomycosis. Curr. Clin. Top. Infect. Dis. **17:** 188–204.

7. SMITHLINE, N., D.A. OGDEN, A.I. COHN & K. JOHNSON. 1967. Disseminated coccidioidomycosis in renal transplant patients. *In* Coccidioidomycosis. Proceedings of the 2nd Coccidioidomycosis Symposium; 1965 Dec 8–10; Phoenix, AZ. L. Ajello, Ed.: 201–206. The University of Arizona Press. Tucson, AZ.

8. HALL, K.A., G.K. SETHI, L.J. ROSADO, *et al.* 1993. Coccidioidomycosis and heart transplantation. J. Heart Lung Transplant. **12:** 525–526.

9. CALHOUN, D.L., J.N. GALGIANI, C. ZUKOSKI & J.G. COPELAND. 1985. Coccidioidomycosis in recent renal or cardiac transplant recipients. *In* Coccidioidomycosis. Proceedings of the 4th Internal Conference on Coccidioidomycosis; 1984 Mar 14–17; San Diego, CA. H.E. Einstein & A. Catanzaro, Eds.: 312–318. The National Foundation for Infectious Diseases. Washington, DC.

10. BRADDY, C.M., R.L. HEILMAN & J.E. BLAIR. 2006. Coccidioidomycosis after renal transplantation in an endemic area. Am. J. Transplant. **6:** 340–345.

11. BLAIR, J.E., B. COAKLEY, A.C. SANTELLI, *et al.* 2006. Serologic testing for symptomatic coccidioidomycosis in immunocompetent and immunosuppressed hosts. Mycopathologia **162:** 317–324.

12. KUSNE, S. & J.E. BLAIR. 2006. Viral and fungal infections after liver transplantation. Part II. Liver Transpl. **12:** 2–11.

13. GEORGE, M.J., D.R. SNYDMAN, B.G. WERNER, *et al.* 1997. Boston Center for Liver Transplantation CMVIG-Study Group. The independent role of cytomegalovirus as a risk factor for invasive fungal disease in orthotopic liver transplant recipients. Am. J. Med. **103:** 106–113.

14. GALGIANI, J.N., N.M. AMPEL, J.E. BLAIR, *et al.* 2005. Coccidioidomycosis. Clin. Infect. Dis. **41:** 1217–1223.

15. BLAIR, J.E. 2006. Coccidioidomycosis in liver transplantation. Liver Transpl. **12:** 31–39.

16. BLAIR, J.E. 2004. Coccidioidal pneumonia, arthritis, and soft-tissue infection after kidney transplantation. Transpl. Infect. Dis. **6:** 74–76.

17. HOLT, C.D., D.J. WINSTON, B. KUBAK, *et al.* 1997. Coccidioidomycosis in liver transplant patients. Clin. Infect. Dis. **24:** 216–221.

18. BLAIR, J.E., S. KUSNE, E.J. CAREY & R.L. HEILMAN. 2007. The prevention of recrudescent coccidioidomycosis after solid organ transplantation. Transplantation **83:** in press.

19. SACHDEV, M.S., J.E. BLAIR, D.C. MULLIGAN & S. KUSNE. 2007. Coccidioidomycosis marked by symptoms of end-stage liver disease in transplant candidates. Transplant. Infect. Dis. In press.

20. TRIPATHY, U., G.L. YUNG, J.M. KRIETT, *et al.* 2002. Donor transfer of pulmonary coccidioidomycosis in lung transplantation. Ann. Thorac. Surg. **73:** 306–308.

21. MILLER, M.B., R. HENDREN, & P.H. GILLIGAN. 2004. Posttransplantation disseminated coccidioidomycosis acquired from donor lungs. J. Clin. Microbiol. **42:** 2347–2349.

22. WRIGHT, P.W., D. PAPPAGIANIS, M. WILSON, *et al.* 2003. Donor-related coccidioidomycosis in organ transplant recipients. Clin. Infect. Dis. **37:** 1265–1269.

23. BLAIR, J.E. & D.C. MULLIGAN. 2007. Coccidioidomycosis in healthy persons evaluated for liver or kidney donation. Transplant Infect. Dis. **9:** 78–82.

24. CHA, J.M., S. JUNG, H.S. BAHNG, *et al.* 2000. Multi-organ failure caused by reactivated coccidioidomycosis without dissemination in a patient with renal transplantation. Respirology **5:** 87–90.
25. MAGILL, S.B., T.M. SCHMAHL, J. SOMMER, *et al.* 1998. Coccidioidomycosis-induced thyroiditis and calcitriol-mediated hypercalcemia in a heart transplantation patient. Endocrinologist **8:** 299–302.
26. VARTIVARIAN, S.E., P.E. COUDRON & S.M. MARKOWITZ. 1987. Disseminated coccidioidomycosis: unusual manifestations in a cardiac transplantation patient. Am. J. Med. **83:** 949–952.
27. BLAIR, J.E. & D.D. DOUGLAS. 2004. Coccidioidomycosis in liver transplant recipients relocating to an endemic area. Dig. Dis. Sci. **49:** 1981–1985.

Coccidioidal Meningitis

PAUL L. WILLIAMS

Department of Adult Medicine, Division of Infectious Diseases, Kaiser Permanente Medical Center, Fresno, California, USA

ABSTRACT: Coccidioidal meningitis affects between 200 to 300 persons annually within the endemic area of the United States, with much larger numbers expected in epidemic years. Because this represents a chronic disease for survivors, several thousand patients may be under treatment at any given time. Epidemiology, background, and diagnosis are reviewed. Azole therapy has replaced intrathecal amphotericin B for induction and maintenance therapy for this disease, given its ease of administration and equivalent efficacy in controlling infection even at the cost of losing the opportunity for cure. Both itraconazole and fluconazole have demonstrated efficacy, but have not been compared in randomized human studies. One of the uses of intrathecal amphotericin B is as "add on" therapy in failing azole regimens without evidence of antagonism. Details of therapeutic approach are reviewed. Approach to diagnosis and management of the two principal potentially life threatening complications, hydrocephalus and vasculitis, is also discussed.

KEYWORDS: *C. immitis*; coccidioidomycosis; meningitis

EPIDEMIOLOGY

In nonepidemic years, approximately 200 to 300 cases of coccidioidal meningitis can be anticipated within the area of endemicity (Lower Sonoran life zone, southwestern United States, Central Valley of California). This disease is also endemic to Northern Mexico and Central and South America, although there are no data in terms of disease burden there. In epidemic years in the United States, such as those following severe dust storms or after prolonged drought seasons followed by significant rainfall, numbers escalate considerably. It is important to note that because of the chronic nature of this infection, with rare cures, several thousand survivors may be under treatment at any given time.

BACKGROUND

Coccidioidal meningitis represents the most severe and frequent manifestation of disseminated disease following initial pulmonary infection. Generally,

Address for correspondence: Paul L. Williams, M.D., The Permanente Medical Group, 2651 Highland Avenue, Selma, CA 93662. Voice: 559-898-6152; fax: 559-898-6190.
traildoc11@yahoo.com

Ann. N.Y. Acad. Sci. 1111: 377–384 (2007). © 2007 New York Academy of Sciences.
doi: 10.1196/annals.1406.037

this follows evident pulmonary disease within weeks or months after initial onset of symptoms. On occasion, this may be the initial presentation of coccidioidal infection without symptomatic pulmonary disease being clinically evident. Dissemination is the result of lymphohematogenous spread from lung to meninges. The principal clinical feature of CNS dissemination is that of a chronic basilar meningitis. Symptomatic presentation is generally characterized by fever, meningeal irritation (meningismus being evident in approximately 50% of cases), headache with occasional sensorial changes (cognitive impairment, personality changes), malaise and nausea with vomiting (often reflecting hydrocephalus, one of the common complications discussed below). Occasionally, overt vasculitis, another common complication, can be the presenting clinical manifestation with overt stroke events, altered mental status or a combination of both being observed. Seizures may accompany this complication (reviewed later).[1–3] More uncommon complications include encephalitis, mass-occupying lesions, brain abscesses and aneurysms complicated by subarachnoid hemorrhages, which are generally fatal.[4,5] Neuroimaging, particularly MRI studies, may be very useful in characterizing disease activity and its complications. Some experienced clinicians have advocated doing baseline MRI studies on all patients at diagnosis.[1]

DIAGNOSIS

Once dissemination to the CNS occurs, it is generally characterized by a basilar meningitis with CSF typically free of circulating organisms (by culture and/or cytospin), implying *Coccidioides immitis* attachment to receptor sites within the meninges. Rarely, *C. immitis* may be observed on cytospin preparation and rarer yet is the observation of mycelia or arthroconidia (most often observed in association with a shunt placed for hydrocephalus management).[6] Diagnosis is therefore generally not based on a positive CSF culture, but rather on a positive CSF serology documenting precipitin (IgM) antibody and IgG antibody by immunodiffusion technique or complement fixing antibody (performed at UC Davis). Unfortunately, no antigen test is commercially available. Generally the CSF complement fixing antibody titer has been used to assess response to therapy. Other abnormal CSF parameters generally present and used to assess therapy response include WBCs (generally elevated in the 100–500 range and predominantly, after disease establishment, lymphocytic), CSF glucose (generally low), and CSF protein generally elevated over 150 mg/dL (reflecting abnormalities induced in CSF circulation by exudative response to infection). Opening pressures are essential to obtain and monitor serially during routine lumbar punctures at clinical follow-up, as persistent elevations over 200 mm H_2O often herald the development of hydrocephalus.

TREATMENT OF COCCIDIOIDAL MENINGITIS

Left untreated, coccidioidal meningitis is nearly always fatal within 2 years.[7,8] Intrathecal amphotericin B was the agent of choice in treating this condition until approximately 1990. It is a problematic treatment modality in terms of patient acceptability, administration complexity and efficacy with only approximately 60% of patients responding. Dosing is variabl, ranging from 0.01 mg initially to a maximum of 1.0 mg in some patients, often accompanied by low-dose hydrocortisone added to decrease arachnoiditis and generally administered intracisternally by experienced clinicians, although others initiate treatment by hyperbaric technique. Details of this therapeutic approach have been recently reviewed.[9] Amphotericin B intrathecal therapy has, as its only advantage, the potential for cure of this disease.

Azole therapy has largely replaced intrathecal amphotericin B for induction and maintenance of therapy for this disease and appears to have reasonable efficacy when compared to intrathecal amphotericin B; it also has improved patient acceptability.[7,8] In initial reports by Tucker et al.[10,11] efficacy in over 60% of patients was reported. This initial experience was followed by a report by Galgiani et al.[12] in which 47 patients with coccidioidal meningitis were initially treated with 400 mg of fluconazole administered orally. Approximately 50% of these patients had previously received other antifungal agents (i.e., intrathecal amphotericin B, oral ketoconazole, miconazole) and had relapsed. Of these patients, 79% responded favorably to the initial dose of fluconazole, with only 21% ($n =$ 10) failing to respond. Of the ten nonresponders, one developed hydrocephalus and required concomitant intrathecal amphotericin B deoxycholate. Two patients were treated daily with 800 mg of fluconazole, with one patient dying of hydrocephalus and the other of complications from AIDS. One required the addition of concomitant intrathecal amphotericin B. Six of the remaining nonresponding patients were treated with 800 mg of fluconazole daily, with four demonstrating a favorable response, one was lost to follow up and thus unevaluable, and one later in the course required the addition of intrathecal amphotericin B resulting in ultimate control of infection.

Tucker et al.[13] reported on eight patients with coccidioidal meningitis who were failing to respond to intrathecal amphotericin B and were begun on itraconazole, between 300 and 400 mg orally daily. Five patients received only itraconazole and demonstrated clinical response. Three patients continued intrathecal amphotericin B initially but were later able to continue on itraconazole monotherapy as they later demonstrated clinical success.

Johnson et al.[14] presented a 1997 abstract describing the response of 44 newly diagnosed patients with coccidioidal meningitis treated with fluconazole alone. Sixteen patients received initial therapy with 400 mg of fluconazole and 28 patents were treated initially with 800 mg of fluconazole. Thirty-one percent of patients treated initially with 400 mg responded, whereas 75% responded at the 800-mg dosage. All patients were evaluated using a Mycosis Study Group (MSG) point system protocol that employed both clinical and CSF parameters. These authors concluded that beginning therapy with fluconazole at a higher dose resulted in improved response.

In a clinical outcome study by the same group, 115 patients with coccidioidal meningitis were studied. Sixty percent of patients were initially given 800 mg of fluconazole, 16% received 400 mg, 7% received some dose of fluconazole with concomitant intrathecal amphotericin B deoxycholate, and 5% received intrathecal amphotericin B deoxycholate as monotherapy. Twelve percent of the patients were lost to follow-up and thus unevaluable. Overall, 52% of evaluable patients responded favorably using the MSG scoring system ($n = 60$), 15% ($n = 18$) were nonresponders, 32% ($n = 29$) died of causes unrelated to coccidoidal infection, and 7% of treated patients could not be evaluated as they were not on therapy sufficiently long to assess efficacy at the time these data were reported in abstract form. Eventually, all of the nonresponders' disease was controlled with a combination of intrathecal amphotericin B in conjunction with oral fluconazole.[15]

After more than a decade of experience with azoles in the treatment of coccidioidal meningitis, the following conclusions can be drawn. These agents are a major advance in treatment as they appear equally efficacious in controlling infection (although they lack the potential for cure), but are not always effective in maintaining remission in the long term, even when administered at higher dosage. In addition, because Azoles are fungistatic agents, lifelong therapy is required.[1,16,17] To date, there have been no head-to-head comparative studies of azoles in humans. There are several animal experiments in rabbits[18] and mice[19] comparing itraconazole versus fluconazole in the treatment of experimental coccidioidal meningitis and these have demonstrated equivalent efficacy (rabbit study) or superior efficacy (mouse study) for itraconazole on a mg/kg basis.

There are three possible variations for induction therapy of coccidioidal meningitis at first diagnosis. Some clinicians favor beginning high-dose azole therapy (i.e., 800 mg or higher of fluconazole) until clinical improvement is established and then decreasing the dose slowly over a period of months to maintenance doses of between 400 and 600 mg daily.[1] Another option is to begin treatments combining intrathecal amphotericin B deoxycholate with an azole to gain rapid control, which theoretically allows for the possibility of cure and possibly decreases the risk of complications (discussed below). In addition, using this format potentially could allow for a lower total intrathecal dose of amphotericin B than what might otherwise be expected if added later to a failing azole regimen.[9] Other physicians still begin with an initial dosage of 400 mg of fluconazole and escalate upward later as necessary.[20] However, no comparative studies have been done to compare these approaches to determine whether they are equal or unequal in effectiveness.

In a failing initial azole regimen, a number of options are available. If fluconazole at high dose is failing, itraconazole can reasonably be substituted and many authorities recommend 600 mg daily in this setting. If the initial regimen is itraconazole, then fluconazole at high dose (800 mg daily or higher) can be substituted. Adding intrathecal amphotericin B deoxycholate to a failing azole regimen is also a viable option, but should be performed by experts experienced with this form of therapy. Intrathecal amphotericin B can be administered either by hyperbaric intralumbar technique, as favored by some, or by standard intracisternal injection in experienced hands, which has the advantage of delivery of the antifungal agent into the basal cisterns, the site of predominant infection. Dose ranges from 0.01 mg at initiation, with some patients tolerating 1 mg, generally administered 3–7 times weekly as tolerated, until control of infection is established.[9]

There have been limited, though favorable, case reports of patients with coccidioidal meningitis and spinal arachnoiditis who were not responding to high-dose fluconazole or a combination of azole and intrathecal amphoteracin B dexoxycholate, or high-dose azole combined with intravenous amphotericin B lipid complex (ABLC), but who subsequently responded to voriconazole.[1, 21, 22] Other potential therapeutic agents include intravenous AmBisome, oral nikkomycin Z, and oral posaconazole, but little to no published human data exists concerning these agents and no conclusions or recommendations can be drawn at this time. An experimental efficacy trial in rabbits comparing intravenous liposomal amphotericin B (AmBisome) to oral fluconazole and intravenously administered amphotericin B deoxycholate demonstrated superior efficacy for AmBisome.[23] Three of eight animals given 15 mg/kg had no detectable residual infection either in the brain or spinal cord. Applicability of these findings to humans with coccidioidal meningitis awaits completion of prospective clinical studies in humans. Although there is a theoretical concern regarding the potential for antagonism between amphotericin B and azoles, in over a decade of experience no significant clinical evidence for antagonism in the treatment of coccidioidal infection has been reported.[8, 12, 24]

COCCIDIOIDAL MENINGITIS–ASSOCIATED COMPLICATIONS

Complications of chronic meningeal inflammation include hydrocephalus and vasculitis with associated stroke events. Hydrocephalus may be either communicating, likely on account of failure of the arachnoid granulations to reabsorb CSF because of exudative "plugging," or noncommunicating on account of obstruction at the level of the third ventricle from meningeal inflammation obscuring the foramina of Luschka and Magendie. Romero and colleagues have summarized the five studies listing the incidence of hydrocephalus. Although the average incidence approximates 40% there is considerable variation.[25] The documentation of hydrocephalus ultimately requires placement of a ventricular peritoneal shunt to relieve CSF pressure.

Radiologic evidence of vasculitis by CT or MRI during the course of this chronic disease occurs in approximately 40% of patients.[26] Complications resulting from stroke may occur in 10–40% of patients, with mortality ranging from 15–70%.[2,3,26] Hydrocephalus may be a marker for the later development of vasculitis. Vasculitis appears to be correlated with the intensity of contiguous meningeal inflammation, which may not always be reflected by analysis of CSF.[27] Proinflammatory molecules appear to be involved, but their relative contributions to the development of vasculitis have not been ascertained.[2,28] Vasculitic complications are the commonest cause of death in patients with CM in the modern era.[20]

The best approach to the management of vasculitis has not been established. Some clinicians have reported success with oral dexamethasone, whereas others have expressed reservation with this approach.[1,2] It is hoped that animal model studies will allow for future clarification of the molecular events responsible and ultimately allow for improved preventive and therapeutic strategies.

SUMMARY

In summary, patients with coccidioidal meningitis frequently present to the attending physician a complex clinical picture that is often not straightforward. These patients require therapeutic "customizing." Managing the various associated complications such as vasculitis can also be challenging. Consultation with an experienced clinician familiar with this disease and its management is advisable. Until a highly effective therapeutic fungicidal agent becomes available for this disease, management will remain potentially confounding and problematic, particularly as it relates to management of vasculitis.

REFERENCES

1. JOHNSON, R.H. & H.E. EINSTEIN. 2006. Coccidioidal meningitis. Clin. Infect. Dis. **42:** 103–107.
2. WILLIAMS, P.L. 2001. Vasculitic complications associated with coccidioidal meningitis. Semin. Respir. Infect. **16:** 270–279.
3. WILLIAMS, P.L., R. JOHNSON, D. PAPPAGIANIS, *et al.* 1992. Vasculitic and encephalitic complications associated with *Coccidioides immitis* infection of the central nervous system in humans: report of 10 cases and review. Clin. Infect. Dis. **14:** 673–682.
4. BANUELOS, A.F., P.L. WILLIAMS, R.H. JOHNSON, *et al.* 1996. Central nervous system abscesses due to *Coccidioides* species. Clin. Infect. Dis. **22:** 240–250.

5. ERLY, W.K., E. LABADIE, P.L. WILLIAMS, *et al.* 1999. Disseminated coccidioidomy-cosis complicated by vasculitis: a cause of fatal subarachnoid hemorrhage in two cases. AJNR Am. J. Neuroradiol. **20:** 1605–1608.

6. HAGMAN, H.M., E.G. MADNICK, A.N. D'AGOSTINO, *et al.* 2000. Hyphal forms in the central nervous system of patients with coccidioidomycosis. Clin. Infect. Dis. **30:** 349–353.

7. GALGIANI, J.N. 1999. *Coccidioides* species. *In* Principals and Practice of Infectious Diseases, 6th edition. G.L. Mandel, J.E. Bennett R. Dolin, Eds.: 3040–3051. Williams and Wilkins. Baltimore, MD.

8. WILLIAMS, P.L. & N.M. AMPEL. 1999. *Coccidioides immitis*. *In* Antimicrobial Therapy and Vaccines. V.L. Yu, T.C. Merigan, S.L. Barriere, Eds. Williams and Wilkins. Baltimore, MD.

9. STEVENS, D.A. & S.A. SHATSKY. 2001. Intrathecal amphotericin in the management of coccidioidal meningitis. Semin. Respir. Infect. **16:** 263–269.

10. TUCKER, R.M., J.N. GALGIANI, D.W. DENNING, *et al.* 1990. Treatment of coccidioidal meningitis with fluconazole. Rev. Infect. Dis. **12** (Suppl. 3): S380-S389.

11. TUCKER, R.M., P.L. WILLIAMS, E.G. ARATHOON, *et al.* 1988. Pharmacokinetics of fluconazole in cerebrospinal fluid and serum in human coccidioidal meningitis. Antimicrob. Agents Chemother. **32:** 369–373.

12. GALGIANI, J.N., A. CATANZARO, G.A. CLOUD, *et al.* 1993. Fluconazole therapy for coccidioidal meningitis. The NIAID-Mycoses Study Group. Ann. Intern. Med. **119:** 28–35.

13. TUCKER, R.M., D.W. DENNING, B. DUPONT, *et al.* 1990. Itraconazole therapy for chronic coccidioidal meningitis. Ann. Intern. Med. **112:** 108–112.

14. JOHNSON, R., J. CALDWELL, H. EINSTEIN, & P. WILLIAMS. 1997. Therapy of coc-cidioidal meningitis in the 90's. Proceedings of the Annual Coccicioidomycosis Study Group Meeting.

15. JOHNSON, R., P.L. WILLIAMS, J. CALDWELL, & H. EINSTEIN. 1999. Coccidioidal meningitis: a decade of experience. Proceedings of the Annual Coccidioidomy-cosis Study Group Meeting.

16. STEVENS, D.A. 2006. Coccidioidal meningitis. Clin. Infect. Dis. **43:** 385; author reply 385–386.

17. DEWSNUP, D.H., J.N. GALGIANI, J.R. GRAYBILL, *et al.* 1996. Is it ever safe to stop azole therapy for *Coccidioides immitis* meningitis? Ann. Intern. Med. **124:** 305–310.

18. SORENSEN, K.N., R.A. SOBEL, K.V. CLEMONS *et al.* 2000. Comparison of flucona-zole and itraconazole in a rabbit model of coccidioidal meningitis. Antimicrob. Agents Chemother. **44:** 1512–1517.

19. KAMBERI, P., R.A. SOBEL, K.V. CLEMONS, *et al.* 2006. Comparison of itracona-zole and fluconazole treatment in a murine model of coccidioidal meningitis. Antimicrob. Agents Chemother. **51:** 998–1003.

20. GALGIANI, J.N., N.M. AMPEL, J.E. BLAIR, *et al.* 2005. Coccidioidomycosis. Clin. Infect. Dis. **41:** 1217–1223.

21. CORTEZ, K.J., T.J. WALSH & J.E. BENNETT. 2003. Successful treatment of coccid-ioidal meningitis with voriconazole. Clin. Infect. Dis. **36:** 1619–1622.

22. PROIA, L.A. & A.R. TENORIO. 2004. Successful use of voriconazole for treatment of *Coccidioides* meningitis. Antimicrob. Agents Chemother. **48:** 2341.

23. CLEMONS, K.V., R.A. SOBEL, P.L. WILLIAMS, *et al.* 2002. Efficacy of intravenous li-posomal amphotericin B (AmBisome) against coccidioidal meningitis in rabbits. Antimicrob. Agents Chemother. **46:** 2420–2426.

24. PEREZ, J.A. Jr., R.H. JOHNSON, J.W. CALDWELL, *et al.* 1995. Fluconazole therapy in coccidioidal meningitis maintained with intrathecal amphotericin B. Arch. Intern. Med. **155:** 1665–1668.
25. ROMEO, J.H., L.B. RICE & I.G. MCQUARRIE. 2000. Hydrocephalus in coccidioidal meningitis: case report and review of the literature. Neurosurgery **47:** 773–777.
26. ARSURA, E.L., R. JOHNSON, J. PENROSE, *et al.* 2005. Neuroimaging as a guide to predict outcomes for patients with coccidioidal meningitis. Clin. Infect. Dis. **40:** 624–627.
27. MISCHEL, P.S. & H.V. VINTERS. 1995. Coccidioidomycosis of the central nervous system: neuropathological and vasculopathic manifestations and clinical correlates. Clin. Infect. Dis. **20:** 400–405.
28. ZUCKER, K.E., P. KAMBERI, R.A. SOBEL, *et al.* 2006. Temporal expression of inflammatory mediators in brain basilar artery vasculitis and cerebrospinal fluid of rabbits with coccidioidal meningitis. Clin. Exp. Immunol. **143:** 458–466.

Deep Solitary Brain Mass in a Four-Month-Old Male with Disseminated Coccidioidomycosis

Case Report

JENNIFER D. NOLT AND FRANCESCA R. GEERTSMA

Children's Hospital Central California, Madera, California, USA

ABSTRACT: Parenchymal brain involvement from disseminated coccidioidomycosis occurs rarely and there are few documented pediatric cases. We report a four-month-old male infant with a cerebellar lesion seen in the brain on computed tomography (CT). *Coccidioides immitis* (*C. immitis*) grew on bronchoscopic fluid samples and serum titers to *C. immitis* were 1:1024. Antifungal treatment was initiated and after 3 months, CT scans demonstrated brain mass resolution and serum titers were decreased.

KEYWORDS: disseminated coccidioidomycosis; brain mass; pediatrics

INTRODUCTION

Disseminated coccidioidomycosis has been a recognized entity since the first documented case of coccidioidomycosis was reported in 1892; it occurs in approximately 1% of those affected with this endemic mycosis. The disease was once thought to be caused by an insect that deposited the fungus into the host through skin lesions.[1] Research has since shown that the primary lesion of coccidioidal granuloma is frequently pulmonary.[2]

About 25% of cases of disseminated coccidioidomycosis involve the central nervous system (CNS)[3] and this dissemination usually occurs within weeks to months after the disease is contracted. Reports of parenchymal CNS lesions are very rare.[4] According to an extensive review conducted by Banuelos[5] in 1996, 39 cases of brain coccidioidal lesions were reported since the first case in 1905, and the youngest age he reported was in a 2-year-old child.[6] Although the child was described as having an intracranial coccidioidal lesion, it is unclear whether the intracranial mass was biopsy-proven for *Coccidioides immitis* (*C. immitis*).

Address for correspondence: Jennifer D. Nolt, CPNP, 110 North Valeria Street, Suite 206, Fresno, CA 93701. Voice: 559-264-2504; fax: 559-264-3707.
jwardpnp@hotmail.com

Ann. N.Y. Acad. Sci. 1111: 385–394 (2007). © 2007 New York Academy of Sciences.
doi: 10.1196/annals.1406.006

Of the intracranial coccidioidal cases reported by Banuelos,[5] meningitis was absent in approximately one third. Serology, culture, and cytology may reveal the etiology, but biopsy of the mass provides the best method for diagnosis when serology and culture are negative for *C. immitis.*[7] In many cases, the risks of a biopsy of an intracranial mass are determined to be too significant to the patient, so pathology of the intracranial mass is not known until an autopsy is performed. Schlumberger[4] described a fatal case of coccidioidal brain lesions in a patient in whom spherules of *C. immitis* were also found in the sputum. In this case, the fungus could not be isolated from the spinal fluid. In addition, Mendel[8] described a fatal case in 1994 of a patient with a coccidioidal brain abscess where the fungus was also identified in the patient's sputum and serum. Cerebral spinal fluid (CSF) studies were normal in this patient.

The location of intracranial coccidioidal masses varies. In further analysis of the cases reported by Banuelos,[5] 10 of the 39 coccidioidal intracranial cases were associated with cerebellar lesions. Additionally, the case reported by Craig[2] involved an adult with a left cerebellar coccidioidal lesion. A case involving a male with diabetes mellitus reported by Rhoden[9] also had a left cerebellar coccidioidal lesion. According to Scanarini,[7] the characteristic lesion of coccidioidomycosis is that of a diffuse granulomatous lesion that encases the brain and may obstruct the flow of CSF. The basiliar region is often involved, especially in cases involving meningitis.

CASE REPORT

Hospital Encounter 1: October 2005

The male infant presented at 23 days of life with fever of 104°F. Past medical history revealed a full-term uncomplicated birth with no known immunodeficiency. The infant resided in Kerman, California with his parents and had no significant travel history. The review of systems was significant for a decreased appetite, 1 day of fever and a generalized rash for 1 week. The infant was admitted for a sepsis work-up. Blood, urine, and CSF cultures were negative. CSF studies were relatively unremarkable [7 white blood cells (WBCs) with 32% lymphocytes and 68% monocytes, 4 red blood cells (RBCs), glucose at 46 mg/dL (normal value 50–80 mg/dL) and protein at 63 mg/dL (normal term infant value 20–170 mg/dL)]. Complete blood count (CBC) and chest X ray (CXR) were normal. C-reactive protein (CRP) was mildly elevated at 2.4 mg/L (normal value less than 2 mg/L). Nasal swabs for respiratory synctial virus and influenza antigens were negative. The infant was treated empirically with intravenous ampicillin sodium (Principen; Apothecon, Inc.; Princeton, NJ, USA) and gentamicin sulfate (Garamycin; Schering; Kenilworth, NJ, USA). On day 3, he was afebrile and discharged home with a presumptive diagnosis of a viral illness.

Hospital Encounter 2: December 2005

The infant presented at 3 months of age for irritability and abdominal distension for 2 weeks. On exam, he was fussy and coughing. CXR was preliminarily read as normal. He was diagnosed with abdominal distension of unknown etiology and sent home. Final CXR report illustrated a left lower infiltrate. The infant was asked to return to the hospital for reevaluation.

Hospital Encounter 3: December 2005

The infant returned for care 5 days later. He had a decreased appetite, cough, fussiness, and congestion. The mother reported that she had a recent upper respiratory tract infection; her symptoms resolved without requiring medical intervention. Faint pulmonary crackles were heard on exam. Repeat CXR illustrated infiltrates in the bilateral lower lobes and the right upper lobe. The infant was admitted for pneumonia. Nasal washes for respiratory synctial virus antigen and pertussis polymerase chain reaction (PCR) were negative. CBC was significant for slight anemia (hemoglobin of 10.5 g/dL). CRP was within normal limits. The infant was treated with intravenous cefuroxime sodium (Zinacef; GlaxoSmithKline; Research Triangle Park, NC, USA). He improved clinically and was discharged home after 2 days with a prescription for cefdinir (Omnicef; Abbott Laboratories; Chicago, IL, USA) and instructions to repeat a CXR in 2 weeks. Follow-up CXR revealed minimal infiltrates at the bilateral lung bases and right apex.

Hospital Encounter 4: February 2006

The infant presented at 4 months of age with fever for 3 days, cough, fussiness, generalized cervical adenopathy, and decreased activity. Social history revealed a possible positive tuberculosis exposure as a cousin visiting the home was found to have an abnormal tuberculosis test and was undergoing medical evaluation. On exam, the infant had a fever of 105.2°F, and was irritable, coughing, and short of breath with bilateral crackles. CBC revealed anemia (hemogloblin of 9.1 g/dL). CRP was elevated at 18.6 mg/dL (normal value less than 2 mg/dL). CSF was unremarkable [4 WBCs with 53% lymphocytes and 47% monocytes, 17 RBCs, glucose at 63 mg/dL (normal value 50–80 mg/dL) and protein at 37 mg/dL (normal value 20–40 mg/dL)]. There was no growth on CSF, blood, or urine cultures. Serum and CSF were preliminarily negative for coccidioidal IgG and IgM by enzyme immunoassay. CXR illustrated bilateral upper lobe and left lower lobe infiltrates with a left pleural effusion. The infant was admitted with the diagnosis of pneumonia and started empirically on ceftriaxone sodium (Rocephin; Roche; Nutley, NJ, USA).

On day 2, there was no clinical improvement. *Coccidioides* serology by enzyme immunoassay revealed a negative IgM and positive IgG. CSF *Coccidioides* tests were negative. Clindamycin phosphate (Cleocin Pediatric; Pharmacia & Upjohn; Kalamazoo, MI, USA) was empirically started.

On day 3, computed tomography (CT) scans were performed both to evaluate the cervical adenopathy and to check for generalized adenopathy. The neck CT revealed left-sided adenopathy. The radiologist performing the CT began the evaluation of the neck at the base of the brain. Incidentally, a right cerebellar density was seen. Because of this incidental finding, a CT of the brain was ordered, which revealed a 5-mm density in the right cerebellar hemisphere that was postulated to be either hemorrhage with edema, tuberculosis, malignancy, or infection (FIG. 1). The chest CT showed mediastinal adenopathy with scattered pulmonary nodules and a left pleural effusion. The abdominal CT illustrated adenopathy, bilateral hydronephrosis, and hydroureters.

The infectious diseases department was consulted. Differential diagnoses included disseminated tuberculosis, malignancy, and disseminated coccid-

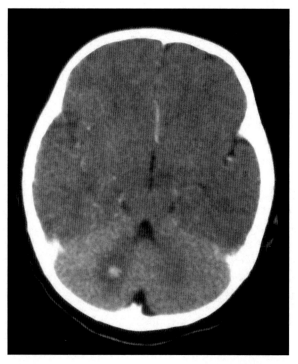

FIGURE 1.   Edema surrounding a 5-mm density in the white matter of the right cerebellar hemisphere.

ioidomycosis. Procedural recommendations included: a tuberculosis skin test, three morning gastric aspirates for acid-fast bacilli (AFB) stain and culture, bronchoscopy with samples sent for cytologic and routine testing, fungal and AFB cultures, and serology for immunodiffusion and complement fixation (CF) antibodies to *C. immitis*. Treatment recommendations included: discontinuing clindamycin phosphate, starting four antituberculosis medications [isoniazid (Nydrazid; Bristol-Myers Squibb; Princeton, NJ, USA), pyrazinamide (Pyrazinamide; Lederle Laboratories; Philadelphia, PA, USA), rifampin (Rifadin; Aventis; Kansas City, MO, USA) and ethambutol hydrochloride (Myambutol; Lederle Laboratories; Philadelphia, PA, USA)], starting oral fluconazole (Diflucan; Pfizer; New York, NY, USA) empirically at 10 mg/kg/day and initiating a public health investigation for the possible tuberculosis exposure.

Serum immunodiffusion antibodies to *C. immitis* were positive at 1:64 by IgG assay and quantity was not sufficient for the CF titer test. Tuberculosis skin test was negative. CXR showed slight improvement. The infant was more energetic and had an improved appetite and decreased cervical adenopathy. Bronchoscopic fluid culture testing revealed *C. immitis*. Ceftriaxone sodium was stopped and oral fluconazole was increased to 15 mg/kg/day. A bone scan to rule out coccidioidomycotic bone involvement revealed renal asymmetry. A cystogram illustrated grade IV vesicouretal reflux, bilateral hydroureter, and mild hydronephrosis. A brain magnetic resonance imaging illustrated a 5–6-mm right cerebellar enhancing lesion without signs of meningeal enhancement or basilar meningitis. A neurosurgical consultation was obtained, but biopsy was deferred on account of the location of the intracranial mass.

The infant was discharged on day 10 with diagnoses of disseminated coccidioidomycosis, cerebellar abscess, and grade IV vesicoureteral reflux. He was to continue oral fluconazole at 15 mg/kg/day and nitrofurantoin (Macrodantin; Procter & Gamble; Cincinnati, OH, USA) as prophylaxis for his urinary reflux. He was to be followed up in the outpatient clinics of the infectious diseases and urology departments.

First Infectious Disease Clinic Follow-Up: March 2006

The infant presented 10 days following discharge. He was taking oral fluconazole at 15 mg/kg/day. The infant's serum titer by CF antibodies to *C. immitis* was at 1:1024. Exam was significant for fever of 102.3°F and congestion, with family members reporting similar symptoms. Repeat CT scans were performed. The CT of the brain revealed a 5-mm focus in the right cerebellar hemisphere with slight edema. The lesion appeared slightly smaller with resolution of the surrounding edema. The neck CT showed stable extensive left neck adenopathy with some nodes suggesting necrosis. The chest CT demonstrated stable mediastinal adenopathy with improving left lung consolidations in the upper and lower lobes. A small consolidation was seen in the right upper and middle lobes and a cavitary lesion was developing in the lingula.

Second and Third Infectious Disease Clinic Follow-Ups: April and May 2006

The infant was clinically well. His serum titer by CF antibodies to *C. immitis* was at 1:256 and immunodiffusion revealed a positive IgG. May's CT of the brain revealed that the right cerebellar lesion was no longer seen. There was a questionable small area of possible enhancement remaining but the remainder of the brain parenchyma was normal (FIG. 2). CXR showed no signs of infiltrate or pneumonia. The infant's parents were instructed to continue to administer fluconazole and attend monthly follow-up appointments.

DISCUSSION

Coccidioidal granuloma tends to occur in individuals who have recently

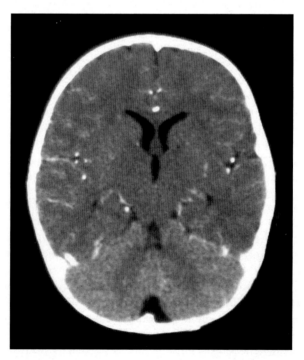

FIGURE 2. Resolution of the right cerebellar lesion leaving a questionable tiny area of enhancement.

arrived in an endemic area and lack specific immunologic protection.[9] In addition, immunosuppression is a known risk factor for disseminated disease. Case reports of coccidioidal intracranial lesions have been reported in patients with AIDS[10] and diabetes.[9,11]

Patients with intracranial coccidioidal lesions can manifest signs and symptoms of their disease via two different pathways. First, patients can undergo a primary pulmonary or skin infection and spread to the CNS, but the symptoms seen represent the primary infection. The second situation involves patients who have CNS involvement and the symptoms expressed reflect the secondary neurological damage from the disseminated focus, but the original focus of infection (e.g., pulmonary or skin infection) is no longer evident. In these instances, the original foci have experienced a remarkable degree of healing.[12] According to Banuelos,[5] the most discrete intracranial granulomas are not due to direct meningeal invasion, but are caused by hematogenous spread from a lung or infected meningeal source. The blood–brain barrier is supposed to protect against intracranial invasion, but barrier dysfunction can occur by hematogenous transport of mediator substances released or activated by damaged or activated cells and result in CNS diseases.[13]

In our patient's case, he presented with generalized symptoms of fever, cough, and cervical adenopathy and through diagnostic tests to identify the etiology behind these symptoms his CNS involvement became apparent. His chest CT illustrated pulmonary nodules that are hypothesized to be the primary coccidioidal focus from which dissemination to the cerebellum occurred; this is supported by the growth of the *C. immitis* culture from the bronchoscopy specimen. A hematogenous origin of the cerebellar abscess is suspected in our patient as there was no evidence of any other contiguous structures being affected. Focal brain lesions are unusual in CNS coccidioidomycosis. Our patient is particularly intriguing given that he had no evidence of meningitis and his presentation was at such a young age. We hypothesize that his disease originated in the lung and traveled hematogenously or via the lymphatics to the cerebellar parenchyma. We further speculate that perhaps his blood–brain barrier defense mechanisms were immature because of his age.

Our patient's intracranial mass was discovered through happenstance. The usual method of evaluating cases of disseminated coccidioidomycosis involves a lumbar puncture to evaluate CSF and a bone scan to evaluate bony structures. Unless a patient presents with neurological symptoms, brain imaging is not routinely performed. Perhaps we should include head imaging in the work-up of all infants suspected of having disseminated coccidioidomycosis in view of the lack of presenting neurological signs and symptoms of CNS infections in this age group.

Optimal treatment of coccidioidal brain lesions is unknown. CT scans of the brain have been shown to be most valuable in the diagnosis and follow-up of patients with brain masses.[14] In our patient's case, he was empirically started on fluconazole to treat for coccidioidomycosis based on positive IgG serology. Amphotericin B (Amphocin; Intermune; Brisbane, CA, USA) was not used because tuberculosis and other etiologies were originally entertained and coccidioidomycosis was not initially highly suspected. Fluconazole was chosen for its excellent penetration into the CSF[15] and continues to be successfully used for patients with coccidioidomycotic brain masses and meningitis. It is unknown whether patients with intracranial coccidioidal masses who undergo surgical brain lesion resection and antifungal treatment will be cured. In our patient's case, risks of surgical intervention were determined to be too significant. Prolonged antifungal treatment, perhaps indefinitely, should be considered in all patients with intracranial coccidioidal lesions because of the high rate of recurrence.[5]

CONCLUSION

Our patient had disseminated coccidioidomycosis. Serum titers by CF antibodies to *C. immitis* were significantly elevated at 1:1024 and culture from a bronchoscopy specimen grew *C. immitis*. It is unusual to grow *C. immitis*. Similar to one-third of the reported intracranial coccidioidal cases, our patient's CSF studies were normal. On imaging, he had extensive pulmonary disease with lymphadenopathy, along with neck adenopathy. CT of the brain demonstrated a right cerebellar mass. Unfortunately, biopsy of the mass could not be undertaken to determine the definitive etiology on account of the surgical risk, but the mass resolved and the infant's *Coccidioides* serologic level decreased with appropriate anticoccidioidal treatment.

Since the infant received little therapy with anti-AFB or antibacterials, it would be difficult to conclude that the etiology of his intracranial lesion could derive from tuberculosis or typical bacteria. One would not have expected bacterial or AFB disease to resolve without extensive medical and/or surgical intervention. In addition, the mass was not believed to be attributable to neurocysticercosis. This diagnosis was considered in the initial differential, but upon further investigation, our patient was not tested for *Cisticercos* sp. because this organism causes abscesses to form in the cerebellar hemisphere and its presentation differs from that experienced by our patient. There are two types of cysts that can develop in the brain from cysticercosis. The less virulent form, *Cysticercus cellulosae,* has a predilection for the dorsolateral subarachnoid space, is often clinically silent, and rarely requires surgical intervention. The more inflammatory form, *Cysticercus racemosus,* has a predilection for the basal subarachnoid cisterns, forms grape-like clusters, and often causes seizures, intracranial hypertension, and hydrocephalus. In cases of active parenchymal neurocysticercosis, neuroimaging studies typically normalize within 1 to 2 years regardless of whether patients receive antiparasitic drugs[16]; this longer period of time required for neuroimaging resolution also argues against this diagnosis in our patient as neuroimaging resolution occurred after only three months.

We feel our patient's clinical course, including his dramatic response to only antifungal therapy taken together with his positive bronchoscopy culture for *C. immitis*, location of the intracranial lesion (which is typical in adults with coccidioidal intracranial abscesses), and evidence of high serology titer antibodies by CF to *C. immitis* (which is indicative of extensive extrapulmonary disease) make a strong case for this infant to have had a deep solitary brain mass due to *C. immitis*.

Our patient continues to thrive without any sign of a neurological abnormality. He is growing well and achieving his developmental milestones. CXR reveals that his pulmonary abnormalities have healed, and serum titers by CF to *C. immitis* continue a slow descent while he continues on oral fluconazole therapy at 15 mg/kg/day. His hydronephrosis and hydroureters, unrelated to the coccidioidomycosis infection, remain stable.

ACKNOWLEDGMENTS

The authors extend their gratitude to Tom Warren for offering his technical support in formatting the CT images.

REFERENCES

1. JACOBSON, H. 1928. Coccidioidal granuloma. Calif. Western Med. **29:** 392–396.
2. CRAIG, W. & M. DOCKERTY. 1941. Coccidioidal granuloma: a brief review with report of a case of meningeal involvement. Minn. Med. **24:** 150–154.
3. COURVILLE, C. & K. ABBOTT. 1938. Pathology of coccidioidal granuloma of the central nervous system and its envelopes. Bull. Los Angeles Neurol. Soc. **3:** 27–41.
4. SCHLUMBERGER, H. 1945. A fatal case of cerebral coccidioidomycosis with cultural studies. Am. J. Med. Sci. **3:** 27–41.
5. BANUELOS, A. *et al.* 1996. Central nervous system abscesses due to *Coccidioides* species. Clin. Infect. Dis. **22:** 240–250.
6. BUCHSBAUM, H. 1977. Clinical management of coccidioidal meningitis. *In* L. Ajello, Ed.: Coccidioidomycosis.:191–199. Symp. Specialists. Miami.
7. SCANARINI, M. *et al.* 1991. Primary intrasellar coccidioidomycosis stimulating a pituitary adenoma. Neurosurgery **28:** 748–751.
8. MENDEL, E. *et al.* 1994. Coccidioidomycotic brain abscess. J. Neurosurg. **80:** 140–142.
9. RHODEN, A. 1946. Coccidioidal brain abscess. Bull. Los Angeles Neurol. Soc. **11:** 80–85.
10. JARVIK, J. *et al.* 1988. Coccidioidomycotic brain abscess in an HIV-infected man. West. J. Med. **149:** 83–86.
11. NAKAZAWA, G. *et al.* 1983. Intracerebellar coccidioidal granuloma. Am. J. Neuroradiol. **4:** 1243–1244.

12. ABBOT, K. & O. CUTLER. 1936. Chronic coccidiodal meningitis. Arch. Pathol. **21:** 320–330.
13. GREENWOOD, J. 1991. Mechanisms of blood-brain barrier breakdown. Neuroradiology **33:** 95–100.
14. RODRIGUEZ, J. *et al.* 1978. The role of computed axial tomography in the diagnosis and treatment of brain inflammation and parasitic lesions: our experience in Mexico. Neuroradiology **16:** 458–461.
15. GALGIANI, J. *et al.* 1993. Fluconazole therapy for coccidioidal meningitis. Ann. Intern. Med. **119:** 28–35.
16. SCHELD, W. *et al.* 2004. Infections of the Central Nervous System. Third edition. Lippincott Williams & Wilkins. Philadelphia.

Comparative Aspects of Coccidioidomycosis in Animals and Humans

LISA F. SHUBITZ

Department of Veterinary Science and Microbiology, Valley Fever Center for Excellence, University of Arizona, Tucson, Arizona 85721

ABSTRACT: *Coccidioides* spp. appear capable of infecting all mammals and at least some reptiles. Development of disease as a result of infection is species-dependent. Dogs seem to have a susceptibility similar to that of humans, with subclinical infections, mild-to-severe primary pulmonary disease, and disseminated disease. Whereas central nervous system disease in humans is typically meningitis, brain disease in dogs and cats takes the form of granulomatous parenchymal masses. Osteomyelitis is the most common form of disseminated disease in the dog, while skin lesions predominate in the cat. Orally administered azole antifungal agents are the backbone of therapy in animals as they are in humans.

KEYWORDS: *Coccidioides*; coccidioidomycosis; dogs; cats; horses; llamas; primates

INTRODUCTION

Coccidioides spp. (*C. posadasii* and *C. immitis*) are fungi that inhabit the semiarid desert soils of the southwestern United States and Mexico, and localized regions of Central and South America. These dimorphic fungi are considered true pathogens and change from their saprophytic mycelial form found in soil into the round, thick-walled spherule/endospore form inside of an animal host. The infection is nearly always acquired by inhalation. *Coccidioides* spp. appear capable of infecting all mammals and at least some reptiles,[1,2] but have not been reported in any avian species to date. Though infection has been documented in a wide variety of animals, clinical states among species range from entirely subclinical to progressive and fatal.

Address for correspondence: Lisa F. Shubitz, D.V.M., Department of Veterinary Science and Microbiology, Bldg. 90, Room 221, University of Arizona, Tucson, Arizona 85721. Voice: 520-626-8198; fax: 520-621-6663.
lfshubit@u.arizona.edu

Ann. N.Y. Acad. Sci. 1111: 395–403 (2007). © 2007 New York Academy of Sciences.
doi: 10.1196/annals.1406.007

ANIMAL SPECIES AFFECTED BY COCCIDIOIDOMYCOSIS

Lesions containing *Coccidioides* spp. have been identified grossly and histologically in domestic livestock (i.e., cattle, sheep, and swine), but these species seldom develop disease.[3,4] Clinical infection is also relatively uncommon in horses, but is well-documented, and most reports are of disseminated disease with a poor outcome.[5–9] Llamas, in contrast, appear quite susceptible to progressive infection.[10,11](personal communications, Drs. B. Page and N. Leverens).

Among exotic animal species, which are mainly from wildlife exhibits in the endemic area, susceptibility is variable, but coccidioidomycosis has been reported as the cause of death among a wide variety of animals. Examples include nondomestic canids, nondomestic felids, bats, wallabies and kangaroos, tapirs, and Przewalski's horses.[2,12–14] Primates as a group appear particularly susceptible and many species, including lemurs, chimpanzees, gorillas, macaques, and baboons,[13,15,16] have been reported to have died from disseminated coccidioidomycosis.

Coccidioidal infection has also been detected in free-ranging wildlife. There have been two reports of coccidioidomycosis in western cougars, in which it caused death.[17,18] It has been reported as an incidental finding in three coyotes in southern Arizona.[19] Necropsy revealed disseminated coccidioidomycosis as the cause of death in a bottlenose dolphin captured emaciated and ill off the southern California coast,[20] and from a sea otter in the same area.[21] Seven free-living California sea lions have also been reported with disseminated coccidioidomycosis based on postmortem examination.[22] Presumably, such marine mammals are inhaling arthroconidia blown from land over the water. To the best of the author's knowledge, *Coccidioides* spp. do not grow in sea water.

Finally, the domestic dog has a frequency and range of coccidioidal illness similar to those in humans. In a recent prospective study, 70% of dogs that became seropositive for *Coccidioides* spp. were without evidence of clinical disease,[23] similar to the 60% rate of subclinical infection reported for humans.[24,25] Domestic cats are diagnosed with coccidioidomycosis much less often than dogs, but commonly are severely ill with disseminated disease at the time of diagnosis.[26]

INCIDENCE OF DISEASE

Estimates of disease incidence in dogs indicate that about 4% of the canine population of Pima, Maricopa, and Pinal counties in Arizona become ill with coccidioidomycosis on an annual basis.[23] Annual infection estimates of 150,000 for humans[25] are not exactly equivalent to the canine data because we do not have canine census data and the human data do not divide the clinically ill from asymptomatic infections. If the data were able to be compared by the number of illnesses per population, the author surmises it would be similar.

Regarding feline coccidioidomycosis, clinical records suggest there are about 50 canine cases for every feline case that is diagnosed. However, a review of necropsy records shows a rate of one feline case for every three canine cases (personal communication, Dr. S. Dial). Two possible reasons for this are that the rate in cats is more similar to that in dogs than is generally seen, or that demise due to undiagnosed coccidioidal infection is more common in cats than dogs, leading to more frequent necropsy to determine the cause of death. The latter suggests that increased awareness about coccidioidomycosis in cats may increase antemortem diagnosis. Also, cats are usually more systemically compromised than dogs when clinical signs are noted and possibly they decompensate and die before a diagnosis can be obtained.

DIAGNOSIS

Serology is the backbone of diagnosis of coccidioidomycosis in animals and uses the same reagents and controls as for serodiagnosis of humans. Since canine serum contains anticomplementary components,[27] agar gel immunodiffusion (AGID) is the test of choice in the dog. AGID is widely used by commercial laboratories for serological testing of all animal species, and is currently the most commonly used test in human serology as well.[28]

A comparison of the relationship between IgG titers and disease in dogs and humans has been reviewed elsewhere.[23] Briefly stated, there is reportedly good correlation in people with severity of illness and the IgG titer, while the canine study showed a significant overlap among subclinically and clinically infected dogs with titers between <1:2 and 1:16. Also, veterinarians and veterinary pathologists observe a clinically relevant but undefined number of seronegative dogs with coccidioidal disease.

Little is known about the relationship of serologic titer and clinical disease in cats. Greene and Troy reported that of 39 cats that were diagnosed with coccidioidomycosis and had serologic performed, all were seropositive.[26] There is no research regarding serologic status of healthy cats, but on the basis of currently available literature, seropositivity appears to correlate with clinically important disease in cats.

CLINICAL MANIFESTATIONS

Like humans, dogs develop primary pulmonary and disseminated coccidioidomycosis. Common clinical signs in dogs with primary coccidioidomycosis are coughing, weight loss, fever, lethargy, and anorexia, similar to the way humans may present.[25] Thoracic radiographs in dogs may reveal pneumonia, ranging from mild interstitial infiltration to extensive miliary pneumonia or consolidation of lung lobes, or only hilar lymphadenopathy may be present, the latter a hallmark of coccidioidomycosis in dogs in the endemic region. Humans have similar radiographic changes, though hilar lymphadenopathy appears to be reported less often in humans than in dogs, and to the author's knowledge there is no equivalent in dogs to the coccidioidal cavity seen in human lungs.[25]

Manifestations of disseminated disease are referable to the system involved. Bone is the most common site of dissemination in dogs.[27,29] Appendicular skeleton is affected more often than axial, whereas in humans with bone involvement, vertebral osteomyelitis is most prevalent.[30] Radiographic lesions of dogs and cats usually have a proliferative component with periosteal new bone formation as well as lysis,[27,31] which is in contrast to the largely lytic nature of coccidioidal bone lesions in humans.[30] Osteomyelitis in primates resembles that in humans with a primarily lytic process.[15,32,33]

Central nervous system disease occurs in dogs and cats, but is usually the result of a granulomatous mass lesion in the brain tissue[34] rather than the meningitis typical in humans; mass lesions of the brain as seen in animals are rare in humans.[35] The majority of dogs with brain disease present with seizures (personal communication, Dr. D. Levesque) while cats appear to have a wider range of manifestations, including incoordination, hyperesthesia, seizures, and behavior changes[26] (personal communication, Dr. S. Dial). Meningitis and infections of the actual spinal cord parenchyma are uncommonly reported in animals, though the inflammation from vertebral osteomyelitis may result in pain and paresis or paralysis. A cat with a spinal cord granuloma was reported in 2005.[36]

Skin lesions are common in people and in cats. More than half of cats diagnosed clinically with coccidioidomycosis have skin lesions.[26] In cats, non-healing lesions that can appear as dermatitis, ulceration, or abscesses that do not respond to antibiotics are often eventually biopsied, cultured, or aspirated, yielding the diagnosis of coccidioidomycosis. Skin lesions are also seen in approximately half of human disseminated cases and are highly variable in presentation.[37] In dogs, skin lesions are most commonly draining tracts that overlie sites of osteomyelitis.[27] Dogs may also develop abscesses involving soft tissue only, but all cutaneous manifestations are fewer in dogs than in people and cats.

Though *Coccidioides* spp. appear to be able to disseminate to virtually any site in the body of animals or humans, pericardial disease in the dog merits discussion. In two different necropsy series, pericardial, epicardial, and myocardial infection was the cause of death in about 25% of cases necropsied (13/46, 13/52) because of either heart failure or sudden death.[29,38] A retrospective study was also published recently in which 17 dogs underwent subtotal pericardiectomy for pericardial/epicardial coccidioidomycosis in a period of 4 years at one surgical referral center in the endemic region (Phoenix, AZ).[39] By comparison, a search of the human literature revealed only 15 reported cases of coccidioidal pericarditis.[40–45] Involvement of the pericardium was also noted in a horse at necropsy,[5] a llama,[11] and in 5/15 cats in a necropsy series (personal communication, Dr. S. Dial).

Recurrence of disease is a relatively common event among people that have had disseminated coccidioidomycosis, averaging 25–35% on the basis of clinical treatment trials of disseminated coccidioidomycosis.[46] Reinfection; with subsequent development of clinical signs, however, is rarely reported in humans and is usually the result of a high-dose laboratory exposure or a change in the immune status of the patient.[47] Though there are no data to differentiate reinfection versus recurrence in dogs, clinical experience confirms that illness can redevelop in dogs anywhere from weeks to years following withdrawal of medication. The percentage of dogs in this category is not known. The opinion of the author is that the bulk of dogs suffer recurrence rather than reinfection, this opinion is based on the paradigm for people and the fact that a case series from a time period prior to the availability of medication reported that 90% of the dogs that died of necropsy-confirmed coccidioidomycosis were 4 years of age or less.[48] This suggests that dogs develop lasting immunity to *Coccidioides* spp. as humans do. Reinfection cannot be ruled out on the basis of current knowledge, however, and dogs might be exposed to sufficiently large doses of the fungal arthroconidia to overwhelm immunity because of their proximity to dirt and their habit of sniffing. In a series of cats, approximately 30% had recurrence of clinical signs after withdrawal of medication,[26] a rate similar to that of people, and the author predicts that recurrence rates in dogs after discontinuing medication is in the same range as that of humans and cats.

TREATMENT

Treatment of coccidioidomycosis in animals is modeled after therapy in humans and uses the same arsenal of antifungal drugs. Economic considerations associated with cost of treatment may affect therapeutic decisions more often in animals than in humans. There are some unique problems with animal species that are not encountered in people. One example is delivery of medication to llamas and alpacas, closely related South American camelid species quite susceptible to disseminated coccidioidomycosis. The animals have a fermentative forestomach orad to the true stomach and intestines, making oral delivery of medication difficult on account of breakdown of drugs before they encounter an absorptive length of gastrointestinal tract. Jugular veins are difficult to access in these species for chronic administration of intravenous antifungal therapy as well. The author was recently involved with the treatment of an alpaca using subcutaneously administered amphotericin B deoxycholate[49] without success. Treatment was then attempted on two alpacas by administering fluconazole compounded into a gel rectally, but therapeutic drug levels could not be attained (<4 μg/mL);

the animals appeared to clinically improve for a time but ultimately succumbed. Recent conversations with two additional veterinarians revealed that they have treated alpacas and llamas with orally administered ketoconazole, itraconazole, or fluconazole (personal communications, Drs. K. Orr and N. Leverens) and obtained clinical improvement, though long-term outcome was not available in these animals. In light of these apparent clinical responses, studies evaluating blood levels of orally administered azoles in llamas and alpacas would be useful.

Acceptability of medication, cost of medication, and follow-up testing for treatment progress are challenges of treating nondomestic species in zoological institutions, where diagnostic testing of animals often requires anesthesia, and medication must be delivered so the animal will readily consume it.

In dogs and cats, oral treatment with ketoconazole, itraconazole, and fluconazole is usually feasible, both physically and economically, and these drugs are the first line of therapy used by veterinarians (unpublished survey results, C.D. Butkiewicz, *et al.*, April 2006). Oral therapy is convenient for owners and avoids the risk of renal complications associated with the use of amphotericin B deoxycholate intravenous infusions. Doses of fluconazole, ketoconazole, and itraconazole published in a veterinary formulary (*Veterinary Drug Handbook*, 4th edition, D. Plumb. Iowa State Press, Ames, Iowa, 2002; see individual drug monographs.) vary from 2.5 mg/kg twice daily to 20 mg/kg twice daily, though these dosages are anecdotal or based upon small case series, nearly all involving treatment of fungi other than *Coccidioides* spp. The range of doses in use by Arizona veterinarians per the aforementioned survey is consistent with the published doses for these drugs, with the exception of a wider range of fluconazole that includes higher-than-published doses. In the author's experience, doses of 10 mg/kg of fluconazole twice daily provide an improved response over 5 mg/kg twice daily in animals that are responding poorly. With more than 85% of Arizona veterinarians surveyed treating canine coccidioidomycosis with fluconazole (unpublished survey results, C.D. Butkiewicz, *et al.*, April 2006), it is clear from the wide range of doses in use that objective studies that include pharmacokinetics and response to fluconazole therapy in dogs are needed to help standardize therapy.

Side effects of oral azole therapy in dogs and cats include reduced appetite, vomiting, diarrhea, and elevation of liver enzymes, similar to side effects of azoles in humans.[50] In addition, a vasculitis that causes sterile, suppurative skin lesions has been documented as an adverse effect of itraconazole in dogs,[51,52] and the author is aware of similar anecdotal cases. An uncommon side effect, particularly of fluconazole, that has been reported to the author by several veterinarians is polyuria/polydipsia that resolves if the medication is stopped. Similar to the alopecia reported in humans,[50] some dogs experience thinning of the hair coat with fluconazole treatment (author's experience and personal communication with other veterinarians). Ketoconazole causes a reversible lightening of the haircoat in dogs (see individual drug monographs in *Veterinary Drug Handbook* referenced above) and may result in reduced fertility of males during treatment.

In recent years, amphotericin B lipid complex (Abelcet, Enzon Pharmaceuticals, Bridgewater, NJ, USA) has been used successfully to treat some canine and feline patients failing treatment with oral azole antifungal drugs (unpublished data, M.E. Matz and L.F. Shubitz). While it is renal-sparing in the majority of animals,[53] economic considerations limit its use. The author has also used subcutaneously administered amphotericin B deoxycholate plus fluconazole with success in a dog with coccidioidomycosis; however, the injection site reactions and failures to respond in other cases have limited therapy by this method even though it, too, is renal-sparing.[49]

In conclusion, there are very few reports in the literature about treatment of coccidioidomycosis in animals and there is a need for research in this area.

SUMMARY

All mammals, including humans, share a common susceptibility to infection with *Coccidioides* spp., though development of disease is variable. Dogs and humans share a rate and range of disease that is highly similar. Notable differences include the type of central nervous system lesions, and the radiographic appearance of osteomyelitic lesions. Therapies for humans and animals involve use of the same drugs, though special challenges can occur in treating animal species.

REFERENCES

1. TIMM, K.I., R.J. SONN & B.D. HULTGREN. 1988. Coccidioidomycosis in a Sonoran gopher snake, *Pituophis melanoleucus affinis*. J. Med. Vet. Mycol. **26:** 101–104.
2. REED, R.E., K.A. INGRAM, C. REGGIARDO & M.R. SHUPE. Coccidioidomycosis in domestic and wild animals. Proceedings, 5th International Conference, National Foundation for Infectious Diseases.: 245–254.
3. MADDY, K.T. 1959. Coccidioidomycosis in animals. Vet. Med. **54:** 233–242.
4. MADDY, K.T. 1954. Coccidioidomycosis of cattle in the southwestern United States. J. Am. Vet. Med. Assoc. **124:** 456–464.
5. DEMARTINI, J.C. & W.E. RIDDLE. 1969. Disseminated coccidioidomycosis in two horses and a pony. J. Am. Vet. Med. Assoc. **155:** 149–156.
6. KRAMME, P.M. & E.L. ZIEMER. 1990. Disseminated coccidioidomycosis in a horse with osteomyelitis. J. Am. Vet. Med. Assoc. **196:** 106–109.
7. ZIEMER, E.L., D. PAPPAGIANIS, J.E. MADIGAN, et al. 1992. Coccidioidomycosis in horses: 15 cases (1975–1984). J. Am. Vet. Med. Assoc. **201:** 910–916.
8. FOLEY, J.P. & A.M. LEGENDRE. 1992. Treatment of coccidioidomycosis osteomyelitis with itraconazole in a horse. J. Vet. Int. Med. **6:** 333–334.
9. LANGHAM, R.F., E.S. BENEKE & D.L. WHITENACK. 1977. Abortion in a mare due to coccidioidomycosis. J. Am. Vet. Med. Assoc. **170:** 178–179.
10. FOWLER, M.E., D. PAPPAGIANIS & I. INGRAM. 1992. Coccidioidomycosis in llamas in the United States: 19 cases (1981–1989). J. Am. Vet. Med. Assoc. **201:** 1609–1614.

11. MUIR, S. & D. PAPPAGIANIS. 1982. Coccidioidomycosis in the llama: case report and epidemiologic survey. J. Am. Vet. Med. Assoc. **181:** 1334–1337.
12. HENRICKSON, R.V. & E.L. BIBERSTEIN. 1972. Coccidioidomycosis accompanying hepatic disease in two Bengal tigers. J. Am. Vet. Med. Assoc. **161:** 674–677.
13. REED, R.E., E.J. BICKNELL & H.B. HOOD. 1984. A thirty-year record of coccidioidomycosis in exotic pets and zoo animals in Arizona. Proceedings, 4th International Symposium on Coccidioidomycosis.: 275–281.
14. TERIO, K.A., I.H. STALIS, J.L. ALLEN, et al. 2003. Coccidioidomycosis in Przewalski's horses (*Equus przewalskii*) J. Zoo Wildl. Med. **34:** 339–345.
15. ROSENBERG, D.P., C.A. GLEISER & K.D. CAREY. 1984. Spinal coccidioidomycosis in a baboon. J. Am. Vet. Med. Assoc. **185:** 1379–1381.
16. BURTON, M., R.J. MORTON, E. RAMSEY & E.L. STAIR. 1986. Coccidioidomycosis in a ring-tailed lemur. J. Am. Vet. Med. Assoc. **189:** 1209–1212.
17. ADASKA, J.M. 1999. Peritoneal coccidioidomycosis in a mountain lion in California. J. Wildl. Dis. **35:** 75–77.
18. CLYDE, V.L., G.V. KOLLIAS, M.E. ROELKE & M.R. WELLS. 1990. Disseminated coccidioidomycosis in a western Couger (*Felis concolor*). J. Zoo Wildl. Med. 200–205.
19. STRAUB, M., R.J. TRAUTMAN & J.W. GREENE. 1961. Coccidioidomycosis in 3 coyotes. Am. J. Vet. Res. **22:** 811–813.
20. REIDARSON, T.H., L.A. GRINER, D. PAPPAGIANIS & J. MCBAIN. 1998. Coccidioidomycosis in a bottlenose dolphin. J. Wildl. Dis. **34:** 629–631.
21. CORNELL, L.H., K.G. OSBORN, J.E. ANTRIM JR. & J.G. SIMPSON. 1979. Coccidioidomycosis in a California sea otter. J. Wildl. Dis. **15:** 373–378.
22. FAUQUIER, D.A., F.M.D. GULLAND, J.G. TRUPKIEWICZ, et al. 1996. Coccidioidomycosis in free-living California sea lions (*Zalophus californianus*) in central California. J. Wildl. Dis. **32:** 707–710.
23. SHUBITZ, L.F., C.D. BUTKIEWICZ, S.M. DIAL & C.P. LINDAN. 2005. Incidence of *Coccidioides* infection among dogs residing in a region in which the organism is endemic. J. Am. Vet. Med. Assoc. **226:** 1846–1850.
24. GALGIANI, J.N., N.M. AMPEL, J.E. BLAIR, et al. 2005. Coccidioidomycosis. Clin. Infect. Dis. **41:** 1217–1223.
25. SNYDER, L.S. & J.N. GALGIANI. 1997. Coccidioidomycosis: the initial pulmonary infection and beyond. Sem. Resp. Crit. Care Med. **18:** 235–247.
26. GREENE, R.T. & G.C. TROY. 1995. Coccidioidomycosis in 48 cats: a retrospective study (1984–1993). J. Vet. Intern. Med. **9:** 86–91.
27. GREENE, R.T. 1998. Coccidioidomycosis. *In* Infectious Diseases in the Dog and Cat. C.E. Greene, Ed.: 391–398. W.B. Saunders. Philadelphia, PA.
28. PAPPAGIANIS, D. 2001. Serologic studies in coccidioidomycosis. Semin. Respir. Infect. **16:** 242–250.
29. REED, R.E. 1956. Diagnosis of disseminated canine coccidioidomycosis. J. Am. Vet. Med. Assoc. **128:** 196–201.
30. DERESINSKI, S.C. 1980. Coccidioidomycosis of bone and joints. *In* Coccidioidomycosis: a Text. D.A. Stevens, Ed.:195–211. Plenum Publishing. New York.
31. JOHNSON, L.R., E.J. HERRGESELL, A.P. DAVIDSON & D. PAPPAGIANIS. 2003. Clinical, clinicopathologic, and radiographic findings in dogs with coccidioidomycosis: 24 cases (1995–2000). J. Am. Vet. Med. Assoc. **222:** 461–466.
32. PAPPAGIANIS, D., J. VANDERLIP & B. MAY. 1973. Coccidioidomycosis naturally acquired by a monkey, *Cercocebus atys*, in Davis, California. Sabouraudia **11:** 52–55.

33. CASTLEMAN, W.L., J. ANDERSON & C.A. HOLMBERG. 1980. Posterior paralysis and spinal osteomyelitis in a rhesus monkey with coccidioidomycosis. J. Am. Vet. Med. Assoc. **177:** 933–934.

34. SHUBITZ, L.F. & S.M. DIAL. 2005. Coccidioidomycosis: a diagnostic challenge. Clin. Tech. Sm. Anim. Pract. **20:** 220–226.

35. KELLY, P.C. 1980. Coccidioidal meningitis. *In* Coccidioidomycosis: a Text. D.A. Stevens, Ed.: 163–193. Plenum Medical Book Company. New York.

36. FOUREMAN, P., R. LONGSHORE & S.B. PLUMMER. 2005. Spinal cord granuloma due to *Coccidioides immitis* in a cat. J. Vet. Int. Med. **19:** 373–376.

37. JACOBS, P.H. 1980. Cutaneous coccidioidomycosis. *In* Coccidioidomycosis: a Text. D.A. Stevens, Ed.: 213–224. Plenum Medical Book Company. New York.

38. SHUBITZ, L.F., M.E. MATZ, T.H. NOON, *et al.* 2001. Constrictive pericarditis secondary to *Coccidioides immitis* infection in a dog. J. Am. Vet. Med. Assoc. **218:** 537–540, 526.

39. HEINRITZ, C.K., S.D. GILSON, M.J. SODERSTROM, *et al.* 2005. Subtotal pericardectomy and epicardial excision for treatment of coccidioidomycosis-induced effusive-constrictive pericarditis in dogs: 17 cases (1999–2003). J. Am. Vet. Med. Assoc. **227:** 435–440.

40. AMUNDSON, D.E. 1993. Perplexing pericarditis caused by coccidioidomycosis. South. Med. J. **86:** 694–696.

41. CHAPMAN, M.G. & L. KAPLAN. 1957. Cardiac involvement in coccidioidomycosis. Am. J. Med. **23:** 87–98.

42. BAYER, A.S., T.T. YOSHIKAWA, J.E. GALPIN & L.B. GUZE. 1976. Unusual syndromes of coccidioidomycosis: diagnostic and therapeutic considerations; a report of 10 cases and review of the English literature. Medicine **55:** 131–152.

43. SCHWARTZ, E.L., E.B. WALDMANN, R.M. PAYNE, *et al.* 1976. Coccidioidal pericarditis. Chest **70:** 670–672.

44. OUDIZ, R., P. MAHAISAVARIYA, S.K. PENG, *et al.* 1995. Disseminated coccidioidomycosis with rapid progression to effusive-constrictive pericarditis. J. Am. Soc. Echocardiography **8:** 947–952.

45. FAUL, J.L., K. HOANG, J. SCHMOKER, *et al.* 1999. Constrictive pericarditis due to coccidioidomycosis. Ann. Thorac. Surg. **68:** 1407–1409.

46. OLDFIELD, E.C., W.D. BONE, C.R. MARTIN, *et al.* 1997. Prediction of relapse after treatment of coccidioidomycosis. Clin. Infect. Dis. **25:** 1205–1210.

47. PAPPAGIANIS, D. 1999. *Coccidioides immitis* antigen. J. Infect. Dis. **180:** 243–244.

48. MADDY, K.T. 1958. Disseminated coccidioidomycosis of the dog. J. Am. Vet. Med. Assoc. **132:** 483–489.

49. MALIK, R., A.J. CRAIG, D.I. WIGNEY, *et al.* 1996. Combination chemotherapy of canine and feline cryptococcosis using subcutaneously administered amphotericin B. Austr. Vet. J. **73:** 124–128.

50. TERRELL, C.L. 1999. Antifungal agents. Part II. The Azoles. Mayo Clinic Proc. **74:** 78–100.

51. PLOTNIK, A.N., E.W. BOSHOVEN & R.A. ROSYCHUK. 1997. Primary cutaneous coccicioidomycosis and subsequent drug eruption to itraconazole in a dog. J. Am. Anim. Hosp. Assoc. **33:** 139–143.

52. LEGENDRE, A.M., B.W. ROHRBACH, R.L. TOAL, *et al.* 1996. Treatment of blastomycosis with itraconazole in 112 dogs. J. Vet. Int. Med. **10:** 365–371.

53. KRAWIEC, D.R., B.C. MCKIERNAN, A.R. TWARDOCK, *et al.* 1996. Use of an amphotericin B lipid complex for treatment of blastomycosis in dogs. J. Am. Vet. Med. Assoc. **209:** 2073–2075.

Diagnosis and Treatment of Ocular Coccidioidomycosis in a Female Captive Chimpanzee (*Pan troglodytes*)

A Case Study

K. HOFFMAN, E. N. VIDEAN, J. FRITZ, AND J. MURPHY

Primate Foundation of Arizona, Mesa Arizona 85277-0027, USA

ABSTRACT: We report here the first documented case of ocular coccidioidomycosis in a chimpanzee (*Pan troglodytes*). In 1996, a 12-year-old female chimpanzee was undergoing treatment with an experimental triazole, BayR3783, for coccidioidomycosis when she was diagnosed with severe conjunctivitis in the right eye. Subsequent development of a coccidioidal granuloma of the ventral conjunctiva and anterior uvea was noted over the next several months, distorting the lens, iris, pupil, and sclera and progressing to uveitis. Treatment with BayR3783 and subconjunctival injections of triamcinolone were successful in reducing the ocular mass, but extensive damage was done to the lens and cornea. This case study provides an interesting comparison to ocular coccidioidomycosis cases observed in both humans and canines.

KEYWORDS: chimpanzee; cocci granuloma; uveitis; primate

BACKGROUND

Coccidioidomycosis is known to occur in humans and domesticated animals (i.e., dogs, cats, horses), as well as in exotic animals housed at zoos, sanctuaries, and research facilities located in the endemic areas.[1] Cases of coccidioidomycosis have been confirmed in a number of species of nonhuman primates, including macaques,[2–4] baboons,[5–6] lemurs,[7] spider monkeys,[4] gorillas,[4,8] and chimpanzees.[4,9–11] The Primate Foundation of Arizona (PFA) is located in an area that is considered endemic for coccidioidomycosis. Since 1975, we have used vaccines and rigorous routine screening in an effort to protect the chimpanzees housed at PFA from coccidioidomycosis, as well as making use of

Address for correspondence: Elaine N. Videan, Ph.D., Primate Foundation of Arizona, P.O. Box 20027, Mesa, AZ 85277-0027. Voice: 480-832-3780; fax: 480-830-7039.
evpfa@wydebeam.com

Ann. N.Y. Acad. Sci. 1111: 404–410 (2007). © 2007 New York Academy of Sciences.
doi: 10.1196/annals.1406.018

multiple treatment modalities in animals that have become ill with this fungal disease. At PFA and elsewhere, the same patterns of involvement seen in humans have been reported in the limited data available for chimpanzees.[9–12] The primary site of infection is predominately pulmonary, with skin lesions, meningeal infections, and skeletal lesions seen only in severe cases.[11–12]

Ocular involvement in disseminated coccidioidomycosis has been documented in humans and domesticated animals.[1,13–15] In humans, ocular coccidioidomycosis is rare and often difficult to diagnose. Ocular coccidioidomycosis in humans is thought to occur in less than 0.5% of those affected with coccidioidomycosis,[13] although, the presence of chorioretinal scars on patients with pulmonary infection suggests that the rate of subclinical ocular involvement may be as high as 10%.[16] The most common sites of ocular involvement include the conjunctiva and cornea (i.e., conjunctivitis, keratoconjunctivitis), sclera (i.e., scleritis, episcleritis), and uvea (i.e., granulomatous iridocyclitis, granulomatous uveitis).[13,17] The patterns of involvement are similar for both dogs[1,14] and cats.[1,15] However, no cases of ocular involvement have been reported for nonhuman primates. We report here the first documented case of ocular coccidioidomycosis in a chimpanzee (*Pan troglodytes*).

CASE STUDY

In December 1993, a 9-year-old female chimpanzee, housed at PFA, was diagnosed with coccidioidomycosis. Symptoms included sneezing, coughing, increased lethargy, and decreased appetite. Diagnosis was through routine biannual serum testing and thoracic radiograph conducted on December 29, 1993. Complement fixation (CF) was 1:64 (FIG. 1) and radiograph interpretation showed a possible hilar lymphadenopathy on the left side. Treatment with an experimental triazole, BayR3783 (10 mg/kg/day), and sodium ascorbate (4 g/day) was initiated on January 1, 1994. The dosage and use of BayR3783 were established on the basis of an experimental protocol already in place at PFA.[18]

After one year (January 1994 through December 1994) of BayR3783 and sodium ascorbate treatment, the CF titers dropped to 1:2 (FIG. 1). However, because of sporadic recurrent upper respiratory symptoms and hilar lesions present upon radiographs taken in December 1995, treatment was continued as above. In March 1996, while still on the above treatment of BayR3783 and sodium ascorbate, the chimpanzee, now 12 years old, was observed rubbing her eye. A physical examination in April 1996 diagnosed her with severe conjunctivitis in the right eye. Fluorescene staining was performed and results were negative. Several drops of Gentacidin Opht were administered while the animal was under anesthesia and amoxicillin (1,000 mg/day) and ibuprofen (800 mg/day) were prescribed, in addition to the concurrent BayR3783 and sodium ascorbate. Additional administration of topical medications was not

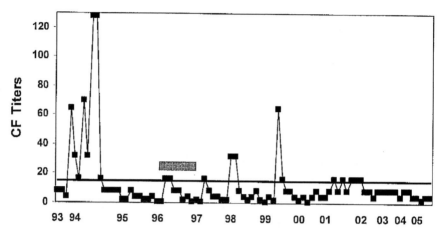

FIGURE 1. CF titers from December 1993 through December 2005. Horizontal line represents CF titer 1:16. Shading represents a period of ocular involvement.

possible without anesthetizing the patient, which was not deemed practical or in the animal's best interest on a daily or regular basis.

In May 1996, a white plaque-like lesion on the ventral half of the sclera and bulbar conjunctiva was visible, and presumptively diagnosed as a coccidioidal granuloma. In addition, the iris was dull, swollen, and distorted in shape, the pupil was fixed and would not dilate well, and the fundus could not be visualized. Per recommendation of a veterinary ophthalmologist, the patient was given a subconjunctival injection of 1 mg triamcinolone and several drops of Gentacidin Opht while under anesthesia. At this time a conjunctival scraping was also taken and the diagnosis revealed suppurative conjunctivitis. The cocci treatment regimen of BayR3783 (10 mg/kg/day) and sodium ascorbate (4 g/day) was continued as above.

In June 1996, the granuloma had grown to a raised mass (approximately 6 mm × 12 mm) and corneal edema was present along with three raised focal white spots on the iris (FIG. 2). Aqueous flare was present and the retina was not visible under examination. While the animal was under anesthesia, an injection of 4 mg triamcinolone was administered and a biopsy of the mass was taken. Histopathological diagnosis of the mass was diffuse ulcerative and chronic lymphocytic, plasmacytic conjunctivitis. Given the ophthalmic examination and histopathological results, the veterinary ophthalmologist diagnosed the patient with uveitis secondary to a coccidioidal granuloma. At this time a semimonthly subconjunctival injection of 4 mg triamcinolone was added, in addition to the BayR3783 (10 mg/kg/day) and sodium ascorbate (4 g/day).

Over the next 3 months (July–September), improvement was seen in both the iris lesions and uveitis. In July 1996, the iris lesions were reduced in size and the semimonthly subconjunctival injections of triamcinolone were decreased to 2 mg. In August 1996, the triamcinolone had decreased, if not eliminated,

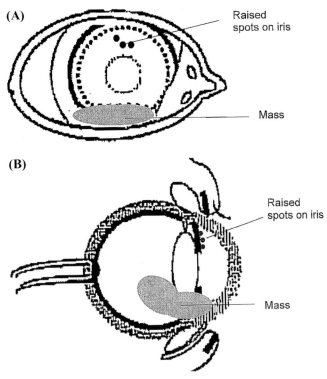

(A)

Raised
spots on iris

Mass

(B)

Raised
spots on iris

Mass

FIGURE 2. Diagram of affected eye June 1996, direct (**A**) and lateral (**B**) views. Black circles represent raised spots on iris and gray areas represent mass.

the periocular inflammation, the fundus was visible, and the iris lesions were disappearing. The granuloma had not changed in size and remained firmly attached to the underlying sclera, distorting the shape of the ventral half of the globe. In September 1996, the focal lesions on the iris were undetectable and the fundus looked normal, whereas the granuloma mass remained unchanged. Cocci treatment during this time, BayR3783 (10 mg/kg/day) and sodium ascorbate (4 g/day), was continued as above.

In October 1996, the iris lesions had disappeared and the granuloma mass had decreased in size (3 mm × 3 mm), but scar tissue appeared to have formed behind the iris, distorting the pupil. Both direct and indirect pupillary light reflexes were absent in the affected eye. The last triamcinolone injection (2 mg) was given while the animal was under anesthesia on October 2, 1996. In November 1996, although the mass had disappeared, extensive damage had occurred to the lens and cornea. Ophthalmic examination at this time revealed a subluxated lens, anterior synechiae of the ventral part of the iris, and corneal scarring.

CURRENT STATUS

This female chimpanzee is currently 21 years old and appears to be in good physical condition. She has periodically exhibited elevated CF titers (FIG. 1) along with small lesions found on thoracic radiographs in 1997, 2001, 2002, and 2005. Because of these, she has undergone additional treatment for coccidioidomycosis. She was treated with ketoconazole from April 1997 to May 1998 and again from October 1998 to June 1999, and with fluconazole from June 1999 to present. However, she has had no other ophthalmic complications.

In February 2006, ophthalmic examination showed anterior lens luxation, a dysconic and fixed iris, and corneal edema as a result of the coccidiodal granuloma in 1996 (FIG. 3). The edema present was mild-to-moderate dorsally progressing to severe ventrally. In addition, a hypermature cataract was evident and both direct and indirect pupillary light reflexes and the dazzle reflex were absent. The veterinary ophthalmologists concluded that she is avisual in her right eye. However, she functions normally within her social group and has suffered no long-term behavioral issues.

CONCLUSION

This case study provides an interesting comparison with ocular coccidioidomycosis cases observed in humans and represents the first documented case of ocular involvement in a nonhuman primate. Ocular coccidioidomycosis is typically either an external disease (i.e., conjunctivitis, scleritis) or an internal disease (i.e., uveitis, iridocyclitis).[14,20] However, cases of conjunctivitis progressing to uveitis have been reported.[21] The corneal and conjunctival

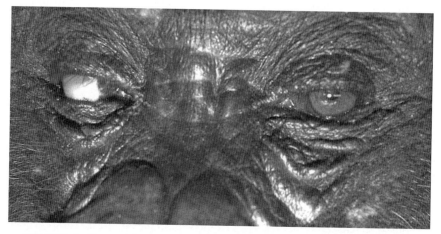

FIGURE 3. Photo of right (affected) and left (nonaffected) eyes, February 2006.

involvement in this case study was most likely a delayed hypersensitivity reaction associated with the pulmonary disease.[14] This may explain why the female chimpanzee in this case study developed the coccidioidal granuloma during a time of relatively low CF titers (FIG. 1). This is supported by reports of ocular coccidioidomycosis documented in humans with no clinical evidence of systemic disease.[13] Additionally, in up to 40% of coccidioidomycosis cases in dogs, it is estimated that ocular involvement may be the only clinical sign.[14] The progression to uveitis is probably due to the rich vascularization of the uvea.[17,20] Finally, the monocular involvement in our case study suggests that additional factors may play a role in development of ocular coccidioidomycosis. Both in humans and in dogs ocular coccidioidomycosis is typically unilateral, affecting only one eye.[14,21] Further research is needed to examine the reasons behind the unilateral involvement and whether other nonhuman primates with coccidioidomycosis exhibit ocular manifestations.

ACKNOWLEDGMENTS

We thank PFA staff members, Kelly Carbone, Erika Evans, and the dedicated carestaff. We further thank the veterinary ophthalmologists who consulted on this project: Drs. Reuben Meredith, Julius Brinkis, and Lisa Felchle. This project was supported in part by the University of Texas, M.D. Anderson Cancer Research Center, as a subcontract (1U42RR15090-05) within the National Institutes of Health Biomedical Research Program.

REFERENCES

1. SHUBITZ, L.F. & S.M. DIAL. 2005. Coccidioidomycosis: a diagnostic challenge. Clin. Tech. Small Anim. Prac. **20:** 220–226.
2. BEAMAN, L., C. HOLMBERG, R. HENRICKSON & B. OSBURN. 1980. The incidence of coccidioidomycosis among nonhuman primates housed outdoors at the California Primate Research Center. J. Med. Primatol. **9:** 254–261.
3. BREZNOCK, A.W., R.V. HENRICKSON, S. SILVERMAN & L.W. SCHWARTZ. 1975. Coccidioidomycosis in a rhesus monkey. J. Am. Vet. Med. Assoc. **167:** 657–661.
4. REED, R.E., E.J. BICKNELL & H.B. HOOD. 1985. A thirty-year record of coccidioidomycosis in exotic pets and zoo animals in Arizona. *In* Proceedings of the 4th International Conference on Coccidioidomycosis. H.E. Einstein & A. Cantanzaro, Eds.: 281–285. National Foundation for Infectious Diseases. Washington, DC.
5. BELLINI, S., G.B. HUBBARD & L. KAUFMAN. 1991. Spontaneous fatal coccidioidomycosis in a native-born hybrid baboon (*Papio cynocephalus anubis/Papio cynocephalus cynocephalus*). Lab. Anim. Sci. **41:** 509–511.
6. JOHNSON, J.H., A.M. WOLF, J.F. EDWARDS, *et al.* 1998. Disseminated coccidioidomycosis in a mandrill baboon (*Mandrillus sphinx*): a case report. J. Zoo Wildl. Med. **29:** 208–213.

7. BURTON, M., R.J. MORTON, E. RAMSAY & E.L. STAIR. 1986. Coccidioidomycosis in a ring-tailed lemur. JAVMA **189:** 1209–1211.
8. HOOD, H.B. & J. GALGIANI. 1985. Treatment of a male lowland gorilla for coccidioidomycosis with Ketoconazole at the Phoenix Zoo. *In* Proceedings of the 4th International Conference on Coccidioidomycosis. H.E. Einstein & A. Cantanzaro, Eds.: 288–291. National Foundation for Infectious Diseases. Washington.
9. HERRIN, K.V., A. MIRANDA & D. LOEBENBERG. 2005. Posaconazole therapy for systemic coccidioidomycosis in a chimpanzee (*Pan troglodytes*): a case report. Mycoses **48:** 447–452.
10. HOFFMAN, K.M., J. FRITZ, S. HOWELL & J. MURPHY. 2001. The effect of coccidioidomycosis on blood chemistry in captive chimpanzees (*Pan troglodytes*). *In* Proceedings of the Annual Coccidioidomycosis Study Group Meeting Vol 45: 4.
11. MARZKE, M., D. HAWKEY & J. FRITZ. 1988. Coccidioidomycosis in the chimpanzee. Am. J. Primatol. **14:** 432.
12. MARZKE, M.W. & C.F. MERBS. 1984. Evidence of hypertrophic pulmonary osteoarthropathy in a chimpanzee, *Pan troglodytes*. J. Med. Primatol. **13:** 135–145.
13. RODENBIKER, H.T. & J.P. GANLEY. 1980. Ocular coccidioidomycosis. Survey Ophthalmol. **24:** 263–290.
14. ANGELL, J.A., R.E. MEREDITH, J.N. SHIVELY & R.L. SIGLER. 1987. Ocular lesions associated with coccidioidomycosis in dogs: 35 cases (1980–1985). JAVMA **190:** 1319–1322.
15. GREENE, R.T. & G.C. TROY. 1995. Coccidioidomycosis in 48 cats: a retrospective study (1984–1993). J. Vet. Int. Med. **9:** 86–91.
16. RODENBIKER, H.T., J.P. GANLEY, J.N. GALGIANI & S.G. AXLINE. 1981. Prevalence of chorioretinal scars associated with coccidioidomycosis. Arch. Ophthalmol. **99:** 71–75.
17. CRUM, N.F., E.R. LEDERMAN, C.M. STAFFORD, *et al.* 2004. Coccidioidomycosis: a descriptive survey of a reemerging disease. Clinical characteristics and current controversies.
18. PAPPAGIANIS, D., B.L. ZIMMER, G. THEODOROPOULOS, *et al.* 1990. Therapeutic effect of the triazole BayR3783 in mouse models of coccidioidomycosis, blastomycosis, and histoplasmosis. Antimicrob. Agents, Chemother. **34:** 1132–1138.
19. KROHNE, S.G. 2000. Canine systemic fungal infections. Vet. Clin. N. Am. Sm. Anim. Prac. **30:** 1063–1090.
20. ZAKKA, K.A., R.Y. FOOS & W.J. BROWN. 1978. Intraocular coccidioidomycosis. Surv. Ophthalmol. **22:** 313–321.
21. MOORTHY, R.S., N.A. RAO, Y. SIDIKARA & R.Y. FOOS. 1994. Coccidioidomycosis iridocyclitis. Ophthalmology **101:** 1923–1928.

State-of-the-Art Treatment of Coccidioidomycosis

Skin and Soft-Tissue Infections

JANIS E. BLAIR

Division of Infectious Diseases, Mayo Clinic, Scottsdale, Arizona

ABSTRACT: Coccidioidomycosis is a fungal infection common in the southwestern United States that is caused by the endemic *Coccidioides* species of fungus. Coccidioidal infections are generally manifested as self-limited respiratory illnesses, but affected patients rarely present with coccidioidomycosis in extrapulmonary locations. Skin and soft-tissue coccidioidomycosis may occur in 15% to 67% of patients with disseminated infection. Skin manifestations of coccidioidomycosis can either be reactive rashes, such as erythema multiforme or erythema nodosum associated with primary pulmonary infection, or they can be the result of extrapulmonary dissemination of the infection to the skin. As many as 90% of persons with disseminated infection to the skin have other extrapulmonary sites of infection, and the presence of coccidioidal skin lesions should prompt an investigation for other extrapulmonary foci of infection. Lymph nodes are a common site of extrapulmonary infection. Nearly every organ system and soft-tissue have been described as infected with Coc*cidioides* species, but subcutaneous abscesses, phlegmon, and sinus tracts are not uncommon and often are themselves the result of coccidioidal infection in neighboring lymph nodes, bones, or joints. A biopsy of the abnormal area is the most direct way to diagnose skin and soft-tissue lesions. Fluconazole and itraconazole are preferred therapeutic agents, and surgical intervention may be required as an adjunctive measure. This article reviews the types and locations of disseminated infections, as well as diagnostic studies and treatment of this difficult-to-treat manifestation of coccidioidomycosis.

KEYWORDS: coccidioidomycosis; dissemination; infection; skin infection; soft-tissue; treatment; antifungal

INTRODUCTION

Coccidioidomycosis results from inhalation of airborne arthrospores of the fungal species *Coccidioides,* which is endemic to the southwestern United

Address for correspondence: Janis E. Blair, M.D., Division of Infectious Diseases, Mayo Clinic, 13400 East Shea Boulevard, Scottsdale, AZ 85259.

Ann. N.Y. Acad. Sci. 1111: 411–421 (2007). © 2007 New York Academy of Sciences.
doi: 10.1196/annals.1406.010

States, northern Mexico, and limited areas of Central and South America.[1] In 60% of healthy persons, coccidioidal infection is asymptomatic, and in 35% to 40% it causes mild-to-severe infections limited to the lungs.[1] About 1% to 5% of all coccidioidal infections are extrapulmonary, primarily resulting from hematogenous dissemination.[2] Persons at risk for disseminated coccidioidomycosis include those of certain racial heritage (e.g., Filipinos or African Americans), and those with diabetes mellitus, or immunosuppressing illnesses, or those receiving immunosuppressive therapy or who are pregnant. The most common locations of extrapulmonary coccidioidomycosis are the skin, lymph nodes, skeleton, and central nervous system.[2]

In this article, I discuss current issues about the anatomy, diagnosis, and treatment of skin and soft-tissue coccidioidomycosis as summarized and presented at the Sixth International Symposium on Coccidioidomycosis, Stanford, California.

Cutaneous Manifestations of Coccidioidomycosis

Cutaneous manifestations are common and can be classified as either of two distinct types: reactive changes associated with primary pulmonary infection or changes resulting from infection of the skin with *Coccidioides* spp. Reactive skin changes do not reflect the presence of coccidioidal organisms within the skin. Skin infection with the fungus is primarily the result of the hematogenous spread of a primary pulmonary infection and rarely the result of primary trauma to the skin.

Reactive Skin Changes

Since the earliest observations of primary coccidioidomycosis, rashes have been commonly observed. In the first few days of illness, an early "toxic erythema"[3] occurs that may be fine, diffuse, erythematous, and macular in about 10% of persons with primary coccidioidal infection; it often resolves before medical evaluation. In contrast, erythema nodosum and erythema multiforme are considered "specific erythemas" of coccidioidomycosis.[3] These two rashes are thought to reflect the development of a hypersensitivity response to the coccidioidal infection. Erythema nodosum is widely regarded as a predictor of good prognosis, although protection is not complete.[3] Antifungal therapy has no role in the treatment of reactive skin changes.

Hematogenous Dissemination into Cutaneous Tissues

The skin is one of the most common sites of disseminated coccidioidal infection,[1] and it is affected in 15% to 67%[4,5] of persons with disseminated infection. As many as 90% of persons with coccidioidal skin lesions also have extrapulmonary infection in other tissues,[2] and the identification of coccidioidal skin lesions should prompt the investigation for other extrapulmonary foci.[1] Skin lesions are most frequently found on the face, scalp, neck, and chest wall.[4] Facial lesions have been associated with meningitis in some patients[5] but not in others.[2]

Papules, nodules, pustules, furuncles, verrucous plaques, abscesses, and ulcerations are all common manifestations[2,4]; however, nodules were the predominant finding in one study.[2] These lesions often lack associated pain or discomfort.[1] Their differential diagnosis includes cutaneous infections with other endemic mycoses, such as histoplasmosis or blastomycosis, cryptococcosis, tuberculosis, sarcoidosis, or primary skin malignancies, such as basal cell or squamous cell carcinomas or other skin problems.[2]

Primary Cutaneous Coccidioidomycosis

Although most cases of cutaneous coccidioidomycosis result from hematogenous dissemination from pulmonary infection, a person may rarely sustain cutaneous trauma that results in a primary cutaneous coccidioidal infection. Precipitating traumatic events include bug bites, injury from cactus thorns or barbed wire, skin trauma while embalming a deceased person who had disseminated coccidioidomycosis, and accidental inoculation by laboratory workers.[6] Diagnostic criteria for primary cutaneous coccidioidomycosis include the following: no history of pulmonary infection, a short incubation period of 1 to 3 weeks from the time of skin trauma to the development of a visible lesion, a chancriform ulcerated nodule or plaque initial lesion, a positive skin test, an initially negative complement fixation serologic test (later positive only in low titer), regional lymphadenopathy or adenitis, sporotrichoid spread, and spontaneous healing within a few weeks if immunocompetent.[7]

Soft-Tissue Coccidioidomycosis

Lymph node infection with *Coccidioides* spp. is common. Hilar and mediastinal lymphadenopathy are involved as the primary infection complex.[4] Distant nodal involvement results from entrapment of the organisms by way of the bloodstream. The supraclavicular and cervical nodes are the most frequently involved extrapulmonary lymph nodes. Involvement of lymph nodes is important for a few reasons. Lymph nodes may contain viable organisms for an extended period of time and act as a source of propagation and dissemination of infection.[4] In addition, necrotic lymph nodes can ulcerate, discharging their infected contents into the surrounding soft-tissues, resulting in phlegmon, abscesses, and sinus tracts.[4] Fine-needle aspiration into the necrotic node may be diagnostic.

Subcutaneous infections, abscesses, and sinus tracts may result from the spread of infection from neighboring lymph nodes, bones, and joints. To be brought under control, these subcutaneous infections may harbor purulent collections that require surgical drainage or débridement.

Nearly every organ system and soft-tissue have been described as infected with coccidioidomycosis, including the upper respiratory structures, intrathoracic, intraabdominal, and pelvic structures, skeletal musculature, and subcutaneous tissues.[1,4,8] In an autopsy study of 95 persons with disseminated coccidioidomycosis, splenic infection was found to be very common, exceeded only by infection of the lymph nodes.[4] Infection was less commonly identified in the kidneys, adrenal glands, liver, skeletal muscles, heart structures, gastrointestinal tract, and genitourinary structures.[4]

Diagnosis

Biopsy of abnormal areas is the most direct way to diagnose skin and soft-tissue lesions. Soft-tissue infections may be accessible to percutaneous aspiration or biopsy, but if not, surgical biopsy may be required.

Histopathology

Granulomatous dermatitis, eosinophilic infiltration, microabscesses, perivascular inflammation, and gummatous necrosis are all manifestations of coccidioidal skin infection.[9] Spherules of *Coccidioides* spp. can be identified with methenamine silver,[9] periodic acid-Schiff[9] stains, or Papanicolaou stains. One study reported skin biopsy histopathologic tests to be diagnostic in 18 of 18 cases.[2]

Diagnostic Imaging

Many patients with disseminated infection are initially evaluated weeks to months after the initial pulmonary infection, and therefore chest radiographs and imaging are often normal. Computed tomography is useful to define the location and extent of soft-tissue infections.

Microbiology

Coccidioides spp. grew from cultures of 60% to 90%[2,9] of biopsy-proven coccidioidal skin infections.[2] This fungus grows easily as a mold on most laboratory media, including bacterial and fungal media. However, although growth can be detected as early as 2 to 3 days, other specimens require a few weeks; therefore, specific fungal cultures are recommended to assure an adequate duration of incubation.

The laboratory personnel should be notified if coccidioidomycosis is suspected, so that additional precautions can be taken to protect laboratory personnel from laboratory-acquired infection, especially in nonendemic areas. Detection of fungal material by nucleic amplification of clinical specimens is being developed but is not currently available in clinical laboratories.

Serology

A number of serologic methodologies can be used to detect anticoccidioidal antibodies, including enzyme immunoassay, immunodiffusion, and complement fixation. Complement fixation provides a quantitative result that declines with therapeutic response over time. Serologic titers range from undetectable to 1:2 to 1:512 (mean, 1:128) in case series of coccidioidal skin infection.[2,9]

Medical Therapy

Soon after its discovery, amphotericin B became the treatment of choice for skin and soft-tissue coccidioidomycosis. It was effective, but no reports of scientific trials were published to characterize its efficacy. Amphotericin, however, is fraught with adverse effects in recipients. It can be delivered systemically only through the intravenous route, and it is often associated with symptoms during the infusion, including but not limited to fever, chills, rigors, headache, nausea, and vomiting. Some of these symptoms can be ameliorated with preinfusion administration of acetaminophen, diphenhydramine hydrochloride, heparin, or meperidine. Over time, renal toxicity and potassium and magnesium wasting become increasingly problematic. The newer lipid formulations of amphotericin are associated with less renal toxicity. Available formulations include amphotericin deoxycholate (0.5–1.5 mg/kg per day or alternate-day administration) or lipid formulations of amphotericin (2.0–5.0 mg/kg per day).[10]

Subsequently, imidazole and azole therapies were developed. For many reasons (e.g., oral formulations, decreased toxicity, and efficacy), these medications have become the treatment of choice. Currently recommended options include oral or intravenous fluconazole 400–800 mg/day and oral or intravenous itraconazole 200 mg two to three times daily. The oral cyclodextrin suspension of itraconazole is available to improve oral absorption.[10] Since absorption of oral itraconazole therapy can be suboptimal in some patients, serum concentrations should be performed to document adequate absorption.

TABLE 1. Summary of prospective trials addressing treatment of skin and soft-tissue coccidioidomycosis

Author (year)	Selection criteria	Medication	Site	N	Response, no. (%)	Relapse, no. (%)	Comment
Stevens (1983)[12]; Stiller et al. (1980)[13]	Failed or relapsed with AMB or refused AMB	Miconazole 200–1,000 mg tid i.v.	S ST	30	12 of 30 (4) response or cure	NR (60)	i.v. dosing limited duration of tx to mean of 45 days
Stevens et al. (1983)[14]	Failed or relapsed with AMB or refused AMB	Ketoconazole ≥200–400 mg/day	S ST	22	20 of 22 (91)	NR (11)	Pts treated for 12-month or 6-month postresolution, whichever was longer
Galgiani et al. (1988)[15]	Culture-proven immunosuppressed pts included	Randomized ketoconazole 400 vs. 800 mg/day	ST	33	21 of 33 (64) overall	NR for this subgroup	Pts not responding at lower dose were offered higher dose
Tucker et al. (1990)[16]	Culture-proven or histology positive	Escalation of doses of itraconazole to 400 mg/day	S ST	21	16 of 21 (76)	3 of 16 (19) responders	Pts treated 12 months total, or 6 months after clinical resolution, whichever was longer
Graybill et al. (1990)[17]	Histology positive, no HIV	Itraconazole 100–400 mg/day	S ST	14	7 of 14 (50)	2 of 7 (29) responders	Daily doses increased if clinical score unimproved 21% at 3 months; minimum at 6 months; no maximum tx duration
Diaz et al. (1992)[18]	Culture positive, no immunosuppressive medication	Fluconazole 100, 200, or 400 mg/day, with dose escalation at investigator's discretion	S ST	2	2 of 2 (100)	1 of 2 (50) treated with 100 mg/day for 12 months	Small cohort was part of larger series of fluconazole-treated deep fungal infections

Continued

TABLE 1. *Continued*

Author (year)	Selection criteria	Medication	Site	N	Response, no. (%)	Relapse, no. (%)	Comment
Hostetler et al. (1994)[19]	Culture or histology proven, not on immunosuppressive therapy, HIV included	SCH 39304 100–200 mg/day, with few at higher doses	S ST	12	NR (0%* at 4 months, 25%[a] at 8 months, 62%[a] at 12 months)	NR	Median tx time was 5 months; product withdrawn from further development
Catanzaro et al. (1995)[20]	Culture or histology positive, no immunosuppressive therapy	Fluconazole 200 mg/day, increased to 400 mg with inadequate response	S ST	22	8 of 19 (42) at 200 mg/day, and 8 of 10 (80) at 400 mg/day[b]	3 of 13 (23)	Pilot study initiated tx at 50 then 100 mg/day; enrollment closed due to insufficient response
Galgiani et al. (2000)[21]	Culture or histology positive; HIV included	Fluconazole 400 mg/day vs. itraconazole 200 mg bid	ST	71 (32 fluconazole, 39 itraconazole)	21 of 32 (66) fluconazole and 29 of 39 (74) itraconazole (P = NS)[c]; 22 of 32 (69) fluconazole and 31 of 39 (79) itraconazole (P = NS)[d]	NR	Randomized double-blind study found no difference in persons with ST infection

AMB = amphotericin B; bid = twice daily; HIV = human immunodeficiency virus; i.v. = intravenous; NR = not reported; pts = patients; S = skin; ST = soft-tissue; tid = three times daily; tx = treatment

[a] Numerator and denominator not provided in the publication.

[b] Of 21 evaluable.

[c] At 8 months.

[d] At 12 months.

TABLE 2. Selected retrospective studies of skin and soft-tissue coccidioidomycosis

Author (year)	Selection criteria	Medication	Site	N	Response, no. (%)[a]	Relapse, no. (%)	Comment
Johnson et al. (1996)[22]	Retrospective review; single public hospital practice (1992–1994)	18 received azole alone, 6 received AMB alone, 16 had both (in sequence), 13 with ST involvement had surgical debridement[b]	S, ST	40	28 of 40 (70)	NR	
Anstead et al. (2005)[11]	Culture confirmed; all available therapy failed	Posaconazole 600–800 mg/day	S	1	1 of 1 (100)	None	1 in case series of 6 pts[c]
Quimby et al. (1992)[9]	Retrospective review; single dermatology practice	AMB, fluconazole, or ketoconazole	S (1 with concurrent ST infection)	6	6 of 6 (100); 1 required 2 antifungal agents in series to achieve control; 1 required surgical debridement of ST infection	1 of 6 (17)	Meningitis developed in 1 on AMB
Crum et al. (2004)[2]	Retrospective review; single military-based practice (1994–2002)	AMB, fluconazole, or itraconazole, either initially or sequentially	S (90% also had ST lesions)	18	16 of 18 (89); tx failed in 2 when concurrent skeletal infection was initially treated with fluconazole	6 of 18 (33)	

AMB = amphotericin B; i.v. = intravenous; NR = not reported; S = skin; ST = soft-tissue; tx = treatment.
[a]Improved or resolved.
[b]Treatment at discretion of individual physician.
[c]Two other patients with skin lesions had concurrent bone lesions and were classified as having "skeletal" infections.

Voriconazole and posaconazole are both recently developed azoles. A few case reports have suggested the efficacy of voriconazole for refractory coccidioidomycosis in select patients. Posaconazole was successful in treating soft-tissue coccidioidomycosis, which had been refractory to other available treatments.[11] However, there have been no published reports of trials delineating the relative efficacy of these newer azoles compared with other azoles or amphotericin. The efficacy of caspofungin for skin or soft-tissue coccidioidomycosis has not yet been studied.

Several trials of coccidioidal treatment have been performed, and the results for the skin and soft-tissue subgroups are delineated in TABLE 1. The earliest studies were performed only in patients who either refused amphotericin treatment or were unable to receive it because of its failure or their intolerance of its adverse effects. These studies demonstrated the difficulty of treating disseminated coccidioidal infection; the overall response rate ranged from 25% to 91%, but the rate of relapse was high in many cases.

Select retrospective studies of the efficacy of treatment for skin and soft-tissue infections are summarized in TABLE 2. No consensus paper exists for the optimal therapy of cutaneous or soft-tissue coccidioidomycosis. Patients with widespread, rapidly progressive lesions, or those who are immunocompromised, should be considered for treatment with amphotericin B.[2] One author has recommended higher doses of treatment for patients of Filipino descent, because the risk of coccidioidal relapse is high in these persons.[2]

For most patients who have uncomplicated coccidioidal skin or soft-tissue infection, fluconazole is the medication of choice, given its good bioavailability, high concentration in dermal appendages, and low toxicity.[2] Itraconazole is also considered a first-line treatment.[10] Treatment is continued for at least 6 to 12 months, depending on the serologic (serial complement fixation titers) and clinical response. Because relapse can be high, retreatment may be needed, and persons with multiple relapses may require lifetime therapy to achieve continuous control.[2] Combination therapy has been used in limited situations with some success, but no data exist to indicate its efficacy.[10]

Surgical Therapy

No trials have clarified if or when surgical débridement is necessary in patients with skin and soft-tissue coccidioidomycosis. Numerous anecdotes highlight the occasional importance of surgical adjuvant treatment to achieve control of infection.[2,8] The consensus of experts in the endemic area indicates that surgical débridement should be strongly considered in patients with a large phlegmon or abscess, with destructive lesions, or with areas that impinge on critical structures, such as the spinal cord or pericardium.[10] My experience has been that the excision of involved skin and subcutaneous tissue results in cure in patients with relapsed infection after successful medical therapy.

CONCLUSION

Although the treatment of disseminated coccidioidomycosis has been studied for decades, and the options for such treatment have broadened considerably, further studies are still needed to effect a cure for all patients who contract this difficult-to-treat infection. The words of Marshall Fiese,[23] penned 50 years ago, still echo true today: "Treatment of disseminated coccidioidomycosis is a subject to be approached with humility. We do not yet have the answer, although research continues apace in several interesting directions."

ACKNOWLEDGMENT

Editing, proofreading, and reference verification were provided by the Section of Scientific Publications, Mayo Clinic.

REFERENCES

1. CHILLER, T.M., J.N. GALGIANI & D.A. STEVENS. 2003. Coccidioidomycosis. Infect. Dis. Clin. North Am. **17:** 41–57.
2. CRUM, N.F., E.R. LEDERMAN, C.M. STAFFORD, *et al.* 2004. Coccidioidomycosis: a descriptive survey of a reemerging disease. Clinical characteristics and current controversies. Medicine (Baltimore) **83:** 149–175.
3. CATANZARO, A. & D.J. DRUTZ. 1980. Primary coccidioidomycosis. *In* Coccidioidomycosis: A Text. D.A. Stevens, Ed.: 139–145. Plenum Medical Book Company. New York.
4. FORBUS, W.D. & A.M. BESTEBREURTEJE. 1946. Coccidioidomycosis: a study of 95 cases of the disseminated type with special reference to the pathogenesis of the disease. Mil. Surg. 653–719.
5. ARSURA, E.L., W.B. KILGORE, J.W. CALDWELL, *et al.* 1998. Association between facial cutaneous coccidioidomycosis and meningitis. West. J. Med. **169:** 13–16.
6. CHANG, A., R.C. TUNG, T.S. MCGILLIS, *et al.* 2003. Primary cutaneous coccidioidomycosis. J. Am. Acad. Dermatol. **49:** 944–949.
7. WILSON, J.W., C.E. SMITH & O.A. PLUNKETT. 1953. Primary cutaneous coccidioidomycosis: the criteria for diagnosis and a report of a case. Calif. Med. **79:** 233–239.
8. CRUM-CIANFLONE, N.F., A.A. TRUETT, N. TENEZA-MORA, *et al.* 2006. Unusual presentations of coccidioidomycosis: a case series and review of the literature. Medicine (Baltimore) **85:** 263–277.
9. QUIMBY, S.R., S.M. CONNOLLY, R.K. WINKELMANN, *et al.* 1992. Clinicopathologic spectrum of specific cutaneous lesions of disseminated coccidioidomycosis. J. Am. Acad. Dermatol. **26:** 79–85.
10. GALGIANI, J.N., N.M. AMPEL, J.E. BLAIR, *et al.* 2005. Coccidioidomycosis. Clin. Infect. Dis. **41:** 1217–1223.
11. ANSTEAD, G.M., G. CORCORAN, J. LEWIS, *et al.* 2005. Refractory coccidioidomycosis treated with posaconazole. Clin. Infect. Dis. **40:** 1770–1776.

12. STEVENS, D.A. 1983. Miconazole in the treatment of coccidioidomycosis. Drugs **26:** 347–354.

13. STILLER, R.L., R. DEFELICE, C. BRASS, *et al.* 1980. Therapy of cutaneous coccidioidomycosis with imadazoles: comparison of results with miconazole and ketoconazole. Proceedings of the 5th International Conference on the Mycoses: Superficial Cutaneous and Subcutaneous Infections. Pan. Am. J. Public. Health **396:** 375–381.

14. STEVENS, D.A., R.L. STILLER, P.L. WILLIAMS, *et al.* 1983. Experience with ketoconazole in three major manifestations of progressive coccidioidomycosis. Am. J. Med. **74:** 58–63.

15. GALGIANI, J.N., D.A. STEVENS, J.R. GRAYBILL, *et al.* 1988. Ketoconazole therapy of progressive coccidioidomycosis: Comparison of 400- and 800-mg doses and observations at higher doses. Am. J. Med. **84:** 603–610.

16. TUCKER, R.M., D.W. DENNING, E.G. ARATHOON, *et al.* 1990. Itraconazole therapy for nonmeningeal coccidioidomycosis: clinical and laboratory observations. J. Am. Acad. Dermatol. **23:** 593–601.

17. GRAYBILL, J.R., D.A. STEVENS, J.N. GALGIANI, *et al.* NAIAD MYCOSES STUDY GROUP. 1990. Itraconazole treatment of coccidioidomycosis. Am. J. Med. **89:** 282–290.

18. DIAZ, M., R. NEGRONI, F. MONTERO-GEI, *et al.* PAN-AMERICAN STUDY GROUP. 1992. A Pan-American 5-year study of fluconazole therapy for deep mycoses in the immunocompetent host. Clin. Infect. Dis. **14**(Suppl 1): S68–S76.

19. HOSTETLER, J.S., A. CATANZARO, D.A. STEVENS, *et al.* 1994. Treatment of coccidioidomycosis with SCH 39304. J. Med. Vet. Mycol. **32:** 105–114.

20. CATANZARO, A., J.N. GALGIANI, B.E. LEVINE, *et al.* NIAID MYCOSES STUDY GROUP. 1995. Fluconazole in the treatment of chronic pulmonary and nonmeningeal disseminated coccidioidomycosis. Am. J. Med. **98:** 249–256.

21. GALGIANI, J.N., A. CATANZARO, G.A. CLOUD, *et al.* MYCOSES STUDY GROUP. 2000. Comparison of oral fluconazole and itraconazole for progressive, nonmeningeal coccidioidomycosis: a randomized, double-blind trial. Ann. Intern. Med. **133:** 676–686.

22. JOHNSON, R.H. & D. CALDWELL. 1996. State of the art lecture: extra-pulmonary nonmeningeal coccidioidomycosis. *In* Coccidioidomycosis. Proceedings of the 5th International Conference on Coccidioidomycosis. 1994, Aug. 24–27. H. Einstein, Ed.: 347–358. National Foundation for Infectious Diseases. Washington DC.

23. FIESE, M.J. 1958. Coccidioidomycosis.: 174–186. Charles C Thomas. Springfield. IL.

State-of-the-Art Treatment
of Coccidioidomycosis Skeletal Infections

JANIS E. BLAIR

Division of Infectious Diseases, Mayo Clinic, Scottsdale, Arizona

ABSTRACT: Coccidioidomycosis is a fungal infection endemic to the southwestern United States. Typically a respiratory illness, coccidioidomycosis can rarely present as extrapulmonary infection. Skeletal coccidioidomycosis occurs in 20% to 50% of disseminated infections. Skeletal coccidioidomycosis is a chronic and progressive infection that eventually results in bone destruction and loss of function and often involves adjacent structures, such as joints, muscles, and tendons and other soft tissues. Sinus tract formation may occur. This infection may be multifocal. Although radiographs, white blood cell count scans, and other imaging methods identify and define relevant abnormalities, histopathologic examination with culture of the involved bone is the only means to confirm the diagnosis. Serologic testing is adjunctive, and complement fixation titers can be evaluated serially to assess response to treatment. A number of studies addressing the efficacy of various antifungal agents have been performed, and the results of these studies as they pertain to skeletal coccidioidomycosis are summarized herein. Among the various studies, response rates ranged from 23% to 100%, but relapse was common. A combination of medical therapy—often, itraconazole or fluconazole—and surgical débridement is often needed to control skeletal coccidioidomycosis. Early diagnosis and treatment are critical to avoid long-term problems with chronically infected bones and joints. Anatomical issues, diagnostic studies, and data related to treatment of this form of extrapulmonary coccidioidomycosis are reviewed in this article.

KEYWORDS: *Coccidioides*; coccidioidomycosis; fungal skeletal diseases

INTRODUCTION

Coccidioidomycosis is a fungal infection endemic to the Lower Sonoran Life Zone in the southwestern United States, northern Mexico, Central America (Guatemala, Honduras, and Nicaragua), and South America (Argentina, Colombia, Paraguay, and Venezuela). The *Coccidioides* species causes an asymptomatic respiratory infection in 60% of persons who inhale the airborne

Address for correspondence: Janis E. Blair, M.D., Division of Infectious Diseases, Mayo Clinic, 13400 East Shea Boulevard, Scottsdale, AZ 85259.

Ann. N.Y. Acad. Sci. 1111: 422–433 (2007). © 2007 New York Academy of Sciences.
doi: 10.1196/annals.1406.000

arthrospores. Another 15% experience mild infections and do not require medical attention. The majority of the remaining infections are symptomatic respiratory infections of sufficient severity that the persons seek medical evaluation and treatment.

Extrapulmonary infections are uncommon (<5% of infections) in coccidioidomycosis and are more likely to occur in persons who are immunosuppressed, such as patients treated with immunosuppressive medication or infected with the human immunodeficiency virus. Extrapulmonary infections can also occur with increased frequency in otherwise healthy persons, including males; persons of certain races, such as Filipino or African American; and women who are pregnant. The most common sites for extrapulmonary coccidioidomycosis include skin, bones and joints, and the meninges.[1]

Bone infection accounts for 20% to 50% of extrapulmonary coccidiodal infections.[2,3] This article discusses current issues in the anatomy, diagnosis, and treatment of skeletal coccidioidomycosis, as presented at the Sixth International Symposium on Coccidioidomycosis, reported in this volume.

ANATOMICAL ISSUES

Extrapulmonary manifestations of coccidioidomycosis are nearly always the result of hematogenous spread after a primary pulmonary infection.[1] By the time a patient experiences overt skeletal involvement, however, the primary pulmonary infection is often resolved and findings on chest radiographs are normal. Skeletal coccidioidomycosis may seemingly develop during the course of medical therapy for the pulmonary infection, but this is likely the result of a preestablished, but heretofore unknown, infection resulting from hematogenous dissemination.[4]

Skeletal coccidioidomycosis is a chronic and progressive infection. It eventually results in bone destruction, loss of function, and involvement of adjacent structures, such as joints, muscles, tendons, and other soft tissues. Sinus tract formation from chronic coccidioidal osteomyelitis may occur.[3]

In coccidioidomycosis, any bone may become involved. However, the axial structures, such as vertebras, skull, sternum, and ribs, are more commonly involved than the appendicular structures, such as the long bones. Often, multiple bones are involved.[5] Symmetrical involvement has been described in some, but not all, series.[3,6]

Axial Skeleton

Both autopsy[5] and clinical series[2,7] demonstrate that the vertebral column is the most common site of skeletal osteomyelitis. Multifocal, noncontiguous vertebral involvement is common[6] (FIG. 1A). Vertebral infection begins in the central cancellous area of the vertebral body and subsequently extends to

(A)   **(B)**

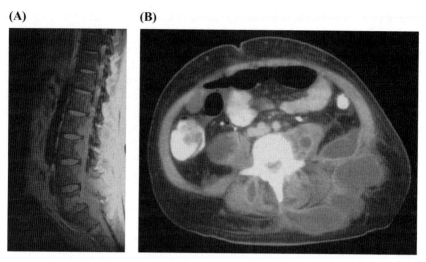

FIGURE 1. Multifocal coccidioidomycosis of the spine. (**A**) Fat suppressed, T2-weighted magnetic resonance imaging with abnormal signal in posterior lower lumbar vertebras and in epidural spaces. Note the relative sparing of the disks. (**B**) Computed tomographic scan showing multiple contiguous soft-tissue abscesses.

the periosteum, later extending into adjacent soft tissue to form paraspinal phlegmon or abscesses[5] (FIG. 1B). A common example of these extensions is coccidioidal psoas abscesses.[5] Collapse of the vertebral bodies may occur and may be asymmetrical.[6] Vertebral coccidioidomycosis may initially be confused with the tuberculous infection Pott's disease.[3,5] Relative sparing of the disks has been seen in vertebral lesions in some series until late into the course of infection; this sparing is not a universal observation.[6,8] Infections of the skull and ribs were not uncommon in one autopsy study, which found that these infections often extend into adjacent structures.[5]

Appendicular Skeleton

Any portion of the bone or periosteum may be involved in the granuloma formation. In one autopsy series, the ends of the long bones, especially at the bony prominences (e.g., tibial tubercle, ankle, acromial process of the scapula, medial aspect of the clavicle), were involved most often.[5] When infection is present in cancellous bone, the overlying articular cartilage and joint may become involved in the infection process.

Joint Infection

In skeletal coccidioidomycosis, joint infection is a less common manifestation than osteomyelitis.[2,7] Synovitis with effusion is the most common clinical presentation, and concurrent osteomyelitis may be present. Although any joint may be infected with *Coccidioides*, the joints of the lower extremities are more often involved than those of the upper extremities, and the knee is the joint most commonly involved. Coccidioidal infection of the knee joint was present in 8 of 10 joint infections in a series of 134 patients with skeletal coccidioidomycosis.[7] Other studies indicate knee involvement in joint infections in 8 of 16,[8] 2 of 8,[9] and 4 of 5[10] cases. Patients with coccidioidal knee synovitis may present without any other manifestation of coccidioidal infection. Typically, arthrocentesis shows exudative synovial fluid and histopathologic examination of the synovium shows granulomatous change and spherules. However, cultures are positive in only about half of the cases.[1] Treatment requires synovectomy and concurrent antifungal therapy.

Tenosynovitis

Coccidioidal involvement of the ligaments and tendons may occur because of direct involvement at the site of a joint infection. Isolated infection of a tendon and its sheath may also occur without infection of adjacent bone or joint.[11] Synovectomy, in combination with antifungal therapy, has been suggested to avoid tendon erosion and rupture.[12]

Muscles

Voluntary muscles adjacent to infected bones and joints may become involved with coccidioidomycosis by direct extension of the infection. Psoas muscle infection is the most common example of this phenomenon.[5] Primary infection of the muscles from hematogenous spread rarely occurs.

DIAGNOSIS

Radiographs are an important tool for the diagnosis of *Coccidioides* infection. Although early infection may cause few or no radiographic abnormalities, a long-standing infection is likely to show single or multiple lytic punched-out, destructive lesions with osteopenia and irregular or ill-defined borders (FIG. 2A). Small bone involvement may have a moth-eaten appearance.[13] A sequestrum is a late and less frequent radiographic finding. Periosteal new bone formation may be seen but is not common. Sclerosis may be seen, although infrequently, if lesions are treated early and healing is allowed to occur.[8] Radiographs with these findings suggest a number of possible infectious and noninfectious processes, including tuberculosis, actinomycosis, other fungal infections (such as blastomycosis or cryptococcosis), bacterial infections, and metastatic carcinoma.[8]

(A)   **(B)**

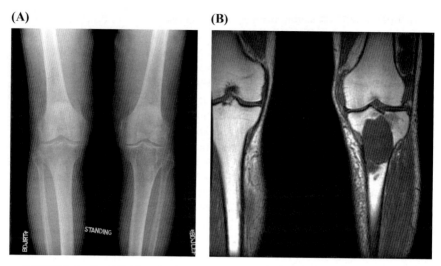

FIGURE 2. Coccidioidomycosis of the left proximal tibia. **(A)** Plain radiograph of both knees. Left tibia has large lytic lesion of coccidioidal skeletal infection. Bone scan (not shown) showed intense uptake at left proximal tibia but no other abnormalities. **(B)** T1-weighted magnetic resonance imaging spin echo sequence. Image of same large, proximal tibial lesion clearly delineates the abnormality.

A whole-skeleton bone scan, together with follow-up plain films of areas under question, allows the clinician to screen for polycentric bony involvement in cases of suspected or known disseminated coccidioidomycosis.[3] In one retrospective series, a technetium Tc[99m] bone scan was 100% sensitive in identifying bony coccidioidomycosis, whereas the findings on plain radiography and magnetic resonance imaging were variable.[10] Computed tomography and magnetic resonance imaging (FIG. 2B) may be used to further define areas of abnormality identified with the bone scan, allowing the surgeon to more precisely plan the surgical débridement.

In bone infections from other microorganisms, white blood cell counts are often normal, and this is also typically true in coccidioidal skeletal infections.[9] Many instances of osteomyelitis or synovitis due to bacterial microorganisms are associated with an elevated erythrocyte sedimentation rate, which also becomes a useful indicator for following the course of the infection. In contrast, the erythrocyte sedimentation rate is often normal in coccidioidal bone infections, and it is a particularly unreliable indicator in patients who are immunosuppressed.[9]

Tubercular osteomyelitis is in the differential diagnosis of osteomyelitis, especially in developing countries where tuberculosis is endemic. A chest radiograph, tuberculin skin test, and serologic testing for coccidioidomycosis may provide important clues to the etiology of the bony infection.

Histopathologic examination with culture of the involved bone, obtained by either percutaneous biopsy or surgical débridement, is the only means to confirm the diagnosis of skeletal coccidioidal infection. Granulomatous inflammation is typical, and spherules are often seen.

Coccidioides species grows well on bacterial and fungal media but may require 5 or more days for growth. If a specimen is incubated on bacterial media alone, the diagnosis may be missed because the organism may not grow before the termination of the culture at 3 to 5 days. By comparison, fungal cultures are generally incubated for several weeks and should be adequate for identifying growth of this organism.

Serologic testing is an adjunctive diagnostic measure, and many methods are available in both local and reference laboratories.[14] Complement fixation (CF) is a quantitative serologic test, and titers generally decrease as the patient receives treatment over time, indicating successful therapy. Since laboratory-to-laboratory variation may occur, serial CF tests performed to follow the response to treatment should be done at the same laboratory.[14] The initial CF titer has been correlated with the likelihood of relapse. In one study, a course of medical therapy alone was more likely to fail in patients with an initial CF titer of 1:128 or greater.[8] Other studies have not identified any correlation.[10]

TREATMENT

Medical Therapy

For many years, amphotericin B was the treatment of choice for coccidioidal skeletal infections. It was used because of its efficacy, and it achieved detectable levels in the joints of treated patients.[8] In a recent retrospective review, amphotericin B was noted to be a superior initial therapy in patients with multiple skeletal lesions or extraskeletal lesions.[3] However, amphotericin B treatment is complicated by a number of problems. The drug is available only as a parenteral agent, and it is often associated with the effects of infusional toxicities, such as fever, chills, headache, nausea, and vomiting. Renal toxicity and potassium and magnesium wasting are also problematic with cumulative doses over time. The newer lipid formulations of amphotericin B are associated with less renal toxicity. Available formulations include amphotericin B deoxycholate (0.5–1.5 mg/kg per day or on alternate days) and lipid formulations of amphotericin B (2.0–5.0 mg/kg per day).[15]

Imidazole and other azole therapies have been studied and developed over the past few decades. In many situations, these agents have become the preferred treatment because of their oral formulations, decreased toxicity, and apparent comparable efficacy. Options for treatment of coccidioidomycosis include oral fluconazole (400–800 mg daily) and oral itraconazole (200 mg 2 or 3 times daily). If the latter is used, serum concentrations of itraconazole should be measured to assess absorption. Oral cyclodextrin suspension of itraconazole improves its absorption.[15]

Newer azoles include voriconazole and posaconazole. A few case reports have suggested that voriconazole is efficacious in the treatment of refractory coccidioidomycosis in select patients. The newly approved posaconazole has been associated with one of two treatment successes in patients with refractory skeletal coccidioidomycosis.[16] However, no trials of the efficacy of these newer azoles in skeletal coccidioidomycosis have been published.

A number of trials on the treatment of coccidioidomycosis have been performed. The results in the skeletal coccidioidomycosis subgroups are summarized in TABLE 1. Many of the earliest studies evaluating nonamphotericin B antifungal medications were performed in patients in whom amphotericin B therapy was unsuccessful, who were intolerant of its side effects, or who simply refused the medication because of side effects. A survey of the coccidioidomycosis treatment trials to date demonstrates the difficulties in treating skeletal coccidioidomycosis; efficacy rates of medications currently available for this condition (i.e., fluconazole and itraconazole) range from a 52% response to an 85% response. A head-to-head comparison of fluconazole versus itraconazole showed a superior response to itraconazole in patients with skeletal infections.[25]

TABLE 2 summarizes the results of selected retrospective studies of skeletal coccidioidomycosis in single-site practices within the endemic area.[2,3,7,8,12,13,16] Patients were treated with one or more azoles or amphotericin B, or both, at the discretion of the treating physician. Whether a patient had surgical débridement was also variable and was determined on an individual basis. The treatment responses (defined by the condition's alleviation or cure) ranged from 50% to 100%. These differences undoubtedly reflect patient demographics, severity of infection, and changes in treatment over time.

The most effective duration of medical therapy for these infections is not known and is generally determined on an individual basis. Many patients continue to take medication to achieve long-term suppression long after the clinical resolution of the infection. Concurrent infection in locations with high rates of recurrence (such as the meninges) may dictate life-long treatment to avoid relapse of infection. Duration of treatment may be guided by clinical response, by CF titers, and by the return to normal values of inflammatory marker levels (such as anemia of chronic disease, serum albumin, erythrocyte sedimentation rate, and C-reactive protein) if original levels were abnormal. If and when a decision is made to discontinue antifungal therapy, close follow-up is recommended to detect early evidence of recrudescent infection.

TABLE 1. Summary of studies or reports addressing the treatment of skeletal coccidioidomycosis

Author/year (Ref.)	Patient selection	Medication	N	Patients Response(%)	Patients Relapse(%)	Comments
Deresinski and Stevens, 1979 (17); Stevens, 1983 (18)	Unsuccessful amphotericin B tx, disease relapse, or pt refusal of amphotericin B tx	Miconazole initiated at 200 mg tid i.v., advanced as needed or tolerated to 1,000 mg tid i.v.	31 (17 joint, 13 bone, 1 tenosynovium)	10/31 (32%) response or cure	5/9 (56%) evaluable responders; relapse more likely in pts with <1 month of tx	i.v. dosing limited duration of tx to a mean of 56 days Arthritis responded better than osteomyelitis
Stevens et al. 1983 (19)	Unsuccessful amphotericin B tx, disease relapse, or pt refusal of amphotericin B tx	Ketoconazole, 200–400 mg/day or greater	16	13/16 (81%)	2/6 (33%) in evaluable pts who completed tx	Pts treated either for 12 months or for 6 months after resolution of infection, whichever was longer
Galgiani et al. 1988 (20)	Positive culture ICH included	Randomized ketoconazole, 400 mg/day vs. 800 mg/day	44	Overall, 10/44 (23%); if treated >6 months, 10/37 (27%)	Not stated for subgroup	Pts in whom the lower dose was successful were offered the higher dose
Tucker et al. 1990 (21)	Positive culture or histologic test	Itraconazole dosage increased to 400 mg/day	25	20/25 (80%)	Not stated for subgroup	Pts treated either for 12 months or for 6 months after resolution of infection, whichever was longer
Graybill et al. 1990 (22)	Positive histologic test, no HIV	Itraconazole, 100–400 mg/day	12	8/12 (67%)	2/6 (33%) in initial responders; overall, 11/12 (92%) if prior clinical failure	Daily doses were increased if clinical score was not improved by 21% at 3 months; minimum tx duration, 6 months; no maximal tx duration
Hostetler et al. 1994 (23)	Positive culture or histologic test; pts not taking immunosuppressive therapy; HIV included	SCH 39304, 100–200 mg/day; few pts received higher doses	17	At 4 months, 7%,[a] at 8 months, 17%,[a] at 12 months, 31%[a]	Not stated	Median tx time, 8 months; product withdrawn from further development
Catanzaro et al. 1995 (24)	Positive culture or histologic test; no immunosuppressive tx	Fluconazole, 200 mg/day; increased to 400 mg/day if inadequate response to lower dosage	14	At 200 mg/day, 6/14 (43%), at 400 mg/day, 6/8 (75%), overall response, 12/14 (86%)	5/10 (50%)	Pilot study initiated tx at 50 mg/day, then 100 mg/day; enrollment closed at these dosages because of insufficient tx response
Galgiani et al. 2000 (25)	Positive culture or histologic test; HIV included	Fluconazole, 400 mg/day vs. itraconazole, 200 mg bid	50 (42 osteomyelitis, 10 joint infection [8 bone and joint involvement])	At 8 months: 7/27 (26%) fluconazole, 12/23 (52%) itraconazole; P = 0.8 At 12 months: 10/27 (37%) fluconazole, 16/23 (70%) itraconazole; P = 0.03	Not stated 0.08	Randomized, double-blind trial

bid = 2 times daily; HIV = infection with human immunodeficiency virus; ICH = immunocompromised host; i.v. = intravenous administration; pt = patient; tid = 3 times daily; tx = treatment.
[a] Numerator and denominator not provided in referenced article.

TABLE 2. Selected retrospective studies of skeletal coccidioidomycosis

Author/year (Ref.)	Patient selection	Medication	Surgery	N	Patients — Response (condition improved or resolved)	Relapse
Iger and Coppola, 1985 (7)	Retrospective review; regional central-California practices, 1977–1983 (7)	Amphotericin B i.v. or by local instillation (afterward, some pts received ketoconazole)	Surgical débridement: 10/11 (91%) bone; 8/12 (67%) joint	23 (11 bone infection, 12 joint infection)	9/11 (82%) bone, 6/12 (50%) joint	Not stated
Bried and Galgiani, 1986 (8)	Retrospective review; single university-based practice; immunosuppressed pts included	Amphotericin B or ketoconazole, or both, sequentially; medical tx alone in 14 pts	Both medical and surgical Tx in 13 pts	24	Of 14 treated medically, 7 (50%) responded; of 13 treated with combined medical and surgical tx, 13 (100%) responded	2/7 (29%) treated medically had relapses after discontinuing Tx
Johnson and Caldwell, 1996 (2)	Retrospective review of single public hospital practice, 1992–1994	Tx at discretion of physician: 10 received azole alone, 6 received amphotericin B alone, 10 received amphotericin B and azole sequentially	15 had surgical débridement	27 (22 bone infection, 5 joint infection)	18/27 (67%)	Not stated
Kushwaha et al. 1996 (9)	Retrospective review; single, primarily pediatric orthopedic practice, 1975–1994	22 received amphotericin B (10 with tx followed by azoles), 2 received azoles alone	24/25 had surgical Tx	25	19/22 (86%)	Not stated
Crum et al. 2004 (3)	Retrospective review; single military-based practice, 1994–2002	Amphotericin B, fluconazole, or itraconazole either initially or sequentially	Surgical evaluation in all pts, with débridement when clinically indicated	28	22/28 (79%); amphotericin B was superior to azole for initial tx ($P = 0.01$), especially if extraskeletal or multiple skeletal lesions were identified	3/28 (11%)
Anstead et al. 2005 (16)	Culture confirmed; all available tx was unsuccessful	Posaconazole, 600–800 mg/day		2 (bone infection)	1/2 (50%)	0/1

i.v. = intravenously; pts = patients; Tx = treatment.

Surgical Management

Surgery is commonly used in conjunction with antifungal therapy in the management of skeletal coccidioidomycosis. Although the condition of some patients clearly improves with medical therapy alone,[10] experienced practitioners in endemic areas have observed that with aggressive débridement and removal of the infected bone or synovium whenever practical, the condition may improve more rapidly and resolve more frequently.[2,8] At a minimum, surgical débridement or stabilization should be strongly considered when a large abscess, a sequestrum, spinal instability, or impingement on a critical organ or tissue is present.[15] Vertebras that are unstable because of extensive infection have been managed with aggressive débridement and instrumentation or bone grafting, or both.[4] Whether newer and more potent antifungal medications will lessen the need for surgical débridement is not known.

CONCLUSION

Early diagnosis and treatment of skeletal coccidioidomycosis are critical to avoid long-term problems with chronically infected bones and joints. When *Coccidiodes* infection is identified, the extent of the infection should be assessed, both within the skeletal system and in other common areas of dissemination, such as skin and soft tissue.

Specific recommendations for antifungal medical treatment for skeletal coccidioidomycosis are lacking. Although in one single comparative trial[25] itraconazole showed superiority over fluconazole for the treatment of skeletal coccidioidomycosis, fluconazole has shown efficacy in other studies. Surgical débridement should be considered as an important adjunctive therapy. Duration of treatment varies and depends on the patient's clinical response and serial serologic and radiographic findings. Close follow-up is recommended after cessation of treatment to identify early evidence of relapse.

ACKNOWLEDGMENT

Editing, proofreading, and reference verification were provided by the Section of Scientific Publications, Mayo Clinic.

REFERENCES

1. CHILLER, T.M., J.N. GALGIANI & D.A. STEVENS. 2003. Coccidioidomycosis. Infect. Dis. Clin. North Am. **17:** 41–57.
2. JOHNSON, R. & D. CALDWELL. 1996. State of the art lecture: extra-pulmonary nonmeningeal coccidioidomycosis. *In* Coccidioidomycosis: Proceedings of the

5th International Conference on Coccidioidomycosis (held Aug. 24–27, 1996). H.E. Einstein & A. Catanzano, Eds.: 347–358. National Foundation for Infectious Diseases. Bethesda, MD.

3. CRUM, N.F., E.R. LEDERMAN, C.M. STAFFORD, *et al.* 2004. Coccidioidomycosis: a descriptive survey of a reemerging disease: clinical characteristics and current controversies. Medicine (Baltimore) **83:** 149–175.

4. WROBEL, C.J., E.T. CHAPPELL & W. TAYLOR. 2001. Clinical presentation, radiological findings, and treatment results of coccidioidomycosis involving the spine: report on 23 cases. J. Neurosurg. **95**(Suppl):33–39.

5. FORBUS, W.D. & A.M. BESTEBREURTJE. 1946. Coccidioidomycosis: study of 95 cases of disseminated type with special reference to pathogenesis of disease. Mil. Surgeon. **99:** 653–719.

6. DALINKA, M.K. & W.H. GREENDYKE. 1971. The spinal manifestations of coccidioidomycosis. J. Can. Assoc. Radiol. **22:** 93–99.

7. IGER, M. & A.J. COPPOLA. 1985. Review of 135 cases of bone and joint coccidioidomycosis. *In* Coccidioidomycosis: Proceedings of the 4th International Conference. H. Einstein, A. Catanzaro, Eds.: 379–389. (held Mar. 14–17, 1984). National Foundation for Infectious Diseases. Washington DC.

8. BRIED, J.M. & J.N. GALGIANI. 1986. *Coccidioides immitis* infections in bones and joints. Clin. Orthop. Relat. Res. **211:** 235–243.

9. KUSHWAHA, V.P., B.A. SHAW, J.A. GERARDI & W.L. OPPENHEIM. 1996. Musculoskeletal coccidioidomycosis: a review of 25 cases. Clin. Orthop. Relat. Res. **332:** 190–199.

10. HOLLEY, K., M. MULDOON & S. TASKER. 2002. *Coccidioides immitis* osteomyelitis: a case series review. Orthopedics. **25:** 827–832.

11. REID, G.D., A. KLINKHOFF, C. BOZEK & J.F. DENEGRI. 1984. Coccidioidomycosis tenosynovitis: case report and review of the literature. J. Rheumatol. **11:** 392–394.

12. IVERSON, R.E. & L.M. VISTNES. 1973. Coccidioidomycosis tenosynovitis in the hand. J. Bone Joint Surg. Am. **55:** 413–417.

13. MCGAHAN, J.P., D.S. GRAVES, P.E. PALMER, *et al.* 1981. Classic and contemporary imaging of coccidioidomycosis. AJR Am. J. Roentgenol. **136:** 393–404.

14. PAPPAGIANIS, D. & B.L. ZIMMER. 1990. Serology of coccidioidomycosis. Clin. Microbiol. Rev. **3:** 247–268.

15. GALGIANI, J.N., N.M. AMPEL, J.E. BLAIR, *et al.* 2005. Infectious Diseases Society of America. Coccidioidomycosis. Clin. Infect. Dis. **41:** 1217–1223. Epub 2005 Sep. 20.

16. ANSTEAD, G.M., G. CORCORAN, J. LEWIS, *et al.* 2005. Refractory coccidioidomycosis treated with posaconazole. Clin. Infect. Dis. **40:** 1770–1776. Epub 2005 May 13.

17. DERESINSKI, S.C., D.A. STEVENS. 1979. Bone and joint coccidioidomycosis treated with miconazole. Am. Rev. Respir. Dis. **120:** 1101–1107.

18. STEVENS, D.A.. 1983. Miconazole in the treatment of coccidioidomycosis. Drugs **26:** 347–354.

19. STEVENS, D.A., R.L. STILLER, P.L. WILLIAMS & A.M. SUGAR. 1983. Experience with ketoconazole in three major manifestations of progressive coccidioidomycosis. Am. J. Med. **74:** 58–63.

20. GALGIANI, J.N., D.A. STEVENS, J.R. GRAYBILL, *et al.* 1988. Ketoconazole therapy of progressive coccidioidomycosis: comparison of 400- and 800-mg doses and observations at higher doses. Am. J. Med. **84:** 603–610.

21. TUCKER, R.M., D.W. DENNING, E.G. ARATHOON, *et al*. 1990. Itraconazole therapy for nonmeningeal coccidioidomycosis: clinical and laboratory observations. J. Am. Acad. Dermatol. **23:** 593–601.

22. GRAYBILL, J.R., D.A. STEVENS, J.N. GALGIANI, *et al*. 1990. NIAID Mycoses Study Group. Itraconazole treatment of coccidioidomycosis. Am. J. Med. **89:** 282–290.

23. HOSTETLER, J.S., A. CATANZARO, D.A. STEVENS, *et al*. 1994. Treatment of coccidioidomycosis with SCH 39304. J. Med. Vet. Mycol. **32:** 105–114.

24. CATANZARO, A., J.N. GALGIANI, B.E. LEVINE, *et al*. 1995. NIAID Mycoses Study Group. Fluconazole in the treatment of chronic pulmonary and nonmeningeal disseminated coccidioidomycosis. Am. J. Med. **98:** 249–256.

25. GALGIANI, J.N., A. CATANZARO, G.A. CLOUD, *et al*. 2000. Mycoses Study Group. Comparison of oral fluconazole and itraconazole for progressive, nonmeningeal coccidioidomycosis: a randomized, double-blind trial. Ann. Intern. Med. **133:** 676–686.

Amphotericin B and Coccidioidomycosis

ROYCE H. JOHNSON AND HANS E. EINSTEIN

Kern Medical Center–Medicine, Bakersfield, California, USA

ABSTRACT: Prior to the 1950s no effective therapy for coccidioidomy-cosis existed. The advent of amphotericin B ushered in the therapeutic era for coccidioidomycosis. Until this time amphotericin B and its lipid congeners have been regarded as the "gold standard" of therapy for se-vere pulmonary and disseminated coccidioidomycosis. The availability of azoles and later triazoles for the past three decades have relegated the amphotericins into a rescue mode, used mainly in widely disseminated cases, azole intolerance, or when there are contraindications to Azoles, such as pregnancy. In meningitis the intrathecal use of amphotericin B is still used frequently by some clinicians alone or with a triazole. The newer lipid preparations, while more expensive, have significantly re-duced toxicity, particularly nephropathy.

KEYWORDS: amphotericin B; coccidioidomycosis; therapy

BACKGROUND

Coccidioidomycosis is a disease caused by thermal dimorphic fungus found throughout the lower Sonoran life zone in the Western Hemisphere. For the first 60 years after the discovery of the disease by Wernicke and Posadas, there were many attempts at therapy (TABLE 1).[1] Some of these therapies had extensive investigations, even in the mid 20th century.[2]

It was not until the mid 1950s that we had the introduction of the first effective antifungal agent, amphotericin B, as the deoxycholate suspension. Coccidioidomycosis was one of the early illnesses against which this agent was measured.[3] There were no comparative or placebo-controlled trials in the modern sense of the term. Amphotericin was the first drug that had demon-strable efficacy in altering the course of severe pulmonary or disseminated coccidioidal infections. The accumulated accounts of such interventions raised amphotericin B deoxycholate to that of the gold standard of therapy for coc-cidioidomycosis.[4]

Address for correspondence: Royce Johnson, Kern Medical Center–Medicine, 1830 Flower Street, Bakersfield, CA 93305. Voice: 661-326-2224; fax: 661-326-2478.
johnsonr@kernmedctr.com

Ann. N.Y. Acad. Sci. 1111: 434–441 (2007). © 2007 New York Academy of Sciences.
doi: 10.1196/annals.1406.019

TABLE 1. Some of the agents proposed for treatment of coccidioidomycosis that have been discarded or inadequately tested

Methyl violet	Bismuth potassium tartrate
Tincture of iodine	Iodobismitol
Potassium bromide	Sodium thiosulfate
Oil of turpentine	Fuadin
Carbolic acid	Sulfonamides
Potassium permanganate	Convalescent serum
Colloidal copper	Penicillin
Gold salts	Streptomycin
Colloidal lead	Streptothrycin
Thymol	Salicil
Potassium iodide	Plumbagin
Creosote	Bacillomycin
Guaiacol	Chloraquin
Antimony and potassium tartrate	Histamine
Roentgen therapy	Chloromycetin
Gentian violet	Protoanemonin
Bismuth	Fradicin
Typhoid vaccine	Thiolutin
Coccidioidin	Rimocidin
Copper sulfate	Candicidin
Lead acetate	Para-amino benzoic acid
Mercury cyanide	Hydroxychloraquin
Novasurol	

AMPHOTERICIN B

Chemistry and Mechanism of Action

Amphotericin B is a natural substance isolated from a strain of *Streptomyces nodosus* obtained from the soil adjacent to the Oronoco River in Venezuela.[5] Amphotericin B is a member of the polyene group of antibiotics that also includes nystatin as the other most commonly recognized member. It has hydrophilic and hydrophobic regions. In its pure form it has very little solubility in aqueous solutions. Because of this latter characteristic, complexing with some other agent is required for clinical administration. The first such agent used was sodium deoxycholate that contains sodium phosphates as buffers (including mono- and dibasic phosphates, sodium hydroxide, and phosphoric acid) and is marketed as lyophilized cake (which may partially reduce to powder after manufacture). This complexing produces micellar structures and presents a fine suspension that appears to the unaided eye to be a solution. This suspension is buffered with sodium phosphate. The exact composition of this buffer has changed over time.

Amphotericin B deoxycholate is marketed as a powder that is resuspended in sterile water. The addition of a salt-containing solution invariably results in precipitation of the amphotericin B. The prepared amphotericin B deoxycholate suspension is compatible with dextrose-containing solutions for intravenous or intrathecal administration and with sterile water for some local applications. The mechanism of action, which is shared in common with other polyenes, is based on the binding of the hydrophobic moiety of the amphotericin B suspension to the spherule's cell membrane ergosterol moiety.[6] This binding produces damage to the cell membrane and, potentially, cell death. Amphotericin B also has the capability of binding to the cholesterol of mammalian cell membranes, which is responsible for a major fraction of its toxic potential. The differential binding of ergosterol and cholesterol is primarily responsible for the therapeutic index of the agent. Amphotericin B may have additional mechanisms of producing toxicity in fungal organisms as well. Immediate human toxicities may be precipitated by amphotericin's ability to bind to one or more toll-like receptors in mammalian cells.[7]

Amphotericin B Deoxycholate: Clinical Utilization

For 50 years amphotericin B deoxycholate has been administered intravenously, intrathecally, intralesionally, intra-articularly and infused into surgical sites.[8] The vast majority of the drug has been used intravenously to treat severe forms of pulmonary and disseminated coccidioidomycosis.

Early in the use of intravenous amphotericin it was discovered that, unlike cryptococcal meningitis, there was an inadequate response in coccidioidal meningitis given by this route.[9]

Intrathecal amphotericin, first by lumbar administration and later by intracisternal injection and Omaya reservoir administration into the ventricle and the cisterna magna, became the treatment of choice for coccidioidal meningitis. It was not until the advent of azole therapy, particularly fluconazole therapy, that a viable alternative to intrathecal amphotericin became available.[10,11]

Intralesional and intra-articular injections were undertaken to increase the local concentration of the drug at the site of the infection and to decrease the systemic toxicity of the drug given intravenously. It is reasonably well-tolerated intralesionally. Intra-articular injection most often resulted in postinjection synovitis, probably based on an inflammatory response to the particles of amphotericin B deoxycholate in a manner similar to that of gout and pseudo-gout. Infusion of surgical sites with amphotericin, particular orthopedic surgical sites, was often undertaken for similar reasons.[12] While local administration has been tried with antibacterials, in most circumstances, this only results in the drug's becoming systemic. In the case of amphotericin, at least theoretically because of its molecular size and its lipotrophic characteristics, it is thought that much of the drug actually may persist where deposited in local tissues.

Newer Amphotericin B Preparations

Multiple attempts have been made to improve upon the early polyene antifungal agents. One of the earliest attempts was the development of a methyl ester of amphotericin B.[13,14] There was significant preclinical and clinical investigation with this agent, which, however, proved to have significant neurotoxicity, causing its further investigation to be abandoned.[15,16] More recently, there have been investigations with a hydrosome preparation of amphotericin B in an animal mode.[17]

Most of the efforts at improving amphotericin B over the last 20 years, however, have been focused on the preparation of amphotericin B with a lipid conjugate. Several preparations have been investigated, three of which came to clinical trials and commercialization. The first of these is amphotericin B colloidal dispersion (ABCD), the second is amphotericin B lipid complex (ABLC), and the third is amphotericin B liposomal (AB-Lip).[18–23]

The principal motivation for the development of additional amphotericin B products is the search for agents that are more efficacious, more tolerable, and less toxic, particularly less nephrotoxic than the gold-standard amphotericin B deoxycholate. The principal acute toxicity of amphotericin B deoxycholate—nausea, vomiting, rigors, fever, hypertension/hypotension, and hypoxia—does appear to be mitigated by the addition of any of the above-mentioned lipid moieties to the amphotericin B suspension. It soon became apparent that the acute toxicities associated with ABCD were not substantially less than those of the deoxycholate preparation.

It is unclear whether chronic complications, such as anemia, anorexia, and cardiomyopathy, are less common with the lipid preparation amphotericins than the deoxycholate.[24,25]

What is clear is that all three of the lipid preparations produce less substantial long-term nephrotoxicity.[26] It is increasingly apparent that amphotericin B lipid preparations are the new "gold standard" of polyene therapy.[27] It should be noted that the newer agents are very costly compared to amphotericin B deoxycholate.

Amphotericin B probably produces renal injury by a variety of mechanisms.[28,29] Early in therapy, one can see a significant rise in creatinine. This is probably based on a poorly understood renal vasoconstriction of the afferent arteriole. This may abate with patience and hydration. The deoxycholate moiety may be nephrotoxic and account for the differential renal toxicity of amphotericin B deoxycholate compared to the lipid congeners.[30] Additional tubular injury clearly occurs, eventuating in hypokalemia and hypomagnesemia and probably less clinically significant, bicarbonate and amino acid loss.[31] Eventually, there is loss of functioning nephrons, and in individuals treated with high doses for protracted periods, significant renal failure requiring hemodialysis can occur. This substantial nephrotoxicity appears to be more common in individuals treated with a higher dose, increased frequency and duration, and in fungal infections other than coccidioidomycosis.[32] Renal toxicity is increased by concomitant administration of other nephrotoxic drugs and in volume-depleted patients. It should also be noted that significant recovery of renal function can occur and this has been noted since the early days of treatment.[33]

Amphotericin B Dosage in Coccidioidomycosis

Amphotericin B deoxycholate can be given in a range from 0.3 mg/kg to 1.5 mg/kg. In coccidioidomycosis, doses from 0.7 to 1 mg/kg have been most usual. Durations of therapy have often been calculated in total doses with this preparation. Doses as small as 1 g total have been felt to have a positive outcome in relatively modest disease. Doses of 3–5 g total have been used in more severe disease. We have used doses as high as 30 g given over more than 6 months, albeit very rarely.

Total durations of therapy with amphotericin B deoxycholate and with the lipid preparations are most frequently 1–4 months. The medication is often given on a daily basis for more severe forms of disease. For patients with milder forms of disease therapy may be initiated on a three times per week basis. As patients with more severe disease improve they are frequently tapered to a three times per week regime.

Amphotericin B lipid complex is usually administered at 5 mg/kg. Amphotericin B liposomal doses can be administered in amounts from 3 to 5 mg/kg. At our institution, the higher dose is more usual.

Medication can be administered in the intensive care unit, a medical/surgical ward, an outpatient infusion center, or with a home health agency. In the latter circumstance, a nurse is usually required to stay with a patient during the infusion process unlike the circumstance with most antibacterial therapy. Premedication with acetaminophen 650–1000 mg and diphenhydramine 50 mg is customary. The addition of pretherapy metoclopramide for individuals that demonstrate vomiting is frequently necessary. The use of small doses of meperidine intravenously for the specific control of rigors is occasionally required.

Careful follow-up of the patient's CBC, electrolytes (including calcium, magnesium, and phosphate) and liver function tests is requisite. Infusion and oral therapy with magnesium and potassium is very frequently required because of renal tubular toxicity. Anemia occasionally requires therapeutic intervention.

It is desirable to have an amphotericin order form that encompasses much of the above information. Additionally, a flow sheet of the therapy and the laboratory values can be advantageous in following patients with what is a relatively complicated therapy over time.

Indications for Amphotericin B in the Azole Era

At Kern Medical Center, amphotericin is used with substantially less frequency than it was before the 1990s. At this time, it is used for severe or rapidly progressive alveolar or miliary pulmonary disease with hypoxia ($PaO_2 \leq$ 70 mmHg on room air). Pregnant individuals with lesser degrees of pneumonia and/or a CF titer $\geq$ 1:16 are usually also placed on amphotericin therapy because azoles are contraindicated in pregnancy. Azoles, and particularly fluconazole, have clearly become the main treatment of uncomplicated primary and disseminated coccidioidomycosis.[34]

Osteoarticular and soft tissue disease is also common circumstance where amphotericin B preparations can continue to be useful. Multifocal disease, particularly involving the axial skeleton, or rapidly progressive diseases are the key reasons for primary amphotericin B. Failed azole therapy with or without surgical intervention is a common indication for an amphotericin B preparation.

The most common reason that intrathecal amphotericin B for using in coccidioidal meningitis at this time is azole failure. Intrathecal amphotericin B is occasionally preferred by patients because of its potential to be less than lifelong, as is required with azoles.[35,36]

SUMMARY

Amphotericin B preparations after 60 years continue to be an important therapeutic modality in coccidioidomycosis. The advent of the lipid preparation drugs, particularly ABLC and AB-Lip, has decreased the toxicity, but not evidently increased the efficacy of amphotericin B therapy. Azole therapy is the predominant therapy for mild and moderate coccidioidal disease; amphotericin is now reserved for use in the most severe forms of the disease as it appears to produce a more rapid response. It is also used in individuals in whom high-dose azole therapy has failed.

REFERENCES

1. FIESE, M.J. 1958. Coccidioidomycosis. 182. Charles C Thomas: Springfield, IL.
2. COHEN, R. 1955. A new fungicide for coccidioidomycosis. Arch. Pediat. **72:** 154–155.
3. WINN, A. 1959. The use of amphotericin B in the treatment of coccidioidal disease. Sem. Mycotic Infect. **3:** 617–635.
4. EINSTEIN, H.E. & R.H. JOHNSON. 1993. Coccidioidomycosis: new aspects of epidemiology and therapy. CID **16:** 349–356.
5. GROLL, H. & T.J. WALSH. 2000. Polyene antifungal agents. *In* Fungal Diseases of the Lung, third edition **17:** 271–277.

6. BRAJTBURG, J. & J. BOLARD. 1996. Carrier effects on biological activity of amphotericin B. Clin. Microbiol. Rev.. **9:** 512–531.

7. SAU, K., *et al.* 2003. The antifungal drug amphotericin B promotes inflammatory cytokine release by a toll-like receptor and CD14-dependent mechanism. J. Biol. Chem. **278:** 37561–37568.

8. STEIN, R.S. *et al.* 1965. Treatment of coccidioidomycosis infection of bone with local amphotericin B suction-irrigation. Case Reports **108:** 161–164.

9. UTZ, J.P. 1967. Recent experience in the chemotherapy of the systemic mycoses. *In* Coccidioidomycosis. L. Ajello, Ed.: 113–117. University of Arizona Press. Tucson.

10. GRAYBILL, J.R. *et al.* 1988. Ketoconazole treatment of coccidioidal meningitis. Ann. N.Y. Acad. Sci. **544:** 488–496.

11. GALGIANI, J.N. *et al.* 2000. Comparison of oral fluconazole and itraconazole for progressive, non-meningeal coccidioidomycosis: a randomized, double-blind trial. Ann. Int. Med. **133:** 676–686.

12. IGER, M. & J. LARSON. 1967. Coccidioidal osteomyelitis. *In* Proceedings of Coccidioidomycosis. (Second Coccidioidomycosis Symposium). L. Ajello, Ed.: 89–92. University of Arizona Press. Tucson.

13. HOEPRICH, P.D. *et al.* 1988. Treatment of fungal infections with semisynthetic derivatives of amphotericin B. Ann. N. Y. Acad. Sci. **544:** 517–546.

14. LAWRENCE, R.M. & P.D. HOEPRICH. 1976. Comparison of amphotericin B and amphotericin B methyl ester: efficacy in murine coccidioidomycosis. J. Infect. Dis. **133:** 168–174.

15. HOEPRICH, P.D. 1982. Amphotericin B methyl ester and leukoencephalopathy: the other side of the coin. J. Infect. Dis. **146:** 173–175.

16. ELLIS, W.G. *et al.* 1982. Leukoencephalopathy in patients treated with amphotericin B methyl ester. J. Infect. Dis. **146:** 125–137.

17. CLEMONS, K.V. *et al.* 2001. A novel heparin-coated hydrophilic preparation of amphotericin B hydrosomes. Curr. Opin. Invest. Drugs **2:** 480–487.

18. GONZALEZ, G.M. *et al.* 2004. Efficacy of amphotericin B (AMB) lipid complex, AMB colloidal dispersion, liposomal AMB, and conventional AMB in treatment of murine coccidioidomycosis. Antimicrob. Agents Chemother. **48:** 2140–2143.

19. CLEMONS, K.V. & D.A. STEVENS. 1992. Efficacies of amphotericin B lipid complex (ABLC) and conventional amphotericin B against murine coccidioidomycosis. J. Antimicrob. Chemother. **30:** 353–363.

20. CLEMONS, K.V. & D.A. STEVENS. 1991. Comparative efficacy of amphotericin B colloidal dispersion and amphotericin B deoxycholate suspension in treatment of murine coccidioidomycosis. Antimicrob. Agents Chemother. **35:** 1829–1833.

21. CLEMONS, K.V. *et al.* 2002. Efficacy of intravenous liposomal amphotericin B (AmBisome) against coccidioidal meningitis in rabbits. Antimicrob. Agents Chemother. **46:** 2420–2426.

22. KOEHLER, A.P. *et al.* 1996. Successful treatment of disseminated coccidioidomycosis with amphotericin B lipid complex. Case Reports **95:** 113–115.

23. ANTONY, S. *et al.* 2003. Use of liposomal amphotericin B in the treatment of disseminated coccidioidomycosis. J. Natl. Med. Assoc. **95:** 982–985.

24. ARSURA, E.L. *et al.* 1994. Amphotericin B-induced dilated cardiomyopathy. Am. J. Med. **97:** 560–562.

25. DANAHER, P.J. *et al.* 2004. Reversible dilated cardiomyopathy related to amphotericin B therapy. J. Antimicrob. Chemother. **53:** 115–117.

26. WALSH, T.J. *et al.* 1999. N. Engl. J. Med. **340:** 764.

27. OSTROSKY-ZEICHNER, L. *et al.* 2003. Amphotericin B: time for a new "gold standard": reviews of anti-infective agents. CID **37**: 415–425.

28. REYNOLDS, E.S. *et al.* 1963. The renal lesion related to amphotericin B treatment for coccidioidomycosis. Med. Clin. North Am. **47**: 1149–1154.

29. IOVINE, G. *et al.* 1963. Nephrotoxicity amphotericin B. Arch. Intern. Med. **112**: 853–862.

30. ZAGER, R.A. *et al.* 1992. Direct amphotericin B-mediated tubular toxicity: assessments of selected cytoprotective agents. Kidney Int. **41**: 1588–1594.

31. MCCHESNEY, J.A. & J.F. MARQUARDT. 1964. Hypokalemic paralysis induced by amphotericin B. **189**: 1029–1031.

32. TAKACS, F.J. *et al.* 1968. Amphotericin B nephrotoxicity with irreversible renal failure. Case Study **59**: 716–724.

33. HOLEMAN, C.W., JR. & H. EINSTEIN. 1963. The toxic effects of amphotericin B in man. Calif. Med. **99**: 90–93.

34. GALGIANI, J.N. *et al.* 2005. Coccidioidomycosis. CID **41**: 1217–1223.

35. SPINELLO, I.M. & R.H. JOHNSON. 2006. A 19-year-old pregnant woman with a skin lesion and respiratory failure. Chest Cardiopulm. Crit. Care J. **130**: 611–615.

36. STEVENS, D.A. & S.A. SHATSKY. 2001. Intrathecal amphotericin in the management of coccidioidal meningitis. Sem. Resp. Infect. **16**: 263–269.

Azole Therapy of Clinical and Experimental Coccidioidomycosis

DAVID A. STEVENS AND KARL V. CLEMONS

Department of Medicine, Santa Clara Valley Medical Center, San Jose, California, USA

California Institute for Medical Research, San Jose, California, USA

Stanford University, Stanford, California, USA

ABSTRACT: The therapy of coccidioidomycosis has been an early target, both experimentally and clinically, for study of new members of the azole class of drugs, because of the recognition that coccidioidomycosis is one of the most difficult mycoses to treat, and because our research group and our collaborators have been eager to pioneer new therapies for this problem pathogen. There have been steady advances in the pharmacologic and antimicrobial properties of this class since the initial introduction of miconazole, and many patients with coccidioidomycosis have benefited. Perhaps the greatest contribution has been the development of well-tolerated oral drugs that make possible prolonged courses of a conveniently administered agent, and perhaps the most impressive advance has been the utility of the agents in coccidioidal meningitis, at least as an adjunct to the polyenes. More potent agents are still required, so that complete biological cure can be attained in meningeal and nonmeningeal coccidioidomycosis.

KEYWORDS: azole antifungals; clinical coccidioidomycosis therapy; experimental coccidioidomycosis therapy

The azole class of drugs acts on fungi by inhibiting the synthesis of ergosterol in the fungal cell membrane. The clinical azole era, in general and the field of coccidioidomycosis specifically began with a "bang" with the testing of miconazole (FIG. 1). In our initial studies,[1] we showed with 26 *Coccidioides* isolates that the mean minimum inhibitory concentration (MIC) was 0.70 mcg/mL (mean; range 0.10–1.70) in the saprobic phase, whereas the endospore phase produced 33–79% lower results. Dosing the drug twice daily, subcutaneously or intramuscularly, could save all animals given a 100% lethal challenge.[1] The MIC values could be exceeded in human blood with tolerable intravenous doses. This led directly to clinical studies,[2] which produced

Address for correspondence: Dr. D.A. Stevens, Department of Medicine, Santa Clara Valley Medical Center, 751 So. Bascom Avenue, San Jose, CA 95128–2699. Voice: 408-885-4303; fax: 408-885-4306. stevens@stanford.edu

Ann. N.Y. Acad. Sci. 1111: 442–454 (2007). © 2007 New York Academy of Sciences.
doi: 10.1196/annals.1406.039

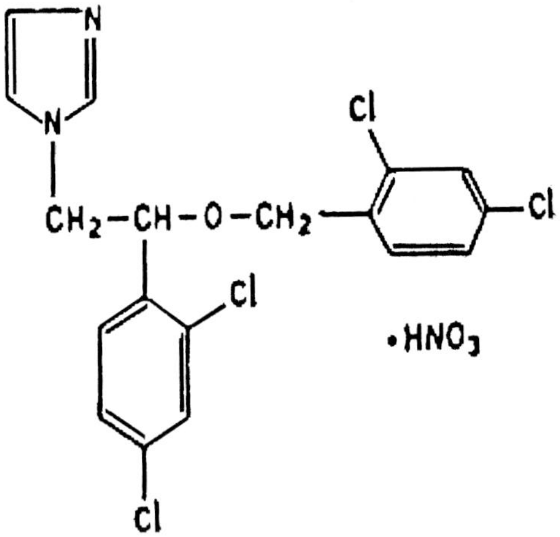

FIGURE 1. Molecular structure of miconazole.

impressive results in chronic and disseminated coccidioidomycosis (TABLE 1). However, as the drug had to be given intravenously, prolonged courses were not convenient. An apparent consequence was that relapse was common (TABLE 2). In addition, there were a number of side effects, mostly attributable to the vehicle needed for this insoluble drug (TABLE 3). It is probable that all members of this class of drugs are teratogenic.

The encouraging results led to the development of the first orally effective antifungal azole, ketoconazole. We showed that oral dosing could protect mice against a lethal coccidioidal challenge, and, with prolonged and twice-daily dosing, could cure some animals (TABLE 4).[3] Clinical studies soon followed, and our own experience in coccidioidomycosis is summarized in TABLE 5.[4] While this experience signaled for the first time that an oral drug would be useful in therapy of coccidioidomycosis, in the prolonged courses that became possible with an oral agent (and were required to treat a chronic disease with granuloma formation and fibrosis, such as coccidioidomycosis), the drug was not without some problematic side effects, particularly when the dose was increased to treat recalcitrant or slowly responding disease (TABLE 6).[5] Particularly problematic were new azole side effects that we described, namely, mammalian steroid

TABLE 1. Responses to miconazole in coccidioidomycosis

	Courses	Mean duration Rx (days)	Response
Skin/soft tissue	33	44.9	40%
Chronic pulmonary	33	43.7	72%
Meningitis	46	71.8	31%
Skeletal	31	56.7	32%

From Stevens.[2] Reprinted by permission.

TABLE 2.   Relapse in responders with coccidioidomycosis to miconazole

	Relapse
Skin/soft tissue	60%
Chronic pulmonary	75%
Meningitis	78%
Skeletal	56%

From Stevens.[2] Reprinted by permission.

blockade.[6] In addition, preclinical testing of modifications of the imidazole molecule was suggesting the possibility of extension of the azole spectrum to other fungal pathogens not reached by ketoconazole, particularly *Cryptococcus* and *Aspergillus*. These two pathogens were becoming of particular interest with the advent of AIDS, and the increase in transplantation as a treatment for organ dysfunction and of immunosuppression for a variety of diseases, respectively.

These factors led to the development of the triazole, itraconazole[7] (FIG. 2), which was free of the prior limitations. We showed the marked susceptibility of *Coccidioides in vitro* to this new agent (FIG. 3). These inhibitory and cidal concentrations were easily exceeded in the blood, at steady state, in patients with coccidioidomycosis[8] (FIG. 4). Most sites of clinical coccidioidomycosis were responsive (FIG. 5), though sometimes long courses were needed to achieve responses (FIG. 6). Although absorption of itraconazole capsules is a problem

TABLE 3.   Miconazole side effects

	Incidence	*n*
Nausea	46%	127
Hyponatremia	46%	91
Anemia	44%	100
Pruritus	36%	104
Phlebitis	35%	105
Thrombocytosis	31%	70

n = number of most recent courses studied for side effect.

TABLE 4.   Ketoconazole in experimental murine coccidioidomycosis

Treatment	Dead/total	Survivors-visceral culture negative[a]
None	25/39	0/14
40 mg/kg qd × 17 d	0/10	0/10
80 mg/kg bid × 46–96 d	0/18	9/18

From Borelli *et al.*[3] Reprinted by permission.
[a] Lung, liver, spleen, and kidney.

TABLE 5. Coccidioidal ketoconazole responses

	Percent response		
	Skeletal	Chronic pulmonary	Skin/soft tissue
Complete	21	12	68
Major	42	41	9
Minor	5	41	14
No or worse	31	6	9
N	19	17	22

From Stevens et al.[4] Reprinted by permission.

in some populations (those with hypochlorhydria, mucositis), it is less so in coccidioidomycosis, where many of the patients are otherwise healthy. The development of a cyclodextrin solution has obviated much of the absorption problem.[9] Side effects were uncommon,[10] and the main pharmacologic/toxicologic problems with the drug were drug–drug interactions.[11] These were of different varieties: drugs which decreased itraconazole levels, either through increasing cytochrome P450 (CYP) enzymes in the liver or by neutralizing the stomach acidity needed for absorption, or agents whose own levels would be increased by itraconazole, to toxic proportions, usually through blockade of CYP enzymes, or, less commonly, by affecting drug efflux. Coccidioidal relapse appeared to be less of a problem than with any prior agent (FIG. 7). Of particular interest, coccidioidal meningitis appeared responsive to oral therapy too (FIG. 8).[12]

The next triazole to be developed was fluconazole (FIG. 9), which had markedly different pharmacologic properties.[13] It is well absorbed after oral administration, and absorption is not pH dependent. It also produced high response rates in various manifestations of coccidioidomycosis, but there appeared to be a problem, especially with the use of low doses, of relapse, even

TABLE 6. Toxicity of ketoconazole

	Percent patients with toxicity	
Toxicity	400 mg/day	800 mg/day
Nausea/vomiting	17	43
Rash	10	10
Pruritus	4	15
Gynecomastia	2	14
Decreased libido	4	14
Menstrual irregularities	6	19
Liver abnormalities	2	5
Alopecia	1	6
Diarrhea	1	10
Miscellaneous	19	33

From Sugar et al.[5] Reprinted by permission.

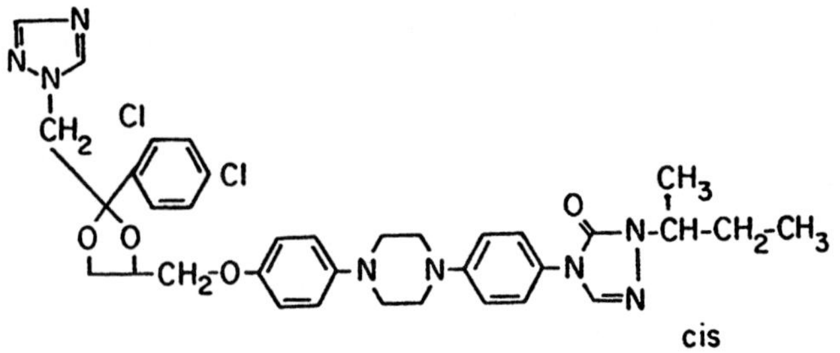

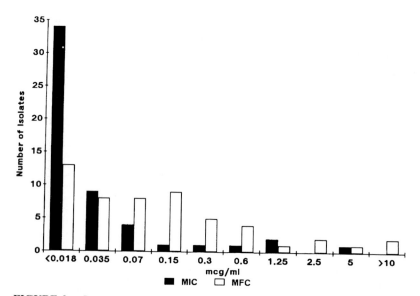

FIGURE 2. Molecular structure of itraconazole (R-51, 211).

after prolonged courses (TABLE 7).[14] Side effects were, similarly to those of itraconazole, infrequent (TABLE 8).[15] Pushing the dose upward, >400 mg/day, in an attempt to achieve a better or faster response, was also more feasible with either triazole than had been the experience with ketoconazole. Response rates in meningeal coccidioidomycosis with fluconazole were similar to those seen with itraconazole (FIG. 10).[16]

It later became apparent with azole therapy of meningeal coccidioidomycosis that the disease was not cured, even in patients whose symptoms cleared and whose cerebrospinal fluid (CSF) normalized.[17] This conclusion came from a follow-up study of 16 responders who stopped therapy, and subsequently,

FIGURE 3. *In vitro* susceptibility of *Coccidioides* to itraconazole. Fifty-three clinical isolates tested in mycelial form by broth dilution. (From Tucker *et al.*[8] Reprinted by permission.)

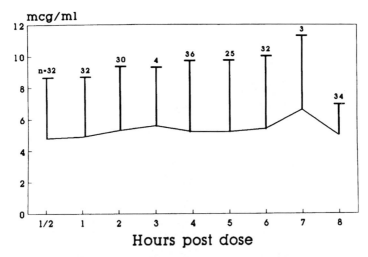

FIGURE 4. Steady-state serum concentrations of itraconazole after capsule administration of 200 mg twice daily to 42 coccidioidomycosis patients. (From Tucker *et al.* Reprinted by permission.)

sooner or later, most relapsed. This stresses the importance of lifetime therapy for coccidioidal meningitis with these agents.

A singular randomized clinical trial has compared itraconazole versus fluconazole for nonmeningeal coccidioidomycosis.[18] The trial was geared for an evaluation of results after 8 months of treatment, at which point the differences were small, but with 4 more months of observation, the superiority

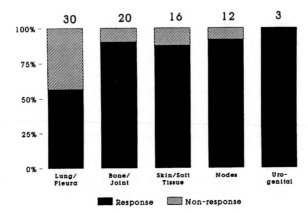

FIGURE 5. Responses at sites of coccidioidal disease to itraconazole. Responses at 81 of the 106 sites were assessable. (From Tucker *et al.* Reprinted by permission.)

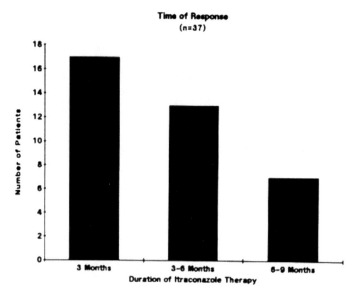

FIGURE 6.   Time to clinical response to itraconazole. (From Tucker *et al*. Reprinted by permission.)

of itraconazole overall became statistically significant, although this was at-tributable almost exclusively to the marked superiority of itraconazole for bone and joint disease (TABLE 9).

In a rabbit model of coccidioidal meningitis, with optimal dosing, itracona-zole and fluconazole produced equivalent results.[19] A recent dose-ranging study[20] in a murine model of coccidioidal meningitis confirmed this, and indi-cated that, on a mg/kg basis, itraconazole is superior to fluconazole for therapy (FIG. 11). Thus, when other pharmacologic issues (e.g., absorption, drug–drug interactions) are not paramount, itraconazole may be clinically utilized more for oral treatment of meningitis.

With respect to the newest azoles, although voriconazole appears to be ac-tive *in vitro*, clinical experience with it in coccidioidomycosis is limited and anecdotal,[21–24] and verification of the activity of the drug in animal models is

TABLE 7.   MSG multicenter trials of fluconazole 200–400 mg in nonmeningeal coccid-ioidomycosis

	n	Response	Relapse
Skeletal	14	86%	50%
Chronic pulmonary	40	55%	39%
Skin/soft tissue	21	76%	23%

From Catanzaro *et al*. Reprinted by permission.

TABLE 8. Possible symptomatic adverse effects of fluconazole

Gastrointestinal	
Diarrhea	2 (2%)
Gastritis	2
Vomiting	2
Nausea	2
Acute gastroenterocolitis	1 (1%)
Constipation	1
Pain	
Arthralgia	2 (2%)
Abdominal	1 (1%)
Arm	1
Pelvic	1
Endocrine or cutaneous	
Hair loss	3 (3%)
Polydipsia	1 (1%)
Dry lips	1
Pruritus	1
Impotence	1
Premature ejaculation	1
Generalized	
Anorexia	3 (3%)
Fatigue	2 (2%)
Malaise	2
Fever	1 (1%)
Weight loss	1
Neurological/muscular	
Headache	3 (3%)
Dizziness	2 (2%)
Weakness	1 (1%)
Weak legs	1
Blurring	1
Tremors	1
Tinnitus	1
Fatal	
Myocardial infarction	1 (1%)
Encephalitis-like symptoms	1
Total = 47 in 25 of the 93 patients	

From Stevens et al.[15] Reprinted by permission.

difficult because of metabolism issues in rodents. There are numerous drug–drug interactions. This drug is discussed by others in this volume. Study of posaconazole (FIG. 12) has been much more intense.[25] The drug is highly active, and fungicidal, *in vitro* against *Coccidioides*, similar to itraconazole (TABLE 10). In animal models it proved >200-fold as potent as fluconazole, on a mg/kg basis, and >10- to 20-fold as potent as itraconazole,[25] and is the only agent studied to date that produces a substantial proportion of cures in the mouse model. Preliminary data from a salvage trial indicates about three

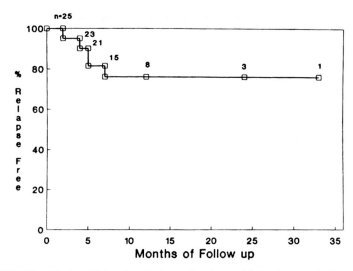

FIGURE 7. Kaplan–Meier plot of relapse-free interval following conclusion of apparently successful response to itraconazole. (From Tucker *et al.* Reprinted by permission.)

fourths of patients with nonmeningeal coccidioidal disease refractory to other agents will respond.

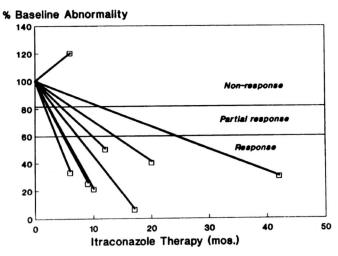

FIGURE 8. Response of coccidioidal meningitis to itraconazole, plotted as end-therapy score compared to pretherapy score. The scoring system aggregates clinical, serological, spinal fluid, etc. parameters of evaluation, and is explained in references 12 and 15. (From Tucker *et al.*[12] Reprinted by permission.)

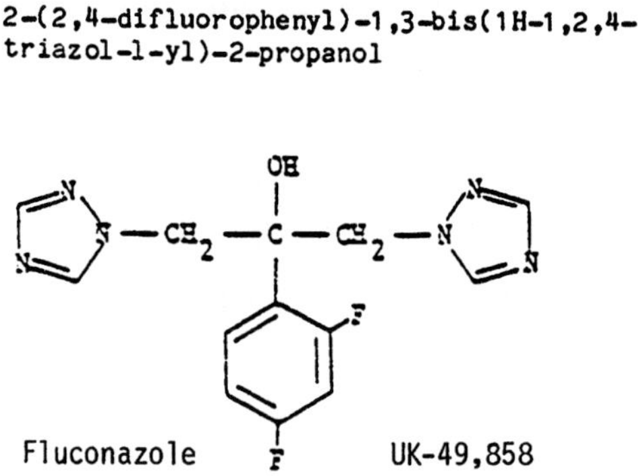

2-(2,4-difluorophenyl)-1,3-bis(1H-1,2,4-
triazol-1-yl)-2-propanol

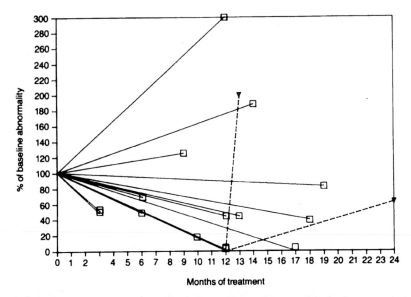

Fluconazole   UK-49,858

FIGURE 9.   Molecular structure of fluconazole.

The animal models used for screening the compounds detailed above as well as other azoles[26] have proven to be a critical crucible for predicting later clinical utility. The clinical trials suggest that cure is achieved only with difficulty, and hence to lessen relapses azole therapy should be given for ≥ 6 months after apparent disease inactivity.[26] The present azoles are well tolerated, making long courses possible. The azoles have proven to be a powerful

FIGURE 10.   Response of coccidioidal meningitis to fluconazole, in the same manner as in FIGURE. 8. The two dotted lines represent change in score for two patients after therapy discontinued. (From Tucker *et al.*[16] Reprinted by permission.)

TABLE 9. Patients who responded to treatment after 12 months according to category of disease

Variable	Fluconazole group n/n	Itraconazole group n/n (%)	P value
All patients	54/94 (57 [47–68])	70/97 (72 [62–81])	0.04
Patients with pulmonary infection	22/35 (63 [45–79])	23/35 (66 [48–81])	>0.2
Patients with soft tissue infection	22/32 (69 [50–84])	31/39 (79 [64–91])	>0.2
Patients with skeletal infection	10/27 (37 [19–58])	16/23 (70 [47–87])	0.03

(From Galgiani et al.[16] Reprinted by permission.)
NOTE: Numbers in square brackets are 95% CIs.

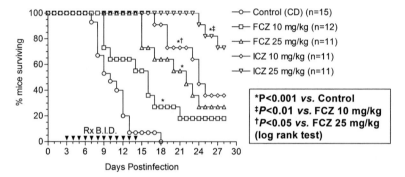

FIGURE 11. Dose comparisons of oral itraconazole (ICZ) and fluconazole (FCZ) on survival in the murine model of coccidioidal meningitis. CD = hydroxypropyl cyclodextrin (control vehicle for the test azoles). (From Kamberi et al.[16] Reproduced by permission.)

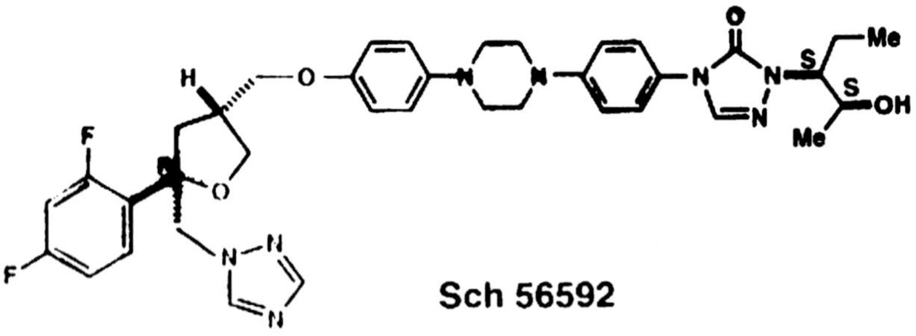

Sch 56592

FIGURE 12. Molecular structure of posaconazole.

weapon in the battle against *Coccidioides*; they are attractive compared to the polyenes because the azoles can be given orally as well as intravenously and because of the toxicity of the polyenes, and there may well be even more pleasant surprises in the future.

TABLE 10. MIC/MFC patient isolates from multicenter coccidioidomycosis study

	≤0.39	0.78	1.56	3.13	6.25	12.5	25	50	100	>100
MIC (μg/mL)										
POS	2	6	3	1	0	0	0	0	0	0
FCZ	0	0	0	0	0	2	5	5	0	0
ITZ	1	8	3	0	0	0	0	0	0	0
MFC (μg/mL)										
POS	2	1	4	4	1	0	0	0	0	0
FCZ	0	0	0	0	0	2	3	5	1	1
ITZ	1	2	5	3	1	0	0	0	0	0

NOTE: The table shows number of isolates at each μg/mL value. MIC indicates minimum inhibitory concentration; MFC = minimum fungicidal concentration; POS = posaconazole; FCZ = fluconazole; ITZ = itraconazole

REFERENCES

1. LEVINE, H.B., D.A. STEVENS, J.H. COBB, *et al.* 1975. Miconazole in coccidioidomycosis. I. Assays of activity in mice and *in vitro.* J. Infect. Dis. **132:** 407–414.
2. STEVENS, D.A. 1977. The role of miconazole in systemic fungal infections. Am. Rev. Resp. Dis. **116:** 801–806.
3. BORELLI, D., J.L. BRAN, J. FUENTES, *et al.* 1979. Ketoconazole, an oral antifungal: laboratory and clinical assessment of imidazole drugs. Postgrad. Med. J. **55:** 657–661.
4. STEVENS, D.A., R.L. STILLER, P.L. WILLIAMS, *et al.* 1983. Experience with ketoconazole in three major presentations of progressive coccidioidomycosis. Am. J. Med. **74:** 58–63.
5. SUGAR, A.M., S.G. ALSIP, J.N. GALGIANI, *et al.* 1987. Pharmacology and toxicity of high dose ketoconazole. Antimicrob. Agents Chemother. **31:** 1874–1878.
6. STEVENS, D.A. 1985. Ketoconazole metamorphosis: an antimicrobial becomes an endocrine drug. Arch. Intern. Med. **145:** 813–815.
7. GANER, A., E. ARATHOON & D.A. STEVENS. 1987. Initial experience in therapy of progressive mycoses with itraconazole, the first triazole in clinical studies. Rev. Infect. Dis. **9:** S77–S86.
8. TUCKER, R.M., D.W. DENNING, E.G. ARATHOON, *et al.* 1990. Itraconazole therapy of nonmeningeal coccidioidomycosis: clinical and laboratory observations. J. Am. Acad. Derm. **23:** 593–601.
9. STEVENS, D.A. 1999. Itraconazole in cyclodextrin solution. Pharmacotherapy **19:** 603–611.
10. TUCKER, R.M., Y. HAQ, D.W. DENNING, *et al.* 1990. The toxicity of itraconazole in 189 patients on chronic therapy. J. Antimicrob. Chemother. **26:** 561–566.
11. TUCKER, R.M., D.W. DENNING, L.H. HANSON, *et al.* 1992. The interaction of azoles with rifampin, phenytoin, and carbamazepine: *in vitro* and clinical observations. Clin. Infect. Dis. **14:** 165–174.
12. TUCKER, R.M., D.W. DENNING, B. DUPONT, *et al.* 1990. Itraconazole therapy of chronic coccidioidal meningitis. Ann. Intern. Med. **112:** 108–112.
13. DIAZ, M., R. NEGRONI, F. MONTERO-GEI, *et al.* 1992. A Panamerican five year study of fluconazole therapy of deep mycoses in the noncompromised host. Clin. Infect. Dis. **14**(Suppl. 1)**:** S68–S76.

14. CATANZARO, A., J.N. GALGIANI, B.E. LEVINE, *et al.* 1995. Fluconazole in the treatment of pulmonary and nonmeningeal coccidioidomycosis. Am. J. Med. **98:** 249–256.
15. STEVENS, D.A., M. DIAZ, R. NEGRONI, *et al.* 1997. Safety evaluation of chronic fluconazole therapy. Chemotherapy **43:** 371–377.
16. TUCKER, R.M., J.N. GALGIANI, D.W. DENNING, *et al.* 1990. Treatment of coccidioidal meningitis with fluconazole. Rev. Infect. Dis. **12:** S380–S389.
17. DEWSNUP, D.H., J.N. GALGIANI, J.R. GRAYBILL, *et al.* 1996. Is it ever safe to stop azole therapy for *Coccidioides immitis* meningitis? Ann. Intern. Med. **124:** 305–310.
18. GALGIANI, J.N., A. CATANZARO, G.A. CLOUD, *et al.* 2000. Randomized double-blind comparison of oral fluconazole and itraconazole for progressive nonmeningeal coccidioidomycosis. Ann. Intern. Med. **133:** 676–686.
19. SORENSEN, K.N., R.A. SOBEL, K.V. CLEMONS, *et al.* 2000. Comparison of fluconazole and itraconazole in the treatment of experimental coccidioidal meningitis in a rabbit model. Antimicrob. Agents Chemother. **44:** 1512–1517.
20. KAMBERI, P., R.A. SOBEL, K.V. CLEMONS, *et al.* 2007. Comparison of itraconazole and fluconazole against coccidioidal meningitis in a murine model. Antimicrob. Agents Chemother. **51:** 998–1063.
21. PROIA, L.A. & A.R. TENORIO. 2004. Successful use of voriconazole for treatment of *Coccidioides* meningitis. Antimicrob Agents Chemother. **48:** 2341.
22. ANTONY, S.J., P. JURCZYK & L. BRUMBLE. 2006. Successful use of combination antifungal therapy in the treatment of *Coccidioides* meningitis. J. Natl. Med. Assoc. **98:** 940–942.
23. PRABHU, R.M., M. BONNELL, B.L. CURRIER, *et al.* 2004. Successful treatment of disseminated nonmeningeal coccidioidomycosis with voriconazole. Clin. Infect. Dis. **39:** e74–e77.
24. CORTEZ, K.J., T.J. WALSH & J.E. BENNETT. 2003. Successful treatment of coccidioidal meningitis with voriconazole. Clin. Infect. Dis. **36:** 1619–1622.
25. LUTZ, J.E., K.V. CLEMONS, B.H. ARISTIZABAL, *et al.* 1997. Activity of the triazole SCH56592 against disseminated murine coccidioidomycosis. Antimicrob. Agents Chemother. **41:** 1558–1561.
26. STEVENS, D.A. 1997. Azoles in the treatment of coccidioidomycosis. *In:* Proceedings of the Centennial Conference on Coccidioidomycosis, H.E. Einstein, D. Pappagianis, A. Catanzaro, Eds.: 255–264. Nat. Foundation for Infectious Diseases.

Poster Presentations

The following list gives the titles and authors of posters presented during the Sixth International Symposium on Coccidioidomycosis. The complete abstract of each presentation is available online at www.vfce.arizona.edu. Abstracts from previous years' meetings are also available.

Efficacy of Ambruticin Analogs Against Coccidioidomycosis in a Mouse Model of Infection
L. SHUBITZ, J. GALGIANI, Z. TIAN, Z. ZHONG, L. KATZ, AND P. TIMMERMANS
University of Arizona

Feline-Disseminated Coccidioidomycosis: A Report of 19 Cases (1995–2003)
M. WESTON AND S. DIAL
University of Arizona

A Proteomic Approach to Valley Fever Vaccine Discovery
J. ROHRBOUGH, T. PENG, J. SIMONS, P. HAYNES, K. ORSBORN, S. JOHNSON, V. WYSOCKI, D. PAPPAGIANIS, AND J. GALGIANI
University of Arizona and University of California, Davis

Coccidioidomycosis in Durango and Coahuila, Mexico:
Clinic Characteristics (1999–2005)
CLODOVEO DE LEÒN-CHAPA, BEATRIZ ALVAREZ-GARCÍA, AND VÍCTOR M. VELASCO-RODRÍGUEZ
Mexican Social Security Institute (IMSS); Specialties Hospital No. 71 of Torreon; and University of Coahuila, Mexico

Skin Reactivity to Coccidioidin in Immunocompromised Patients in a Northern City of Mexico: Research's Advances
BEATRIZ ALVAREZ-GARCÍA AND VÍCTOR M. VELASCO-RODRÍGUEZ
Mexican Institute of Social Security (IMSS); Specialties Hospital No. 71 of Torreon; and University of Coahuila, Mexico

Ann. N.Y. Acad. Sci. 1111: 455–459 (2007). © 2007 New York Academy of Sciences.
doi: 10.1196/annals.1406.051

Identification of *Coccidioides immitis* and *C. posadasii* by Sequencing of its Regions of rDNA
K. TINTELNOT, G. S. DE HOOG, A. ANTWEILER, H. LOSERT, M. SEIBOLD, M. BRANDT, A. H.G. GERRITS VAN DEN ENDE, AND M. C. FISHER
Robert Koch Institut, Berlin, Germany; Centraalbureau voor Schimmelcultures, Utrecht, the Netherlands; Centers for Disease Control and Prevention, Atlanta, GA, USA; and Department of Disease Epidemiology, Imperial College, London, UK

Antifungal Susceptibility Profiles of *Coccidioides immitis* and *Coccidioides posadasii* from Endemic and Non-Endemic Areas
R. RAMANI AND V. CHATURVEDI
Wadsworth Center, Albany, NY

Antigen Detection in Coccidioidomycosis: Detection of a Cross-Reactive Antigen in the Second-Generation Histoplasma Antigen Assay
L. WHEAT, T. KUBERSKI, A. MYERS, M. DURKIN, AND P. CONNOLLY
Miravista Diagnostics

Identification and Molecular Cloning of a pH-Sensitive Protein-Like Protein Present in the Coccidioidal Vaccine T27 K
S. JOHNSON AND D. PAPPAGIANIS
University of California, Davis

Coccidiomycosis in Former Manganum Miners
G. NEMSADZE
Tkibuli District Hospital

Some Like it Hot: Differences in Thermotolerance of *Coccidioides* Species
B. BARKER, S. STATT, J. GALGIANI, AND M. ORBACH
University of Arizona

Genomic Analyses of Coccidioides, a Cryptically Sexual Human Pathogenic Fungus
B. BARKER AND S. KROKEN
University of Arizona

Expression of *Coccidioides posadasii* Chitin Synthases of *Coccidioides posadasii* During Mycelial and Spherule Growth
M. A. Mandel, J. N. Galgiani, and M. J. Orbach
University of Arizona

Fourteen-Year Experience with Coccidioidomycosis at Kern Medical Center
Shehla Baqi, Greti Petersen, Amina Haggag, Ryan Cabatbat, Maisara Rahman, Hans Einstein, and Royce Johnson
Kern Medical Center, Bakersfield, CA, USA

The Coccidioidomycosis Infection in Mexico: A National Project to Determine its Prevalence
A. Aroch-Calderón, L.R. Castañón-Olivares, and R. Laniado-Laborin
Hospital General de Tijuana, Isesalud, SS. Tijuana, Mexico

Clinical Experience with a Urinary Antigen Test in Patients Suspected of Having a *Coccidioides immitis* Infection
T. Kuberski and L. J. Wheat
Miravista Diagnostics

Public Health Surveillance for Coccidioidomycosis In Arizona
S. Anderson, A. Vossbrink, and D. Engelthaler
Arizona Department of Health Services

Population-Based Epidemiological Study of Greater Tucson, Arizona
J. Tabor and M. O'Rourke
University of Arizona

Disparities in Testing Practices for *Coccidioides* among Patients with Community-Acquired Pneumonia: Metropolitan Phoenix, 2003–2004
D. C. Chang, B. J. Park, L. A. Burwell, K. Wannemuehler, S. Anderson, D. Engelthaler, and S. K. Fridkin
Centers for Disease Control, Atlanta, GA, USA

Outbreak of Coccidioidomycosis in a State Prison—California, 2005
J. Yuan, C. Wheeler, R. Chapnick, R. Kanan, D. Vugia, and J. Mohle-Boetani
California Department of Health Services; Centers for Disease Control and Prevention; and California Department of Corrections and Rehabilitation

Epidemiological Profile of Coccidioidomycosis Infection in a Predominantly Hispanic Population
MIGUEL ANGEL PEÑA-RUIZ, ZUBER D. MULLA, ANDRES ESCOBAR, AND ARMANDO MEZA
Texas Tech University Health Sciences Center, El Paso, TX, USA; University of Texas School of Public Health at Houston, El Paso Regional Campus

Antigenicity, Safety and Efficacy of a Recombinant Coccidioidomycosis Vaccine in Cynomolgus Macaques (*Macaca fascicularis*)
N. W. LERCHE, D. PAPPAGIANIS, S.M. JOHNSON, J. L. YEE, J. N. GALGIANI, AND R. F. HECTOR
California National Primate Research Center, University of California Davis; University of Arizona; and University of California, San Francisco

Case Study: Diagnosis and Treatment of Ocular Coccidioidomycosis in a Female Chimpanzee (*Pan Troglodytes*)
K. HOFFMAN, J. FRITZ, E. VIDEAN, AND J. MURPHY
Primate Foundation of Arizona, Mesa, Arizona, USA

Case Report of a Deep Solitary Brain Mass in a Four-Month-Old Male with Disseminated Coccidioidomycosis
J. NOLT AND F. GEERTSMA
Children's Hospital Central California, Madera, California, USA

Identification of a Cu/Zn-Superoxide Dismutase in the Coccidioidal T27 K Vaccine using Proteomic Methods
J. LUNETTA, S. JOHNSON, AND D. PAPPAGIANIS
University of California, Davis

Characteristics of the Protective Subcellular Coccidioidal 27 K Vaccine
D. PAPPAGIANIS, S. JOHNSON, AND J. LUNETTA
University of California, Davis

Nuclear Labeling of *Coccidioides posadasii* with GFP
L. LI, E. KELLNER, J. GALGIANI, AND M. J. ORBACH
University of Arizona

Identification of a Class I 1,2-Alpha-Mannosidase Protein in the Coccidioidal T27 K Vaccine using Immunoproteomic Methods
J. LUNETTA, S. JOHNSON, AND D. PAPPAGIANIS
University of California, Davis

Coccidioides Species in Mexico
L.R. CASTAÑÓN-OLIVARES, D. GUEREÑA-ELIZALDE, M.R. GONZÁLEZ-MARTÍNEZ, A. F. LICEA-NAVARRO, G.M. GONZÁLEZ-GONZÁLEZ, AND A. AROCH-CALDERÓN
Facultad de Medicina, Universidad Nacional Autónoma de México; Laboratorio de Soluciones Genómicas; Unidad Médica de Atención Especializada No.71, Instituto Mexicano del Seguro Social; Biotecnología Marina, Centro de Investigación Científica y de Educación Superior de Ensenada; and Facultad de Medicina, Universidad Autónoma de Nuevo León

Modeling Ecological Niche of *Coccidioides* sp. in Baja California, Mexico
R. BAPTISTA, M. RIQUELME, AND A. HINOJOSA
Faculty Science UABC/Microbiology Department CICES

Index of Contributors